KV-467-915

Numerical Modeling of the Global Atmosphere in the Climate System

edited by

Philip Mote
JISAO/SMA Climate Impacts Group,
University of Washington,
Seattle, WA, U.S.A.

and

Alan O'Neill
NERC Centre for Global Atmospheric Modelling,
University of Reading,
Department of Meteorology,
Earley Gate, Reading, United Kingdom

Kluwer Academic Publishers

Dordrecht / Boston / London

Published in cooperation with NATO Scientific Affairs Division

Proceedings of the NATO Advanced Study Institute on
Numerical Modeling of the Global Atmosphere in the Climate System
Castelvecchio Pascoli, Italy
May 25-June 5 1998

A C.I.P. Catalogue record for this book is available from the Library of Congress.

ISBN 0-7923-6301-9 (HB)
ISBN 0-7923-6302-7 (PB)

Published by Kluwer Academic Publishers,
P.O. Box 17, 3300 AA Dordrecht, The Netherlands.

Sold and distributed in North, Central and South America
by Kluwer Academic Publishers,
101 Philip Drive, Norwell, MA 02061, U.S.A.

In all other countries, sold and distributed
by Kluwer Academic Publishers,
P.O. Box 322, 3300 AH Dordrecht, The Netherlands.

Printed on acid-free paper

Printed in the Netherlands.

Numerical Modeling of the Global Atmosphere in the Climate System

NATO Science Series

A Series presenting the results of activities sponsored by the NATO Science Committee. The Series is published by IOS Press and Kluwer Academic Publishers, in conjunction with the NATO Scientific Affairs Division.

A. Life Sciences — IOS Press
B. Physics — Kluwer Academic Publishers
C. Mathematical and Physical Sciences — Kluwer Academic Publishers
D. Behavioural and Social Sciences — Kluwer Academic Publishers
E. Applied Sciences — Kluwer Academic Publishers
F. Computer and Systems Sciences — IOS Press

1. Disarmament Technologies — Kluwer Academic Publishers
2. Environmental Security — Kluwer Academic Publishers
3. High Technology — Kluwer Academic Publishers
4. Science and Technology Policy — IOS Press
5. Computer Networking — IOS Press

NATO-PCO-DATA BASE

The NATO Science Series continues the series of books published formerly in the NATO ASI Series. An electronic index to the NATO ASI Series provides full bibliographical references (with keywords and/or abstracts) to more than 50000 contributions from international scientists published in all sections of the NATO ASI Series.
Access to the NATO-PCO-DATA BASE is possible via CD-ROM "NATO-PCO-DATA BASE" with user-friendly retrieval software in English, French and German (WTV GmbH and DATAWARE Technologies Inc. 1989).

The CD-ROM of the NATO ASI Series can be ordered from: PCO, Overijse, Belgium

Contents

PREFACE

Part 1. Preliminary considerations

1. Weather and Climate GCMs: A memoir   K. Miyakoda

1	Introduction	1
2	Phillips' GCM experiment	2
3	Numerical integration of GCMs	3
4	Subgrid-scale processes	8
5	Climate GCMs	11
6	Long range forecasts: extending the frontier	13
7	Future issues of GCM forecasts	18
8	Epilogue	21

2. The GCM as a Dynamical System: Implications for numerical simulations

                                                              J.-F. Royer

1	Introduction	29
2	Formal Description of a Climate Model as a Dynamical System	29
3	Initial Conditions	32
4	Model Parameters and Boundary Conditions	34
5	Observation and Storage of Model Results	36
6	Analysis of Model Results	37
7	Defining the Sensitivity of a GCM	42
8	The Different Categories of Climate Simulations	46
9	The Different Steps in Running a GCM Experiment	50

3. Analysis and Verification of Model Climate   G.J. Boer

1	Introduction	59
2	Climate Analysis and Verification	60
3	The Basic Equations of Climate and Climate Analysis	60
4	Data Considerations	65
5	A Hierarchy of Budget Equations and Climate Statistics	66
6	Model Results	71
7	Concluding Comments	80

4. Statistical Treatment of Model Output   M. Déqué

  1   Introduction 83
  2   The Mean 84
  3   The Variability 87
  4   Conclusion 102

5. Use of Simplified Atmospheric Models   J. Thuburn

  1   Introduction 105
  2   Shallow water models 106
  3   Zonally averaged models 109
  4   3D models with highly simplified parametrized physics 112
  5   Stratosphere-mesosphere models 114
  6   Conclusion 117

6. Designing a GCM Experiment : Fundamentals of the planning process
                                                              P.W. Mote

  1   Introduction 119
  2   Control Runs 119
  3   Changing a Model 120
  4   Designing an Experiment 121
  5   Checking for Errors 124
  6   Summary 125

Part 2. Components of an Atmospheric General Circulation Model

7. Numerical Approximations for global Atmospheric GCMs
                                        D.L. Williamson and R. Laprise

  1   Introduction 127
  2   Finite Difference Approximations 130
  3   Spectral Method 158
  4   Semi-Lagrangian Method 171
  5   Orographic and Other Stationary Forcings 184
  6   Finite Element Approximations 198
  7   Reduced Grids 198
  8   Vertical Coordinates 200
  9   Example of an AGCM 204

8. Boundary Layer Processes   N. McFarlane

  1   Introduction 221
  2   Basic Features of the Atmospheric Boundary Layer 222
  3   Cloudy Boundary Layers 235

9. Moist Convection   **N. McFarlane**
- 1   Basic Considerations 239
- 2   Conditional Instability 240
- 3   Approaches to Parameterization 245

10. Clouds and Cloud Water Prediction   **J.-J. Morcrette, Ch. Jakob, and J. Teixeira**
- 1   Clouds: What is their role in the climate system? 263
- 2   Cloud fraction 264
- 3   A quick history of cloud representation at ECMWF 265
- 4   Validation of cloud products 267
- 5   Concluding remarks 276

11. Radiation   **J.-J. Morcrette and S.A. Clough**
- 1   Introduction 281
- 2   Sensitivity of the ECMWF model to the representation of the longwave radiative transfer 283
- 3   Validation 287
- 4   Perspectives 291

12. Gravity-Wave Drag   **N. McFarlane**
- 1   General aspects of the parameterization problem 297
- 2   Basic Gravity Wave Dynamics 298
- 3   Parameterization of Orographic Gravity-wave drag 302
- 4   Non-orographic waves 305
- 5   Effects of Parameterized Gravity-wave Drag 306
- 6   Uncertainties 317

13. Land Surface Processes and Hydrology   **J.-F. Royer**
- 1   Introduction 321
- 2   Water and Energy Budgets at the Surface 322
- 3   Heat and Water in the Soil 331
- 4   Vegetation 334
- 5   Snow Cover 339
- 6   Soil Hydrology 340
- 7   Sensitivity Experiments 345
- 8   Conclusion 346

14 Atmospheric Chemistry and Aerosol Dynamics   **J. Feichter**
- 1   Introduction 353
- 2   Budgets and Cycles 354
- 3   Transport of trace species 356
- 4   Sources and sinks 360
- 5   Chemistry and aerosol micro-physics 367
- 6   Evaluation 369

Part 3. Applying GCMs

15. Atmospheric Data Assimilation   A. O'Neill
1   Introduction 375
2   What is data assimilation? 376
3   What are the benefits of data assimilation? 377
4   How is data assimilation done? 378
5   Concluding remarks 385
6   Supplement: the (error) covariance matrix S_x 385

16. Seasonal Predictions   P.W. Mote, E.S. Sarachik, and M. Déqué
1   Introduction 387
2   Three Types of Predictions 390
3   Predicting Global Anomalies by Predicting ENSO 393
4   A Multiyear Multimodel Experiment: PROVOST 397

17. Regional Models   M. Déqué
1   Introduction 403
2   Resolution impact 404
3   Limited Area Model 408
4   Variable resolution model 410
5   Other solutions 414
6   Conclusion 415

18. Toward a Complete Model of the Climate System   B.A. Boville
1   Introduction 419
2   Constraints Imposed by Coupling 421
3   Ocean General Circulation Models 425
4   Sea Ice Models 428
5   NCAR CSM-1 Coupled Solution, Equilibrium Forcing 430
6   1%/year CO_2 Increase Experiment 435
7   Computer Hardware 438
8   Summary 440

19. Climate Model Intercomparison   G.J. Boer
1   Introduction 443
2   Model Intercomparison 444
3   Atmospheric Models 446
4   Coupled Models 452
5   Model Sensitivity 457
6   Concluding Remarks 462

20. Paleoclimate Modeling   P. Valdes
1   Introduction 465
2   Motivation 466
3   Types of Climate Model 470
4   Case Studies 476

5   Summary 484

21. Simulating Future Climate   **G.J. Boer**
1   Introduction 489
2   International Aspects 490
3   Simulating Historical and Future Climate 492
4   Climate Change in the 20th Century 495
5   Simulating Future Climate Change 498
6   Climate Impact, Adaptation, and Mitigation 501
7   Summary 502

Index   **505**

PREFACE

Numerical modeling of the global atmosphere has entered a new era. Whereas atmospheric modeling was once the domain of a few research units at universities or government laboratories, it can now be performed almost anywhere thanks to the affordability of computing power. Atmospheric general circulation models (GCMs) are being used by a rapidly growing scientific community in a wide range of applications. With widespread interest in anthropogenic climate change, GCMs have a role also in informing policy discussions. Many of the scientists using GCMs have backgrounds in fields other than atmospheric sciences and may be unaware of how GCMs are constructed.

Recognizing this explosion in the application of GCMs, we organized a two-week course in order to give young scientists who are relatively new to the field of atmospheric modeling a thorough grounding in the basic principles on which GCMs are constructed, an insight into their strengths and weaknesses, and guidance on how meaningful numerical experiments are formulated and analyzed. Sponsored by the North Atlantic Treaty Organization (NATO) and other institutions, this Advanced Study Institute (ASI) took place May 25–June 5, 1998, at Il Ciocco, a remote hotel on a Tuscan hillside in Italy. In addition to lectures, which form the basis for the chapters in this book, the ASI featured a computer modeling project in which the students designed GCM experiments and ran a state-of-the-art GCM (the NCAR CCM3) on rented Sun workstations at Il Ciocco, demonstrating the point made above that GCMs can be run almost anywhere.

The first part of this book describes some general considerations in planning and executing an experiment with a GCM. The second part describes the inner workings of a GCM, including the dynamical core and the most common parameterizations. Finally, the third part describes some of the present applications of GCMs. As the second part of the title, "in the climate system", points out, atmospheric GCMs (which were once nearly synonymous with "climate models") now form but one component of climate system models.

This book assumes that the reader is familiar with some basic principles of atmospheric dynamics. Several good books on the subject are available, including "Atmospheric Science" by J.M. Wallace and P.V. Hobbs, and "Introduction to Dynamic Meteorology" by J.R. Holton (both published by Academic Press).

Previous books on the subject of atmospheric modeling include the excellent, though now somewhat dated, text on "Numerical Prediction and Dynamic Meteorology" by G.J. Haltiner and R.T. Williams (John Wiley and Sons, 1980), and the comprehensive "Climate System Modeling," edited by K. Trenberth (Cambridge University Press, 1992). Readers will find that Part II of this book provides more recent surveys of numerical techniques covered in part in those and other books on the subject, but to our knowledge Parts I and III largely contain topics not covered in other books.

Acknowledgements

The success of the NATO ASI rested on the shoulders of several people: the other committee members, Byron Boville and Jean-François Royer; the NATO Science Committee; Simon Walmsley, who provided computers and support; and Byron Boville, Jim Rosinski, and David Williamson, whose cheerful and expert assistance with the NCAR CCM3 contributed immeasurably to the success of the projects. We also thank Anne Pinnock for handling a great many tedious details of both the ASI and the book, and Stefano Nativi for his able linguistic assistance in innumerable ways during the ASI.

We are also indebted to many people who contributed to this book. The students at the ASI suggested helpful revisions to the chapters, and are acknowledged individually by the authors. Roger Brugge provided invaluable editorial assistance by converting many of the chapters into LaTeX for uniformity. Further editorial assistance was provided by Anne Pinnock, Katya Partan, and Nathaniel Mote.

WEATHER AND CLIMATE GCMS

A Memoir

K. MIYAKODA
George Mason University / COLA
4041 Powder Mill Rd., Suite 302
Calverton MD 20705 USA

1. Introduction

The genesis of general circulation models (GCMs) took place at the Geophysical Fluid Dynamics Laboratory (GFDL) in Princeton, New Jersey. I spent much of my career there and my office was a superb observation deck from which to watch progress in GCM activities. I came to the U.S. in 1961, first worked at GFDL in 1963, and settled down permanently in 1965. The innovative director of GFDL, Joseph Smagorinsky, had studied numerical weather prediction in its early stage at the Institute of Advanced Study (IAS) at Princeton under Charney.

From the beginning, there have been two types of GCMs: weather models (e.g., Charney *et al.* [15]), which produce operational forecasts, and climate models (e.g., Phillips [81]), with which scientists investigate fundamental processes. I always felt that weather prediction aims at a narrow target, while climate modeling aims at wide targets. Lennart Bengtsson often commented that GCMs tend to become more precise and accurate if they are used for practical forecasts. On the other hand, if GCMs were not exposed to an academic environment, their utility and flexibility would not expand.

This essay is my recollection of the development of GCMs. Although my research focused on weather prediction (especially long-range), rather than on climate modeling, I worked at GFDL for 30 years, observing the climate research of my colleagues. Most papers quoted here refer to work before 1982, when the World Meteorological Organization (WMO) began a project on long-range forecasts (described in section 6.2) and coincidentally just before the 1982–83 El Niño occurred.

P. Mote and A. O'Neill (eds.), Numerical Modeling of the Global Atmosphere in the Climate System, 1–27.
© 2000 *Kluwer Academic Publishers. Printed in the Netherlands.*

2. Phillips' GCM experiment

The starting point of long-range forecasts and climate studies using GCMs is the research of Norman Phillips [81], which took place before I emigrated, but I read his paper as a student in Tokyo. Smagorinsky [100] characterized this study as "a monumental experiment," and he wrote that a new era had been opened by this experiment. Phillips built a two layer quasi-geostrophic atmospheric model and carried out a 31 day run of this model. This is the first GCM run in history, and it was done at IAS using the first automatic computer.

The model's domain was hemispheric between a pole and the equator, and the east-west extension was chosen sufficiently large to accommodate one large baroclinic wave with periodic lateral boundary conditions at both ends. The grid sizes were: $\Delta x = 375$ km and $\Delta y = 625$ km, and the west to east domain corresponded to $16\Delta x$. Net radiation and latent heat processes were empirically specified by a heating function.

Phillips started his integration with an isothermal atmosphere at rest and ran the model with a time step $\Delta t = 1$day. The meridional gradient in net heating gradually built up a meridional gradient in temperature with a meridional circulation that consisted of a single weak direct cell. In the second stage, a random number was introduced to give rise to a perturbation in the geopotential field. The time step was then reduced to 1 hour. Subsequently disturbances of wavelength of ~ 6000 km were produced, and the flow patterns tilted westward with height, the wave moving eastward. Meanwhile, horizontal transport of zonal momentum was directed toward the mid-latitudes, creating a jet of 80 m s^{-1} at the 250 mb level. The three-cell structure emerged between the equator and the pole.

Phillips analyzed the result and recognized that the solutions resemble the observational evidence about the atmospheric general circulation. The baroclinic instability theory, which had been established before 1956, is in a linear framework, and since waves that are small enough to behave linearly are probably too small to detect, the validity of baroclinic instability theory was difficult to prove with observations. Phillips' experiment imitated observations so well with the inclusion of nonlinearity that it was possible to investigate the connection between nature's complexity and the linear theories. The experiment indicated that unstable baroclinic waves are generated when the vertical wind shear of the basic flow reaches a critical value, as is expected. The disturbances transport heat in the expected direction. The secondary effect is to transport momentum in such a way as to maintain the jet against frictional dissipation, and thus the westeries are established in the middle latitudes. All these results completely agree with what the general circulation theory had postulated. As Lewis [42] documented, this was the first breakthrough in research on climatology. Charney mentioned [83] that this was one of three remarkable achievements in the IAS at Princeton, others being the first implementation of a computer for numerical weather prediction, and the realization of cyclone development with a $2\frac{1}{2}$-level model.

It was surprising that Phillips was successful in running the model without any serious computational instability for such a long time. The major reason is that he treated relatively mild weather. The isothermal initial state gave way to more realistic conditions after 30 days, but the calculation could not continue indefinitely because truncation errors grew and computational instabilities finally destroyed the calculation.

3. Numerical integration of GCMs

3.1. FINITE DIFFERENCE METHOD

The period from 1955 to 1963 represents the struggle to unravel the cause of computational instability and the search for techniques to overcome these instabilities. Climate simulations were modest then. For example, at the first international conference on numerical weather prediction in Tokyo I presented a paper [63] entitled "500 hour forecasts" (not 500 days). The issue we faced in longer integrations was how to reduce the truncation error, because it was thought essential to obtain an accurate solution for a long run. This objective turns out to be wrong. The question is, which is preferred: an accurate solution without stability (my obsession), or a less accurate solution with stability.

Akio Arakawa [1] presented a method to obtain a stable solution by using the so-called Arakawa Jacobian. This was really a breakthrough for long time integration of the barotropic vorticity equation. The principle is simple. Energy and enstrophy (squared vorticity) are conserved within the framework of the finite difference equations. These two squared quantities can be conserved for all time. In other words, the integration of the equation is computationally stable, because two quantities are bounded, and therefore, all quantities are bounded. This method was applied to the two-level GCM of the University of California at Los Angeles (UCLA), which was referred to as the Mintz-Arakawa model. Now GCMs could be integrated as long as desired.

As just mentioned, I was obsessed by the notion of an accurate numerical solution. In fact, accuracy versus stability was a controversial issue among the applied mathematicians. If one approximates a differential Jacobian using finite differences, the resulting formula consists of the main term and the remaining terms of higher order. The question is whether the remaining terms should be ignored. The Arakawa approach is to prevent production of spurious energy and enstrophy first, and if so desired, the grid size should be made smaller. In those days, I had the impression that the mathematicians at the Courant Institute, New York University, were critical of the Arakawa Jacobian, or at best unenthusiastic. In fact, the famous text book of Richtmyer and Morton [87] did not include the Arakawa Jacobian at all. For example, the Lax-Wendroff method [40] (a product of the Courant Institute) lets the remaining term serve as a damping term. Sometimes this term appears as a "nonlinear viscosity" (see section 4), and it becomes large when the deformation of the flow pattern is large.

Stimulated by the idea of the Arakawa Jacobian, Kirk Bryan [13] at GFDL developed a scheme that conserves a single squared quantity, whereas Arakawa's scheme conserved two squared quantities. Conservation of one quantity is sufficient from the standpoint of computational stability. The GCMs at GFDL ran with Bryan's scheme ([102], [46]), with the application of nonlinear viscosity. Our group at GFDL used this model for extended range forecasts (to two weeks) with observed initial conditions. The results were very encouraging [67] and were presented at the first meeting of the Global Atmospheric Research Programme in 1967. Bryan's scheme was also applied efficiently to an irregular grid by Kurihara and Holloway [36]. The implication of irregular grids was important, because the convergence of meridians at the poles of the globe presented a formidable difficulty (see also Chapter 7, sections 1–3). This problem was soon ameliorated by the introduction of the "polar filtering technique," which allows the time step to remain large even though Δx gets small (see equation 2.14 of Chapter 7). Our group used Kurihara's irregular grid model for global forecasts from observed initial conditions [64]. This turned out to be the first global prediction, and was successful in forecasting a blocking event ten days ahead, a result that impressed European meteorologists immensely.

The Arakawa Jacobian is an excellent scheme, because the solution is not only stable but the resulting patterns are noise-free, compared with, for example, Bryan's scheme. However, the Arakawa scheme is only applicable to the case in which the advection term is of Jacobian form. In other words, the advective velocity $\bar{V}$ is divided into two components by the Helmholtz theorem, i.e., the rotational component, $\hat{k} \times \nabla \psi$, and the divergent component, $\nabla \chi$; if $\bar{V}$ is represented only by $\hat{k} \times \nabla \psi$, then the Jacobian scheme can be utilized. For divergent (three-dimensional) flow an extended Arakawa Jacobian is needed. Arakawa and Lamb [2] achieved both an energy and potential enstrophy conservation scheme for the general flow, and applied it to the three dimensional UCLA model (Figure 1). In Yugoslavia, Mesinger and Janjic (see [60]) developed a similar scheme. Their scheme also can have energy and potential enstrophy conservation with the ψ-component, but not with ψ and χ together. The finite difference alogorithm had now become very sophisticated and complicated, so it became more difficult for generalists (graduate students, for example) to follow this path without deep involvement.

3.2. SPECTRAL METHOD

We now turn to the development of the spectral method; for details on its application to GCMs today, see chapter 7, section 3. The Fourier method can guarantee

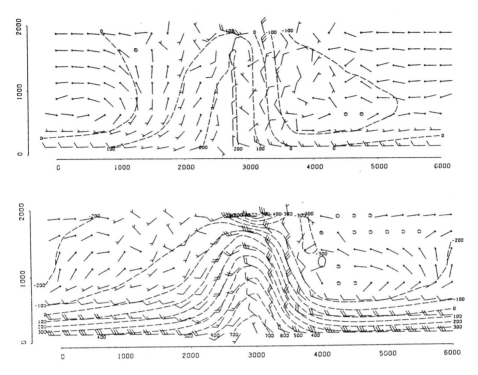

Figure 1. Solutions of the shallow water equations over a steep mountain as represented by flow vectors and streamlines (upper) by the simpler scheme, i.e., energy and potential enstrophy conserving with only purely horizontal non-divergent flow, and (lower) by the new method, i.e., conserving with general divergent flow. (After [2]. Reproduced by permission of the American Meteorological Society.)

the conservation of kinetic energy and enstrophy. For the global atmosphere, the spherical harmonic functions Y_{mn} are more appropriate than Fourier functions (Figure 2). In fact, it is another merit of the spherical harmonic method that the treatment at the poles is straightforward; for example, the "polar filtering" method used in the finite difference scheme for increasing the time step is not needed. The activity along this line has been going on since the end of 1950's ([98], [82], [35], [3]). The spectral method is more suitable for an atmospheric model than for an oceanic model, because of the difficulties posed by lateral boundaries (although some regional models use spectral methods; see Chapter 17).

A problem with spectral methods is the representation of discontinuous fields, for example, the distribution of rainfall [89]. If the Fourier transform is applied to a delta-function, the resulting distribution shows a number of ripples around the main rainfall region, which is called the "Gibbs phenomenon." Furthermore, the wave produced by local rainfall is spread over the whole world instantly, because of the modal nature of the basis functions. The reason for these two undesirable

6

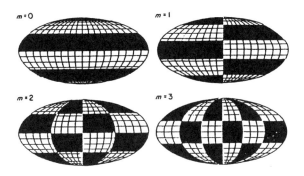

Figure 2. Spherical harmonic functions, $Y_{mn}(\lambda, \mu)$. For the associated Legendre functions P_{mn} see Chapter 7, Figure 3.2.

features of the spectral method is that any finite number of terms in a Fourier series is insufficient to resolve the rainfall profile. In this case, a finite difference scheme may have a slight advantage, because although it produces severe truncation error, the resulting rainfall pattern is not so bad, compared with the Fourier method.

Another fundamental difficulty with spectral methods is the treatment of non-linear terms, which are products of two or more variables. A product of two variables then consists of a product of different Y_{mn}; for this reason, the spectral scheme in which all calculations take place in spectral space is referred to as the "Interaction Coefficient method." Orszag [77] and Eliasen *et al.* [23] developed a new method, the "Transform method," which simplifies the nonlinear calculation (see also [44]). In this method, the mathematical process is switched flexibly between the spectral and the grid calculation, depending upon which is convenient. For example, the horizontal advection term in the x-coordinate consists of two variables, i.e., $A = u$ and $B = \partial \zeta / \partial x$, where u is the speed of the advective flow and ζ is the vorticity. In the Transform method, the product AB is obtained at grid points, and this set of values is decomposed into Fourier series.

Bill Bourke, a young nuclear physicist from Melbourne, Australia, went to Montreal in 1969, and worked on the application of the Transform method to a shallow water spherical model under the advice of A. Robert. (More detail on shallow water models can be found in Section 2 of Chapter 5 and Section 3.4 of Chapter 7.) In contrast to the approach of Orszag, the basic dependent variables Bourke used were vorticity ζ and divergence D (u and v are the so-called pseudo-scalars and are not appropriate for the expansion with the spherical harmonics [88]). I met Bourke at Princeton in 1971, and when he asked me my opinion of his work, I encouraged him to continue it. In retrospect he had already gone quite far at that time, and soon completed his shallow water model [10]. This model was subsequently distributed to a number of groups around the world and formed the basis for the baroclinic spectral models that those groups developed, each within the next few years. For example, the Bureau of Meteorology in Australia developed a baroclinic spectral model in 1974, GFDL in 1974, University of Reading in 1975, and the Prediction Unit in Canada in 1976. The Australian and Canadian bureaus of meteorology implemented operational numercal weather prediction systems based on the transform method in early 1976. McAvaney *et al.* [56], an Australian research group, conducted a pioneering GCM integration utilizing the spectral method and semi-implicit time integration (see Chapter 7, section 4.5), and their work showed that this method is indeed inexpensive.

While the major efforts of GCM development were, for the following decade or so, directed to the development of physics, the next substantial advance in numerical methods came with the development of the semi-Lagrangian method by Robert [90] and Bates and McDonald [4], with contributions by Harold Ritchie. In the semi-Lagrangian method, the advection terms are replaced by the Lagrangian transport. As a result, the semi-Lagrangian method has several advantages over previous methods: (1) no more computational restriction on the time step, Δt, in the sense of the Courant-Friedrich-Lewy restriction (see Chapter 7, section 2.1); (2) the same framework can be used for the treatment of advection in all three dimensions, as opposed to doing a grid calculation in the vertical and a spectral calculation horizontally; (3) no more Gaussian grid restriction for the latitudinal direction; and (4) no concern about aliasing error. As a result, a drastic reduction of grid-points near the poles is possible. The semi-Lagrangian scheme can be applied equally well in a spectral or a finite-difference GCM, and has brought a dramatic change in the computational aspects at the European Centre for Medium-Range Weather Forecasts (ECMWF).

In 1982, Robert called me from Montreal, suggesting that we use the semi-Lagrangian scheme, but I declined the offer. The time required for re-coding a model to use the semi-Lagrangian scheme for all prognostic variables (i.e., the wind vector (u, v, ω), the temperature T and the moisture q) with this scheme is considerable, putting the scheme beyond the reach of most research groups. Proliferation of this scheme is not as easy as that of the spectral method in 1972–76. In 1998, ECMWF and the Canadian Meteorological Centre are the only operational centers that use this method, together with a small model in the Irish Meteorological Service. Semi-Lagrangian advection of q has become relatively commonplace, but few research groups (the National Center for Atmospheric Research (NCAR) being a notable exception; see Chapter 7) have switched to fully semi-Lagrangian dynamics, and it is an open question how many research groups will make the switch.

Despite the fact that a fully Lagrangian scheme precisely conserves advected quantities, the conservation of advected quantities by the semi-Lagrangian scheme has not been guaranteed. The water vapor, for example, is not positive definite so far. Recently, however, a new scheme has been developed [45] in which the mass, enthalpy and momentum are conserved. In the future, operational centers will continue to refine the resolution, and therefore, they will use the semi-Lagrangian methods, which are cheaper at high resolution than other methods.

3.3. VALIDITY OF GCM SOLUTIONS

Finally I return to my old concern about the accuracy of the GCM solution. As I mentioned earlier, the GCM calculation produces a large amount of truncation error, which grows with time. Is there a point at which the truncation error is so large that the GCM solution becomes insignificant by comparison, and therefore, invalid? If the problem is steady state, such as the flow pattern around an airplane's wing, the solution is significant. However, there are problems with the case of long-range forecasts and climate simulation. Our wish and belief are that the whole solutions of GCM, including the truncation error, are determined by the governing physics and dynamics of the GCM, and therefore, they should be significant. Besides, it is a common practice to produce multiple realizations of GCM integration by changing the initial condition and averaging the ensemble of solutions [41].

4. Subgrid-scale processes

In Section 3.1, I mentioned the "nonlinear viscosity" implicitly contained in some finite-difference schemes. Although nonlinear viscosity is no longer widely used, it formed an important step in the development of GCMs. Besides, the parameterization of subgrid-scale cumulus convection in the atmosphere and mesoscale eddies in the ocean are still areas of active research.

4.1. SMAGORINSKY'S NONLINEAR VISCOSITY

Smagorinsky [99] applied an explicit nonlinear viscosity scheme to his GCM, and the scheme scheme came to my notice when I joined GFDL. However, if you read his paper, you cannot find the derivation of this scheme anywhere. I borrowed his personal notes and made a copy of them. After his retirement, he was invited by a group of turbulence experts to talk about his updated view of subgrid-scale eddy viscosity [101].

In his formula, the coefficient of this viscosity is assumed to be proportional to the resolvable velocity deformation, and the final result is that the eddy viscosity is proportional to the square of the grid-size. Lilly was a member of GFDL before 1963, and he appeared to be sufficiently impressed by Smagorinsky's work on nonlinear viscosity that he continued working on it after he moved to NCAR, as we shall see shortly. In GFDL, this scheme was used in the GCM until a switch to the spectral method.

4.2. TURBULENCE CLOSURE SCHEME

In the town of Princeton, there were several scientists who were working on turbulence theories in various research facilities, for example, Mellor, Donaldson, Lewellen, and Herring. At Pennsylvania State University, not far from Princeton, similar activities were going on (involving, for example, Panofsky, Tennekes, Lumley and Wyngaard). It is not surprising, therefore, that the second-order turbulence closure model was derived by Mellor [57] and Mellor and Yamada [59], based on the closure assumptions of Rotta and Kolmogorov. George Mellor is a unique, independent and bright professor at Princeton University, who was involved in models and experiments on a neutral, rotationless turbulent boundary layer [58].

Lilly continued his work on nonlinear viscosity at NCAR, extending the treatment to three-dimensional turbulence. He later presented a paper [43] showing that Smagorinsky's formulation is exactly what is needed to cascade resolvable-scale turbulence energy to the scale of the inertial subrange. In Lilly's group, Deardorff worked mainly on boundary layer turbulence and published his turbulence theory [19]. Ironically, the latter is the direct descendant of Smagorinsky's or can be traced back even to von Neumann and Richtmeyer.

The equations derived by Mellor in Princeton and by Deardorff in Colorado are very similar to each other, except that the characteristic parameters for the closure assumption are different, as I documented in a side-by-side comparison [65]. The conceptual difference in the two turbulence schemes is the definition of turbulence, with Mellow using the ensemble average and Deardorff using the spatial average. Mellor and Yamada use the turbulent length scale l as the characteristic length scale, while Deardorff uses the grid size, Δs. As a consequence, Mellor and Herring had to propose an empirical formula for the length scale l in terms of other quantities.

These theories are appealing, because the formulation of eddy viscosity in the Earth's planetary boundary layer agrees with the hypothesis on the nocturnal jet proposed by Blackadar [7]. This hypothesis stresses the decrease of vertical mixing inside the boundary layer at night, and as a result, the inertial oscillation of wind is not weakened by vertical mixing, but remains strong. Bill Bonner [9] mentioned that major wet episodes over the central U.S. in summer occur in association with the strong low level nocturnal jet (see also [28],[78]). The low level jets bring Gulf Coast moisture efficiently into the central continent (Figure 3); note the correspondence between the jet strength (right hand panels) and rainfall in the central U.S. One of the questions is, however, that the large scale moisture transport takes place only in a certain synoptic situation. Anyway, if you believe the turbulence closure approach, Smagorinsky's nonlinear viscosity should be reasonable, though the details remain to be refined, such as the magnitude of coefficients and the balance with some neglected terms.

One of the merits of the turbulence closure approach is that the formulation

10

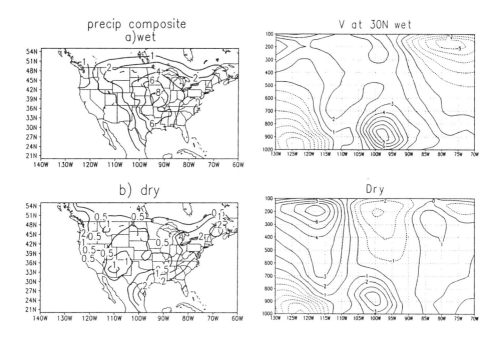

Figure 3. (left) Composite of rainfall for (a) wet events (contours 1, 2, 4, 6 and 8 mm day^{-1}) and (b) dry events (contours 0.5, 1, 2 and 3 mm day^{-1}). (right) Composite of the vertical profile of meridional wind at 30° N for (a) wet and (b) dry events; contour interval is 1 m s^{-1}. (After [68]. Reproduced by permission of the American Meteorological Society.)

and the coefficients are based on the "similarity paradigm", and that the coefficients should be determined by universal constants. Accordingly, the framework should be applicable equally to the atmosphere as well as the ocean, and even to Jupiter's atmosphere without changing the values of the constants. If a bulk method, as opposed to the turbulence closure method, is to be used for economical reasons, it should be consistent with the original multi-level framework. But this is just my personal feeling; in practice, one could be flexible to adjust the method to the observational evidence and also consider efficiency. So far as the horizontal eddy coefficients are concerned, the necessity of including the Reynolds eddy term varies, depending on the numerical scheme. Whether it is the Arakawa scheme or the Bryan scheme or the spectral scheme, the constants vary, depending upon the degree of production of deformation. In other words, the magnitude of the eddy viscosity in the free atmosphere or ocean is not so large in nature, but the deformation in finite difference calculations requires a large amount of damping. For the spectral GCM, therefore, it is a general attitude (or policy) to avoid the nonlinear formula for the horizontal Reynolds term, because there is no strong requirement. Linear formulae such as $-K_1\nabla^4\zeta$ and $-K_2\nabla^4 D$ are often used, where K_1 and K_2 are constants.

4.3. LARGE EDDY SIMULATION

One of the interesting developments from Smagorinsky's subgrid-scale eddy viscosity is the Large Eddy Simulation (LES) Studies, started by Deardorff [18] and Lilly [43]. The turbulence approach is usually framed in terms of a hierarchy of scales. The subgrid-scale transfer terms in the planetary boundary layer can be examined by using the turbulence eddies in a model of the next scale up in the hierarchy. The grid-scale turbulence in the next level of the hierarchy can be investigated by even finer grid resolution. If schemes or coefficients lower in the hierarchy are found to be inadequate, they should be corrected accordingly. This correcting process really happened, but this was successful only to a limited extent. The most crucial subgrid-scale physics in atmospheric GCMs is the "cumulus parameterization." The LES research on this process has been carried out (e.g., [116], [104]), but the conclusion has not yet been reached. The reason for the difficulty is that computers are not large enough to accommodate several cumuli simultaneously with a grid-size of, say, 1 km.

5. Climate GCMs

5.1. VARIETY OF CLIMATE GCMS

The construction of climate GCMs became active particularly after the development of the Arakawa Jacobian, and it gained further momentum after the proliferation of spectral models. Many different GCMs have emerged with different treatments of physical processes, such as the radiation, cumulus convection, planetary boundary layer processes, lateral diffusion, and, later, clouds. Overviews of the present state of these subjects can be found in Chapters 8–14.

The activity associated with climate GCMs had spread to several centers around the U.S. in the 1960's, including GFDL, UCLA, NCAR, and GISS (Goddard Institute for Space Studies, New York, NY). In those years, the main objectives for using climate GCMs were to simulate the observed climate of the atmosphere (e.g., [62], [102], [49]), and to explore the "sensitivity" of climate. Sensitivity studies (see Chapter 2, or [21]) are performed by changing some aspect of the physics or atmospheric consitituents. One of the typical examples is to change the concentration of CO_2 in the atmosphere (see Figure 4). It had been shown [33] that the atmospheric concentration of CO_2 was steadily increasing, and Manabe and Wetherald ([52], [53]) were perhaps the first to model the effects of such an increase using a GCM. These works suggested that climate change of significant magnitude may occur in future, with warming in most of the troposphere and cool-

12

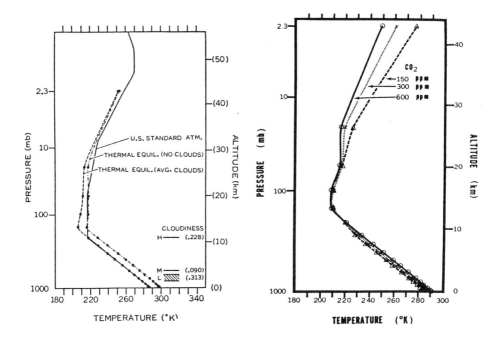

Figure 4. (left) Global mean temperature profile of the atmosphere in radiative-convective equilibrium (RCE, dashed) and the U.S. standard atmosphere (solid). (right) RCE temperature profile for various values of atmospheric CO_2 concentration in units of parts per million by volume. (After [51], [52]. Reproduced by permission of the American Meteorological Society.)

ing in the stratosphere. (It is interesting to note that they also did a reduced-CO_2 experiment.)

From their simple beginnings, climate GCMs have expanded their territories to new frontiers. They have been broadened to study the middle atmosphere, paleoclimate (Chapter 20), and atmospheric chemistry (Chapter 14). Climate system models (Chapter 18) include detailed models of processes in the ocean, on the land surface, in the biosphere and the cryosphere.

Paleoclimate offers useful lessons for the issue of increasing CO_2, because of the large variations in atmospheric CO_2 experienced in the past. Paleoclimatic records of temperature were extracted from Greenland and Antarctic ice cores and from ocean sediment cores; they contain evidence of striking changes in concentration of atmospheric CO_2, in particular during the period from about 18,000 to 9,000 years ago [38]. Based on the data published by CLIMAP [17], Gates [24] and Manabe and Broccoli [47] simulated past climate with GCMs.

5.2. CONSTRUCTION OF OCEAN GCMS

The first ocean GCMs were built by Bryan [12] at GFDL, Takano [105] at the University of Tokyo, and Sarkisyan [92] in the Soviet Union. In contrast to the acceptance of GCMs by meteorologists, the ocean GCM was not so popular among oceanographers. The main reason may have been the lack of marine observations, for which reason there had been no attempts at ocean forecasting. Another reason is that the leaders in oceanography considered GCMs as a technical exercise, as opposed to a scientific tool. I recall the discussion in 1957 with a well-known oceanography professor at the University of Tokyo. I proposed that he organize a committee to purchase or build a scientific computer at the university. He opposed this proposal, telling me that there were more important issues in oceanography. It is interesting to note that at least one pioneer in ocean GCMs, Bryan, earned his PhD not in oceanography but in meteorology.

The first attempt to couple an ocean GCM with an atmospheric GCM was carried out in a highly idealized geometry by Manabe and Bryan in 1969 [48]. They used different time steps for the integration of the atmosphere and ocean GCMs ("asynchronous integration") for the sake of economy, because they started the integration with the ocean at rest.

At present, the oceanographic model activities are divided into at least three categories. The first category is for forecasts of El Niño on the time-scale of 1–6 years. It requires the 60°N-60°S extension of the Pacific basin with 1°–2° grid resolution, and a depth of at most 1 km. The grid resolution is 1°–2° for the entire domain, but at least 1/3° grid resolution meridionally for the equatorial Pacific basin to resolve the oceanic Kelvin and Rossby waves. The salinity is not important. This kind of model emerged after 1983 [80]. The second category is for the simulation of the world ocean circulation including mesoscale eddies in the time-scale of 10–30 years, which requires at least 1/6° grid resolution, and a world ocean of 2 km depth [94]. The currents are driven by wind, and therefore, the salinity variability may not be essential. However, the treatment of high-latitude sea-ice could be important. It can be argued that, with an appropriate parameterization of mesoscale eddies (e.g., [25], and Chapter 18), 1/6° grid resolution may not be required. The third category is for the thermohaline circulation in the time-scale of 60–500 years [50], which requires 1° grid resolution, and a whole global ocean, and a depth of at least 1 km with 4 km desirable. The salinity is essential. See section 7.

6. Long range forecasts: extending the frontier

As weather GCMs developed, interest grew in making long-range forecasts (LRF). At first these consisted simply of longer runs of weather GCMs with no changes in their formulation or in the lower boundary conditions, but as the goals changed, the approach did too. Forecasts on time scales of one month or less appear to require an adequate ensemble of initial conditions, while for seasonal (3 months or more) predictions this subtlety of initial conditions appears to matter less [11] than accurate forecasts of the lower boundary conditions. Chapter 16 outlines the present state of seasonal predictions; here I describe the events that led to seasonal predictions, focusing first on the two-week to one-month timescale.

6.1. EXTENDED FORECASTS

My group at GFDL began a project of extending weather forecasts to two weeks in 1965, then we began extending forecasts to one month in 1975. Unlike the operational approach at ECMWF, which has been very successful in making 10 day forecasts, I thought that GFDL should take a different course closer to basic research. In fact, we had already started a study of utilizing the observed sea surface temperature (SST) for forecasts, but we found that SST anomalies are only effective after 20 days and we simply used climatological SST. This point was quite controversial with Jerome Namias, who was a pioneer on LRF and has stressed the importance of SST ([70], [72]), even insisting that SST influenced the atmosphere after only two days.

One of our major objectives was to forecast "blocking," an important phenomenon in the middle- and high-latitudes ([86], [26]). Blocking has a longer time scale than midlatitude cyclones and consequently is of interest in extended forecasts. Progress in understanding the dynamics of blocking has been slow.

The approaches to one-month forecasts differ from the approaches to daily weather forecasting. For one thing, one-month forecasts extend beyond the limit of predictability (see Section 1.2 of Chapter 16). Therefore, to evaluate the success of a one-month forecast, it is not appropriate to compare the observed day-to-day weather patterns with those forecasted by models. More appropriate comparisons are outlined in Chapter 16. Another difference from the daily forecasts is that multiple forecasts, or ensembles, were used earlier: in 1981. Shukla [95] presented three realizations for one-month GCM runs, concluding that the similarity among the runs was still stronger than would be the case if all information had been lost and therefore that there was still some skill. My group showed three ensemble forecasts for eight January cases [66]. Unfortunately our GCM was not good enough to obtain accurate forecasts. However, if the systematic error of the particular GCM is removed after the end of a forecast, the skill of prediction is substantially improved, and the monthly forecasts are not bad.

One way to evaluate forecasts on the largest scale is to compare the teleconnection patterns of the 500 hPa height field between the model and observations. Wallace and Gutzler [108] identified five categories of teleconnection patterns, such as the PNA (Pacific/North American), which was originally noted by Namias in 1969 [71]. These teleconnection patterns have an approximate relation with the

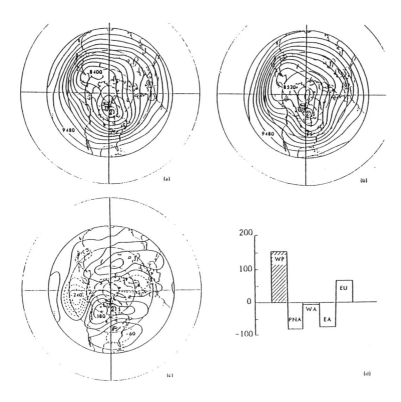

Figure 5. Geopotential height at 200 hPa for (a) negative PNA and (b) positive PNA; contour interval 120 m. (c) The difference of geopotential height between the positive and negative PNA pattern; contour interval 60 m. (d) An example of teleconnection index at 500 mb in five categories for March 1965.

leading EOFs (empirical orthogonal functions), indicating that these patterns are coherent modes of variability (see Figure 5). Blackmon *et al.* [8] showed that, if the geopotential height distribution is low-pass filtered to retain frequencies less than $(10 \text{ days})^{-1}$, there remain stationary and low-frequency planetary and some synoptic-scale waves. This means that no baroclinically unstable wave is left in these patterns, and therefore, the filtered patterns may be predictable.

6.2. WMO COMMITTEE ON LRF

The transition from extended forecasts to long-range forecasts (LRF) was slow to occur for several reasons: the state of knowledge and techniques of extended forecasts was insufficiently advanced to perform LRF; statistical LRF had poor skill, which discouraged further efforts; and LRF faced competition for able scientists from the growing field of global warming research, as exemplified by the renaming of NOAA's long-range forecast section as the "Climate Analysis Center." Nevertheless, according to a survey by the World Meteorological Organization (WMO) survey in 1978, 34 countries were carrying out scientific work on LRF; 10 meteorological services were regularly issuing one-month forecasts, while 22 services issued seasonal forecasts. A WMO committee on LRF was formed in 1980 with Gilman as the chairman, and I was a member of this group [112].

From the standpoint of the GCM, the technical capability has proceeded steadily, as described above, but the predictability limit has always bothered us. Coughlan, an Australian on the WMO committee, presented a brief report on some success of monthly and seasonal forecasts, referring to a review written by Nicholls [75]. LRF in Australia was not simply an academic exercise, but a very practical one as well because of the direct and profound influence of El Niño-Southern Oscillation (ENSO). Therefore, their experience has been the driving force of the LRF committee.

The committee planned a large meeting on LRF at Princeton [113]. I was assigned to co-chair the meeting with Gena Kurbatkin of the Soviet Union. Participants reached a consensus on several recommendations; highest priority was assigned, first to the improvement of atmospheric GCMs, and second to the development of air-sea-land coupled GCMs.

6.3. THEORY AND PREDICTION OF ENSO

The influence on the WMO's committee of the Australian viewpoint (that ENSO was of fundamental importance to LRF) was, it turns out, prescient. ENSO clearly plays an important role in the long-range evolution of the ocean and atmosphere and, as will be seen in Chapter 16, the current approach to seasonal forecasts rests on the crucial step of forecasting ENSO itself. I describe here some of the developments before 1982.

Sir Gilbert Walker first described the Southern Oscillation—the atmospheric counterpart to El Niño—in 1924 [107]. Dynamical theories developed much later, and initially focused on the ocean [115] and the atmosphere [55] separately, but already in 1969 Bjerknes [6] recognized that ENSO is inherently a coupled ocean-atmosphere phenomenon (see the reviews by Philander [79] and more recently Neelin et al. [74] and Latif et al. [39]). The first time I heard about El Niño was from Bjerknes himself in 1967, and since then it has been one of my group's subjects (e.g., [91]). In 1983, the new term "ENSO" was proposed, reflecting the atmospheric as well as oceanic perspective.

A connection between El Niño and the extratropics was inferred by Bjerknes in 1961 [5] and was extensively studied by 1982 ([6], [91], [31], [29], [34], [110]). According to the hypothesis of Horel and Wallace [29], precipitation induced by SST anomalies in the tropical Pacific acts as a source of heating in the atmosphere. This heating excites an atmospheric Rossby wavetrain [30] that propagates into

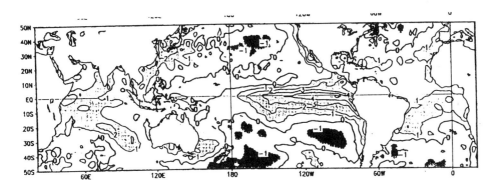

Figure 6. SST anomalies for January 1998. Contour interval is 1°C. (After an NCEP report.)

higher latitudes along (roughly) a great circle ray path.

From the standpoint of LRF, propagation of a Rossby wavetrain alone is not sufficient for making good forecasts. The inherent predictability of the extratropical atmosphere is an important aspect. In fact, high potential predictability was speculated in the tropics rather than in the extratropics, but tropical disturbances such as easterly waves and tropical depressions, to say nothing of the much larger extratropical disturbances, can disturb the El Niño signals and prevent accurate LRF for the extratropics. Concerning this problem, Charney and Shukla [16] suggested that since the large scale monsoon circulation is dynamically stable and since the (slowly evolving) lower boundary conditions exert significant influence on the time averaged monsoon flow, the monsoon circulation must be potentially predictable. Shukla [96] speculated further, based on his GCM experiment, that although the tropical disturbances are less predictable, the equilibration of their amplitudes is quite rapid and the upscale-cascade process (small to large eddies) is weak, due to the smallness of the Coriolis parameter there. In other words, interaction of large-scale overturnings with the tropical disturbances is not strong enough to detract from the predictability of the planetary scale circulation.

6.4. OUTSTANDING EL NIÑOS

A strong El Niño event in 1972–73 stimulated research on the subject by both meteorologists and oceanographers ([114], [111], [84], [85], [61]) using observations, theory, and GCMs. With this preparation, the 1982–83 El Niño created a great deal of enthusiasm toward understanding the phenomenon. The 1997–98 El Niño brought very warm SST (comparable to that for 1982–83; see Figure 6) to the eastern Equatorial Pacific and was associated with a warm winter to the eastern U.S. and heavy rains and mudslides to the western U.S.

In order to utilize the information of ENSO for the forecasts of middle latitudes, it is a prerequisite for a GCM to have good teleconnection capability from the equatorial Pacific to the U.S. David Straus at COLA (Center for Ocean-Land-Atmosphere Studies) investigated the teleconnections for four GCMs: those of COLA, CPC (Climate Prediction Center), NCEP (National Center for Environmental Prediction) and NCAR [97]. Each group carried out simulations for three months, specifying the observed SST over the globe. The forecasts were done for about 15 winters from 1982 to 1997. The ensemble sizes are different for each group, ranging from 5 to 10. Straus performed SVD (singular value decomposition) analysis [109], using winter mean SST on one side and 500 hPa anomalies of analyzed geopotential height on the other side. He then regressed the time series of simulated seasonal mean height anomalies for each GCM on the time series of the first SVD mode and obtained the variance patterns for the four GCMs (Figure 7). The transmission of the El Niño effect from the equatorial Pacific to the mid-latitudes turns out to be so delicate that various atmospheric GCMs gave different solutions. This figure indicates a very important point; that is, even given perfect knowledge of the future state of ENSO, the details as to how the midlatitudes will respond are still murky, and more accurate GCMs may be needed. At this stage, it is not clear why the models give such different results.

7. Future issues of GCM forecasts

Before concluding my retrospective look at the development of GCMs, I wish to make a few remarks about developments in ENSO prediction and other topics since 1982. Modelers developed coupled air-sea models, with oceanographers taking the lead in investigating El Niño. For example, Cane and Zebiak [14] built a simple coupled model and made successful predictions, though their model is too simple to be classified as a GCM. Schopf and Suarez [93] generated El Niño processes with a coupled GCM and, based on the results, proposed the "delayed oscillator" theory for ENSO. For this timescale, the oceanic part sets the slow part of the processes, and the atmospheric part acts as a quick linear response.

Sensitivity studies of CO_2 described in section 5 (and in Chapter 21) have now gained overwhelming support by the public. The IPCC has issued every year a (International Panel on Climate Change) has issued every year a projection of the global trend of the atmosphere temperature related to the scenario of increasing CO_2 and aerosols, which has been calculated by several GCMs around the world. Associated with this report, one issue is that the GCM projections for IPCC contain some caviats; for example, the GCM runs lack a proper ENSO, many use "flux correction", and all use uncertain estimates of future greenhouse gas and aerosol concentrations and distributions. One thing we can do about the first two of these caviats is the stepwise extension of forecast range and the confirmation of the realizability of GCM simulation for all time-scale variations.

The extension of forecast range begs the question, What lead time will people find interesting? My view is that people would not care too much for the fore-

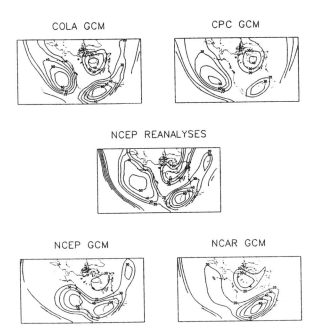

COLA GCM   CPC GCM

NCEP REANALYSES

NCEP GCM   NCAR GCM

Figure 7. 500 hPa geopotential height variances explained by tropical SST in terms of SVD analysis. GCM simulations of 4 groups, i.e., COLA, CPC, NCEP, and NCAR are presented. The pattern in the middle is the NCEP Re-analysis, which is regarded as the reference. Contours represent the percentage. (After [97].)

casts on time scales longer than El Niño, because the ambiguity of forecasts is so large. However, they may care about the climate-change time scale (20–100 years) which is the focus of the IPCC reports.

7.1. INTERANNUAL VARIATIONS

The periodicity of ENSO is about 4 years with the warm phase, El Niño, and the cold phase, La Niña, at both extrema. In order to unravel the mechanism of the teleconnection phenomena associated with ENSO, a coupled ocean-atmosphere GCM (see Chapters 16 and 18) is needed, in which the ocean GCM has the characteristics described in the first category in section 5. Of course, they already exist, but they should be refined so that the ENSO simulation is more accurate. The simulation of various monsoons should be correct as well, because these monsoons are very likely to be in the relation of cousin with El Niño; the trigger could be common for both monsoons and El Niño. The oscillation is not regular, which is one of the key aspects for this phenomenon. If the time variation is made clear,

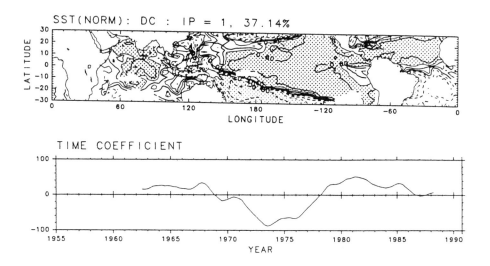

Figure 8. EOF 1 for the decadal component of the tropical ocean SST between 30°S and 30°N. This mode can account for 37.14% of the total variance of the decadal component. (top) the distribution of SST; contour interval is 0.2° C, and (bottom) time series of the coefficients. (After [32]; reproduced by permission of the Meteorological Society of Japan.)

the knowledge would contribute appreciably to more reliable El Niño forecasts as well.

7.2. INTERDECADAL VARIATIONS

Several authors (e.g., [73], [76], [106]) have noted an abrupt shift in the Pacific Ocean in the late 1970s. For example, Namias *et al.* [73] studied interannual variability of North Pacific SST anomalies. Kachi and Nitta [32] studied interannual and interdecadal variations in the global atmosphere and ocean, focusing on SST (see Figure 8), sea-level pressure, surface winds, and 500 hPa geopotential height. Variations at the interdecadal time-scale are dominant over the entire Pacific Ocean, whereas variations at the ENSO time-scale are dominant only over the equatorial Pacific.

Nakamura *et al.* [69] found that the strongest variability of SST associated with this slow oscillation (> 7 years) is located around the subarctic front in the Pacific at 42°N. They also found that the SST is increasing over the tropical central and eastern Pacific, and is decreasing in the mid-latitudes of North and

South Pacific. The winter PNA pattern has been more intense recently, compared to the period before 1970's. The interdecadal variation of the ocean in the northern Pacific may be proceeding independently of tropical processes.

In order to simulate this variability, which may involve the thermohaline "conveyer belt" circulation, an ocean GCM in the second category (see Section 5.2) is needed. There have been some attempts at simulation with an ocean GCM of this category (e.g., [22]) but to treat the problem properly requires long simulations with a coupled atmosphere-ocean GCM. Such a simulation could lead to an extraordinary increase in our knowledge of the world ocean circulation.

7.3. HALF CENTURY VARIATION

Manabe and Stouffer [50] investigated the results from a fully coupled atmosphere-ocean climate model for realistic ocean basin geometry. They showed that there is a solution that represents Atlantic overturning and North Atlantic Deep Water formation. On the other hand, another solution does not include North Atlantic sinking, as had been predicted by Stommel in 1961 [103]. Figure 9 shows the surface air temperature difference between the two climatic equilibria (see [27]). Further study [20] has revealed that this model has irregular oscillations of the thermohaline circulation with a time scale of approximately 50 years. The irregular oscillation appears to be driven by density anomalies in the ocean, and is triggered by nearly random surface buoyancy forcing of heat and water fluxes. The sea surface temperature fluctuations associated with this variations have a spatial pattern that bears resemblance to a pattern of observed interdecadal variability [37].

8. Epilogue

This essay is my recollection of progress and events prior to 1982. However, without the 1997–98 El Niño, confidence in the utility of LRF would not be established. Now people on the street know about the existence of El Niño and of LRF. Perhaps, they don't care whether the LRF is monthly or seasonal forecast, but they are concerned about whether the gradual increase in the intensity of El Niño from 1972–73 to 1997–98 is due to natural variability or due to global warming. Perhaps it is too early to say.

Although climate GCMs have not been my research subject, I worked at GFDL for 30 years and observed the climate research of my colleagues based on climate GCMs, particularly Manabe, Bryan, Mahlman (middle atmosphere), Holland (ocean), Williams (planetary atmospheres), Philander (El Niño/La Niña), and Sarmiento (atmospheric trace constituents). Manabe's works have been really fantastic. He used a variety of methods to stay on the cutting edge: "flux adjustment", "moist convective adjustment", "asynchronous integration", "bucket soil

22

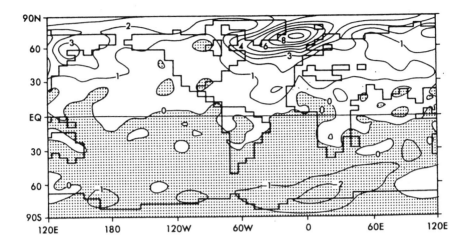

Figure 9.   Difference in surface air temperature (degrees C) between two solutions in a coupled air-sea model, the state with North Atlantic sinking minus the state without. (After [50]. Reproduced by permission of the American Meteorological Society.)

moisture", to name a few. The essence of his GCM approach has been simplicity. Perhaps this is an important principle for a first generation modeler or model user.

Acknowledgements

I wish to express my gratitude to Drs. Joe Smagorinsky, Akio Arakawa, George Mellor, Bill Bourke, Suki Manabe, Kingtse Mo, Hisashi Nakamura, and David Williamson, who reviewed this memoir and gave me many valuable suggestions and comments. I thank Drs. Jim Kinter and Uma Bhatt, COLA, and Drs. Jassim Al-Saadi and Evi Schuepbach, participants in the NATO ASI, who checked the logic, structure, and English of this essay. I also wish to express my gratitude to Dr. David Straus, who kindly showed me his research results, and to Mr. Curt Steinmetz, who patiently helped me to prepare this manuscript.

References

1.   Arakawa, A. (1966) Computational design for long-term numerical integration of the equations of fluid motion: two-dimensional incompressible flow. Part 1, *Comp. Phys.* **1,** 119–143.

2.   Arakawa, A. and Lamb, V.R. (1981) A potential enstrophy and energy conserving scheme for the shallow water equation. *Mon. Wea. Rev.* **109**, 18–36.

3.   Baer, F. and Platzman, G.W. (1961) A procedure for numerical integration of the spectral vorticity equation, *J. Meteor.* **18**, 393–401.

4.   Bates, J.R. and McDonald, A. (1982) Multiply-upstream, semi-Lagrangian advective schemes: Analysis and application to a multilevel primitive equation model, *Mon. Wea. Rev.* **112**, 1831–1842.

5.   Bjerknes, J. (1961) El Niño study based on analysis of ocean surface temperatures 1935-57, *Bull. Inter-Amer. Trop. Tuna Comm.* **5**, 219–303.

6.   Bjerknes, J. (1969) Atmospheric teleconnections from the equatorial Pacific, *Mon. Wea. Rev.* **97**, 165–172.

7.   Blackadar, A.K. (1957) Boundary layer wind maxima and their significance for the growth of nocturnal jet, *Bull. Amer. Meteor. Soc.* **38**, 283–290.

8.   Blackmon, M.L., Wallace, M., Lau, N.-C., and Mullen, S.L. (1977) An observational study of the Northern Hemisphere wintertime circulation, *J. Atmos.Sci.* **34**, 1040–1053.

9.   Bonner, W.D. (1968) Climatology of the low level jet, *Mon. Wea. Rev.* **96**, 833–850.

10.   Bourke, W. (1972) An efficient, one level, primitive-equations spectral model, *Mon. Wea. Rev.* **100**, 683–689.

11.   Brankovic, C. and Palmer, T.N. (1997) Atmospheric seasonal predictability and estimates of ensemble size, *Mon. Wea. Rev.* **125**, 859–874.

12.   Bryan, K. (1963) A numerical investigation of a nonlinear model of a wind-driven ocean, *J. Atmos. Sci.* **20**, 594–606.

13.   Bryan, K. (1966) A scheme for numerical integration of the equations of motions on an irregular grid free of non-linear instability, *Mon. Wea. Rev.* **94**, 39–40.

14.   Cane, M.A. and Zebiak, S.E. (1986) A theory for El Niño-Southern Oscillation, *Science* **228**, 1085–1087.

15.   Charney, J.G., Fjortoft, R., and von Neumann, J. (1950) Numerical integration of the barotropic vorticity equation, *Tellus* **1**, 38–54.

16.   Charney, J.G. and Shukla, J. (1981) Predictability of monsoons, in J. Lighthill and R. Pearce, Eds., *Monsoon Dynamics*, Cambridge University Press, Cambridge.

17.   CLIMAP Project Members (1976) The surface of the ice-age Earth, *Science* **191**, 1131–1136.

18.   Deardorff, J.W. (1970) Convective velocity and temperature scales for the unstable planetary boundary layer and Rayleigh convection, *J. Atmos. Sci.* **27**, 1211–1213.

19.   Deardorff, J.W. (1973) The use of subgrid transport equations in a three-dimensional model of atmospheric turbulence, *ASME Publication,* 1–10.

20.   Delworth, T., Manabe, S., and Stouffer, R.J. (1993) Interdecadal variations of the thermohaline circulation in a coupled ocean-atmosphere model, *J. Climate* **6**, 1993–2011.

21.   Dickinson, R.E. (1987) Climate sensitivity, *Adv. In Geophys.* **28A**, 99–129.

22.   Dukowicz, J.K. and Smith, R.D. (1994) Implicit free-surface method for the Bryan-Cox-Semtner model, *J. Geophys. Res.* **99**, 7991–8014.

23.   Eliasen, E., Machenhauer, B., and Rasmussen, E. (1971) On a numerical method for integration of the hydrodynamical equations with a spectral representation of the horizontal fields, *Tech. Rep.,* Copenhagen University, Copenhagen.

24.   Gates, W.L. (1976) Modeling the ice age climate, *Science* **191**, 1138–1144.

25.   Gent, P.R. and McWilliams, J.C. (1990) Isopycnal mixing in ocean circulation models, *J. Phys. Oceanogr.* **20**, 150–155.

26.   Green, J.S.A. (1977) The weather during July 1976: some dynamical considerations of the drought, *Weather* **32**, 120–128.

27.   Held, I.M. (1993) Large-scale dynamics and global warming, *Bull. Amer. Met. Soc.* **74**, 228–241.

28. Helfand, H.M. and Schubert, S.D. (1995) Climatology of the Great Plains low-level jet and its contribution to the continental moisture budget of the United States, *J. Climate* **8**, 784–806.

29. Horel, J.D. and Wallace, J.M. (1981) Planetary-scale atmospheric phenomena associated with the interannual variability of sea surface temperature in the equatorial Pacific, *Mon. Wea. Rev.* **109**, 813–829.

30. Hoskins, B.J. and Karoly, D.J. (1981) The steady linear response of a spherical atmosphere to thermal and orographic forcing. *J. Atmos. Sci.* **38**, 1179–1196.

31. Julian, P.R. and Chervin, R.M. (1978) A study of the Southern Oscillation and Walker Circulation phenomenon, *J. Atmos. Sci.* **106**, 1433–1451.

32. Kachi, M. and Nitta, T. (1997) Decadal variations of the global atmosphere-ocean system, *J. Meteor. Soc. Japan* **75**, 657–675.

33. Keeling, C.D. (1958) The concentration and isotopic abundances of atmospheric carbon dioxide in rural areas, *Geochemica et Cosmochica Acta* **13**, 322–334.

34. Keshavamurty, R.N. (1982) Response of the atmosphere to sea surface temperature anomalies over the equatorial Pacific and the teleconnections of the Southern Oscillation, *J. Atmos. Sci.* **39**, 1241–1259.

35. Kubota, S. (1960) Surface spherical harmonic representation of system of equations for analysis, *Papers Meteor. Geophys.* **10**, 145–166.

36. Kurihara, Y. and Holloway, J.L. (1967) Numerical integration of a nine-level global primitive equations model formulated by the box method, *Mon. Wea. Rev.* **95**, 509–530.

37. Kushnir, Y. (1993) Interdecadal variations in North Atlantic sea surface temperature and associated atmospheric conditions, *J. Climate* **7**, 141–157.

38. Kutzbach, J.E. (1985) Modeling of paleoclimates, *Advance in Geophys.* **28A**, 159–196.

39. Latif, M., Anderson, T., Barnett, T., Cane, M., Kleeman, R., Leetma, A., O'Brien, J., Rosati, A., and Schneider, E. (1998) A review of the predictability and prediction of ENSO, *J. Geophys. Res.* **103**, 14,375–14,393.

40. Lax, P. and Wendroff, B. (1960) Systems of conservation laws. Communications on *Pure and Applied Mathematics* **13**, 217–237.

41. Leith, C.E. (1974) Theoretical skill of Monte Carlo forecasts, *Mon. Wea. Rev.* **102**, 409–418.

42. Lewis, J.M. (1998) Clarifying the dynamics of the general circulation: Phillips's 1956 experiment, *Bull. Amer. Meteor. Soc.* **79**, 39–60.

43. Lilly, D.K. (1967) The representation of small-scale turbulence in numerical simulation experiments, *Proc. IBM Sci., Computing Symp. Environmental Sci.*, IBM Form No. **320–1951**, 195–210.

44. Machenhauer, B. (1979) The spectral method. Numerical methods used in atmospheric models, *GARP Publ. Series* **11**, No. 17, 121–275.

45. Machenhauer, B. (1997) The spectral method, *Numerical methods used in atmospheric models* **11**, GARP Publ. Series No. 17, pp. 121–275.

46. Manabe, S., Smagorinsky, J., and Strickler, R.F. (1965) Simulated climatology of a general circulation model with a hydrologic cycle, *Mon. Wea. Rev.* **93**, 769–798.

47. Manabe, S., and Broccoli, T. (1985) A comparison of climate model sensitivity with data from the last glacial maximum, *J. Atmos. Sci.* **42**, 2643–2651.

48. Manabe, S. and Bryan, K. (1969) Climate calculations with a combined ocean-atmosphere model, *J. Atmos. Sci.* **26**, 786–789.

49. Manabe, S., Hahn, D.G., and Holloway, J.L. (1974) The seasonal variation of the tropical circulation as simulated by a global model of the atmosphere, *J. Atmos. Sci.* **31**, 43–83.

50. Manabe, S. and Stouffer, R.J. (1988) Two stable equilibria of a coupled ocean-atmosphere model, *J. Climate* **1**, 841–866.

51. Manabe, S. and Strickler, R.F. (1964) Thermal equlibrium of the atmosphere with convective adjustment, *J. Atmos. Sci.* **21**, 361–385.

52. Manabe, S. and Wetherald, R.T. (1967) Thermal equlibrium of the atmosphere with a given distribution of relative humidity, *J. Atmos. Sci.* **24**, 241–259.
53. Manabe, S. and Wetherald, R.T. (1975) The effect of doubling the CO_2 concentration of the climate of a general circulation model, *J. Atmos. Sci.* **32**, 3–15.
54. Manabe, S. and Wetherald, R.T. (1983) CO_2 and hydrology, *Adv. in Geophys.* **28A**, 131–156.
55. Matsuno, T. (1966) Quasi-geostrophic motions in equatorial areas, *J. Meteor. Soc. Japan* **2**, 25–43.
56. McAvaney, B.J., Bourke, W.P., and Puri, K.K. (1978) A global spectral model for simulation of the general circulation, *J. Atmos. Sci.* **35**, 1557–1583.
57. Mellor, G.L. (1973) Analytic prediction of the properties of stratified planetary surface layers, *J. Atmos. Sci.* **30**, 1061–1069.
58. Mellor, G.L. (1985) Ensemble average, turbulence closure, *Adv. in Geophys. 28B*, 345–358.
59. Mellor, G.L. and Yamada, T. (1974) A hierachy of turbulence closure models for planetary boundary layers, *J. Atmos. Sci.* **31**, 1791–1806.
60. Mesinger, F., and Arakawa, A. (1976) Numerical methods used in atmospheric models, *GARP Publication Series* **1**, WMO/ICSU Joint Organizing Committee, Geneva.
61. Miller, F.R. and Laurs, R.M. (1975) The El Niño of 1972–73 in the eastern tropical Pacific Ocean, *Bull. Inter-Amer. Trop. Tuna Comm.* **16**, 403–448.
62. Mintz, Y. (1965) Very-long-term global integration of the primitive equations of atmosphereric motion: an experiment in climate simulation, *WMO Tech. Notes* no. 66, pp 141–167; and *Amer. Meteor. Soc. Monograph* **8**, no. 30., pp 30–36 (1968).
63. Miyakoda, K. (1960) A trial of 500 hour barotropic forecast, In: *Proc. Internat. Sym. In Numer. Wea. Pred. In Tokyo,* Nov. 7–13, 1960, pp. 221–240.
64. Miyakoda, K., Moyer, R.W., Stambler, H., Clarke, R.H, and Strickler, R.F. (1971): A prediction experiment with a global model of the Kurihara-grid, *J. Meteor. Soc. Japan* **49**, 521–536.
65. Miyakoda, K. and Sirutis, J. (1977) Comparative integrations of global models with various parameterized processes of subgrid-scale vertical transports. Description of the parameterisations, *Contrib. Atmos. Phys.* **50**, 445–487.
66. Miyakoda, K. and Sirutis, J. (1985) Extended range forecasting, *Advances in Geophysics* **28B**, 55–85.
67. Miyakoda, K., Smagorinsky, J., Strickler, R.F., and Hembree, G.D. (1969) Experimental extended predictions with a nine-level hemispheric model, *Mon. Wea. Rev.* **97**, 1–76.
68. Mo, K.C., Paegle, J.N., and Higgins, R.W. (1997) Atmospheric processes associated with summer floods and droughts in the central United States, *J. Climate* **10**, 3028–3046.
69. Nakamura, H., Lin, G., and Yamagata, T. (1997) Decadal climate variability in the North Pacific during the recent decades, *Bull. Amer. Meteor. Soc.* **78**, 2215–2225.
70. Namias, J. (1962) Influences of abnormal surface heat sources and sinks on atmospheric behavior, in *Proc. Internat. Sym. In Numer. Wea. Pred., Tokyo,* Meteor. Soc. Japan, pp. 615–627.
71. Namias, J. (1969) Seasonal interactions between the North Pacific Ocean and the atmosphere during the 1960's, *Mon. Wea. Rev.* **97**, 173–192.
72. Namias, J. (1976) Negative ocean-air feedback systems over the North Pacific in the transition from warm to cold seasons. *Mon. Wea. Rev.* **104**, 1107–1121.
73. Namias, J., Yuan, X., and Cayan, D.R. (1988) Persistence of North Pacific sea surface temperature and atmospheric flow patterns, *J. Climate* **1**, 682–703.
74. Neelin, J.D., Battisti, D.S., Hirst, A.C., Jin, F.-F., Wakata, Y., Yamagata, T., and Zebiak, S.E. (1998) ENSO theory, *J. Geophys. Res.* **103**, 14,261–14,290.
75. Nicholls, N. (1980) Long-range weather forecasting, value, status, and prospects, *Rev. Geophys. Space Phys.* **18**, 771–788.

76. Nitta, T. and Yamada, S. (1989) Recent warming of tropical sea surface temperature and its relationship to the Northern hemisphere circulation, *J. Meteor. Soc. Japan* **71**, 367–375.

77. Orszag, S.A. (1970) Transform method for the calculation of vector-coupled sums: Application to the spectral form of the vorticity equation, *J. Atmos. Sci.* **27**, 890–895.

78. Paegle, J., Mo, K.C., and Paegle, J.N. (1996) dependence of simulated precipitation on surface evaporation during the 1993 United States summer floods, *Mon. Wea. Rev.* **124**, 345–361.

79. Philander, S.G.H. (1990) *El Niño, La Niña, and the Southern Oscillation*, Academic Press, San Diego.

80. Philander, S.G.H. and Seigel, A.D. (1984) Simulation of El Niño of 1982–1983, in J.G.J. Nihoul (ed.),*Coupled-Atmosphere Models*, Elsevier Oceanography Series No. 40, pp. 517–541.

81. Phillips, N.A. (1956) The general circulation of the atmosphere: a numerical experiment, *Quart. J. Roy. Meteor. Soc.* **82**, 123–164.

82. Platzman, G.W. (1960) The spectral form of the vorticity equation, *J. Meteor.* **17**, 635–644.

83. Platzman,G.W. (1990) The atmosphere—A challenge, Charney's recollection, in R. Lindzen, E. Lorenz, and G.W. Platzman (eds.), *The atmosphere—A challenge. The science of Jule Gregory Charney,* American Meteorological Society, Boston, pp. 11–88.

84. Quinn, W.H. (1974) monitoring and predicting El Niño invasions, *J. Appl. Meteor.* **13**, 825–830.

85. Ramage, C.S. (1975) Preliminary discussion of the meteorology of the 1972–73 El Niño, *Bull. Amer. Meteor. Soc.* **56**, 234–242.

86. Rex, D.R. (1950) Blocking action in the middle troposphere and its effect upon regional climate. I. An aerological study of blocking action, *Tellus* **2**, 169–211.

87. Richtmeyer, R.D. and Morton, R. (1967) Difference method for initial value problem, 2nd edition, Interscience Publ., New York.

88. Robert, A. (1966) The integration of a low-order spatial form of the primitive meteorological equations, *J. Meteor. Soc. Japan* **44**, 237–245.

89. Robert, A. (1968) The treatment of moisture and precipitation in atmospheric models integrated by the spectral method, *J. Appl. Meteor.* **7**, 730–735.

90. Robert, A. (1982) A semi-Lagrangian and semi-implicit numerical integration scheme for the primitive meteorological equations, *J. Meteor. Soc. Japan* **60**, 319–325.

91. Rowntree, P.R. (1972) The influence of tropical east Pacific Ocean temperature on the atmosphere, *Quart. J. Roy. Meteor. Soc.* **98**, 290–321.

92. Sarkisyan, A.S. (1966) *Fundamentals of the theory and calculation of ocean currents* (in Russian), Gidrometeoizdat, Moscow.

93. Schopf, P.S. and Suarez, M.J. (1988) Vacillation in a coupled ocean-atmosphere model, *J. Atmos. Sci.* **45**, 549–566.

94. Semtner, B. and Chervin, R.M. (1988) A simulation of the global ocean circulation with resolved eddies, *J. Geophys. Res.* **93**, 15,502–15,775.

95. Shukla,J. (1981) Dynamical predictability of monthly means, *J. Atmos. Sci.* **38**, 2547–2572.

96. Shukla, J. (1984) Predictability of time averages: Part II: The influence of the boundary forcings, in D.M. Burridge and E. Källen (eds.) *Problems and prospects in long and medium range weather forecasting*, Springer-Verlag, Berlin, pp 155–200.

97. Shukla, J., Anderson, J., Baumhefner, D., Brankovic, C., Chang, Y., Kalnay, E., Marx, L., Palmer, T., Paolino, D., Ploshay, J., Schubert, S., Straus, D., Suarez, M., and Tribbia, J. (2000) Dynamical seasonal prediction, *Bull. Amer. Meteor. Soc.*, under review.

98. Silberman, I.S. (1954) Planetary waves in the atmosphere, *J. Meteor.* **11**, 27–34.

99. Smagorinsky, J. (1963) General circulation ecxperiments with the primitive equations: I, The basic experiment, *Mon. Wea. Rev.* **91**, 99–164.

100. Smagorinsky, J. (1983) The beginning of numerical weather prediction and general circulation modeling. Early recollections, *Adv. Geophys.* **25**, 3–37.

101. Smagorinsky, J. (1993) Some historical remarks on the use of nonlinear viscosities, in B. Galperin and S.A. Orszag (eds.) *Large Eddy Simulation of Complex Engineering and Geophysical Flows, Proc. Internat. Workshop in LES, St. Petersburg, FL,* Dec. 19–21, 1990, Cambridge University Press, pp. 1–34.

102. Smagorinsky, S., Manabe, S., and Holloway, J.L. (1965) Numerical results from a nine-level general circulation model of the atmosphere, *Mon. Wea. Rev.* **93**, 727–768.

103. Stommel, H. (1961) Thermohaline convection with two stable regimes of flow, *Tellus* **13**, 224–230.

104. Sui, C.H., Li, X., and Lau, K.M. (1998) Radiative-convective processes in simulated diurnal variations of tropical oceanic convection, *J. Atmos. Sci.* **55**, 2345–2357.

105. Takano, K. (1974) A general circulation model of the world ocean, Numerical simulation of weather and climate, Tech. Report No. 8, Dept. of Meteorology, UCLA, 47 pp.

106. Trenberth, K.E. (1990) Recent observed interdecadal climate changes in the Northern Hemisphere, *Bull. Amer. Met. Soc.* **71**, 988–993.

107. Walker, G.T. (1924) Correlation in seasonal variations of weather, IX. A further study of world weather, *Memoirs of the India. Meteor. Depart.* **24**, 275–352.

108. Wallace, J.M. and Gutzler, D.S. (1981) Teleconnections in the geopotential height field during the Northern Hemisphere winter, *Mon. Wea. Rev.* **109**, 784–812.

109. Wallace, J.M., Smith, C., and Bretherton, C. (1992) Singular value decomposition of wintertime sea surface temperature and 500-mb height anomalies, *J. Climate* **5**, 561–576.

110. Webster, P.J. (1982) Seasonality in the local and remote atmospheric response to sea surface temperature anomalies, *J. Atmos. Sci.* **39**, 41–52.

111. Wooster, W.S. and Guillen, O. (1979) Characteristics of El Niño in 1972, *J. Mar. Res.* **32**, 387–404.

112. WMO (World Meteorological Organization) (1980) Report of the informal meeting of experts on long-range forecasting. Progress on short-, medium-, and long-range weather prediction research, Geneva, 1–10.

113. WMO (1982) The WMO-CAS/JSC expert study meeting on long-range forecasting, Public. Ser. No.1, pp238.

114. Wyrtki, K. (1973) Teleconnections in the equatorial Pacific Ocean, Science, 180, 66–68.

115. Wyrtki,K. (1975) El Niño — the dynamic response of the equatorial Pacific Ocean to atmospheric forcing, *J. Phys. Oceanogr.* **5**, 572–584.

116. Yamazaki, M. (1996) Study on CISK and cloud system, in A. Sumi (ed.), *Evolution from Meteorolgical to Climatological Dynamics* (in Japanese), Center for Climate System Research, Tokyo, pp 75–96.

THE GCM AS A DYNAMICAL SYSTEM

Implications for numerical simulations

J.-F. ROYER
Météo-France, CNRM
42 Av. G. Coriolis, 31057 Toulouse Cedex 1, France

1. Introduction

A general circulation model (GCM) can be viewed as a piece of software for generating pseudo-climatological series. However, it is conceptually different from the statistical climatological generators [102] [41], in that it tries to reproduce in detail the dynamical and physical mechanisms operating in the climate system. Following Lorenz [53], who proposed to formulate climatic change as a mathematical problem, it is a good starting point to attempt to describe formally the atmospheric system and its numerical representation in the framework of dynamical system theory, in order to bring out more clearly the underlying mathematical structure. Dynamical system theory has in fact been developed as a unifying language for the description of similar nonlinear phenomena appearing in very diverse scientific disciplines [71] [101] .

2. Formal Description of a Climate Model as a Dynamical System

A GCM represents a portion $\mathcal{M}$ of the climate system; $\mathcal{M}$ can be defined precisely by specifying its spatial boundary $\mathcal{B}$. This boundary is, in general, permeable, and the fluxes of matter, momentum, and energy that cross it have to be specified as boundary conditions. In the case of a global GCM, the model encompasses the whole lower atmosphere, thus eliminating lateral boundaries, while in the case of a Limited Area Model (LAM) or Regional Climate Model (RCM) [28], the fluxes through the lateral boundaries of the chosen region have to be provided (see Chapter 17). The upper boundary is, in general, some high atmospheric level through which fluxes of matter can be neglected, while fluxes of energy are limited to radiative exchange with the outside interplanetary space, both by incoming

29

P. Mote and A. O'Neill (eds.), Numerical Modeling of the Global Atmosphere in the Climate System, 29–58.
© 2000 *Kluwer Academic Publishers. Printed in the Netherlands.*

and reflected shortwave solar radiation and by outgoing longwave infrared radiation emitted by the earth-atmosphere system [45] (see Chapter 11). The lower boundary is usually set at, or below, the earth's surface. In the case of a coupled atmosphere-ocean GCM, this lower boundary will be at some prescribed ocean depth below the thermocline in the case of an ocean-mixed layer model, or eventually at the bottom of the ocean in the case of a full dynamical ocean model (as discussed in Chapter 18). In a purely Atmospheric GCM (AGCM) the lower boundary is generally specified just at the ocean surface, and the fluxes are computed by prescribing the sea surface temperature. Over the continents, the lower boundary is usually set at some depth inside the soil in order to include as part of the AGCM the strong interactions between the upper layers of the soil and the atmosphere (see Chapter 13). Such interactions are acting even at short time scales, which makes it preferable to treat the upper layers of the ground as interactively coupled to the atmospheric boundary layer.

The collection of all the relevant dynamical and physical variables necessary to describe uniquely the internal state of the portion of the earth-atmosphere system contained within the boundary $\mathcal{B}$ are called state variables and are symbolized by the infinite-dimensional vector $\mathcal{X}$. The boundary conditions, which describe the influence of the outside universe upon the system, will be considered as an "external forcing" represented by a vector of control variables $\mathcal{U}_t = \mathcal{U}(t)$, the evolution of which needs to be specified completely as a function of time t in order to determine uniquely the future evolution of the system $\mathcal{M}$. The operator describing the physical laws governing the evolution of the system is denoted by a functional $\mathcal{F}$, which symbolizes in a compact notation various kinds of mathematical operations like partial differentiation, convolution, and integration, as well as various nonlinear operators.

The continuous mathematical model, which is specified by an appropriate choice of variables and equations representing the known physical laws of the earth-atmosphere system, can be cast symbolically into the form

$$\frac{d\mathcal{X}}{dt} = \mathcal{F}(\mathcal{X}, \mathcal{U}_t). \tag{1}$$

This underlying abstract model has to be truncated in some way in order to transform it from an infinite-dimensional problem into a finite- (though huge-) dimensional approximation, for which a numerical representation and solution can be sought by numerical methods (as discussed in Chapter 7). After either a discretization in space and time or a projection onto a finite set of basis functions, the derived numerical model can be represented as

$$X_{p+1} = F(X_p, U_p) \tag{2}$$

where the vector $X_p = X(t_p)$ is now an N-dimensional vector of real numbers representing the complete collection of all the model prognostic variables that are sufficient to determine the internal state of the model. The number N of prognostic variables is also called the number of degrees of freedom of the model. The coordinates of the state vector X can represent various things like the values of a physical variable at a grid point, or more abstract concepts like the real and imaginary components of spectral coefficients in a decomposition of horizontal fields into spherical harmonics at some vertical level. Besides the prognostic variables, the model will also contain certain diagnostic variables Y, but because (by definition) these variables can be computed by a diagnostic relationship $Y = G(X, U)$ from the prognostic variables, they can be (and indeed have been) eliminated from the model's formal definition written above.

In mathematical language, a GCM represented by the above equation is an "iterated mapping" operating in the N-dimensional vector state space X (also called "phase space"). An iteration represents a time-step of the model that transforms the model state X_p representing the atmosphere at time t_p to its new state X_{p+1} at time $t_p + \delta t$, where δt is the time step. It is well-known that the time step cannot be chosen arbitrarily and is limited by stability constraints to smaller and smaller values when the spatial resolution is increased (see Chapter 7). The successive positions of the state vectors X_p generate what is called an orbit or trajectory in state space. When the external conditions are held constant, the control vector U is independent of time and the mapping is said to be autonomous, which means that the orbit of X at all future times is completely determined once given an initial position X_0 for starting the iterations. This kind of idealized simulation is sometimes applied in so-called perpetual month (usually January or July) simulations using a daily average constant insolation and fixed surface conditions [106].

For such an autonomous system, only the choice of the initial starting point X_0 and of the model parameters determines the whole subsequent evolution of the system and all its statistical properties. The study of simple autonomous mappings has been the object of intensive mathematical investigations [32] [101]. It has been shown that the orbits are generally attracted to some restricted limiting subset of the phase space, which is called the attracting set or "attractor" of the dynamical system. The attractor can be very simple like a point attractor representing a steady state of the system, or a closed one-dimensional line representing a periodic phenomenon, or a torus in the case of 2 periods, or more complex manifolds with complex topologies called "strange attractors" necessitating the use of "fractal" geometries [22].

There can be multiple attractors in a system and the phase space can be divided into "basins of attraction," where the division is determined by the attractor to which an orbit converges from the given initial point. Some of the initial points can generate a diverging orbit that fails to remain bounded in phase space, i.e., it is attracted to infinity. This situation is referred to in the GCM jargon as "divergence" or "explosion" or "blow up" of the model. Though frequently indicating an inappropriate choice of time step or an error in model specification or coding,

32

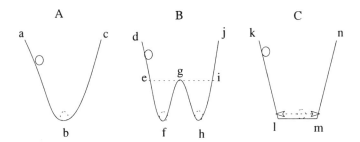

Figure 1. Illustration of the attractor for the case of a small solid ball rolling in a one-dimensional landscape under the influence of gravity and surface friction. In configuration A (a point attractor) the ball, after a number of damped oscillations, will ultimately settle at point b. In B (a multiple attractor), the ball will end up either at f or h according to its initial position with respect to the basins of attraction efg or ghi. In C (line attractor) the ball will stop anywhere on the segment between l and m if the bottom is assumed perfectly horizontal.

such divergence could also, in principle, occur with a perfectly sound model if the initial condition were chosen improperly outside the basin of attraction of the finite attractors of the model, for example, if specifying physically unrealistic initial values.

3. Initial Conditions

The possible existence of multiple attractors makes the choice of the initial condition far from trivial. As the location of the attractor of the model is not known initially, it may not be easy to start from a point close to it. Several methods are applied in practice. One method can be to start from a stable solution like, for example, an atmosphere at rest with no horizontal gradients [86]. Though this initial condition is usually remote from the attractor it provides a rather smooth starting point, and it has usually been found that the physical processes in the GCM are effective enough to restore the horizontal gradients and set the atmosphere in its usual motions within a time span of a few months. This short "spinup" time for the atmosphere is a consequence of the lack of inertia of the atmospheric processes, which gives them a rather short memory of their previous state. Processes at the land surface (in particular cumulative processes like those that depend on the storage of water below the ground) have a larger inertia [83] and therefore often need several years before reaching their equilibrium values, when started from unrealistic initial values such as empty reservoirs or saturated ones. The specification of initial values of such long-memory processes can exert a long-lasting influence on the subsequent simulation, and should be carefully taken care of in the design of a numerical experiment.

A good method for starting in the vicinity of the model attractor is to start close to an observed atmospheric state. This method is compulsory in the case of numerical weather prediction, and suitable methods have been developed for incorporating observations in what is referred to as the "analysis" of atmospheric fields. Sophisticated methods of interpolation try to use in an optimal way the information contained in observations by consideration of their statistical structure in space and time. As the different fields are dealt with separately, they can be "unbalanced" with respect to the dynamical equations, resulting in undesirable oscillations in the first iterations of the model. Methods of "initialization" that project the model states onto a lower-dimensional subset have been developed to start from a more balanced state (see Chapter 15). New variational methods, which iteratively adjust the model trajectory towards the observations by making use of the linear tangent equations, are a promising approach but entail a high computational cost [11] . Despite all the progress that is being made to make the best possible use of observations for the specification of the model's initial state, it is not obvious that the use of an actual atmospheric state is the best method for choosing an initial condition for a climate simulation. In fact, the attractor of the real climate system is likely to be different from the attractor of the model, though climate modelers devote a significant part of their time to reducing model systematic errors and strive to improve the model in all possible ways so as to mimic as accurately as possible the observed climate behavior. As long as the two attractors are not completely identical, the initial part of the trajectory will continue to remain more or less outside of the model attractor, and thus largely influenced by the transient and nonstationary attracting dynamics, rather than by the natural long-term stationary dynamics prevailing on the attractor itself.

Although this transient phase (in which the initial portion of the trajectory is drawn toward the model's attractor) is the focus of interest in short-range numerical weather forecasting and is an essential part in extended-range forecasting, the transient phase needs to be left out when analyzing properly the "climate" of the model, precisely because of its transient nature. The usual practice in climate simulations is to discard from subsequent analysis an initial portion of the simulation known as "spin-up", waiting until the model has had time to stabilize and adjust itself to its boundary conditions. How long one should wait for that is not always obvious nor easy to define. The bulk of the atmosphere by itself has a rather short inertia and adjusts in less than two months to external forcings, but some of the surface processes like the soil reservoirs can take up to several years. For simulations of the current climate, it may not be necessary to wait for the adjustment of the soil reservoirs provided they have been initialized at realistic values. Conversely, in climate change scenarios or in other sensitivity experiments in which a possible change of the surface hydrology can be expected, this long adjustment time can become an essential issue and may require discarding a much longer transient spin-up period before arriving at a climate in statistical equilibrium.

In fact, in simulations in which the external forcings are not stationary in time, like climate change scenarios in which the concentration of greenhouse gases is gradually increased, the notion of an equilibrium climate becomes less relevant, but the problem of specifying an initial state remains crucial, particularly for the slowly evolving part of the system such as the deep oceans.

Perhaps the simplest method, when observed initial conditions are not required, is just to start from a state of the same model obtained from some previous simulation. Even though in a new simulation changes are generally made to some of the model parameters or boundary conditions, the previous model state will usually incorporate the same model constraints and thus provide a good starting point giving a smooth initial trajectory, not too remote from the new attractor set. The limitations of this frequently used method is that, unless an interpolation or truncation is applied to the previous model fields, one is restricted to using the same resolution of the GCM.

4. Model Parameters and Boundary Conditions

Among the control variables that have been denoted by the M-dimensional vector U_t, some are in fact fixed parameters that have been included here for notational convenience. Fundamental physical constants, for instance, can be included in this list. Though this will in practice seldom be the case, they could in principle be changed if more precise determination was obtained in the future. Some of the environmental parameters such as the rotation rate of the earth, its orbital parameters [2], the orography of the earth's surface, and the sea level are in fact slowly evolving, but on such a long time-scale that they can be considered as fixed values during the course of usual GCM simulations. Their values may, however, need to be adjusted when simulating past climate conditions. Some other parameters are not really constant in time, like the total solar radiance (improperly called the "solar constant") or the atmospheric concentration of carbon dioxide, but their temporal fluctuations (and spatial fluctuations in the case of carbon dioxide) are usually assumed small enough to be neglected. The model parameters, which really deserve to be called "control" parameters, are some empirical parameters introduced in the equations to represent phenomena that cannot be explicitly resolved by the model, such as the large variety of physical processes taking place at spatial scales smaller than the model grid spacing. The methods of parameterization of these unresolved subgrid-scale processes are most of the time based on elementary and rudimentary physical or statistical assumptions that introduce various empirical parameters, the values of which are so uncertain that they become "tuning" parameters, such as critical humidity thresholds for cloud formation or mixing lengths for turbulent transfer. They need to be adjusted by looking at the model outputs and tuned progressively so as to bring the model to a more desirable behavior. A large part of the "art" of large-scale modeling in the model validation is devoted to trying to estimate the best choice of these empirical tuning parameters, by combining critical analysis of model deficiency and experienced deductions about their possible origin.

In most simulations, some of the control variables do change as a function of time, and the mapping representing the model is then called nonautonomous. The variations of some of the external variables are periodic in time, like the variations of the astronomical parameters that determine the insolation angles: the rotation of the Earth and its revolution around the sun. These two fundamental periods in the solar forcing are reflected in many climatological parameters that exhibit more or less regular diurnal or annual variations. The annual cycle in particular is prominent in several boundary conditions like surface temperatures, sea ice cover and vegetation cover. Moreover, besides their smooth response to the annual solar forcing, these external systems can exhibit more complex interannual variations and longer term trends, due to their interaction with the atmospheric circulation, and to their own complex nonlinear dynamics. The fact that the model is nonautonomous means that some of the control parameters can be varied during the course of an integration.

We are thus confronted with the problem of specifying these external parameters, and since in principle they could be specified arbitrarily they can be used to influence or control the behavior of the model. The object of sensitivity experiments is to investigate experimentally how the model behavior is influenced by the control parameters. Time series of the control parameters U_t can be specified from various sources of observations. In the case of modeling past observed conditions, some of the external parameters can be estimated from past observations (see Chapter 20). In the Atmospheric Model Intercomparison Project (AMIP) [26], for example, the sea surface temperatures (SSTs) where specified by the monthly average analyses over the period 1979–1988 (Chapter 19). Some other properties such as surface vegetation cover can also be deduced from satellite observations.

The nonautonomy of the atmospheric GCMs becomes a more serious issue when the model is used to make forecasts, either on the seasonal time scale, or even more so, for climate scenarios for the next century. In order to define the experiment, time series of the forcing U_t have to be anticipated in some manner, either by physical, statistical, or other kinds of argument. For instance, in the case of seasonal forecasting, a simple persistence or a statistical regression of the initial SST anomalies superposed on the seasonal climatological cycle can be considered as a first reasonable approximation for a time span of a few months ahead but will generally fail on longer time scales. In fact, the most rational way to progress is to try to make the model as autonomous as possible by including prognostically in the model the parts of the climatic system that vary on the time scale considered, so that the external environment outside the model can then be treated as nearly constant during the integration period considered. This can be done by coupling atmospheric GCMs to oceanic GCMs, sea-ice models, and vegetation models, so that the time-dependent external parameters in the boundary conditions of the atmospheric model will then become internal parameters of the coupled system (see Chapter 18).

5. Observation and Storage of Model Results

The trajectory of the state vector in phase space at all the successive time steps of the simulation $p = 0, ..P$, together with the time history of the control variables U_p is the most complete description of the model behavior. However the volume of information represented by the collection of all the P points in N-dimensional space is so huge that it cannot all be stored. Furthermore, as the time steps are rather close together in time compared to the time-scales of the processes of interest (since their choice is dictated from stability constraints and not from an application perspective), the information contained in successive iterations is mostly redundant or irrelevant. In fact, since the model is deterministic, any portion of a particular trajectory can be reconstructed simply by reiterating the mapping from a starting point. The most practical and economical way of reconstructing this fine-scale information when needed is simply to rerun the model for the time-span of interest, starting from a model state vector X_{p0} stored at a given time t_{p0}. Such a stored model state, together with the corresponding control parameter set U_{p0} is called a "restart." It is usually prudent to store restarts at regular time intervals (maybe once a month) in order to be able to rerun the model over such time spans.

Another problem with the model state vector X is that the model variables stored in it (i.e., the model prognostic variables), though convenient to formulate the mathematical equations and compute their numerical solution, may not be the most appropriate from a physical point of view. A good example: the spectral coefficients of the streamfunction of the horizontal wind vector on a model sigma surface (see Chapter 7). In most cases it can be appropriate to use the model internal variables for computing diagnostic quantities having a more transparent and meaningful physical interpretation. What is then needed is a kind of "observing system" to sample the model results at regular intervals [72] [92]. The "observations" collected need not be directly the components of the state vector X_p, but can be a subset of derived variables computed from the model internal state. Some of the variables can be diagnostic variables (for instance, the geopotential height computed from the temperature by means of the hydrostatic equations) interpolated on suitable horizontal or vertical grids (e.g., constant pressure surfaces). Sampling the instantaneous values at regular time intervals can be satisfactory for variables, such as upper air data, that do not exhibit short time-scale variations. For other quantities, such as precipitation or various other fluxes of matter and energy, accumulation over a time interval is more relevant. This leads to the introduction of a vector of output variables Z distinct from the state variables X

$$Z_{p+1} = H(Z_p, X_p, U_p) \tag{3}$$

and of a sampling scheme to store the results at specified time intervals (for example once a day, or more frequently if diurnal cycles are to be studied). The process of running the model generates thus a T-dimensional time series.

6. Analysis of Model Results

A GCM simulation generates a trajectory in phase space, which is projected by the observing and sampling scheme into a trajectory in observation space Z. After discarding an initial transient phase, the trajectory follows the attractor of the model. For climate modeling purposes the interest is to determine the shape and characteristics of the attractor corresponding to the chosen control parameters. The particular trajectory generated in a single climate run is a single realisation of a one-dimensional section of the attractor. One possible method to improve the sampling of the attractor is to generate several trajectories by using different initial conditions, in order to generate an ensemble of simulations. An attractor can be a very complex folded surface in a multidimensional space, and sophisticated methods for the analysis of the geometrical and topological characteristics have been developed in the context of low-dimensional nonlinear systems [70] [82] (see also Chapter 5). However, the application of these tools for high-dimensional systems is not straightforward and requires prohibitively long time series, and therefore much simpler statistical analyses are more relevant.

The first global characteristic of the attractor is simply its location in phase space. The single characteristic that best summarizes the location of an object in a multidimensional space is the center of mass of the object. The center of mass of the trajectory generated in a model run corresponds to the temporal mean of the time series of the different output variables. This explains the prevalence of the computation of time averages for the analysis of climate simulation. The display and interpretation of the center of mass of the trajectory is far from simple because it corresponds to fields of several climate variables, most of them with three spatial dimensions. The next characteristic of the attractor would be its moments of inertia about the center of mass. This corresponds to computing the variance-covariance matrix of the time series of the model variables. The problem of representation becomes even far more complex, and some method of decomposition of this matrix is necessary to bring out its dominant structures. The simplest representation consists in extracting only the diagonal vector of this matrix, which represents the temporal variance of the time series. This variance is a measure of the range of the extent in each direction of observation space of the portion of the attractor spanned by the simulated trajectories. Other standard mathematical methods involve the use of singular decomposition of the variance-covariance matrix to find the eigenvectors of this matrix; these methods include Principal Component Analysis (PCA), also known as Empirical Orthogonal Functions (EOFs) in the meteorological literature [5] [68]. An example of the application of EOFs is given in Chapter 4. The advantage of these decompositions is that they yield a series of vectors, ordered in decreasing order to their contribution to the total variance, in the N-dimensional phase space. These vectors can be displayed as 3-D fields by the visualization techniques used for the time averages. These vectors represent the principal axes of the multidimensional ellipsoid of inertia of the phase space attractor. When the variance decreases rapidly with the order of the eigenvector, an efficient reduction of the dimension of the output space can be obtained by retaining only the first eigenvectors that represent a large fraction of the total variance. Though eigenvectors are very useful for a compression of the information, their physical interpretation is rendered difficult by the orthogonality constraint imposed by their mathematical definition. Furthermore, these eigenvector methods are seldom applied to the whole observation space, but most of the time to some subspace of it corresponding to some selected variable of interest in order to reduce the dimensionality of the problem.

Whatever the particular statistics S chosen to summarize the results of an experiment, they will be computed by applying some well defined operator S (for instance a simple arithmetic mean) to the model output variables Z_p stored over a subensemble T of time indices p defining the portion(s) of the attractor over which one wishes to perform the sampling (for instance all the indices corresponding to a chosen month in a multi-year run, for the computation of a climatological monthly mean). This can be expressed formally as:

$$S_T = S(Z_p; \ p \in T) = S_T(X_0, U_P) \qquad (4)$$

This notation expresses in condensed form that the climate statistics S_T are totally determined by two conditions:

1. the vector of initial conditions X_0 and
2. the collection of controls applied during the whole simulation U_P (at least up to the maximum index p_{max} contained in the averaging interval T: i.e., $U_P = \{U_p; p = 0, p_{max}\}$) representing the boundary conditions of the experiment.

The distinction between initial conditions X_0 and boundary conditions U_P is fundamental for the formulation of climate simulations because the two kinds of conditions play very different roles, as will be shown in the following paragraphs.

6.1. SENSITIVITY TO THE INITIAL CONDITIONS

The climate of the model, which is a characterization of the attractor, should ideally be independent of the particular initial condition X_0 chosen to start the simulation. However, if we start another simulation from a different initial condition X_0' in a very close neighborhood of X_0 the exponential increase of small perturbations, even though it subsequently has to slow down and saturate when the perturbations reach a large enough amplitude, will ultimately lead to utterly different trajectories, which will sample in different ways the attractor of the model, even if using the same prescribed boundary conditions U_P. This will result in a different value S_T' of the statistics. However, both values S_T and S_T' are equally likely, and the difference between them can be considered a random quantity produced by the slightly different choice of starting point. In other words, this difference is only sampling "noise," comparable to measurement error in a physical experiment. The appearance of this error is caused by the fact that the statistics S_T being computed over a limited set T are only an estimate of limited accuracy for sampling the underlying asymptotic properties of the attractor that are of interest.

The parallel with physical measurements may be enlightening. The output of a climate simulation can be considered as the equivalent of a single measurement on a physical system. A measurement is useless unless it is accompanied by an indication of its accuracy, and a basic practice is to repeat the measurements several times so as to be able to quantify its accuracy. This does not eliminate the possibility of systematic errors in the instrument, but it gives a measure of the random errors in the measurement process. By analogy with the laboratory practice of repeating a measurement, GCM experiments can be run several times with different initial conditions, thus generating an "ensemble" of simulations. This methodology, though seldom used in the past due to constraints in computer resources, is progressively becoming more usual.

The accuracy to which the attractor is estimated can be improved either by increasing the size of T (the length of the simulation) so as to have a more accurate measurement, or by repeating the experiment several times with different initial conditions so as to have several independent measurements that can be combined to reduce the measurement error. Physicists usually combine the two approaches, trying to improve the accuracy of their instrument and repeating the measurements several times in order to bracket the correct value [57]. The same should be recommended for numerical simulations. It may be useful to think in terms of statistical estimation and consider the outcome of a numerical simulation as a single sample drawn randomly from a population of possible outcomes. The population in the case of climate simulation is the infinite ensemble of values $S_T(X_0)$ that would be obtained by performing a simulation with identical controls U_P for all possible initial states X_0 belonging to the same basin of attraction. This of course is a purely theoretical notion but it permits statistical reasoning to be applied to the results of an experiment. This population might then be qualified by standard statistical measures such as the low-order moments of its distribution. The numerical experiments are then seen as samples taken from the true population, which one should try to use as efficiently as possible for estimating parameters of the parent population by applying classical statistical estimation theory.

In practice, such an estimation can be done by rerunning the same numerical simulation from a number of initial states. It is comparatively easy to generate any number J of neighboring initial conditions to the initial field X_0 by adding to it small random perturbations $\delta_j X_0$. The rapid growth of the perturbations that is the hallmark of the unstable atmospheric dynamics will then generate trajectories that will have rapidly diverging evolutions. Provided the initial growth phase of the perturbations is discarded from the set T the statistics S_{T_j} will appear as independent samples. An estimation of the population mean $\overline{S_T}$ will then simply be given by the arithmetic mean operator:

$$\tilde{S}_T = (1/J) \sum_{j=1}^{J} \mathcal{S}_T(X_0 + \delta_j X_0, U_T) \qquad (5)$$

A measure of uncertainty will be obtained by computing the covariance matrix of the S_T. In fact, running multiple experiments should be considered as a standard procedure in climate simulations, as it is a prerequisite for applying the powerful techniques of statistical analysis and estimation. In the past, modelers have been reluctant to apply this methodology systematically on account of its high cost in computer time, but the continued progression at an exponential rate of the computing power (with a doubling time of 18 months according to Moore's law) should make this approach more feasible.

6.2. SENSITIVITY TO THE CONTROL PARAMETERS

The solution of a dynamical system, and hence of a GCM simulation, can change quantitatively, and even qualitatively, when the controls (using this as a general term encompassing the model parameters and the boundary conditions) are varied. The aim of climate modeling is precisely to discover how the behavior of the model (assumed to have some degree of similarity with the real climate) will respond to changes of some of the control parameters. In formal representation the aim of a sensitivity experiment is to study the variation of the climate statistics S_T in response to a change in the set of control parameters U_T. The most elementary sensitivity experiment consists simply in performing two simulations: one which will be considered as a "reference," or "control," simulation using some standard parameters U_T, and a second one using "modified" or "perturbed" parameters U_T'. It does not seem to be an important issue whether the second simulation starts from the same initial condition X_0 as the first simulation, or from a different one X_0', since in any case the two simulations will most likely have divergent evolution when the different parameters are applied. The difference between the statistics S_T and S_T' is generally considered as the model response to the perturbation $\delta U_T = U_T' - U_T$.

However, as the former discussion on the role of the initial conditions has shown, the result of a single run represents just one single realization from the parent population. The aim of a sensitivity experiment is to discover how the parent populations are affected by changing the experimental setup (U_T). Then it seems intuitively clear that simply drawing one individual from each population is insufficient to make powerful deductions about the difference between the populations, unless one has some prior notion of the possible intrapopulation variations. In fact, in the absence of a formal measure of uncertainty, the interpretation of such simple sensitivity tests has usually relied on the prior experience of the climate modeler. The modeler, through previous experiments or knowledge of the real climate variability, develops a subjective notion of the sampling variability that could be expected in a run, and uses this subjective measure to qualify what differences seemed sufficiently large to be ascribed to the perturbation. This explains why most of the initial sensitivity experiments with GCMs have used large perturbations in the boundary conditions, in order to bypass a discussion of the statistical significance of the response. A few examples of such large perturbations are:

— removal of orography [44] [55] [34]
— removal of sea ice cover [69] [81]
— variations of the solar radiation [99] [29] [80]
— changes in orbital parameters [37] [79]
— large-scale thermal pollution [96]
— doubling or quadrupling of carbon dioxide [56] [97]
— ice age conditions [103]
— Cretaceous climate [1]
— nuclear winter [12] [73]
— global perturbations of SST by ± 2 K [7] [17] [75]
— El Nino SST anomalies [78] [85]
— completely dry or wet continents [84]
— extensive snow cover anomalies [20]
— large extensive albedo anomaly [8] [49]
— total deforestation [51] [74] [91]

In all these cases the main response is usually so large and obviously exceeding the usual climate variance, that a formal statistical discussion may not be necessary. In most of these sensitivity experiments the forcing has not only a large magnitude, but also a fairly simple and localized shape, which helps in the interpretation of the results. The analysis of the results in such experiments is centered on the coherence of the response, both spatially and among different fields, and on the plausible chain of interactions between the mechanisms through which the initial forcing can extend its influence. These experiments are not conceived as precise measurements of the climate sensitivity, but rather as crude explorations of the possible atmospheric feedback mechanisms that could be triggered by the external forcings, and the possible repercussions of the atmospheric response on climate. The above methodology is an attempt to explore the parameter space by going to some of its most remote boundaries that may still have a physical relevance. There is a similarity with laboratory experiments in which materials are submitted to extreme conditions in order to test their resilience and domain of stability, to locate phase changes, and possibly to discover other new interesting physical properties.

However, even in these extreme cases, there may be areas where the response is weaker and thus cannot be ascertained to result from the influence of the forcing or to be a mere sampling fluctuation, and statistical testing creeps back into the picture again. The necessity of testing the statistical significance of the response was emphasized early and several relevant statistical methods have been proposed [52] [95]. However sophisticated the statistical method may be for extracting the "signal" due to the change of the control parameters from the noise due to sampling fluctuations, the basic rule is that the weaker the forcing and the response, the larger has to be the overall volume of data processed either by longer simulations or larger ensembles. Recent publications show that it is becoming more and more usual to perform simulations of at least ten years and use ensembles of several of them, so as to increase the statistical significance. This is becoming increasingly necessary as the forcings applied in sensitivity experiments are becoming more and more realistic, and thus of smaller magnitude and greater complexity. As the response is weaker a more careful statistical analysis of the response is required, as discussed in Chapter 4. Ensemble simulations can be required even for defining precisely the amount of interannual variance of external origin forced by observed monthly means SSTs and thus potentially predictable [35], compared to the amount of variance of internal origin arising from the instabilities of the atmospheric dynamics, and therefore unpredictable.

7. Defining the Sensitivity of a GCM

The aim of numerical simulations is to establish the relationship between parameter vectors U_T and climate statistics S_T so as to know how the climate will respond to changes in external conditions. Owing to the huge dimensionality of both the parameter space and the climate space, it will never be possible to explore parameter space completely. Even in low dimensional systems, the exploration of parameter space has proved extremely complex due to the richness of behavior exhibited by nonlinear systems. The fascinating structures of the famous Mandelbrot set, which gives in the complex plane the set of points which have a finite attractor for a simple quadratic mapping, is a good illustration of the fact that there are regions in parameter space where a small change in a control parameter can lead to utterly different behavior of the dynamical system, which is referred to as the phenomenon of "bifurcation" [38]. The aim of the qualitative description of a dynamical system is to locate the set of bifurcation points so as to partition the parameter space into regions in which the dynamical system exhibits different asymptotic behaviors. This goal should also apply to climate simulations, but we are still a long way from it. The recurrence of ice ages and interglacial periods, with rather abrupt transitions between them on the palaeoclimatic time scale, seems to be evidence that the climate system as a whole can bifurcate between two different regimes, but the precise role of the atmosphere, ocean, cryosphere, biosphere interactions is not yet clear. It seems likely that the essential nonlinearities might be located rather in the long term components of the coupled system such as the ocean thermohaline circulation, the ice accumulation processes, or the carbon cycle, rather than in the atmosphere, which seems to respond most of the time in a smooth way to changes in boundary conditions, at least in most GCM simulations.

Explorations with GCMs, though growing in number, have sampled only a rather small number of scattered points in parameter space, and most of the time not in a systematic enough way to establish whether sudden transitions can occur at certain values of the control parameters. There are only few simulations that have actually tried to vary a control parameter over a range of values, and more experiments of this kind are needed. A good example of an attempt at a systematic exploration of some region of parameter space is provided by Milly and Dunne [62] who varied the water capacity of the land surface from 1 cm to 240 cm in a set of seven experiments. The response of the hydrologic cycle, though large, was found to be a rather smooth function of the control parameter.

In those regions of parameter space where the model responds in a continuous manner to variations of the forcing, it is possible to define a local sensitivity. We make the hypothesis that there is a well defined mapping C that for each control U_P, gives a well defined value of the expectation $\overline{S_T}$ of the climate statistics of interest. In this part we shall drop the index T to simplify the notation, keeping in mind that the statistics S refer to a specified time interval, and that a time dependence may be included in the forcing U.

$$\overline{S} = C(U) \qquad (6)$$

If the system is regular enough for the mapping to be continuous, then it can be approximated by a Taylor expansion limited to first order terms in the vicinity of a reference state U_0:

$$\delta\overline{S} = \overline{S} - \overline{S_0} = \left.\frac{\partial C}{\partial U}\right|_{U_0} \delta U \qquad (7)$$

We can define the linear tangent approximation $s = \mathcal{L}(u)$ with $u = U - U_0$ to

$C(U)$ about a reference value S_0 obtained with forcing U_0 as

$$s = \mathcal{L}(u) = \left.\frac{\partial C}{\partial U}\right|_{U_0} u \tag{8}$$

The Jacobian matrix of the partial derivatives of the components of $\overline{S}$ with respect to each component of U completely defines the sensitivity to variations in control parameters about the reference state, and the Jacobian can be called a sensitivity matrix. However, the dimension of this matrix ($\dim(S) \times \dim(U)$) will in general be too large for the matrix to be practically useful. One possible way to reduce its dimensionality is to consider only a subspace of forcings along a limited number of predefined directions in parameter space consisting of linear combinations of forcings with fixed shapes. Suppose we have defined a reduced number J of elementary forcing variations $u_j = \delta U_j$ which are of special interest. The forcing could be a single parameter (CO_2 concentration for example), or a specified field (typical El Nino anomaly, or uniform SST perturbation), or a combination of several forcings (CO_2 and expected SST response pattern), or even forcing with specified time variation (periodic forcing, forcing increasing at a specified rate). We shall call them forcing patterns. Considering small perturbations λu_j colinear to a chosen forcing pattern u_j, we can write a Taylor expansion truncated at first order in the vicinity of a reference state $S_0 = C(U_0)$:

$$C(U_0 + \lambda u_j) - C(U_0) = \Lambda_j \lambda + \mathcal{O}(\lambda^2) \tag{9}$$

in which

$$\Lambda_j = \frac{dC(U_0 + \lambda U_j)}{d\lambda} \tag{10}$$

is the derivative of $C(U)$ along direction u_j at the reference point U_0. This vector Λ_j which has the same dimension as the statistics S can be considered as the linear sensitivity of the statistics S_0 with respect to the forcing pattern u_j. Knowledge of this sensitivity allows one to compute a linear approximation $\mathcal{L}$ of the variation of S in response to forcings proportional to u_j, and the approximation is all the more accurate the smaller the forcing:

$$\mathcal{L}(U_0 + \lambda u_j) = S_0 + \Lambda_j \lambda = S + \mathcal{O}(\lambda^2). \tag{11}$$

If several forcing patterns are operating simultaneously in a forcing $u = \sum_j \lambda_j u_j$, and if the individual forcings are small enough for the response to remain linear, then a good representation (in the least square sense) of the true response will be given by the linear tangent application:

$$\mathcal{L}(U_0 + u) = S_0 + \sum_j \Lambda_j \lambda_j = S + \sum_j \mathcal{O}(\lambda_j^2). \tag{12}$$

The above relationship is a tautology, stating that if the system is linear in a neighborhood of a reference state, then the superposition principle applies, and the response to a linear combination of forcings is a linear combination of the responses to the individual forcings. The difficulty is to determine precisely to what extent in parameter space the response can be considered linear, an answer which can only be provided by numerical experimentation with gradually increasing forcings (possibly of different signs). An inherent underlying hypothesis in a linear approximation is that the different forcings are acting independently and additively on the response. As the forcings are increased in magnitude, higher order terms in the Taylor expansion can be expected to grow, thus strengthening the influence of the second order terms, which represent the mutual interactions of the different forcing patterns.

The idea of linear forcing patterns has been applied recently to study the response to SST anomalies by using principal component analysis of the SST field as forcing patterns [94]. This is certainly a reasonable strategy since the leading eigenvectors by definition maximize the amount of variance captured by each of them, and contain the bulk of information on SST variations. Computing the climate response to the first components gives the approximate linear response to any SST field from its projection on the first components. Furthermore, the accuracy of the linear approximation can be tested by performing the simulation with the full SST field.

There remains, however, the problem of how to compute the sensitivity vector Λ_j, which has been defined only in a mathematical sense by taking the limit of a vanishingly small perturbation. An approximation to Λ_j can be computed by taking a finite difference

$$\Lambda_j = \frac{S_j - S_0}{\lambda} \tag{13}$$

in which S_j is a value of the statistics S computed in a simulation with the forcing $U_0 + \lambda u_j$, and S_0 a value computed in a simulation with the reference forcing U_0.

The inevitable presence of residual noise due to sampling fluctuations is a fundamental limitation that excludes the use of small perturbations that would generate a response smaller than the noise level. The problem is amplified by the fact that the finite difference, which is used to approximate the derivative, doubles the amount of variance due to the noise. On the other hand, the use of large perturbations to reinforce the signal heightens the danger of the interference of nonlinearities, so that there is a real dilemma in choosing the appropriate amplitude of perturbations for defining a linear sensitivity. Here again, the quest for accuracy requires ensembles of simulations. Since multiple simulations are needed, the question can be raised of the best strategy for choosing them. Suppose we have decided to perform K simulations for determining the sensitivity to a chosen perturbation pattern u_j. A straightforward application of the finite difference formula would suggest that we divide the simulations between the $K/2$ simulations with the reference control U_0 and the $K/2$ simulations with the perturbed control $U_0 + \lambda u_j$, so as to reduce simultaneously the estimation errors of both. However, an alternative strategy is possible: instead of rerunning multiple simulations with the same forcing, we could perform each simulation with a different forcing $U_0 + \lambda_k u_j$ using K different values (positive or negative) of the factors λ_k. This would have the advantage of scanning the response of the system over a range of values of the forcing, and the resulting K values S_{jk} could be used to assess the possible nonlinearities of the response. Having K values with different sampling error ϵ_k,

$$S_{jk} = \mathcal{C}(U_0 + \lambda_k u_j) + \epsilon_{jk} \tag{14}$$

$$S_{jk} = S_0 + \Lambda_j \lambda_k + \epsilon'_{jk}, \tag{15}$$

it is natural to apply a least squares regression technique to estimate the unknown coefficients Λ_j, and compute them from a linear regression of the S_{jk} on the λ_k. If K is sufficiently large, one may even try nonlinear regression by including quadratic and higher order terms in λ_{jk}. In this reasoning we have used K perturbations proportional to a single forcing pattern u_j for a fixed index j, but in fact the formulas are general and apply as well if the K perturbations apply to different forcing patterns. In that case (15) becomes

$$S_k = S_0 + \sum_j \Lambda_j \lambda_{jk} + \epsilon'_k. \tag{16}$$

There are now $J+1$ quantities to determine and this can be done by ordinary least squares, provided $K > J + 1$. This simple regression method is a suitable first approach for determining the sensitivity to given patterns of perturbation.

8. The Different Categories of Climate Simulations

8.1. VALIDATION EXPERIMENTS

The first kind of experiment is simply to try to get as realistic a simulation of the present climate as possible. This is referred to as "model validation" (see Chapter 3). Since the aim is to reproduce the current mean climate and its variability the choice of boundary conditions is in principle straightforward: one should simply use the most realistic available data for setting up the boundary conditions of a validation experiment. However, there still may be some nontrivial issues. Usually standard boundary condition fields for the current climate will be provided as part of the model package. The main fields that are usually specified are:

- the land-sea mask, or a fractional coverage giving land/ocean proportions for each model grid box;
- the mean height of the orography or, for a spectral model the spectral components of the orography, which may have been adjusted in order to reduce negative altitudes over the oceans due to Gibbs oscillations;

— various fields describing the properties of the land surface, like surface albedo, roughness height, vegetation type and fractional coverage, type and depth of the active soil layer, thermal and hydraulic properties, and variance of subgrid orography. This list can be more or less extensive according to the degree of sophistication and complexity of the land-surface parameterization scheme;

— the sea surface temperature (SST) field, and sea-ice fractional cover.

Among the boundary conditions some are fixed fields (land-sea mask, orography, soil type) while others may undergo seasonal and interannual variations (SSTs, vegetation). The largest seasonal variations are found in SST and sea-ice, and there are several possibilities for their specification. In perpetual month simulations all the boundary conditions are kept fixed to their initial value, and thus a single field of SST is required corresponding to the time of the year at which the insolation conditions have been set (for example, a perpetual January simulation). Since, however, the seasonal variations are a fundamental characteristic of climate, most of the simulations used for validation of GCMs include the annual cycle. The usual way of taking into account the seasonal variations of the SST field is to provide monthly mean SST maps and to interpolate between successive monthly means to obtain a smoothly varying set of daily values. There are two possible strategies for choosing the monthly mean SST fields: either to take climatological monthly mean SSTs established over a number of years, or to use the observed monthly SST observed in a particular year. The first GCM simulations applied the first method using compiled climatologies, at a time when global SST analyses were neither performed routinely nor easily available. The advantage of using climatological SSTs was that even the short (a few months) GCM simulations that were possible then could be assumed to represent a mean climate and compared with the observed atmospheric climatologies. However, this assumption is valid only if the atmospheric response to SSTs is linear. The second strategy does not make this assumption and is better for multiyear simulations since it allows one to include the interannual variability. This approach has been applied in the Atmospheric Model Intercomparison Project (AMIP) [26], in which the monthly mean SST analyses performed at COLA-CAC over the 10 year period 1979-1988 were provided for use in the different models taking part in the intercomparison. In coupled atmosphere-ocean models the SSTs are computed by the ocean model and transmitted to the atmospheric model at regular intervals. A mean seasonal variation can also be prescribed for changes in surface characteristics (albedo, leaf area index) depending on the seasonal evolution of vegetation. Satellites can also provide useful information such as albedo or monitoring of the state of the vegetation cover (leaf area index) and its seasonal and interannual variations [3]. In a typical reference experiment the climate simulated by the GCM is compared to the observed climate. The aim of the comparison is to ascertain which features of climate are reproduced realistically and which need improvements. Validation of a model is a very important exercise that measures the credibility of a model and is a prerequisite before using the model for other applications. A new version of a climate model should not be released for use without a documentation of the changes made in it and a validation atlas displaying the main verifying fields. Several objective statistical techniques can be useful to assist in quantifying the agreement (or the lack of it!) between model and observed climates.

8.2. SENSITIVITY EXPERIMENTS

The aim of a sensitivity experiment is to study the influence on the model climate of specified modifications in the parameters or boundary conditions on the model. The standard procedure consists in performing at least two simulations, a "control" (or reference) simulation and one or (hopefully) several "perturbed" (or modified) simulations. The results of the perturbed simulation(s) are then compared with those of the reference simulation so as to identify the response of the model to the imposed changes.

The choice of the perturbation is very much dependent on the particular phenomenon that is under investigation and the previous empirical or theoretical understanding one has of it. The aim of experiments can be exploration guided by some previous knowledge of similar phenomena, or the verification of hypotheses and theories that could explain the phenomena in question. As the climatic system is a very complex and multidimensional system, there is actually a plethora of factors that could have an influence on the climate directly or indirectly through various feedback mechanisms, and an essential task is to distinguish those that have the largest impact from those that are of secondary and marginal importance.

The design of a sensitivity experiment should therefore start with a critical reflection on some physical mechanism, including a review of the relevant literature, and an identification of the weak and uncertain points needing clarification. The design should culminate in the formulation of a testable hypothesis or a precise question to be answered by the sensitivity experiment.

Sensitivity experiments can have different goals:

- Process studies: trying to improve the understanding of the role and feedback of a particular physical process in the climate system. This can be very useful to assess the importance of this process, how it could be better calibrated, or to determine how much further effort should be devoted in improving its representation in the GCM. Among the processes that have received the most attention are:

  - radiative processes [65] [98] [23]
  - cloud processes [50] [100]
  - convection [19] [10] [74]
  - land surface processes [24]
  - model resolution [58] [89]
  - horizontal diffusion [48][88]

- numerical algorithms [76]

− Assessment of the potential impact of external forcings: to determine what are the factors outside the atmosphere that can influence its behavior. In order to build a physical theory of climate and climate change, it is important to be able to determine the extent to which climate is controlled by external parameters and to determine its sensitivity to changes in those parameters. Many qualitative theories have been imagined to explain the variations of past climates, putting emphasis on one or another factor, but most have neither been proved nor disproved by lack of suitable data for validation. GCMs can offer a testbed for such theories and a motivation to put them in quantitative form. Almost all the boundary conditions can have an influence on climate, could play some part in a theory of climate change, and have been used in sensitivity experiments and extensively studied:

- orography [39] [66]
- insolation intensity or distribution [47] [80] [14]
- rotation rate of the earth [104] [40]
- atmospheric composition [6] [61] [63]
- aerosols [4] [9] [42] [27]
- volcanic eruptions [30]
- sea surface temperatures [90] [46] [64] [87] [21]
- sea ice and snow cover [81] [77]
- albedo and other land-surface properties [8] [60] [105]
- vegetation distribution [18]

8.3. CLIMATE SIMULATIONS

The aim of climate simulations is to compute with as much realism as possible the climate conditions that could occur under a realistic change of external conditions that could have occurred in the past (paleoclimatic reconstructions) or could occur in the future (climate scenarios). As there may be uncertainties in the actual external conditions that should apply, some plausible hypotheses are introduced. As an example for paleoclimatic reconstructions, the astronomical parameters are usually known with a good accuracy over the last million years or so, and reconstructions of the SST patterns, height of the ice caps, and vegetation distribution have been made for certain key periods like the last ice age. For more ancient periods such as the Cretaceous (see Chapter 20), much less is known and hypotheses have to be introduced about the atmospheric composition and distribution of continents [1]. Coupled ocean-atmosphere simulations are required when the SST distribution has not been reconstructed from observation, in which case integrations of several thousand of years are required for the deep ocean circulation to adjust to the altered external conditions (Chapter 18). The aim of the climatic reconstruction is to check whether the simulated climate is consistent with the scattered paleoclimatic evidence [15]. An analysis of discrepancies can indicate in which direction the hypotheses made in the simulation should have to be modified. The GCM simulations in this context give a coherent framework that can help to interpret the paleoclimatic data [43] [59].

In the case of future climate scenarios, some hypotheses are made on the evolution of the composition of the atmosphere or the land-surface cover under the influence of human activities. As it is necessary in this context to forecast the response of ocean temperatures, coupled atmosphere-ocean models are required to compute the mutual adjustment of the atmospheric and oceanic circulations and their possible feedbacks. Several such scenarios have been performed and are at the basis of the evaluation of future climate changes by the IPCC [36]. Most of the published scenarios use a "flux correction" method to inhibit excessive drift of the coupled model climate [13] [67]. Recently some models have been calibrated well enough to run in control and increasing CO_2 simulations without using flux correction ([25] [31] [33]; see Chapter 18). A possibility for refining and understanding scenarios is to make additional "time-slice" (or "snapshot") simulations with an atmospheric GCM for specific periods in the future (for example at the time of doubled CO_2) using as boundary conditions the SSTs and ice limits computed from such a coupled model simulation [54]. This gives the possibility of making several simulations in order to assess better the statistical variability of the simulated climate, or to use higher resolution models to refine the regional aspects [16]. Moreover, if there is a control run with constant CO_2, one can use time-slice simulations to correct part of the drift in a coupled model run by calculating the difference between the simulation with increasing CO_2 and the control run, and superposing this difference on the observed climatology, instead of using directly the SST field of the coupled model [93].

9. The Different Steps in Running a GCM Experiment

As a conclusion it may be useful to recall the different steps involved in the realization of a numerical simulation, as this practical and somewhat trivial aspect of the problem is seldom discussed in scientific publications. We have seen that the GCM can be considered as just a piece of computer software that implements the theoretical mapping procedure presented in the preceding sections. The GCM needs for its operation a certain number of control parameters, which are usually stored and provided in various input files of the required format, and a "restart" file representing the initial values of the model internal state. During its operation the model iterates its basic time step and generates new values of its internal state at a new time $t + \delta t$. Periodically (usually at least once a day) an output procedure is activated to store in output files results describing the model state, mainly in

Initialization   **integration**   **Postprocessing**

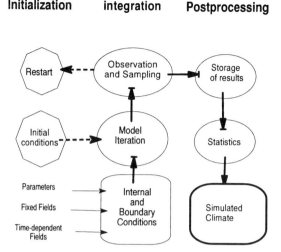

Figure 2.   The different steps of a GCM simulation

the form of instantaneous or time-averaged fields of meteorological parameters of interest. After the model has been run, a post-processing procedure is activated in order to compute, from the model output files, various statistics that condense and summarize the large volume of data generated in the simulation, and displays them in graphical form convenient for their interpretation.

Having in mind this general scheme, displayed in figure 2, we now proceed to examine in more detail the different phases involved in running a GCM model simulation.

9.1. DESIGN

The first and essential phase is of course the precise definition of the conditions of the experiment that will be performed. At this planning stage a certain number of basic and crucial choices have to be made:

- choice of model configuration, including model vertical and horizontal resolution
- choice among different parameterizations (if available) and parameters for parametrizations
- precise definition of the initial and boundary conditions that will be used for the experiment.

It is important to incorporate in the design phase the material constraints (computer resources, time span available) that can limit the scope of the simulation, and to try to keep a reasonable balance between the scientific interest of the simulation and its overall cost in human and computer time.

9.2. PREPARATION

Once the experimental setup has been precisely defined, the different files necessary for model integration are prepared. This can involve changing model parameters, or even recompiling (part of) the model code if the code itself has to be changed. Then files describing the boundary conditions consistent with the chosen model resolution have to be prepared or modified. The boundary conditions provide a description of the environment of the modeled atmosphere, such as the surface geography (land-sea mask, orography, surface radiative, thermal or hydrological properties). Finally, one must prepare an initial condition, from which the model can start its integrations, as well as files and storage space for model output.

It can never be too strongly recommended to check as far as possible the prepared input fields by all means available (printing, plotting or more elaborate graphical outputs) in order to identify possible initial setup errors that could ruin the whole simulation. Once all the input fields are ready, one should attempt to start the model and run it for a short duration in order to verify that all is operating smoothly and generating meaningful results.

9.3. INTEGRATION

When all the necessary input files have been produced and checked, the model integration can begin. Usually the model will be integrated in a single run for only a limited number of days, chosen to satisfy the constraints of the computer operating system, such as maximum duration for a single job. This basic process will have to be repeated, either manually or automatically, until the chosen duration of the experiment has been reached. This integration phase can take quite a while, according to the speed of the operating computer, the number of other users, and the ambitions of the modeler. Care should be paid in the planning stage not only to the overall computer cost of the simulation, but also to the expected real time duration of the simulation, taking into account the expected availability of the computer resources. This recommendation is particularly relevant for models running on large shared (super)computer systems on which the resources are usually distributed among a large number of users, particularly for projects that need to be completed in a very limited time span. While the model run is progressing, it does no harm to sample and check regularly some of the model results with the aim of detecting possible anomalies in the model behavior, so as to avoid a waste of computer time if the simulation seems to be going awry.

9.4. POSTPROCESSING

The next, more interactive phase, is the postprocessing of the model output in order to bring out potentially interesting results. Usually several tools are available along with the model code to process the model output files and extract various standard statistics, such as temporal means (over a month, a season or a year), or spatial means (zonal, areal or global averages) and plot them in different ways, and it is a good idea to use them to look at the general features of the simulation. More elaborate statistical and dynamical analysis procedures may be available to study various aspects of the simulation, as set forth in Chapter 4. Very often, after looking at the results produced by these tools, the modeler will feel like adapting them or developing new methods, in order to bring out more clearly the specific results expected from the simulation.

9.5. INTERPRETATION

Finally, the last phase is the interpretation of the results, which should culminate in an adequately descriptive report on the new findings and ideas brought about by the simulation. This implies a critical examination of the different plots produced in the previous phase, their comparison with other published or unpublished experiments, and an effort to provide a rational explication, with the help of relevant theoretical support.

Acknowledgments: *The author thanks David Stephenson for useful advice and suggestions, Oleg Makarynskyy, Kathleen Schreiber and Jeffrey Yin for their comments that helped to improve the manuscript.*

References

1. Barron, E.J. and Washington, W.M. (1984) The role of geographic variables in explaining paleoclimates: Results from Cretaceous climate model sensitivity studies, *J. Geophys. Res.*, **89**, 1267–1279.
2. Berger, A., Imbrie, J., Hays, J., Kukla, G., and Saltzman, B., Eds. (1984) *Milankovitch and Climate. Understanding the Response to astronomical forcing. Part I*, NATO ASI Series - C: Mathematical and Physical Sciences, 126. D.Reidel Publishing Company, Dordrecht.
3. Bony, S., Le Treut, H., Duvel, J.P., and Kandel, R.S. (1992) Satellite validation of GCM-simulated annual cycle of the Earth radiation budget and cloud forcing, *J. Geophys. Res.*, **97**, 18,061–18,081.
4. Boucher, O. (1995) GCM estimate of the indirect aerosol forcing using satellite- retrieved cloud droplet effective radii, *J. Clim.*, **8**, 1403–1409.
5. Brunet, G. (1994) Empirical normal-mode analysis of atmospheric data, *J. Atmos. Sci.*, **51**, 932–952.

54

6. Cariolle, D., Lasserre-Bigorry, A., Royer, J.F., and Geleyn, J.F. (1990) A general circulation model simulation of the springtime Antarctic ozone decrease and its impact on mid-latitudes, *J. Geophys. Res.*, **95**, 1883–1898.

7. Cess, R.D., Potter, G.L., Blanchet, J.P., Boer, G.J., Del Genio, A.D., Déqué, M., Dymnikov, V., Galin, V., Gates, W.L., Ghan, S.J., Kiehl, J.T., Lacis, A.A., Le Treut, H., Li, Z.X., Liang, X.Z., McAvaney, B.J., Meleshko, V.P., Mitchell, J.F.B., Morcrette, J.J., Randall, D.A., Rikus, L., Roeckner, E., Royer, J.F., Schlese, U., Scheinin, D.A., Slingo, A., Sokolov, A.P., Taylor, K.E., Washington, W.M., Wetherald, R.T., Yagai, I., and Zhang, M.H., (1990) Intercomparison and interpretation of climate feedback processes in 19 general circulation models, *J. Geophys. Res.*, **95**, 16,601–16,615.

8. Charney, J.G., Stone, P.H., and Quirk, W.J. (1975) Drought in the Sahara: A biogeophysical feedback mechanism, *Science*, **187** 434–435.

9. Coakley, Jr., J.A. and Cess, R.D. (1985) Response of the NCAR Community Climate Model to the radiative forcing by the naturally occurring tropospheric aerosol, *J. Atmos. Sci.*, **42**, 1677–1692.

10. Colman, R.A. and McAvaney, B.J. (1995) Sensitivity of the climate response of an atmospheric general circulation model to changes in convective parameterization and horizontal resolution, *J. Geophys. Res.*, **100**, 3155–3172.

11. Courtier, P. (1997) Variational methods, *J. Meteor. Soc. Japan*, **75**, 211–218.

12. Covey, C., Thompson, S.L., and Schneider, S.H. (1985) "Nuclear winter": A diagnosis of atmospheric general circulation model simulations, *J. Geophys. Res.*, **90**, 5615–5628.

13. Cubasch, U., Hasselmann, K., Höck, H., Maier-Reimer, E., Mikolajewicz, U., Santer, B.D., and Sausen, R. (1992) Time-dependent greenhouse warming computations with a coupled ocean- atmosphere model, *Climate Dyn.*, **8**, 55–69.

14. Cubasch, U., Voss, R., Hegerl, G.C., Waszkewitz, J., and Crowley, T.J. (1997) Simulation of the influence of solar radiation variations on the global climate with an ocean-atmosphere general circulation model, *Climate Dyn.*,**13**, 757–767.

15. de Noblet, N.I., Prentice, I.C., Joussaume, S., Texier, D., Botta, A., and Haxeltine, A. (1996) Possible role of atmosphere-biosphere interactions in triggering the last glaciation, *Geophys. Res. Lett.*, **23**, 3191–3194.

16. Déqué, M., Marquet, P., and Jones, R.G. (1998) Simulation of climate change over Europe using a global variable resolution general circulation model, *Climate Dyn.*, **14** 173–189.

17. Déqué, M. and Royer, J.F. (1991) GCM response of the mean zonal surface heat and water budgets to a global sea surface temperature anomaly, *Dynam. Atmos. Oceans*, **16**, 133–146.

18. Dickinson, R.E. and Henderson-Sellers, A. (1988) Modelling tropical deforestation: A study of GCM land-surface parametrizations, *Quart. J. Roy. Meteor. Soc.*, **114**, 439–462.

19. Donner, L.J., Kuo, H.L., and Pitcher, E.J. (1982) The significance of thermodynamics forcing by cumulus convection in a general circulation model, *J. Atmos. Sci.*, **39**, 2159–2181.

20. Douville, H. and Royer, J.F. (1996) Sensitivity of the Asian summer monsoon to an anomalous Eurasian snow cover within the Meteo-France GCM, *Climate Dyn.*, **12**, 449–466.

21. Druyan, L.M. and Hastenrath, S. (1994) Tropical impacts of SST forcing: A case study for 1987 versus 1988, *J. Clim.*, **7**, 1316–1323.

22. Falconer, K. (1990) *Fractal Geometry: Mathematical Foundations and Applications*, John Wiley & Sons, Inc., Chichester.

23. Fouquart, Y., Bonnel, B., and Ramaswamy, V. (1991) Intercomparing shortwave radiation codes for climate studies, *J. Geophys. Res.*, **96**, 8955–8968.

24. Garratt, J.R. (1993) Sensitivity of climate simulations to land-surface and atmospheric boundary-layer treatments - A review, *J. Clim.*, **6**, 419–449.

25. Garric, G., Stephenson, D.B., and Terray, L. (1997) A global coupled atmosphere sea ice upper ocean simulation, *C. R. Acad. Sci., Paris, Ser. II Fasc. A*, **324**, 529–536.

26. Gates, W.L. (1992) AMIP: The Atmospheric Model Intercomparison Project, *Bull. Amer. Meteor. Soc.*, **73**, 1962–1970.

27. Genthon, C. (1992) Simulations of desert dust and sea-salt aerosols in Antarctica with a general circulation model of the atmosphere, *Tellus*, **44B**, 371–389.

28. Giorgi, F. and Mearns, L.O. (1991) Approaches to the simulation of regional climate change: A review, *Rev. Geophys.*, **29**, 191–216.

29. Gordon, H.B. and Davies, D.R. (1977) The sensitivity of response of climatic characteristics in a two-level general circulation model to small changes in solar radiation, *Tellus*, **29**, 484–501.

30. Graf, H.F., Kirchner, I., Sausen, R., and Schubert, S. (1992) The impact of upper-tropospheric aerosol on global atmospheric circulation, *Ann. Geophysicae*, **10**, 698–707.

31. Gregory, J.M. and Mitchell, J.F.B. (1997) The climate response to CO2 of the Hadley Centre coupled AOGCM with and without flux adjustment, *Geophys. Res. Lett.*, **24**, 1943–1946.

32. Guckenheimer, J. and Holmes, P. (1983) *Nonlinear Oscillations, Dynamical Systems and Bifurcations of Vector Fields*, Springer-Verlag, Berlin.

33. Guilyardi, E. and Madec, G. (1997) Performance of the OPA/ARPEGE-T21 global ocean-atmosphere coupled model, *Climate Dyn.*, **13**, 149–165.

34. Hahn, D.G. and Manabe, S. (1975) The role of mountains in the South Asian monsoon circulation, *J. Atmos. Sci.*, **32**, 1515–1541.

35. Harzallah, A. and Sadourny, R. (1995) Internal versus SST-forced atmospheric variability as simulated by an atmospheric general circulation model, *J. Clim.*, **8**, 474–495.

36. Houghton, J.T., Meiro Filho, L.G., Callander, B.A., Harris, N., Kattenberg, A., and Maskell, K., Eds. (1996) *Climate Change 1995 - The Science of Climate Change*, Cambridge University Press, New York. Contribution of WGI to the Second Assessment Report of the Intergovrnmental Panel on Climate Change.

37. Hunt, B.G. (1982) The impact of large variations of the Earth's obliquity on the climate, *J. Meteor. Soc. Japan*, **60**, 309–318.

38. Iooss, G. and Joseph, D.D. (1980) *Elementary Stability and Bifurcation Theory*, Springer-Verlag, New York.

39. Jarraud, M., Simmons, A.J., and Kanamitsu, M. (1988) Sensitivity of medium range weather forecasts to the use of an envelope orography, *Quart. J. Roy. Meteor. Soc.*, **114**, 989–1025.

40. Jenkins, G.S. (1993) A general circulation model study of the effects of faster rotation rate, enhanced CO2 concentration, and reduced solar forcing: Implications for the faint young sun paradox, *J. Geophys. Res.*, **98**, 20,803–20,812.

41. Johnson, G.L., Hanson, C.L., Hardegree, S.P., and Ballard, E.B. (1996) Stochastic weather simulation: Overview and analysis of two commonly used models, *J. Appl. Meteor.*, **35**, 1878–1896.

42. Joussaume, S. (1990) Three-dimensional simulations of the atmospheric cycle of desert dust particles using a general circulation model, *J. Geophys. Res.*, **95**, 1909–1942.

43. Jouzel, J., Koster, R.D., Suozzo, R.J., and Russell, G.L. (1994) Stable water isotope behavior during the Last Glacial Maximum: A general circulation model analysis, *J. Geophys. Res.*, **99**, 25,791–25,801.

44. Kasahara, A., Sasamori, T., and Washington, W.M. (1973) Simulation experiments with a 12-layer stratospheric global circulation model. I. Dynamical effect of the Earth's orography and thermal influence of continentality, *J. Atmos. Sci.*, **30**, 1229–1251.

45. Kiehl, J.T. and Trenberth, K.E. (1997) Earth's annual global mean energy budget, *Bull. Amer. Meteor. Soc.*, **78**, 197–208.

46. Kitoh, A. (1991) Interannual variations in an atmospheric GCM forced by the 1970-1989 SST. Part I: Response of the tropical atmosphere, *J. Meteor. Soc. Japan*, **69**, 251–269.

47. Kutzbach, J.E. and Gallimore, R.G. (1988) Sensitivity of a coupled atmosphere/mixed layer ocean model to changes in orbital forcing at 9000 years BP, *J. Geophys. Res.*, **93**, 803–821.

48. Laursen, L. and Eliasen, E. (1989) On the effects of the damping mechanisms in an atmospheric general circulation model, *Tellus*, **41A**, 385–400.

49. Laval, K. (1986) General circulation model experiments with surface albedo changes, *Climatic Change*, **9**, 91–102.

50. Le Treut, H. and Laval, K. (1986) Sensitivity of the climate simulated by a GCM to the parameterization of stratus clouds, *Ann. Geophys. B*, **4**, 377–384.

51. Lean, J. and Rowntree, P.R. (1997) Understanding the sensitivity of a GCM simulation of Amazonian deforestation to the specification of vegetation and soil characteristics, *J. Clim.*, **10**, 1216–1235.

52. Livezey, R.E. (1985) Statistical analysis of general circulation model climate simulation: Sensitivity and prediction experiments, *J. Atmos. Sci.*, **42**, 1139–1149.

53. Lorenz, E.N. (1970) Climatic change as a mathematical problem, *J. Appl. Meteor.*, **9**, 325–329.

54. Mahfouf, J.F., Cariolle, D., Royer, J.F., Geleyn, J.F., and Timbal, B. (1994) Response of the Meteo-France climate model to changes in CO2 and sea surface temperature, *Climate Dyn.*, **9**, 345–362.

55. Manabe, S. and Terpstra, T.B. (1974) The effects of mountains on the general circulation of the atmosphere as identified by numerical experiments, *J. Atmos. Sci.*, **31**, 3–42.

56. Manabe, S. and Wetherald, R.T. (1975) The effects of doubling the CO2 concentration on the climate of a general circulation model, *J. Atmos. Sci.*, **32**, 3–15.

57. Mandel, J. (1964) *The statistical analysis of experimental data*, Dover Publications, Inc., New York, reprint.

58. Marshall, S., Roads, J.O., and Oglesby, R.J. (1997) Effects of resolution and physics on precipitation in the NCAR Community Climate Model, *J. Geophys. Res.*, **102**, 19,529–19,541.

59. Masson, V. and Joussaume, S. (1997) Energetics of the 6000-yr BP atmospheric circulation in boreal summer, from large-scale to monsoon areas: A study with two versions of the LMD AGCM, *J. Clim.*, **10**, 2888–2903.

60. Meehl, G.A. (1994) Influence of the land surface in the Asian summer monsoon: External conditions versus internal feedbacks, *J. Clim.*, **7**, 1033–1049.

61. Meehl, G.A. and Washington, W.M. (1993) South Asian summer monsoon variability in a model with doubled atmospheric carbon dioxide concentration, *Science*, **260**, 1101–1104.

62. Milly, P.C.D. and Dunne, K.A. (1994) Sensitivity of the global water cycle to the water holding capacity of land, *J. Clim.*, **7**, 506–526.

63. Mitchell, J.F.B. (1983) The seasonal response of a general circulation model to changes in CO2 and sea temperatures, *Quart. J. Roy. Meteor. Soc.*, **109**, 113–152.

64. Molteni, F., Ferranti, L., Palmer, T.N., and Viterbo, P. (1993) A dynamical interpretation of the global response to equatorial Pacific SST anomalies, *J. Clim.*, **6**, 777–795.

65. Morcrette, J.J. and Geleyn, J.F. (1985) On the influence of different radiation parametrizations on model-generated radiation fields, *Quart. J. Roy. Meteor. Soc.*, **111**, 565–585.

66. Mullen, S.L. (1994) The impact of an envelope orography on low-frequency variability and blocking in a low-resolution general circulation model, *J. Clim.*, **7**, 1815–1826.

67. Murphy, J.M. (1995) Transient response of the Hadley Centre coupled ocean-atmosphere model to increasing carbon dioxide. 1: Control climate and flux adjustment, *J. Clim.*, **8**, 36–56.

68. Navarra, A. (1993) A new set of orthonormal modes for linearized meteorological problems, *J. Atmos. Sci.*, **50**, 2569–2583.

69. Newson, R.L. (1973) Response of a general circulation model of the atmosphere to removal of the Arctic ice-cap, *Nature*, **241**, 39–40.
70. Nicolis, C., and Nicolis, G. (Eds.) (1987) Irreversible Phenomena and Dynamical Systems Analysis in Geosciences, *NATO ASI Series - C: Math. Phys. Sci.*, **192,** D. Reidel Publishing Company, Dordrecht.
71. Nicolis, G., and Prigogine, I. (1989) *Exploring complexity an introduction*, W. H. Freeman and Company, New York.
72. Phillips, T.J., Gates, W.L., and Arpe, K. (1992) The effects of sampling frequency on the climate statistics of the European Centre for Medium-Range Weather Forecasts, *J. Geophys. Res.*, **97**, 20,427–20,436.
73. Pittock, A.B., Walsh, K., and Frederiksen, J.S. (1989) General circulation model simulation of mild nuclear winter effects, *Climate Dyn.*, **3**, 191–206.
74. Polcher, J. (1995) Sensitivity of tropical convection to land surface processes, *J. Atmos. Sci.*, **52**, 3143–3161.
75. Randall, D.A., Cess, R.D., Blanchet, J.P., Chalita, S., Colman, R., Dazlich, D.A., Del Genio, A.D., Keup, E., Lacis, A., Le Treut, H., Liang, X.Z., McAvaney, B.J., Mahfouf, J.F., Meleshko, V.P., Morcrette, J.J., Norris, P.M., Potter, G.L., Rikus, L., Roeckner, E., Royer, J.F., Schlese, U., Sheinin, D.A., Sokolov, A.P., Taylor, K.E., Wetherald, R.T., Yagai, I., and Zhang, M.H. (1994) Analysis of snow feedbacks in 14 general circulation models, *J. Geophys. Res.*, **99**, 20,757–20,771.
76. Rasch, P.J. and Williamson, D.L. (1991) The sensitivity of a general circulation model climate to the moisture transport formulation, *J. Geophys. Res.*, **96**, 13123–13137.
77. Rind, D., Healy, R., Parkinson, C., and Martinson, D. (1995) The role of sea ice in 2 x CO2 climate model sensitivity. Part 1: The total influence of sea ice thickness and extent, *J. Clim.*, **8**, 449–463.
78. Rowntree, P.R. (1972) The influence of tropical east Pacific Ocean temperatures on the atmosphere, *Quart. J. Roy. Meteor. Soc.*, **98**, 290–321.
79. Royer, J.F., Chauvin, F., and Pélissier, G. (1994) Sensitivity of the atmospheric circulation to solar irradiance variations, In E. Nesme-Ribes (ed.) *The Solar Engine and its Influence on Terrestrial Atmosphere and Climate*, Springer-Verlag, Berlin.
80. Royer, J.F., Déqué, M., and Pestiaux, P. (1984) A sensitivity experiment to astronomical forcing with a spectral GCM: Simulation of the annual cycle at 125000 BP and 115000 BP, In A. Berger et al. (eds.), *Milankovitch and Climate. Understanding the Response to astronomical forcing. Part II*, D.Reidel Publishing Company, Dordrecht, pp. 733–763.
81. Royer, J.F., Planton, S., and Déqué, M. (1990) A sensitivity experiment for the removal of Arctic sea ice with the French spectral general circulation model, *Climate Dyn.*, **5**, 1–17.
82. Schertzer, D., and Lovejoy, S., Eds. (1991) *Non-Linear Variability in Geophysics: Scaling and Fractals*, Kluwer Academic Publishers, Dordrecht.
83. Serafini, Y.V. (1990) The time scale of land surface hydrology in response to initial soil moisture anomalies a case study, *Tellus*, **42A**, 390–400.
84. Shukla, J. and Mintz, Y. (1982) Influence of land-surface evapotranspiration on the Earth's climate, *Science*, **215**, 1498–1501.
85. Shukla, J. and Wallace, J.M. (1983) Numerical simulation of the atmospheric response to equatorial Pacific sea surface temperature anomalies, *J. Atmos. Sci.*, **40**, 1613–1630.
86. Simmonds, I. (1985) Analysis of the "spinup" of a general circulation model, *J. Geophys. Res.*, **90**, 5637–5660.
87. Soman, M.K. and Slingo, J. (1997) Sensitivity of the Asian summer monsoon to aspects of sea-surface- temperature anomalies in the tropical Pacific Ocean, *Quart. J. Roy. Meteor. Soc.*, **123**, 309–336.

58

88. Stephenson, D.B. (1995) The impact of changing the horizontal diffusion scheme on the northern winter climatology of a general circulation model, *Quart. J. Roy. Meteor. Soc.*, **121**, 211–226.

89. Stephenson, D.B., Chauvin, F., and Royer, J.F. (1998) Simulation of the Asian summer monsoon and its dependence on model horizontal resolution, *J. Meteor. Soc. Japan*, **76**, 237–265.

90. Stephenson, D.B. and Royer, J.F. (1995) GCM simulation of the Southern Oscillation from 1979-88, *Climate Dyn.*, **11**, 115–128.

91. Sud, Y.C., Walker, G.K., Kim, J.H., Liston, G.E., Sellers, P.J., and Lau, W.K.M. (1996) Biogeophysical consequences of a tropical deforestation scenario: A GCM simulation study, *J. Clim.*, **9**, 3225–3247.

92. Thuburn, J. (1991) Data sampling strategies for general circulation models, *Quart. J. Roy. Meteor. Soc.*, **117**, 385–397.

93. Timbal, B., Mahfouf, J.F., Royer, J.F., Cubasch, U., and Murphy, J.M. (1997) Comparison between doubled CO2 time-slice and coupled experiments, *J. Clim.*, **10**, 1463–1469.

94. Trzaska, S., Moron, V., and Fontaine, B. (1996) Global atmospheric response to specific linear combinations of the main SST modes. Part I: Numerical experiments and preliminary results, *Ann. Geophysicae*, **14**, 1066–1077.

95. von Storch, H. and Zwiers, F.W. (1988) Recurrence analysis of climate sensitivity experiments, *J. Clim.*, **1**, 157–171.

96. Washington, W.M. and Chervin, R.M. (1979) Regional climatic effects of large-scale thermal pollution: Simulation studies with the NCAR general circulation model, *J. Appl. Meteor.*, **18**, 3–16.

97. Washington, W.M. and Meehl, G.A. (1983) General circulation model experiments on the climatic effects due to a doubling and quadrupling of carbon dioxide concentration, *J. Geophys. Res.*, **88**, 6600–6610.

98. Weare, B.C. (1988) A comparison of radiation variables calculated in the UCLA general circulation model to observations, *J. Clim.*, **1**, 485–499.

99. Wetherald, R.T. and Manabe, S. (1975) The effects of changing the solar constant on the climate of a general circulation model, *J. Atmos. Sci.*, **32**, 2044–2059.

100. Wetherald, R.T. and Manabe, S. (1988) Cloud feedback processes in a general circulation model, *J. Atmos. Sci.*, **45**, 1397–1415.

101. Wiggins, S. (1990) *Introduction to Applied Dynamical Systems and Chaos*, Springer-Verlag, New York.

102. Wilks, D.S. (1992) Adapting stochastic weather generation algorithms for climate change studies, *Climatic Change*, **22**, 67–84.

103. Williams, J., Barry, R.G., and Washington, W.M. (1974) Simulation of the atmospheric circulation using the NCAR global circulation model with Ice Age boundary conditions, *J. Appl. Meteor.*, **13**, 305–317.

104. Williamson, D.L. (1988) The effect of vertical finite difference approximations on simulations with the NCAR Community Climate Model, *J. Clim.*, **1**, 40–58.

105. Xue, Y.K. (1997) Biosphere feedback on regional climate in tropical north Africa, *Quart. J. Roy. Meteor. Soc.*, **123**, 1483–1515.

106. Zwiers, F.W. and Boer, G.J. (1987) A comparison of climates simulated by a general circulation model when run in the annual cycle and perpetual modes, *Mon. Wea. Rev.*, **115**, 2626–2644.

ANALYSIS AND VERIFICATION OF MODEL CLIMATE

G. J. BOER
Canadian Centre for Climate Modelling and Analysis
Atmospheric Environment Service
University of Victoria
Victoria, B.C., V8W 2Y2   CANADA

1. Introduction

Atmospheric general circulation models (AGCMs) are used for weather forecasting, short-range climate forecasting, and are major components of coupled climate models. The evolution of the CCCma model is typical. An increasingly sophisticated AGCM is part of an evolving sequence of climate models which are applied to increasingly complex simulations of different aspects of climate. An AGCM alone may be used to simulate and investigate the atmospheric climate and climate perturbations that arise with specified sea-surface temperatures (SSTs) and ice boundaries (e.g., [2]). Such studies include the simulation of current climate, local and remote effects of El Niño SST anomalies, short range climate forecasting, atmospheric budgets and balances, and related studies. An atmospheric GCM coupled to a mixed layer ocean component is used to simulate equilibrium climate change due to increasing greenhouse gases, aerosols and other perturbations to the climate, including paleoclimate simulations ([3], [12], Chapter 20). Coupling an atmospheric GCM to a full ocean GCM permits investigation of the time evolution of climate change in the past and the projection of climate change into the future as discussed in Chapter 21.

A climate model, like the proverbial chain, is only as strong as it weakest link. The cycle of model development is sketched in Figure 1 where "analysis and verification" investigates the ability of the model to simulate current climate. The ability to simulate current climate is taken as a basic indication of the model's ability to simulate future climates. The identification of deficiencies in the simulation points the way to needed model improvement.

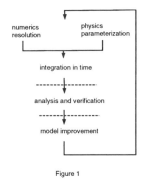

Figure 1

The comprehensive assessment reports of the Intergovernmental Panel on Climate Change ([4], [5]) summarize the state of climate change research. Each report contains a section on model "verification" to put the ability of models to simulate future climate change in context. The international effort to "intercompare" climate model results (Chapter 19) is also a part of the evolutionary process of understanding and improving models through the analysis and verification of model climate.

P. Mote and A. O'Neill (eds.), Numerical Modeling of the Global Atmosphere in the Climate System, 59–82.
© 2000 *Kluwer Academic Publishers. Printed in the Netherlands.*

2. Climate Analysis and Verification

The "climate" is often defined as "the average of the weather" and characterized in terms of basic surface quantities such as the average temperature and precipitation rate. A more complete description of climate involves the equations governing the three-dimensional climate system and the distribution of the statistics of the quantities involved. A climate model simulates the hour to hour evolution of the atmosphere (and the ocean in the case of a coupled model) by numerically integrating the governing equations forward in time. The resulting time-evolving solution is not expected to mirror the evolution of the observed system, which would imply the ability to make forecasts indefinitely into the future, but rather to simulate a "possible" evolution under current "external conditions" (the solar constant, the composition of the atmosphere, the shape of the earth, etc.). Such a solution should, however, have the same *climate statistics* as the real system. The verification of a climate model is thus in terms of the statistics of the solution. Typically, first and second order moments (means, variances and covariances) are considered but there are a bewildering array of other possible quantities to study. Means and variances arise naturally as terms in averaged versions of the governing equations and the associated budget equations discussed below. Since the verification is in terms of the statistics of the solution, sampling considerations are involved. Unlike the case for weather forecasting, however, generally accepted "measures of skill" for climate models have yet to be developed.

Model verification asks several questions: (1) can models reproduce the current climate of the atmosphere (ocean, land, cryosphere, etc.); (2) do models correctly incorporate important climate mechanisms; (3) can models reproduce "perturbed" climates (e.g. solar changes, volcanic effects); (4) are the climate changes projected by models "reliable"; and (5) how do we improve models? There are three broad categories of verification, namely: (1) the *morphology of climate* as given by the spatial distribution and structure of the means, variances, and covariances (and possibly other statistics) of basic climatic parameters; (2) *budgets, balances, and "cycles"* in the climate system where the generation or source of some quantity, its transport, transformation, and ultimately its sink is considered (e.g. the hydrological cycle, the budget of angular momentum, the atmospheric energy cycle); and (3) *process studies of climate*, which investigate particular aspects of the climate system such as the monsoons, blocking, convective processes etc. We concentrate here on the first two of these after first discussing the basic equations of climate and climate analysis.

3. The Basic Equations of Climate and Climate Analysis

3.1 THE PRIMITIVE EQUATIONS

The so called "primitive" equations for the atmosphere, with pressure as the vertical coordinate and using the usual meteorological notation, are written as

$$\frac{dV}{dt} + fk \times V + \nabla\phi = F \tag{1}$$

$$C_p\frac{dT}{dt} - \omega\alpha = Q \tag{2}$$

$$\frac{dq}{dt} = S \tag{3}$$

$$\nabla \cdot V + \frac{\partial\omega}{\partial p} = 0 \tag{4}$$

$$\frac{\partial\phi}{\partial p} = -\alpha \tag{5}$$

$$p\alpha = RT \tag{6}$$

and embody Newton's second law applied to a fluid, the first law of thermodynamics, conservation of moisture, conservation of mass, hydrostatic balance, and the gas law respectively. Here the horizontal velocity V, temperature T, and specific humidity q, are basic *prognostic* variables while the geopotential ϕ, the vertical "pressure velocity" $\omega = dp/dt$, and the specific volume α, may be obtained from them. The terms on the right-hand sides of equations (1-3) represent parameterized physical processes as is discussed below. Note that $\nabla\phi$ is the equivalent of the pressure gradient in these equations.

The kinematic boundary conditions

$$\omega = 0, p = 0 \text{ and } \omega_s = \frac{\partial p_s}{\partial t} + V_s \cdot \nabla p_s \quad p = p_s \tag{7}$$

apply at the top of the atmosphere and at the surface. In pressure coordinates, the total derivative may be written in the advective and flux forms respectively as

$$\frac{dX}{dt} = \frac{\partial X}{\partial t} + V \cdot \nabla X + \omega\frac{\partial X}{\partial p} = \frac{\partial X}{\partial t} + \nabla \cdot XV + \frac{\partial}{\partial p}X\omega \tag{8}$$

which involve quadratic terms. The domain of definition, in the usual spherical polar coordinate system of latitude and longitude, is

$$0 \le \lambda \le 2\pi, -\pi/2 \le \varphi \le \pi/2, 0 \le p \le p_s(\lambda, \varphi, t)$$

and the element of mass is

$$dm = a^2\cos\varphi \, d\lambda \, d\varphi \, dp/g.$$

These equations form the basis for the analysis of model results and of observed data. A similar set of equations applies to the ocean with an equation for salinity replacing that for moisture.

3.2 THE VERTICAL COORDINATE

The primitive equations are usually derived using cartesian coordinates with geometric height as the vertical coordinate but may be recast with an arbitrary vertical coordinate provided that it is a single-valued monotonic function of height (hydrostatic balance is

62

usually assumed). As discussed in Chapter 7, some scaled pressure coordinate, such as $\sigma = p/p_s$, is almost invariably used in global atmospheric models since the bottom boundary condition becomes $\dot{\sigma} = d\sigma/dt = 0$, at $\sigma = 1$ and is therefore easy to treat numerically. The two styles of vertical coordinate are illustrated in Figure 2.

Pressure is the preferred vertical coordinate for *analysis*, however, because: (1) the governing equations adopt a particularly simple form (resembling that for an incompressible fluid), (2) pressure surfaces are quasi-horizontal which aids physical interpretation, (3) advective and flux terms are quadratic, and (4) data from global analyses are routinely available on a set of "standard pressure levels".

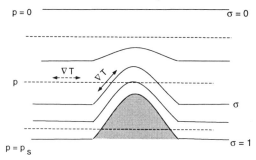

Figure 2. Pressure and scaled pressure vertical coordinates

One disadvantage of pressure as the vertical coordinate is the comparatively complicated lower boundary condition (7) which implies that fields may be multiply connected where topography pierces the pressure surface. While this is ignored here in order to simplify the discussion, Boer [1] provides a formalism that treats this difficulty for calculations in pressure coordinates. A scaled pressure vertical coordinate is usually avoided for analysis because: (1) the resulting equations are more complex (especially the pressure gradient term), (2) σ-surfaces are not horizontal so that "horizontal" and "vertical" gradients of quantities are intertwined as indicated by the horizontal temperature gradients in the figure, (3) flux quantities have the form of triple products, (4) there is no uniformity among models in the use of the vertical coordinate, and (5) there are no "standard" model levels.

3.3 RESOLVED AND PARAMETERIZED QUANTITIES

There are a number of steps in passing from the continuous equations to a working global model. These steps include: (1) the *discretizing* of the equations and the choice of resolution for the model, (2) the *parameterization* of unresolved physical processes, and (3) the *integration* of the resulting equations by stepping forward in time. The left-hand sides of equations (1-3) involve the resolved *dynamics* of the model while the right-hand sides involve the parameterized *physics* of the model.

To make this more explicit, consider the prototype equation

$$\frac{\partial X}{\partial t} + V \cdot \nabla X + \omega \frac{\partial X}{\partial p} = \frac{\partial X}{\partial t} + \nabla \cdot XV + \frac{\partial}{\partial p} X\omega = S \qquad (9)$$

for some variable X. This equation embodies the left-hand-side dynamics, which are a feature of the prognostic equations for V, T, q, as well as a right-hand-side physical "source/sink" term (in fact the prototype equation has just the form of the moisture equation but it could easily be extended to encompass the additional dynamical terms seen in the other equations). However, a numerical model can only represent, and an observations system can only observe, a certain range of scales. We write $X = X_r + X_u$, for the resolved and unresolved components. Applying a "resolution" or "truncation" operator, $R(X) \rightarrow X_r$, which retains only resolved scales, to (9) gives

$$\frac{\partial X_r}{\partial t} + V_r \cdot \nabla X_r + \omega_r \frac{\partial X_r}{\partial p} = \{\nabla \cdot (X_r V_r - (XV)_r)\} + \{\partial(X_r \omega_r - (X\omega)_r)/\partial p\}$$

$$+ \{(S(X))_r - S_r(X_r)\}$$

or, rewriting the three terms in braces on the right-hand-side,

$$\frac{\partial X_r}{\partial t} + V_r \cdot \nabla X_r + \omega_r \frac{\partial X_r}{\partial p} = S_H(X_r) + S_V(X_r) + S_r(X_r) + E_D + E_S.$$

Here the dynamical quantities on the left-hand-side of the equation involve only the *resolved* variables and the effect of the *unresolved* horizontal and vertical dynamical transports and their interactions with the resolved scales (the first two terms in braces) are approximated or *parameterized* as expressions S_H and S_V (usually a form of diffusion) acting on the resolved quantities which now appear as terms on the right-hand-side of the equation. The source/sink term S is parameterized as a function of the resolved variables and returns values at the resolved scale. E_D and E_S are the errors that arise in the approximate treatment of the dynamical and physical source/sink terms respectively. Note that these *parameterization* errors are independent of the additional *numerical approximations* that are made to cast the equations into finite difference (or other) numerical form for their numerical solution.

Replacing the differential operators by numerical approximations (indicated by a caret) adds a *numerical* error E_N, whence the prototype equation, discretized over a grid of points, with $X_{ijk}^\tau = X(\lambda_i, \varphi_j, p_k, \tau \Delta t)$, is written symbolically as

$$\frac{\hat{\partial}}{\partial t} X_{ijk}^\tau + V_{ijk}^\tau \cdot \hat{\nabla} X_{ijk}^\tau + \omega_{ijk}^\tau \frac{\hat{\partial}}{\partial p} X_{ijk}^\tau = S_{ijk}^\tau + S_{H_{ijk}}^{\ \tau} + S_{V_{ijk}}^{\ \tau} + E_D + E_S + E_N. \quad (10)$$

It is important to note that while numerical errors E_N will clearly decrease with resolution, it is not necessarily the case that the parameterization errors E_D, E_S will also decrease. This is the reason that an increase in model resolution does not guarantee an overall improvement in simulated climate. Parameterization error should also decrease with increasing resolution if only because more scales are resolved so their effects do not need to be parameterized, but this need not be the case if the parameterization has been formulated for one resolution but is applied at a different resolution.

Finally, the parameterized physical terms on the right-hand-side of (10) are typically written as the divergence of a vertical flux f (of momentum, moisture or energy) in the form

$$S + S_V + S_H = -g\frac{\partial f}{\partial p} + S_H \approx -g\frac{\partial f}{\partial p}. \quad (11)$$

Since S_H is typically smaller than $S+S_V$ it is often neglected for analysis (but *not* for prognostic calculations).

The global model, consisting now of the discretized version of (1-6) in the manner of (10), is integrated forward in time over many timesteps and provides evolving three

64

dimensional fields of the prognostic variables X^{τ}_{ijk} together with the right-hand-side quantities $S^{\tau}_{ijk}, S_{H\,ijk}^{\tau}, S_{V\,ijk}^{\tau}$. The primary source/sink term S may have several components. For the thermodynamic equation (2), $Q^{\tau}_{ijk} = Q_{sw} + Q_{lw} + Q_{lh} + Q_{con} + Q_f$ represents the heating associated with short- and long-wave radiative effects, latent heat release, convective heat exchange, and the dissipation of kinetic energy into heat. This is in addition to the parameterized subgrid horizontal and vertical heat transports represented by $Q^{\tau}_{H\,ijk}$ and $Q^{\tau}_{V\,ijk}$. For an atmospheric model, $X^{\tau}_{ijk} \Leftrightarrow (V, \omega, T, q, \phi)$ are explicitly resolved dynamical quantities, while $S^{\tau}_{ijk} \Leftrightarrow (F, Q, S)$ (and their components) are parameterized physical processes.

3.4 BUDGET EQUATIONS

Equations (1-6) embody a set of conservation/budget principles that provide a focus for the analysis of many aspects of climate and climate model behaviour. Basic budgets include those of:

- *Mass*, governed by the continuity equation
-
- *Moisture* (or other constituent)

$$\frac{\partial q}{\partial t} + \nabla \cdot qV + \frac{\partial}{\partial p}q\omega = S \approx -g\frac{\partial}{\partial p}(m_q + m_c) \qquad (12)$$

where the parameterized subgrid flux of moisture in any form (vapour, liquid and/or solid) is approximated by the gradient of the vertical flux $m_q + m_c$ of water substance in vapour and condensed form. Water vapour in the atmosphere is a source of "latent" heat which is realized when the vapour condenses and precipitates out. The moisture budget becomes the latent energy budget when multiplied by the latent heat of vaporization L.

- *Angular momentum* $M = a\cos\varphi(\Omega a\cos\varphi + u)$

$$\frac{\partial M}{\partial t} + \nabla \cdot MV + \frac{\partial}{\partial p}M\omega + \frac{\partial}{\partial\lambda}\phi = a\cos\varphi F_\lambda \approx -g\,a\cos\varphi\frac{\partial\tau_\lambda}{\partial p} \qquad (13)$$

where τ is the vertical component of the stress. $\qquad (14)$

- *Kinetic energy* $k = \frac{1}{2}V \cdot V$

$$\frac{\partial k}{\partial t} + \nabla \cdot kV + \frac{\partial}{\partial p}k\omega + V \cdot \nabla\phi = V \cdot F \approx -gV \cdot \frac{\partial\tau}{\partial p} \qquad (15)$$

- *Thermodynamic and potential energy*

$$\frac{\partial}{\partial t}C_pT + \nabla \cdot (C_pT + \phi)V + \frac{\partial}{\partial p}(C_pT + \phi)\omega - V \cdot \nabla\phi = Q_n \qquad (16)$$

where $Q_n = Q - Q_f$ is the "non-frictional" heating. The term $V \cdot \nabla\phi$ represents the dynamical conversion between the two forms of energy (parcels of air gain kinetic energy as they are accelerated down the pressure gradient) in (15-16). The kinetic, thermodynamic and potential, and latent energy equations may be added together in various ways. The sum of (15-16) is often used in a modified form as the "dry static" energy equation (dry since the latent energy of the moisture is not included, and static when k is neglected in comparison with the other energy terms).

- *Total energy*

$$\frac{\partial}{\partial t}(k + C_p T + Lq) + \nabla \cdot (k + C_p T + Lq + \phi)V + \frac{\partial}{\partial p}(k + C_p T + Lq + \phi)\omega$$

$$= Q + LS \approx -g\frac{\partial}{\partial p}\{R_{sw} + R_{lw} + Lm_q + h\} \qquad (17)$$

is the sum of the kinetic, internal, potential, and latent energies. The right-hand-side term $Q \approx -g\partial(R_{sw} + R_{lw} - Lm_q + h)/\partial p$ involves the fluxes of shortwave (R_{sw}) and longwave (R_{lw}) radiation, the latent energy of water vapour, and sensible heat. When k is neglected in the total energy equation it becomes the "moist static" energy equation.

- *The atmospheric energy cycle*

$$\frac{\partial A}{\partial t} = G - C \qquad \frac{\partial K}{\partial t} = C - D \qquad (18)$$

attempts to measure the "basic rate of working" of the atmosphere as a whole by estimating the rate of *generation* G of available potential energy A (the part of the thermodynamic and potential energy that may be converted to kinetic energy), its *conversion* to kinetic energy C, and subsequent *dissipation. D*. The expressions for these quantities and their evaluation from observed and model data is discussed in Section 6.6.

4. Data Considerations

There are four basic types of data available for verification of atmospheric climate models. They are: (1) *satellite data*, especially top of the atmosphere (TOA) radiative fluxes but including derived quantities; (2) *surface data* based mainly on station and ship observations; (3) *objective analyses* (and reanalysis) of the basic meteorological quantities (V, T, q, ϕ) on standard pressure surfaces and a grid of points over the globe; and (4) *derived* quantities, basically the parameterized terms on the right-hand-sides of (1-3), which are not observed directly but are obtained by calculation. In practice the verification of model climate is considerably hampered by a disconcerting lack of knowledge of some of the most basic climate parameters of the real system.

There is no central archive for climate-related data, and modelling groups tend to collect information as and when it becomes available. The internet is a valuable source of climate data. Information on the U.S. National Center for Environmental Prediction (NCEP) reanalysis [6] may be found on the NCEP web site. Other reanalyses have been performed by the European Centre for Medium-Range Weather Forecasts and by the Goddard Laboratory for Atmospheres. Specialized data collections are available from the Program for Climate Model Diagnosis and Intercomparison (PCMDI), which collects data sets for model verification. Similarly, information on satellite data from the

Earth Radiation Budget Experiment (ERBE), the International Satellite Cloud Climatology Project (ISCCP) and other special data sets such as those of the Global Precipitation Climatology Project (GPCP) may be accessed via the internet.

5. A Hierarchy of Budget Equations and Climate Statistics

We develop a hierarchy of budget equations and climate statistics under averaging and integration of various kinds applied to the mass, angular momentum, moisture, and total energy budgets of the atmosphere. Analyses of this general kind have a lengthy history ranging from the very early efforts to understand atmospheric budgets and balances based on surface data, and then on relatively sparse radiosonde data, as described by Lorenz [7], to efforts to include all parts of the globe ([9], [10], [11]), to more recent calculations based on data from global reanalyses [13].

5.1 AVERAGING

Climate statistics are formally "ensemble averages over multiple realizations of a stocastic process" but in practice are time averages or a mixture of time and ensemble averages (e.g., an average over 30 Januarys serves to specify the mean January climate). The implicit assumption is that the climate reflects an underlying regularity behind the irregular unsteady and non-repeating "weather" that is so characteristic of the system. The idea is to separate the flow into two parts: the climate is that part determined by the large scale "external" forcing of the system (the solar constant, the rotation of the earth, and so on) while the remaining component is "internally generated" and consists of the irregular "eddies" that comprise, in simple terms, the weather.

This approach is familiar from turbulence studies where time/ensemble averaging, indicated by an overbar, separates the flow into *mean* and *transient eddy* components

$$X = \bar{X} + X' \qquad \overline{XV} = \bar{X}\bar{V} + \overline{X'V'}. \qquad (19)$$

This decomposition is illustrated for temperature in Figure 3 which displays the mean temperature structure $\bar{T}$ and the standard deviation $\sqrt{\overline{T'^2}}$ (as a measure of the vigour of the transient eddies) on the 850 hPa pressure level for December-February (DJF) from NCEP reanalysis data.

Averaging the prototype equation (9) gives an equation for the climate variable $\bar{X}$ as

$$\frac{\partial}{\partial t}\bar{X} + \bar{V} \cdot \nabla \bar{X} + \bar{\omega}\frac{\partial}{\partial p}\bar{X} = \bar{S} - \left(\nabla \cdot \overline{X'V'} + \frac{\partial}{\partial p}\overline{X'\omega'}\right). \qquad (20)$$

where $\partial \bar{X}/\partial t$ is formally the rate of change of the ensemble average of X, but in practice is usually a climatological time average so that this term is usually small. The effect of the eddies appears in the covariances between variables that arise in the non-linear dynamical terms in the original equations. The average effect of the eddies is to transport momentum, moisture and heat as part of the climatic balance.

The decomposition into mean and transient eddies has some similarity to the decomposition into resolved and unresolved components of Section 2.4. If it were possible to represent the eddy covariances as functions of the mean quantities (via some parameterization or "closure") then the equations for the climate could be written down directly. This view motivated early meteorologists and still finds its proponents in "simplified" energy-balance climate models where the covariance terms in the temperature equation are represented as diffusion terms.

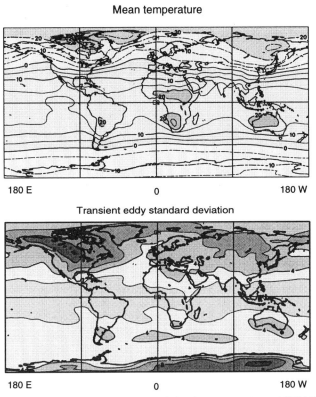

Mean temperature

Transient eddy standard deviation

Figure 3. NCEP climate statistics for temperature at 850 hPa

Suffice it to say that, for the atmosphere at least, this approach has been largely abandoned and the effect of the eddies on the mean climate is obtained by simulating the full flow and performing the decomposition as a diagnostic step. In this diagnostic case, it is often useful to group the mean and eddy components together in the flux form of the equation as

$$\frac{\partial}{\partial t}\overline{X} + \nabla \cdot (\overline{X}\,\overline{V} + \overline{X'V'}) + \frac{\partial}{\partial p}(\overline{X}\,\overline{\omega} + \overline{X'\omega'}) = \overline{S}. \qquad (21)$$

The strong north-south gradient in climate parameters, in contrast to east-west variations, suggests a further zonal averaging, $[x] = \frac{1}{2\pi}\int_o^{2\pi} x\,d\lambda$, whence the decomposition is into *mean meridional, standing eddy,* and *transient eddy* components,

$$X = [\overline{X}] + \overline{X}^* + X' \qquad [\overline{Xv}] = [\overline{X}][\bar{v}] + [\overline{X}^*\bar{v}^*] + [\overline{X'v'}]. \qquad (22)$$

For the temperature field of Figure 3, $\overline{T}^* = \overline{T} - [\overline{T}]$, and $\overline{T}^*$ is the deviation of the

68

temperature in the east-west from the zonal average and, since it is a feature of the time mean field, is a standing structure or "standing eddy".

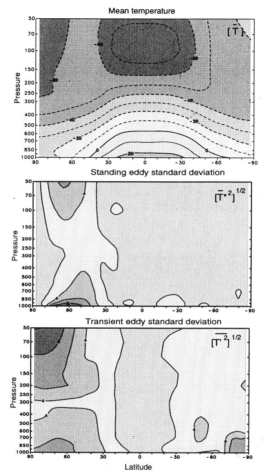

Figure 4. Measures of mean, standing eddy, and transient eddy temperature structures for DJF

The prototype equation becomes

$$\frac{\partial}{\partial t}[\bar{X}] + \frac{[\bar{v}]}{a}\frac{\partial}{\partial \varphi}[\bar{X}] + [\bar{\omega}]\frac{\partial}{\partial p}[\bar{X}] = [\bar{S}]$$

$$- \frac{1}{a\cos\varphi}\frac{\partial}{\partial \varphi}([\bar{X}^*][\bar{v}^*] + [\overline{X'v'}])\cos\varphi - \frac{\partial}{\partial p}([\bar{X}^*][\bar{\omega}^*] + [\overline{X'\omega'}]) \quad (23)$$

where the eddy terms appear on the right-hand-side as eddy effects on the zonal and time-averaged climate. Again, if these terms could be ignored or parameterized, a direct equation for the zonally averaged climate would be available.

We use the decomposition for diagnostic purposes, most often in the flux form

$$\frac{\partial}{\partial t}[\bar{X}] + \frac{1}{a\cos\varphi}\frac{\partial}{\partial\varphi}([\bar{X}][\bar{v}] + [\bar{X}^*][\bar{v}^*] + [\overline{X'v'}])\cos\varphi$$

$$+ \frac{\partial}{\partial p}([\bar{X}][\bar{\omega}] + [\bar{X}^*][\bar{\omega}^*] + [\overline{X'\omega'}]) = [\bar{S}]. \qquad (24)$$

The decomposition under time and zonal averaging, $T = [\bar{T}] + \bar{T}^* + T'$, is used to reveal aspects of the latitudinal/height distribution of climate quantities as illustrated in Figure 4, which displays the temperature structure $[\bar{T}]$ and the standard deviations $\sqrt{[\bar{T}^{*2}]}$ and $\sqrt{[\overline{T'^2}]}$ as measures of the standing and transient eddies.

Finally, vertical integration $\int_o^{p_s} \bar{X} dp/g$ concentrates attention on the geographical distribution of the quantity, while averaging over an area, $\langle X \rangle = \frac{1}{A}\int X dA$, often the entire globe, directs attention to basic global balances.

The mean, variance, and covariance statistics that flow from equations (1-6) and the budget equations (12-18) under averaging, as illustrated by the prototype equations (22, 25), are displayed in Table 1. These are the set of *basic climate statistics*.

Table 1. Means, variances and covariances for atmospheric climate

$\bar{u}$	$\overline{u'u'}$	$\overline{u'v'}$	$\overline{u'\omega'}$	$\overline{u'T'}$	$\overline{u'q'}$	$\overline{u'\phi'}$	$\bar{F}_\lambda$	$\overline{u'F'}_\lambda$
$\bar{v}$		$\overline{v'v'}$	$\overline{v'\omega'}$	$\overline{v'T'}$	$\overline{v'q'}$	$\overline{v'\phi'}$	$\bar{F}_\varphi$	$\overline{v'F'}_\varphi$
$\bar{\omega}$			$\overline{\omega'\omega'}$	$\overline{\omega'T'}$	$\overline{\omega'q'}$	$\overline{\omega'\phi'}$		
$\bar{T}$				$\overline{T'T'}$			$\bar{Q}$	$\overline{Q'T'}$
$\bar{q}$					$\overline{q'q'}$		$\bar{R}$	
$\bar{\phi}$						$\overline{\phi'\phi'}$		

There is a similar array under time and zonal averaging with entries $[\bar{X}]$, $[\bar{X}^*\bar{Y}^*]$, $[\overline{X'Y'}]$.

5.2 A HIERARCY OF BUDGET EQUATIONS UNDER AVERAGING

Figure 5 gives a schematic sense of the kinds of quantities of interest for the verification of the climate of an atmospheric model, namely: (1) the flow of moisture, energy, and momentum across the top and bottom of the atmosphere (TOA and BOA); (2) the morphology of basic climate quantities $\bar{u}, \bar{v}, \bar{\omega}, \bar{T}, \bar{q}, \bar{\phi}, \overline{cloud}...$, symbolized as $\bar{X}$; (3) the distribution of the eddy statistics of Table 1, symbolized as $\overline{XV}, \overline{X'V'}$, which represent fluxes and transports in the body of the atmosphere; (4) the right-hand-side source/sink terms F, Q, R symbolized as $\bar{S}$ (although, in practice these are poorly know and typically given much less attention than the other quantities); and (5) additional surface fields such as mean sea level pressure (mslp) and surface air temperature $\bar{T}_s$.

70

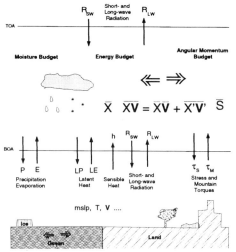

Figure 5. Basic climate balances and statistics.

The connection to the budget equations is indicated formally below where typical forms of the budget equations arise under the various averaging operations. The dimensionality of the terms in the prototype equation (9) is reduced by a sequence of operations resulting in different *levels* of budget equations. Since the time rate of change term is typically small for climatological averages, as noted in conjunction with (20), it is omitted to good approximation to give the *balance equations for climate*. Applying various averaging/integration operators gives the balance equations under:

- *time-averaging*

$$\nabla \cdot \overline{XV} + \frac{\partial}{\partial p}\overline{X\omega} = -g\frac{\partial}{\partial p}\bar{f} \qquad \text{level 3 } (\lambda,\varphi,p)(25)$$

- *time and zonal averaging*

$$\frac{1}{a\cos\varphi}\frac{\partial}{\partial\varphi}[\overline{Xv}]\cos\varphi + \frac{\partial}{\partial p}[\overline{X\omega}] = -g\frac{\partial}{\partial p}[\bar{f}] \qquad \text{level 2 } (\varphi, p) (26)$$

- *time averaging + vertical integration*

$$\nabla \cdot \int \overline{XV}dp = \nabla \cdot \boldsymbol{H} = \bar{f}_T - \bar{f}_S \qquad \text{level 2 } (\lambda,\varphi) (27)$$

- *time and zonal averaging + vertical integration*

$$\frac{1}{a\cos\varphi}\frac{\partial}{\partial\varphi}\{\int[\overline{Xv}]dp/g\}\cos\varphi = \frac{1}{a\cos\varphi}\frac{\partial}{\partial\varphi}H_\varphi = [\bar{f}_T] - [\bar{f}_S] \text{ level 1 } (\varphi) \ (28)$$

- *time averaging, vertical integration and global area averaging*

$$\langle\bar{f}_T\rangle - \langle\bar{f}_S\rangle = 0. \qquad \text{level 0 } (.) \quad (29)$$

This last equation represents the overall balance across the bottom and the top of the atmosphere. The flux terms $(\overline{XV}, \overline{X\omega})$ in these equations have the form of covariances under time and zonal averaging and are further decomposed into mean and eddy components as discussed previously.

The hierarchy of budget equations produce results with the corresponding dimensionality. *Level 0* budgets involve single numbers, *level 1* budgets are functions of latitude only, *level 2* budgets are functions of (latitude, longitude) or (latitude, pressure). Finally, *level 3* budgets are cumbersome because each term is a function of three spatial dimensions (latitude, longitude, pressure).

6. Model Results

We first look at the morphology of the basic quantities suggested by Figure 5 and compare various observed and modelled results. For the example used here we have the results of a recent simulation with a new version of the CCCma atmospheric GCM, referred to as GCM3, with T47L32 spectral dynamics and a 96x48 physics grid. Atmospheric means, variances and covariances from the model for DJF are compared with a 10-year climatology based on the NCEP reanalysis. Then, as examples of budgets under averaging, we consider the vertically integrated and zonally averaged energy transport and the atmospheric energy cycle.

6.1 TOP OF ATMOSPHERE FLUXES, CLOUDS, AND CLOUD FORCING

The way the climate system "processes" the input solar radiation from the sun into the balancing outgoing longwave radiation determines the distribution of temperature and other climate quantities. Increasing the concentration of greenhouse gases and particles in the atmosphere affects these streams of radiation and leads to the kind of climate change discussed in Chapter 21. The processes that affect the flow of radiation in the system are, therefore, particularly important.

The shortwave radiative flux of energy across the top of the atmosphere is the difference between the incoming radiation from the sun and that reflected by the system. The reflected solar is measured in terms of the planetary albedo which is the ratio of these two terms. The reflected solar radiation is greatly affected by the amount and nature of the cloudiness in the atmosphere. While clouds interact strongly with radiation, and while they are generally thought to be the cause of the greatest uncertainty in modelling climate and climate change, it is notoriously difficult to measure their distribution and optical properties. An alternative approach is to measure the *direct radiative effect* of the clouds or what is termed the *cloud forcing* or *cloud radiative forcing*. The TOA shortwave cloud forcing $CF_{sw} = R_{sw} - R_{sw}^{(nc)}$ is the difference between the radiative flux in a grid square with clouds and in the same grid square with the same atmospheric and surface conditions but with "no clouds" indicated as *(nc)*. For satellite observations, this depends on averaging clear and cloudy pixels in a grid square and taking the difference. In a model, the quantity is usually calculated by repeating the radiation calculation for a gridsquare with clouds removed but with no other change. For solar radiation, the effect of the clouds is to cool the system by reflecting radiation back to space so shortwave cloud forcing is typically negative.

The same considerations apply for the flux of thermal radiation across the top of the atmosphere by the outgoing longwave radiation (OLR) term R_{lw}. The longwave cloud forcing $CF_{lw} = R_{lw}^{(nc)} - R_{lw}$, generally acts to warm the system by shielding the surface since the OLR from the tops of cold clouds is less than that from the warm surface. The zonal average of the planetary albedo and of the OLR are shown in Figure 6 for the model and from the Earth Radiation Budget Experiment (ERBE) satellite data. These are very basic radiative terms in the budget of the model, and they deserve and receive considerable attention from climate modellers. Figure 6 also shows the shortwave and longwave cloud radiative forcing for the model compared with results from the International Satellite Cloud Climatology Project (ISCCP). Measuring cloud forcing is technically more difficult than measuring the TOA radiative fluxes because of the problems associated with finding (truly) cloud free regions adjacent to cloudy regions, sampling, etc.

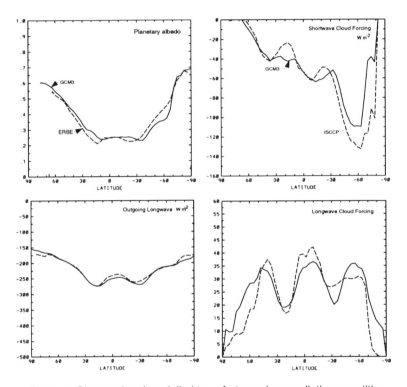

Figure 6. Observed and modelled top-of-atmosphere radiation quantities.

Despite the difficulty of measuring cloud, it is a primary meteorological variable and the fractional distribution of cloudiness from the model is displayed in Figure 7 as a zonal cross-section. The total amount of cloud from the model and from two observation-based data sets is shown in Figure 8. The cloud distribution can be readily seen to be connected to the radiative quantities in the previous figures. The geographical distribution of net cloudiness is also available but is not shown. There is reasonably good coherence between model and observations but the diagrams also indicate that global verification data may also have a considerable range of uncertainty.

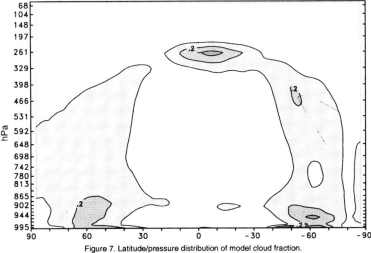

Figure 7. Latitude/pressure distribution of model cloud fraction.

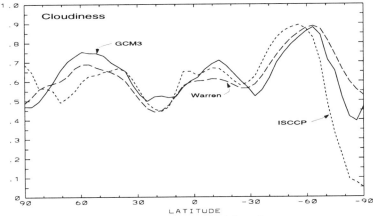

Figure 8. Total cloud amount from model and observations.

6.2 SURFACE ENERGY FLUXES

Figure 5 illustrates the fluxes of energy across the surface and Figure 9 displays zonally averaged values for December-February. No observed values are plotted. There are observationally-based estimates over the oceans, usually obtained by applying empirical flux formulae to mean temperatures and other surface based climate information, but values for land are generally unavailable. Figure 9 illustrates some general features of the surface energy balance. The net flux of energy is into the surface in the summer (southern) hemisphere and out of the surface in the winter (northern) hemisphere.

74

Solar radiation outweighs longwave radiation from the surface over most of the globe
to provide a net radiative energy input which
is maximum in the summer hemisphere.
Latent heat flux (which includes evaporation
and transpiration) is the main surface cooling
mechanism. The sensible heat flux also generally acts to cool the surface (but may warm
the surface in the polar regions) but is considerably smaller in magnitude than latent
heat flux. Taken together, surface energy
fluxes are particularly important to surface
climate. They are the energetic link between
the atmosphere and the underlying land,
ocean, and sea-ice so are a key element in a
successful coupled model.

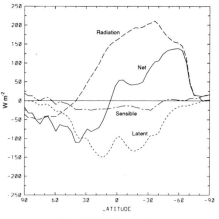

Figure 9. Modelled surface energy fluxes.

6.3 SURFACE CLIMATE

The surface climate is characterized by the distributions of temperature, precipitation,
and mean sea-level pressure. Observed and modelled values are shown in Figure 10.

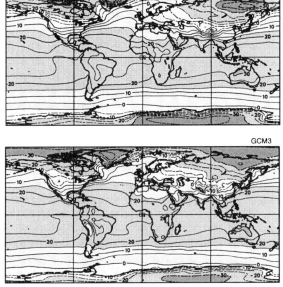

Figure 10a. Observed and modelled surface air temperature (° C).

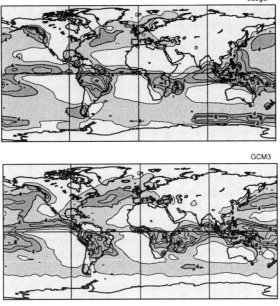

Figure 10b. Observed and modelled precipitation (mm/day).

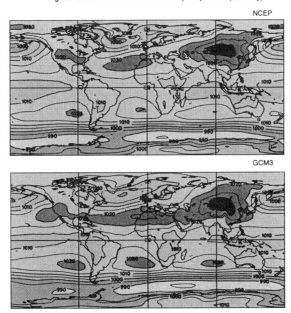

Figure 10c. Observed and modelled mean sea-level pressure (hPa).

Surface temperature and precipitation rate are of immense practical importance and reflect the energy and hydrological balances. Sea-level pressure gives information on surface wind. Simulated values from GCM3 generally agree well with those observed.

76

6.4 THE ZONALLY AVERAGED STRUCTURE OF THE ATMOSHERE

The vertical structure of atmospheric climate for DJF under time and zonal averaging is illustrated in Figure 11. Model results for temperature, moisture, zonal wind, and

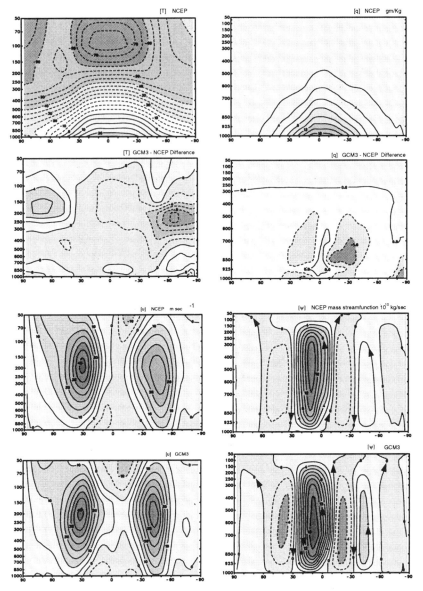

Figure 11. Modelled and observed climate distributions of temperature, specific humidity, zonal wind, and meridional streamfunction.

streamfunction are compared with results from the NCEP reanalysis. The meridional mass streamfunction, $[\overline{\psi}] = 2\pi a\cos\varphi \cdot \int_{0}^{p} [\overline{v}]dp/g$, incorporates information on both $[\overline{v}]$ and $[\overline{\omega}]$. The temperature and specific humidity differences from the NCEP values are shown since otherwise differences would be difficult to see. Given the strong horizontal and vertical gradients in the atmosphere, the differences represent a rather small displacement of the modelled and observed temperature and moisture patterns. The cold summer pole is a common or "systematic deficiency" in many models (Chapter 19) for which a ready explanation is not available. The correspondence of the other zonal fields with the NCEP values is quite reasonable.

The climate is not characterized only by means of quantities but also by variances, covariances, and other statistics. The variance of the winds is the eddy kinetic energy $k_E = \frac{1}{2}[\overline{V}^* \cdot \overline{V}^*] + \frac{1}{2}[\overline{V' \cdot V'}]$ shown in Figure 12. If anything, the GCM is more active and contains more eddy kinetic energy than the observed system. This represents a change from earlier versions of this, and other similar models, where the eddy kinetic energy was weaker than that observed. The speculation at the time was that this was a consequence of the lower horizontal resolution of earlier models but this is not the case for GCM3; improved parameterizations of physical processes are the more likely reason for the increase in model k_E.

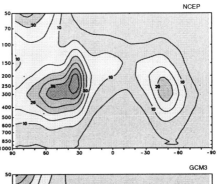

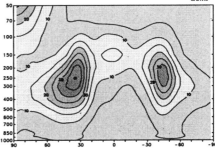

Figure 12. The kinetic energy of the eddies (m s^{-2}).

Covariances of wind components with other quantities represent transports by the eddies. Figure 13 gives the standing and transient eddy fluxes of temperature, $[\overline{T}^*\overline{v}^*]$, $[\overline{T'v'}]$, that arise in the heat budget equation under time and zonal averaging as discussed in Sections 5.1 and 5.2. Model results are compared to values obtained from the NCEP reanalysis. Standing eddy transports are most important in the northern hemisphere where land/sea contrasts and topography force standing structures. Transient eddy transports are important in both hemispheres.

6.5 ENERGY TRANSPORT IN THE ATMOSPHERE

The time averaged, zonally averaged, and vertically integrated energy equation

$$\frac{1}{a\cos\varphi}\frac{\partial}{\partial\varphi}H_\varphi\cos\varphi = [\Delta\overline{R}_N + L\overline{E} + \overline{h}] \tag{30}$$

is obtained by applying the averaging and integration operations of Section 5.1 to the energy equation (17) to obtain an equation of the form of (28). The right-hand-side

78

source/sink of energy comprises the net radiative difference between the top and bottom of the atmosphere and the latent and sensible heat fluxes into the atmosphere.

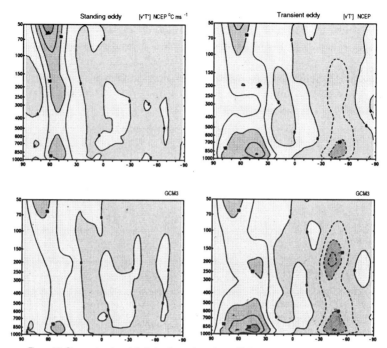

Figure 13. Observed and modelled standing and transient eddy heat transport covariance terms.

In (31) the net northward energy flux is

$$H_\varphi = H_M + H_S + H_T = \int (C_p[\bar{T}] + L[\bar{q}] + [\bar{\phi}])[\bar{v}]dp/g$$

$$+ \int (C_p[\bar{T}^*\bar{v}^*] + L[\bar{q}^*\bar{v}^*] + [\bar{\phi}^*\bar{v}^*])dp/g$$

$$+ \int (C_p[\overline{T'v'}] + L[\overline{q'v'}] + [\overline{\phi'v'}])dp/g$$

which is written in terms of its mean, standing eddy, and transient eddy components. We plot the related *transport* $\mathcal{T} = 2\pi a \cos\varphi H$ which includes transports of internal, potential, and latent energy as well as a pressure work term. Although the result is a function only of latitude (i.e., it is a level 1 diagnostic) the transport involves the temperature, moisture, and momentum equations and hence most of the covariances in Table 1. A large part of the energy transport is accomplished by the eddies as measured by these covariances.

The poleward heat transport in the atmosphere is a basic climate quantity and indicator of atmospheric behaviour. Figure 14 shows the components of the energy transport from the model, and Figure 15 compares each term with the NCEP results to indicate that the model is reproducing not only the net transport but also the appropriate mean and eddy transport components and mechanisms.

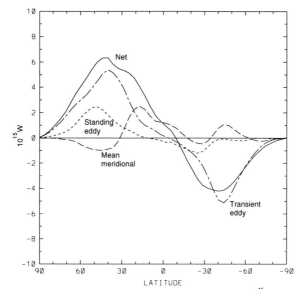

Figure 14. Energy transport in the model atmosphere (10^{15}W).

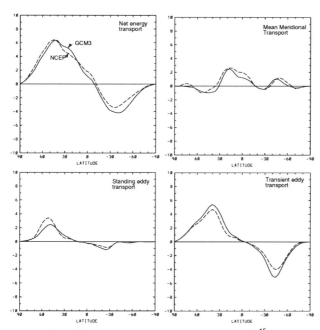

Figure 15. Observed and modelled energy transport components (10^{15}W).

6.6 THE ATMOSPHERIC ENERGY CYCLE

As discussed in section 2.4, the atmospheric energy cycle measures the "basic rate of working" of the system by estimating the rate of generation of available potential energy, its conversion to kinetic energy, and subsequent dissipation. Decomposing the equations into zonal and eddy components under time and zonal averaging gives

$$\frac{\partial A_Z}{\partial t} = G_Z - C_Z - C_A \qquad \frac{\partial K_Z}{\partial t} = C_Z - C_K - D_Z \tag{31}$$

$$\frac{\partial A_E}{\partial t} = G_E - C_E + C_A \qquad \frac{\partial K_E}{\partial t} = C_E + C_K - D_E \tag{32}$$

where for $X = [\bar{X}] + \bar{X}^* + X' = [\bar{X}] + X_E$ and $[X] = \langle X \rangle + [X]^+$

$$A = A_Z + A_E = \frac{1}{2} C_p \int \gamma [\bar{T}]^{+2} dm + \frac{1}{2} C_p \int \gamma [\overline{T_E}]^2 dm \tag{33}$$

$$K = K_Z + K_E = \frac{1}{2} \int [\bar{V}]^2 dm + \frac{1}{2} \int [\overline{V_E}]^2 dm \tag{34}$$

The generation and dissipation terms are

$$G = G_Z + G_E = \int \gamma [\bar{Q}]^+ [\bar{T}]^+ dm + \int \gamma [\overline{Q_E T_E}] dm \tag{35}$$

$$D = D_Z + D_E = -\int [\bar{V}] \cdot [\bar{F}] dm - \int [\overline{V_E \cdot F_E}] dm \tag{36}$$

and the conversion terms are

$$C = C_Z + C_E = -\int [\bar{\omega}]^+ [\bar{\alpha}]^+ dm - \int [\overline{\omega_E \alpha_E}] dm \tag{37}$$

$$C_K = -\int a\cos\varphi \left\{ \left([\overline{u_E v_E}] \frac{1}{a} \frac{\partial}{\partial \varphi} + [\overline{u_E \omega_E}] \frac{\partial}{\partial p} \right) \left(\frac{[\bar{u}]}{a\cos\varphi} \right) \right.$$

$$\left. + \left([\overline{v_E v_E}] \frac{1}{a} \frac{\partial}{\partial \varphi} + [\overline{v_E \omega_E}] \frac{\partial}{\partial p} - [V_E \cdot V_E] \frac{\tan\varphi}{a} \right) \left(\frac{[\bar{v}]}{a\cos\varphi} \right) \right\} dm \tag{38}$$

$$C_A = -\int C_p \left(\frac{\Theta}{T} \right) \left([\overline{v_E T_E}] \frac{1}{a} \frac{\partial}{\partial \varphi} + [\overline{T_E \omega_E}] \frac{\partial}{\partial p} \right) \left(\left(\frac{T}{\Theta} \right) \gamma [\bar{T}]^+ \right) dm . \tag{39}$$

The atmospheric energy cycle involves many of the means, variances and covariances of Table 1. Note that the generation and dissipation terms contain covariances with the parameterized heating and dissipation terms on the right-hand sides of the equations. These terms are not observed and, although estimates may be available as derived quantities as part of a reanalysis, these values depend heavily on model parameterizations. In diagnostic calculation they are typically obtained as budget residuals as done here. Figure 16 interprets the terms in the energy cycle and gives values from the model and from NCEP data. The model's energy cycle is much like that of the real atmosphere although it is slightly more "active" in that the basic rate of working of the modelled atmosphere exceeds that inferred from the observations. Since the terms in the energy cycle are

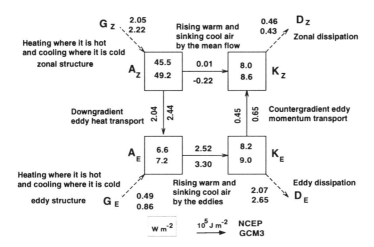

Figure 16. Observed and modelled terms in the atmospheric energy cycle
(upper values from NCEP reanalysis and lower values from GCM3).

obtained as integrals over the mass of the atmosphere, a considerable amount of information is condensed into a few numbers.

7. Concluding Comments

The analysis and verification of model climate, as described here, is based on the budget
equations for the atmosphere under various levels of averaging and integration. The flow
is decomposed into mean and eddy components under averaging, and the resulting array
of means, variances, and covariances that arise in the governing equations constitute the
basic climate statistics. The geographical distribution of these statistics and the associated budgets and balances represent two streams of model verification. A third stream of
verification, of particular climate processes in models, is not followed here. Early results
from a recent version of the CCCma atmospheric general circulation model are compared with observation-based data from satellite, from surface networks, and from the
NCEP reanalysis.

The analysis and verification of climate model behaviour is a complex task that must
always be incomplete since there are very many variables, processes, and timescales that
could be considered. Although there is no readily obtained "measure of goodness" for
model climate, there is a continuing evolution of the level of effort and sophistication
that is applied to the analysis of model results and to model verification. This, together
with improved climatic data from reanalyses and other sources, will certainly advance
the task.

Acknowledgements

Thanks to John Fyfe and Steve Lambert for their helpful comments.

References

1. Boer, G.J. (1984) Diagnostic equations in isobaric coordinates, *Mon. Wea. Rev.* **110**, 1801-1820.

2. Boer, G.J., McFarlane, N.A., Laprise, R., Henderson, J.G., and Blanchet, J.-P. (1984) The Canadian Climate Centre spectral atmospheric general circulation model, *Atmos.-Ocean* **22**, 397-429.

3. Boer, G.J., McFarlane, N.A., and Lazare, M. (1992) Greenhouse gas-induced climate change simulated with the CCC second-generation general circulation model, *J. Clim.* **5**, 1045-1077.

4. IPCC (1990) Climate Change, the IPCC Scientific Assessment, *Cambridge University Press*, Cambridge.

5. IPCC (1995) Climate Change 1995, the Science of Climate Change, *Cambridge University Press*, Cambridge.

6. Kalnay, E., et al. (1996) The NCEP/NCAR 40-year reanalysis project, *Bull. Am. Meteor. Soc.* **77**, 437-471.

7. Lorenz, E.N. (1967) *The nature and theory of the general circulation of the atmosphere*, WMO Publication 218, WMO, Geneva.

8. McFarlane, N.A., Boer, G.J., Blanchet, J.-P., and Lazare, M. (1992) The Canadian Climate Centre second generation general circulation model and its equilibrium climate, *J. Climate* **5**, 1013-1044.

9. Newell, R.E., Kidson, J.W., Vincent, D.G., and Boer, G.J. (1972) *The general circulation of the tropical atmosphere and interactions with extratropical latitudes, Vol.* 1. MIT Press, Cambridge, Mass.

10. Newell, R.E., Kidson, J.W., Vincent, D.G., and Boer, G.J. (1974) *The general circulation of the tropical atmosphere and interactions with extratropical latitudes, Vol.* 2. MIT Press, Cambridge, Mass.

11. Peixoto, J.P and Oort, A. (1992) *Physics of Climate*, American Institute of Physics, New York.

12. Reader, M.C. and Boer, G.J. (1998) The modification of greenhouse gas warming by the direct effect of sulphate aerosols, *Climate Dynamics* **14**, 593-608.

13. WMO (1998) Proceedings of the first WCRP international conference on reanalysis. WCRP report WCRP-104, WMO/TD-No. 876.

STATISTICAL TREATMENT OF MODEL OUTPUT

M. DEQUE
Météo-France CNRM
42 Avenue Coriolis
F-31057 Toulouse
France

1. Introduction

An atmosphere General Circulation Model (GCM) can be seen as a generator of meteorological data from a statistical point of view. It is particularly well suited for statistical treatment, since contrary to observed data it produces homogeneous and continuous series. Spatial sampling is generally regular, and the length of the series is limited only by the compromise between the computer availability and the degree of complexity of the model (the horizontal resolution being the major limiting factor).

The number of statistical tools that are developed to analyze the output of a GCM is limited only by the imagination of the modelers. However, just applying a tool to a model run is generally of little use, except for testing the robustness of the tool against the size of the series. Two types of study can be done with a given tool. The most important consists in applying the tool to a model simulation and to a series of observed data. This approach is called validation. One can see how far the similarity between the true atmosphere (or ocean) and its "numerical equivalent" is. Often a series of meteorological analyses (or reanalyses) is used, since this is suitable for comparison with the model output. The second approach consists in applying the tool to two model simulations performed under different conditions, e.g. with current and doubled CO_2 rates. We learn from this comparison the impact produced by the change on a particular aspect of the climate. A prerequisite is that the aspect we investigate is correctly reproduced in the control simulation, so that the first approach (i.e. validation) must be conducted as a first step. In this Chapter, we develop a variety of techniques for analyzing GCM output, while Chapter 3 constructs a framework for evaluating the performance of a GCM against observations.

P. Mote and A. O'Neill (eds.), Numerical Modeling of the Global Atmosphere in the Climate System, 83–104.
© 2000 *Kluwer Academic Publishers. Printed in the Netherlands.*

84

2. The Mean

2.1. MODEL OUTPUTS

The mean state of a model is often called its climatology. In nearly all the modeling studies, it is measured by its first-order moment. The median, which is less sensitive to outliers, is not used because of its lower statistical accuracy and the computer cost of its evaluation. Qualities often averaged are 500 hPa geopotential height, mean sea-level pressure, precipitation, and 850 hPa temperature. Surface or screen level temperature is also used, but it needs an elevation correction when it is compared with observed data or model data using another orography (e.g. with a higher truncation). Traditionally, a vertical gradient of 6.5 K/km is taken.

The other fields often considered are the top of atmosphere radiation fluxes (shortwave and longwave) for which satellite verification data are available and which measure how balanced the energetics are. At the surface, the evaporation and the run-off fields permit evaluation of the hydrological balance. The longwave and shortwave radiation, the latent and sensible heat fluxes, permit evaluation of the thermal budget. However, it is difficult to have a global coverage of reliable observed climatology for these fields. The cloudiness is sometimes measured by the percentage of cloud cover, but the arithmetic average with time and on the vertical is not justified by physical considerations (non-linear impact), and so-called cloud forcing (CRF) is preferred. It consists simply of subtracting from any radiation flux (SW or LW, surface or top) the corresponding radiation flux calculated without any cloud.

$$CRF = F - F_{clear} \tag{1}$$

When the sky is clear, this cloud forcing is zero. This forcing is negative in the shortwave (the clouds reduce the solar radiation) and positive in the longwave (the clouds have a greenhouse effect). Since it is an extensive variable (an energy flux), the arithmetic average is justified.

The other mean diagnostics traditionally calculated from a GCM output are the zonal averages as a function of latitude and pressure. The fields of temperature and zonal wind are mostly used. The mass transport is often used as a vertical-zonal stream function. It can be used to display the location and intensity of the Hadley and Ferrell cells. It is defined as:

$$\Psi = \frac{2\pi a \cos \varphi}{g} \int < V > dp \tag{2}$$

where φ is the latitude, a the earth radius, g the gravity, and $< V >$ the zonal mean of meridional velocity. The integral is computed as a function of pressure.

There are many other diagnostics (e.g. zonal averages of fluxes at the surface calculated separately over land and over ocean), but the above-mentioned ones correspond to a standard model verification. One of the most useful additional fields is the soil moisture at root level. In the case of a CO_2 doubling, it is a key parameter to determine possible regions of drought.

2.2. TESTS

At a given gridpoint, for a given month (or a given season like DJF or JJA), a model gives a numerical value. The problem is to decide whether this value can be accepted as compatible with the present observed climate (in the case of a control simulation), or whether it can be accepted as different from a control experiment (in the case of a perturbation simulation). This problem is addressed by statistical tests. Indeed, there is little chance of recovering the same numerical value if the simulation is repeated even with a small change, like using another initial condition or another compiler for the model. The reference observed value is also uncertain, owing to the choice of the averaging period, or owing to inaccuracies in the measurements.

The standard test is the Student test (or t-test). It is more than one century old, and the traditional reference in meteorology is [8]. Let $\overline{X}$ and $\overline{Y}$ be the averages of the two samples and S_p^2 be the pooled estimate of variance:

$$S_p^2 = \frac{1}{n + m - 2} \left[\sum_{i=1}^{n} (X_i - \overline{X})^2 + \sum_{j=1}^{m} (Y_i - \overline{Y})^2 \right] \qquad (3)$$

Then the ratio (called the t-value):

$$\frac{\overline{X} - \overline{Y}}{\sqrt{\left(\frac{1}{n} + \frac{1}{m}\right) S_p^2}} \qquad (4)$$

is a random variable following the Student distribution with $n + m - 2$ degrees of freedom. There is no analytical expression of the cumulated distribution function in general, but tables or numerical routines which relate a t-value to a probability for a given number of degrees of freedom are widely available. When n and m are greater than 10, the critical value at the 95% confidence level (two-sided) is close to 2.

The test assumes that you have averaged a certain number of independent samples of two Gaussian variables (different years can be considered as independent samples), and that the variance of the two samples is identical.

This test is robust and easy to implement. The assumption of equality of variances is one drawback. If two climates are very different, the variances are also different, and there is a risk that the test accepts equalities of the means. To avoid this, a Fisher test (F-test) can be applied first. This test is based on the calculation of the variances of the two samples (assumed to be Gaussian). The ratio of the unbiased estimates follows a Fisher distribution with $n - 1$ and $m - 1$ degrees of freedom. As for the Student law, tables and numerical routines are available. However, the F-test is not very robust with short (i.e. less than 30 year) samples, and maps of F-values are often very noisy.

Another limitation is that the samples must be Gaussian. We have generally no way to test this hypothesis, but it is reasonable when we work with monthly or seasonal means. There exist non-parametric tests, e.g. tests based on ranking like the Mann-Whitney test [11], but they are less robust with short samples.

The major drawback with the t-test comes from the fact that it is not applied to a single variable, but to every gridpoint. If a threshold of 95% is chosen, a map with 8192 gridpoints (T42 global grid) will show about 400 points as significant, just by chance. Moreover, since neighboring gridpoints are not independent, these 400 points will be gathered in a few regions. Such a geographical organization may imply that there is a significant impact in those regions. On the other hand, if the model response is a dipole with positive and negative values (this is often the case with 500 hPa height or MSL pressure), the area between the two poles will appear as not significant, since the mean normalized difference is less than the threshold.

For these reasons, the t-test must be used as a tool for verification of an already identified pattern, rather than for an investigation of possible patterns.

There are more powerful tests that allow testing of two fields at once, instead of gridpoint by gridpoint. Unfortunately, knowing that the two vectorial means are different does not reveal in which region the differences appear.

The Hotelling test [11] is the generalization of the t-test to two vector samples of size k. The generalization of (3) is:

$$V_p = \frac{1}{n + m - 2} \left[(n - 1) V_{xx} + (m - 1) V_{yy} \right] \tag{5}$$

where V_{xx} and V_{yy} are the empirical covariance matrices of the two samples. The true covariances are assumed to be identical. The generalization of (4) is:

$$\frac{nm(n + m - k - 1)}{(n + m)(n + m - 2)} \left(\overline{X} - \overline{Y} \right)^T V_p^{-1} \left(\overline{X} - \overline{Y} \right) \tag{6}$$

A similar approach consists in testing each empirical orthogonal function (EOF) in turn, instead of gridpoint by gridpoint, and to reconstruct the pattern with only the significantly affected EOF. This kind of approach can be found in [32].

Another way to test the difference between two fields is to count the fraction of gridpoints above a given threshold. If the number of gridpoints above the 95% significance level is much larger than 5%, then the fields are accepted as significantly different, although we cannot state where the differences are significant.

3. The variability

Atmospheric behavior is not steady, and its fluctuations about the mean state are of great interest, in particular in the activities linked to weather prediction. When a GCM is designed, the modelers try to adjust their empirical coefficients in order to get an acceptable mean climate, but the quality of the transient features is a result of the realism of the algorithms (feedbacks correctly taken into account) rather than a result of the efficiency of the tuning.

Whereas a single parameter (the arithmetic average in time) is used to study the mean state, there exist many approaches to measuring the variability. We will present some of the approaches most commonly used in the meteorological literature. We have mentioned in the last section the Fisher test, but we will not present in the following section methods to test the variability, since they are complex and not very robust in meteorological application. The method we recommend is to split the sample and produce several independent estimates of the parameter. If these estimates agree, then there can be a high degree of confidence in the estimate based on the whole sample. There exists a more sophisticated method, called cross-validation, e.g. in [10], which consists in extracting a moving subsample (the window), calculating the optimal function with the data outside the subsample (the training data), and calculating the quality criterion with the data inside the subsample (the test data). This method is widely used in predictive statistics.

3.1. SECOND-ORDER MOMENTS

Calculating a variance map (or rather a standard deviation map, since the units are more convenient) is the natural step after calculating an average map. There are several ways to proceed. One can calculate for each month or season the variance of the daily data, one can calculate the variance of the monthly or seasonal mean data, or one can calculate the bulk variance. Fortunately, the statistical equivalent of Huygens' theorem shows that the bulk variance is the sum of the mean intra-month variance:

$$\frac{1}{n}\sum_{y=1}^{n}\left(\frac{1}{30}\sum_{d=1}^{30}X_{yd}^2 - \left(\frac{1}{30}\sum_{d=1}^{30}X_{yd}\right)^2\right) \tag{7}$$

and of the inter-month variance (this is also true for the season):

$$\frac{1}{n}\sum_{y=1}^{n}\left(\frac{1}{30}\sum_{d=1}^{30}X_{yd}\right)^2 - \left(\frac{1}{n}\sum_{y=1}^{n}\left(\frac{1}{30}\sum_{d=1}^{30}X_{yd}\right)\right)^2 \tag{8}$$

Generally, the annual cycle is not included in the variance. This can be done either by restricting the period to a calendar month or season, or by subtracting an estimated mean annual cycle to the data prior to computing the variance. The separation between intra-month (i.e. high frequency) and inter-month (i.e. low frequency) is generalized by further time filtering.

The second-order moments are not only the variances, but also the covariances. A particular category of covariances is the meridional transport, which is the covariance between the meridional velocity (V) and another conservative field (such as temperature, specific moisture, geopotential height, zonal velocity or kinetic energy). The vertical transport is a similar quantity, but calculated with the vertical velocity in pressure coordinate (ω). The reason for this is that when averaging a conservation equation with time, the mean sources and sinks due to diabatic processes must include a term which can be written as follows, using the continuity equation:

$$S = \frac{1}{a \cos \varphi} \frac{\partial U X}{\partial \lambda} + \frac{1}{a} \frac{\partial V X}{\partial \varphi} + \frac{\partial \omega X}{\partial p} \qquad (9)$$

where X is the conservative variable and λ the longitude. When averaged over a longitude circle, the first term of (9) is canceled. Similarly, when averaging in an atmosphere column, the third term is canceled. The time-vertical-zonal means of diabatic sources, which are difficult to evaluate directly in model data and almost impossible to evaluate with observed data, are simply the meridional divergence of the meridional transport. Oort and Rasmusson [24] propose separating this transport into transport by the mean zonal circulation, the stationary eddies, and the transient circulation. This last contribution can in turn be separated into transient cells and transient eddies. Noting the time average with overbar, the time departure with prime, the zonal average with square brackets and the zonal deviations with star, the decomposition may be written as:

$$\left[\overline{V X} \right] = \left[\overline{V} \right] \left[\overline{X} \right] + \left[\overline{V^* X^*} \right] \left[\overline{V' X'} \right] \qquad (10)$$

3.2. SPACE-TIME SPECTRA

After a first look at the bulk variance, it is interesting to look at the time variability in different time scales. The basic decomposition of a time series X_t in Fourier coefficients is:

$$X_t = \sum_{k=0}^{T-1} \alpha_k \exp \frac{itk2\pi}{T} \quad \text{with} \quad \alpha_k = \frac{1}{T} \sum_{t=1}^{T} X_t \exp \frac{-itk2\pi}{T} \qquad (11)$$

The coefficient α_k is the contribution of the sinusoidal wave with period T/k to the temporal variations of X. Then, $|\alpha_k|^2$ is the variance of this sinusoidal wave. The distribution of the variance as a function of k is called the variability spectrum. In order to get a stable estimate of the variances, it is necessary to split the time series, calculate several independent values for $|\alpha_k|^2$, and average them. This explains why, with a 10-year sample, one can theoretically calculate the variance of the 5-year period oscillations, but in practice two years is the longest period one can estimate.

If one needs to estimate long period variances with a short sample, the solution is to make the hypothesis that the series is obtained by a Markovian process. Then from the low-order time autocorrelations, an analytic spectrum can be calculated. This method is known as the Maximum Entropy Method [28].

Generally, the variance of a single frequency is poorly estimated, and one prefers to average the variances in a band, since the variances are additive (the sum for $k = 0, T - 1$ of the variances is simply the variance of X for $t = 1, T$). But a similar result can be obtained with a time filtering of X before calculating the variance:

$$\tilde{X}_t = \sum_{k=-l}^{l} \beta_k X_{t+k} \tag{12}$$

Indeed, the above convolution is equivalent, in Fourier space, to a multiplication. Classical time filters have been proposed in meteorology by Blackmon [3]. The spectrum is divided into high-pass (less than 2 days), medium-pass (2 to 6 days) and low-pass (greater than 6 days). More recently, Doblas-Reyes and Déqué [12] have proposed a flexible method to choose the coefficients β for any window.

The above methods address the variability of a scalar series. One can study the variability of the gridpoints, irrespective of the connections between the neighbors. This is not at all suited for propagative phenomena. The Fourier decomposition can be multidimensional, i.e. applied to time and one or two spatial coordinates. If the direction of propagation is known, then the space-time spectral analysis proposed by Hayashi [16] can be performed. The Fourier transform of a series $X(x, t)$, where x is the spatial coordinate (generally the longitude), is:

$$Q(k, \omega) = \frac{1}{T} \sum_{t=1}^{T} \exp(-i2\pi\omega t) \frac{1}{L} \sum_{x=1}^{L} X(x, t) \exp(-ikx) \tag{13}$$

The expectation of the squared modulus of $Q(k, \omega)$ is noted $E(k, \omega)$ and called the two-sided spectrum. Westward (eastward) propagation corresponds to negative (positive) value of the time frequency ω. By convention the zonal wavenumber k is always positive. When displayed as a function of k and ω, E often exhibits a maximum along a line of constant ratio ω/k, which corresponds to the phase velocity.

A further analysis can be performed by considering only the positive values of the frequency. The propagation is eastward when $E(k, \omega) < E(k, -\omega)$. One can then define the propagating spectrum:

$$PR(k, \omega) = |E(k, \omega) - E(k, -\omega)| \tag{14}$$

the stationary spectrum:

$$SR(k, \omega) = 2\min(E(k, \omega), E(k, -\omega)) \tag{15}$$

and the total spectrum:

$$T(k, \omega) = PR(k, \omega) + SR(k, \omega) \tag{16}$$

A simpler way to look at the propagation phenomena which occur along a single direction is the Hovmoeller diagram, which consists simply of plotting the isolines of $X(x, t)$ as a function of x and t. But this simple method does not allow the synthesis of long data series.

3.3. PRINCIPAL COMPONENTS

The literature on Principal Component Analysis (PCA) in meteorology started with Lorenz [19] who popularized the term Empirical Orthogonal Function (EOF). Many papers on applications, extensions, generalizations, and significance tests have been published in the meteorological literature, and only a limited selection is given here. The principle of the EOF decomposition is more than one century old, but was made feasible only with numerical computers. The basic idea is to compress the information by making a rotation in the phase space, and projecting the data on the axes which exhibit the highest temporal variance.

In practice, the PCA is no longer used for data compression, since it takes a lot of computer time and implies a loss of information. It has been used in numerical prediction to reduce the number of variables, e.g. [4]. Nowadays, the primary use of the PCA is to obtain patterns that people try to identify as intrinsic variability modes. The secondary use is to reduce the number of degrees of freedom of the original data before performing another statistical application (e.g. regression or factor analysis), so that the robustness is increased: it is usual to deal with 20 EOFs instead of 8000 gridpoints when the sample size (or the equivalent number of independent data) is less than 50.

The calculation of the first EOFs of a sample $X(x, t)$, where x is the spatial coordinate and t the temporal coordinate (more generally, x may identify a feature of a variable X and t an individual in the statistical sample), consists in evaluating the covariance matrix $V(x, y)$:

$$V(x, y) = \frac{1}{T} \sum_{t=1}^{T} \left(X(x, t) - \overline{X}(x) \right) \left(X(y, t) - \overline{X}(y) \right) \sqrt{a(x)a(y)} \tag{17}$$

where $a(x)$ is the spatial weight of the gridpoint x. This matrix may be huge (8000×8000 for a global field in standard GCM resolution). In fact, algebraic manipulations allow $T \times T$ arrays to be handled instead (see [10]). This matrix is symmetrical, so it has T orthogonal eigenvectors C_i, and T real positive eigenvalues λ_i. The eigenvalue λ_i corresponds to the variance of the projection c_i of X on the direction C_i. For this reason, the eigenvalues are ranked in decreasing order. After normalization, c_i yields an index γ_i:

$$\gamma_i(t) = \frac{1}{\sqrt{\lambda_i}} \sum C_i \left(X(x,t) - \overline{X}(x) \right) \sqrt{a(x)} \qquad (18)$$

corresponding to the pattern:

$$\Gamma_i(t) = \sqrt{\gamma_i} C_i(x) / \sqrt{a(x)} \qquad (19)$$

Then the expansion of X in EOFs is:

$$X(x,t) = \overline{X}(x) + \sum_{i=1}^{T} \gamma_i(t)\Gamma_i(x) \qquad (20)$$

Only the first terms ($i = 1, k$) of this expansion are retained for further analysis. The variance of the term discarded is $\sum_{i=k+1}^{T} \lambda_i$, so that the efficiency of the decomposition in k EOFs is often measured by the ratio $\sum_{i=1}^{k} \lambda_i / \sum_{i=1}^{T} \lambda_i$. One advantage of the EOF expansion is that the indices γ_i and γ_j are uncorrelated for $i \neq j$, so that the evolution of pattern Γ_i can be considered as independent of the evolution of Γ_j.

A variant of the method consists in using the correlation matrix instead of the covariance matrix. This is equivalent to normalizing each gridpoint value by the local standard deviation. This is useful when mixing different fields in a single PCA, or when the domain of analysis contains regions with high variability and regions with low variability, such as 700 hPa height over the tropics and the midlatitudes.

One can also apply a PCA to time-filtered data. Figure 1 shows the result of a PCA on 10 winters of 500 hPa height (from ECMWF analyses) with a low-pass filter (greater than 6 days). The time evolution of the first three Principal Components (PCs), also called factor scores (Figure 1a), shows a slow time evolution which is not completely explained by the time filter (about one change of sign per month). The spatial distribution of the first three EOFs, also called factor loadings (Figure 1b), shows large-scale patterns. In particular, the first EOF corresponds to the well-known North Atlantic Oscillation (NAO) pattern. The percentage of variance of the three PCs is 18%, 14% and 9%.

Figure 2 shows the EOFs with a band-pass filter (2-6 days). One can see that EOF1 and EOF2 are similar, but with an eastward phase shift. It is then difficult to admit that the two patterns are independent. This raises the problem of degeneracy

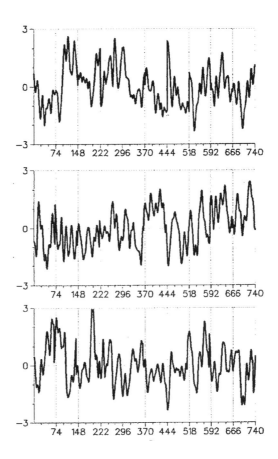

Figure 1a. First three Principal Components (dimensionless) of 500 hPa height in a 10-winter analysis of low-pass filtered data as a function of time (day).

in PCA, a problem that occurs when two (or more) consecutive PCs have very close variances. Indeed, the percentage of variance of the three PCs in this example is 12%, 11% and 6%. With degenerate EOFs, any linear combination (rotation) of the PCs and the respective EOFs is also a solution to the diagonalization. The way to solve the degeneracy problem is to use another criterion to separate the eigenvectors.

Sometimes a problem arises from the fact that the opposite boundaries of the domain are correlated (too small a domain): then the structure of the EOFs is imposed by the size of the domain (its aspect ratio), and the EOFs have little physical meaning. When the domain is much larger, another problem arises: different independent phenomena have similar variances and are artificially coupled. Horel [17]

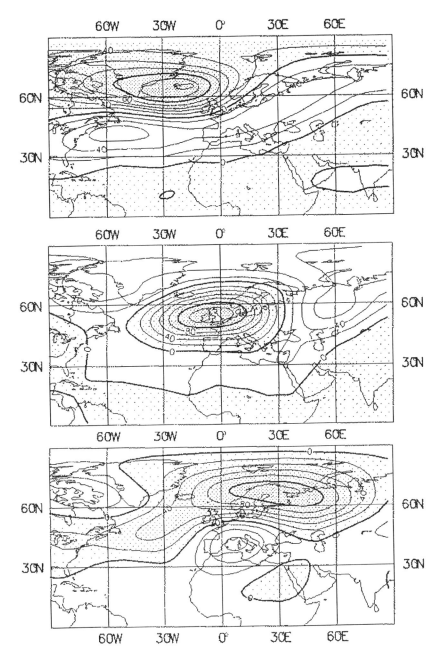

Figure 1b. First three Empirical Orthogonal Functions (m) of 500 hPa height in a 10-winter analysis of low-pass filtered data.

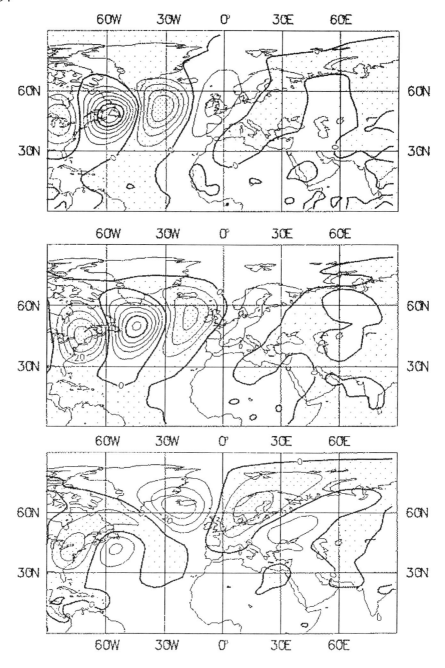

Figure 2. As Figure 1b for band-pass filtered data.

has studied this last phenomenon and proposes the so-called VARIMAX rotation. A third reason for the degeneracy is the presence of moving patterns. This is the case for Figure 2, and the solution is the complex PCA.

The rotated EOF analysis is a family of methods designed for solving the degeneracy problem, or more generally for emphasizing a statistical property other than the maximization of the variance. The orthogonality of the EOFs is lost, but owing to the orthogonality of the rotation the decorrelation of the PCs remains. (20) is replaced by a new expansion.

$$X(x, t) = \overline{X}(x) + \sum_{i=1}^{k} \hat{\gamma}_i(t)\hat{\Gamma}_i(x) + \ldots \tag{21}$$

with:

$$\hat{\gamma}_i = \sum_{j=1}^{k} R_{ij}\gamma_j \quad \hat{\Gamma}_i = \sum_{j=1}^{k} R_{ji}\Gamma_j, \quad \text{with} \sum_{h=1}^{k} R_{ih}R_{hj} = \delta_{i=j} \tag{22}$$

R is an orthonormal matrix (i.e. a rotation matrix in the phase space).

Any choice of the matrix R, driven by statistical considerations, leads to a new set of EOFs and PCs. Richman [27] discusses the rotations that lead to simplified structures for the EOFs. The most widespread method is the VARIMAX.

The principle of the VARIMAX is to maximize the spatial variance of the squared rotated EOFs:

$$w = \sum_{i=1}^{k} \left[\left(\sum_x \hat{\Gamma}_i^4(x)a(x) \right) - \left(\sum_x \hat{\Gamma}_i^2(x)a(x) \right)^2 \right] \tag{23}$$

Indeed, the PCA corresponds to maximizing the spatial average of the squared EOFs, which is identical to $\sum_{i=1}^{k} \lambda_i$. However, when you get a pattern with a lot of local positive and negative maxima, the average of the squared EOFs is large, but the variance is small. When the structure is simple, and consists of a localized maximum, the variance is larger. The practical difference with PCA is that the algorithm is not progressive. That is, if you rotate the first three EOFs, the first two rotated EOFs are different from the ones you get by rotating the first two EOFs only. As a consequence, one has to decide *a priori* the number of EOFs to rotate. The algorithm consists in rotating the EOFs by pairs, so that w increases, and stopping when this increase becomes negligible.

The complex EOF analysis (or CPCA) has been designed to generalize the expansion of (20). Each individual component of the EOF expansion is able to represent a seesaw phenomenon, such as the NAO, but the centers of action cannot move, so that one cannot represent a pressure low traveling across the Atlantic. But if we generalize (20) as:

$$X(x, t) = \overline{X}(x) + \sum_{j=1}^{T} \gamma_j(t) \Gamma_j(x) \cos\left(\omega_j(t) + \Omega_j(x)\right) \tag{24}$$

then we have the basis of the CPCA. The generalization of the j-th EOF is $\Gamma_j(x) \exp(i\Omega_j(x))$ and of the j-th PC is $\gamma_j(x) \exp(i\omega_j(x))$. Two equivalent methods yield the same result. Barnett [1] adds to the real X field a complex field calculated as the Hilbert Transform of X with respect to time. The Hilbert Transform consists in taking each wave of the Fourier decomposition in quadrature, e.g. $\cos \omega t$ yields $\sin \omega t$. Then the covariance of the complex field $X + i HT(X)$ is calculated and diagonalized. As a Hermitian matrix, it has orthogonal eigenvectors and real eigenvalues. This approach is easier to understand, but more complicated to carry out than the one proposed by Wallace and Dickinson [35]. These authors propose diagonalizing the cospectral matrix, which is a complex Hermitian matrix, integrated with respect to frequency in a frequency band. This latter method offers the advantage of naturally including a time filter.

The interpretation of the results of a CPCA is a little more difficult than for a PCA, since we have two fields and two time series. See [18] for a wider review. The field $\Gamma_j(x)$ is the spatial amplitude of the signal. It shows the region where the phenomenon is active. The field $\Omega_j(x)$ is the phase of the signal. Its gradient shows the direction and velocity (when the time frequency is known) of the propagation. The series $\gamma_j(t)$ indicates when the phenomenon is active. The series $\omega_j(t)$ must be considered carefully. If it cannot be fitted by a linear function of t (modulo 2π), then the complex PC does not correspond to a propagation phenomenon but to a statistical artifact. If such a fit is possible, the linear coefficient corresponds to the time pulsation of the phenomenon. Note that this pulsation may be different between two large periods: the frequency may increase or decrease during some events.

Figure 3 illustrates the CPCA in the case of the 500 hPa height presented above. Figure 3a shows, for the first three CPCs of the low-pass filtered data, the amplitude and the phase as a function of time during 10 winters. Figures 3b and 3c show, for the first two CEOFs of the band-pass filtered data, the amplitude, the propagation direction (horizontal gradient of the phase), and full CEOF. The last field is displayed as a vector with real part eastwards and imaginary part northwards.

The percentage of variance of the first three CPCs of the low-pass filtered data is 17%, 13% and 11%, and the periods are 60, 40 and 30 days. In the case of the band-pass filtered data, the percentage of variance of the first three CPCs is 23%, 10% and 6%, and the periods are close to 4 days.

The success of the EOF decomposition in meteorology comes from the fact that they can be linked to physical aspects, since they correspond to normal modes of simplified models [23]. However, their statistical properties make them useful

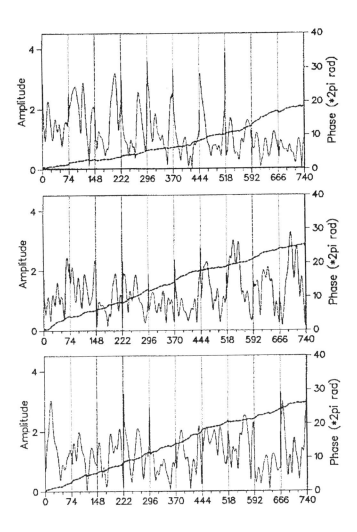

Figure 3a. First three Complex Principal Components of 500 hPa height in a 10-winter analysis of band-pass filtered data as a function of time (day). The solid line is the amplitude (dimensionless) and the dash line the phase (unit: cycle).

in classification problems [2], provided that the choice of the number of EOFs retained is reasonable. (See [25] for selection rules.) The extended EOF (EEOF) technique is simply the application of the EOF analysis to several fields instead of a single one [38]. From a theoretical point of view, there is no difference. From a practical point of view, dealing with n fields at the same time multiplies the size of the matrix to diagonalize by n^2. A first reduction of the number of degrees of freedom by an EOF filtering of the individual fields before performing the EEOF

98

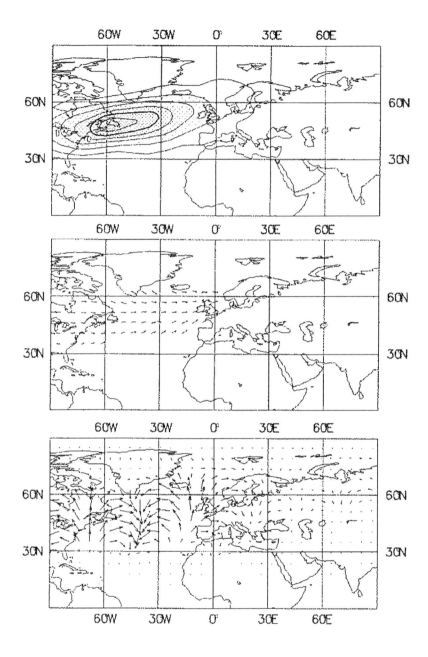

Figure 3b.   First Complex EOF of 500 hPa height in a 10-winter analysis of band-pass filtered data. The upper panel is the amplitude (m), the middle panel the phase gradient, and the bottom panel the complex eigenvector (m) with the real axis along the latitude circle and the imaginary axis along the longitude circle.

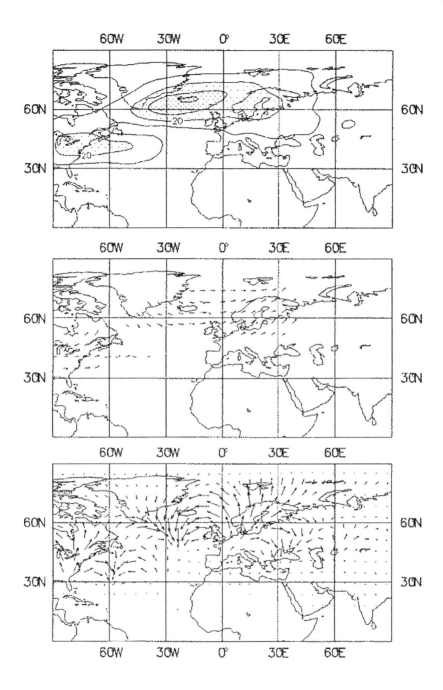

Figure 3c. As Figure 3b for the second complex EOF.

analysis solves the problem of size. One can use different fields, like Sea-Surface Temperature (SST) and wind stress, or the same field at regular lags. In the first case, it is essential to use the correlation matrix, not the covariance matrix. The multichannel singular spectrum analysis [26] consists of using lagged EOFs as an input for an EOF analysis.

3.4. CANONICAL ANALYSIS

In the previous section, we have seen that a field (or a multi-field in the case of EEOF) can be expanded onto a new base of functions. Here we consider two fields, X and Y, on a different spatial domain but during the same time interval, and look for two bases of function:

$$X(x,t) = \overline{X}(x) + \sum_{i=1}^{t} \gamma_i(t)\Gamma_i(x) \qquad (25)$$

$$Y(y,t) = \overline{Y}(y) + \sum_{i=1}^{t} \xi_i(t)\Xi_i(y) \qquad (26)$$

A review of the different methods is given in [5]. The most classical is the Canonical Correlation Analysis (CCA) for which we maximize the correlation between γ_i and ξ_i ([22], [10], and many others). This method consists of diagonalizing a matrix, the eigenvalues being the squared correlations. The correlations are ranked in decreasing order. In fact, only the first two or three are actually investigated. To each canonical correlation corresponds a pair of maps $\Gamma_i(x)$, $\Xi_i(y)$ and a pair of time series $\gamma_i(t)$, $\xi_i(t)$. One interesting property is that $\gamma_i(t)$, $\xi_j(t)$ are uncorrelated ($\gamma_i(t)$, $\gamma_j(t)$ and $\xi_i(t)$, $\xi_j(t)$ are uncorrelated by construction), so that the pairs can be considered as independent phenomena.

A problem that arises with this technique is that one can get canonical correlations of 99.99% associated with components which explain 0.01% of the variance and are just statistical artefacts. The reason for this is the much larger number of degrees of freedom than number of independent events. One method of avoiding this consists of pre-filtering the two fields by EOF expansions, with 10 to 20 EOFs in both cases. This has the additional advantage that the canonical components explain a non-negligible fraction of the variance of the respective fields. Another method is the Singular Value Decomposition (SVD).

In fact, the SVD decomposition is a mathematical technique which can be applied to any matrix. When applied to the covariance matrix between X and Y, this method yields components which maximize their covariance (instead of correlation for CCA). The covariance being the product of the correlation by the standard deviations, there is a compromise between the fraction of variance explained by the components and the correlation. This technique is more recent in meteorology than the CCA [37]. It has the disadvantage that $\gamma_i(t)$, $\xi_j(t)$ are not uncorrelated for $i \neq j$, but this is harmless when the study is restricted to the first pair of components. As for PCA, one can get simpler structures by rotation of the patterns [7].

3.5. MISCELLANEOUS

We examine here statistical techniques which are popular in meteorology, but cannot be directly linked to the family of the PCA or of the CCA. One simple, easy to implement, and rather robust method of investigating the teleconnection patterns is the method described by Wallace and Gutzler [36]. At each gridpoint of a field is associated the largest negative correlation among all the gridpoints of the domain (the largest positive correlation has no interest since it is found among the neighbors). The map of this parameter, called teleconnectivity, allows one to identify the pairs of gridpoints which exhibit the strongest values. For each pair, the correlation map of one of the two points with all the points of the domain allows better documentation of the pattern.

The time autocorrelation has been mentioned above in the section devoted to spectral analysis. One can use it to synthesize the time behavior of a series of indices by an autoregressive process [39]. The generalization of this technique leads to the Principal Oscillation Pattern (POP) analysis [33].

This method consists of diagonalizing the autocorrelation matrix at a given lag. Since the matrix is not symmetrical, the eigenvalues and eigenvectors are complex. Thus each mode is characterized by two maps and two time series. The time evolution of a mode is a chain of patterns, for example:

$$\mathcal{P}_1 \longrightarrow \mathcal{P}_2 \longrightarrow -\mathcal{P}_1 \longrightarrow -\mathcal{P}_2 \longrightarrow \mathcal{P}_1 \longrightarrow \ldots \tag{27}$$

There are two time constants: the real part of the eigenvalue describes the persistence of the pattern, the imaginary part describes the period of the above cycle. Unlike PCA or CCA, there is no way of ranking the patterns. This method can be used for statistical prediction [34]. It can be extended to Fourier space by the complex POP analysis [6].

The methods based on covariance or correlation assume that the link between two fields is linear. One can avoid this constraint, at the expense of robustness, by considering composite averages. This method consists simply in averaging a field only when a second variable follows a certain relation. This is nothing more than conditional expectation. An example of this is given by the composite maps during the ENSO periods [14].

The study of persistence may be carried out with non-linear methods, as in [13]. These authors consider a phenomenon as persistent when a variable remains above a given threshold for more than a given duration. When studying the blocking phenomenon, the use of a dynamical index is more efficient, since it can be applied to an instantaneous field, like a 10-day forecast [29]. This index, based

102

on the North-South gradient of 500 hPa height, has become very popular in GCM validation [30].

The identification of weather regimes, in observed as in simulated data, is an important issue for the description as well as the prediction of climate. Some methods, as in [21] look for local maxima of the density in the phase space. This is only possible with a severe reduction of the dimension by EOFs. Vautard [31] proposes identifying the regimes by minimization of their large-scale tendency. Mo and Ghil [20] identify them by a cluster analysis. In fact, there are so many criteria and methods in cluster analysis that the subject is not exhausted, the main limitation coming from the "short" size (less than 40 years) of observed samples.

4. Conclusion

We have examined a certain number of statistical techniques used in GCM post-processing. Most of the literature references are related to an application to observed data. The reason is that the scientific papers must interest the largest number of readers, and the properties of the real atmosphere are more instructive than the properties of a particular GCM. But all the techniques are applicable to GCM outputs.

Many techniques have been developed to provide an optimal statistical answer to a specific question. For example, the expansion of a field in components which maximize the predictability at a given range has led to a method introduced in [9], but unused since that time.

It is expected that the progress in the representation of the interannual variability by coupled GCMs will allow the use of long simulations as a substitute for observed series for statistical investigations (e.g. analog prediction). Indeed, we have to wait 50 years before getting a 100-year global observed sample. Moreover, this sample will include inhomogeneities in the measurements and trends owing to anthropogenic changes. It is thus essential that modelers check carefully all the aspects of the variability in the long simulations they produce, as in AMIP [15] or in the next AMIP2. Even though (and this is an advantage rather than a disadvantage) it is not feasible to "tune" a model in order to get the best variability, there is constant progress in the realism of the simulated variability. The validation of this aspect of a model is also a way to increase our confidence in the response of the climate to anthropogenic changes in the next century.

Acknowledgements

Thanks are due to F.J. Doblas-Reyes for his help in preparing the manuscript. This work was partly supported by the Commission of the European Union (Contract ENV4-CT97-0497).

References

1. Barnett, T.P. (1983) Interaction of the monsoon and Pacific trade wind system at interannual time scales, Part I: the equatorial zone, *Mon. Wea. Rev.* **111**, 756-773.
2. Barnston, G.A. and Livezey, R.E. (1987) Classification, seasonality and persistence of low-frequency atmospheric circulation patterns, *Mon. Wea. Rev.* **115**, 1083-1115.
3. Blackmon, M.J. (1976) A climatological spectral study of the 500 mb geopotential height of the northern hemisphere, *J. Atmos. Sci.* **33**, 1607-1623.
4. Bodin, S. (1974) The use of empirical orthogonal functions in quasi-geostrophic numerical prediction models, *Tellus* **26**, 582-592.
5. Bretherton, C.S., Smith, C. and Wallace, J.M. (1992) An intercomparison of methods for finding coupled patterns in climate data, *J. Climate* **5**, 541-560.
6. Bürger, G. (1993) Complex Principal Oscillation Pattern analysis, *J. Climate* **6**, 1972-1986.
7. Cheng, X. and Dunkerton, T.J. (1995) Orthogonal rotation of spatial patterns derived from Singular Value Decomposition analysis, *J. Climate* **8**, 2631-2643.
8. Chervin, R.M. and Schneider, S.H. (1976) On determining the statistical significance of climate experiments with General Circulation Models, *J. Atmos. Sci.* **33**, 405-412.
9. Déqué, M. (1988) 10 day predictability of the Northern Hemisphere winter 500 mb height by the ECMWF operational model, *Tellus* **40A**, 26-36.
10. Déqué, M. and Servain, J. (1989) Teleconnections between tropical Atlantic sea surface temperatures and mid-latitude 50 kPa heights during 1964-1986, *J. Climate* **2**, 929-944.
11. Der Megreditchian, G. (1992) *Le traitement statistique des données multidimensionnelles. Application à la météorologie.* CEPADUES Ed., Toulouse.
12. Doblas-Reyes, F.J. and Déqué, M. (1998) A flexible band-pass filtering procedure applied to midlatitude intraseasonal variability, *Mon. Wea. Rev. in press.*
13. Dole, R.M. and Gordon, N.D. (1983) Persistent anomalies of the extratropical Northern Hemisphere wintertime circulation: geographical distribution and regional persistence characteristics, *Mon. Wea. Rev.* **111**, 1567-1586.
14. Fraedrich, K. (1994) An ENSO impact on Europe? A review, *Tellus* **46A**, 541-552.
15. Gates, L.W. (1992) AMIP: the Atmospheric Model Intercomparison Project, *Bull. Amer. Meteor. Soc.* **73**, 1962-1970.
16. Hayashi, Y. (1982) Space-time spectral analysis and its application to atmospheric waves, *J. Meteor. Soc. Japan* **60**, 156-171.
17. Horel, J.D. (1981) A rotated Principal Component Analysis of the interannual variability of the Northern Hemisphere 500 mb height field, *Mon. Wea. Rev.* **109**, 2080-2092.
18. Horel, J.D. (1984) Complex Principal Component Analysis: theory and examples, *J. Clim. Appl. Meteor.* **23**, 1660-1673.
19. Lorenz, E.N. (1956) Empirical Orthogonal Functions and statistical weather prediction, *Scientific report 1*. Dept. of Meteorology, Massachusetts Institute of Technology, Cambridge, Massachusetts, USA.
20. Mo, K. and Ghil, M. (1987) Cluster analysis of multiple planetary flow regimes, *J. Atmos. Sci.* **44**, 877-901.
21. Molteni, F., Sutera, A. and Tronci, N. (1988) The EOFs of the geopotential eddies at 500 mb in winter and their probability density distributions, *J. Atmos. Sci.* **45**, 3063-3080.
22. Nicholls, N.(1987) The use of canonical correlation to study teleconnections, *Mon. Wea. Rev.* **115**, 393-399.
23. North, G.R. (1984) Empirical Orthogonal Functions and normal modes, *J. Atmos. Sci.* **41**, 879-887.
24. Oort, A.H. and Rasmusson, E.M. (1971) Atmospheric circulation statistics, *NOAA Prof. Paper* **5**.

104

25. Overland, J.E. and Preisendorfer, R.W. (1982) A significance test for Principal Components applied to a cyclone climatology, *Mon. Wea. Rev.* **110**, 1-4.

26. Plaut, G. and Vautard, R. (1994) Spells of low-frequency oscillations and weather regimes in the northern hemisphere, *J. Atmos. Sci.* **51**, 210-236.

27. Richman, M.B. (1986) Rotation of Principal Components (Review Article), *J. Climatol.* **6**, 293-335.

28. Thomson, D.J. (1982) Spectrum estimation and Harmonic Analysis, *Proc. IEEE* **70**, 1055-1096.

29. Tibaldi, S. and Molteni, F. (1990) On the operational predictability of blocking, *Tellus* **42A**, 343-365.

30. Tibaldi, S., d'Andrea, F., Tosi, E. and Roeckner, E. (1997) Climatology of Northern Hemisphere blocking in the ECHAM model, *Climate Dyn.* **13**, 649-666.

31. Vautard, R. (1990) Multiple weather regimes over the North Atlantic: analysis of precursors and successors, *Mon. Wea. Rev.* **118**, 2056-2081.

32. von Storch, H. and Roeckner, E. (1983) Verification of General Circulation Models applied to the Hamburg University GCM. Part I : test of individual climate states, *Mon. Wea. Rev.* **111**, 1965-1976.

33. von Storch, H. and Xu, J. (1990) Principal Oscillation Pattern analysis of the tropical 30- to 60-day oscillation. Part I: definition of an index and its prediction, *Climate Dyn.* **5**, 175-190.

34. von Storch, H. and Baumhefner, D.P. (1991) Principal Oscillation Pattern analysis of the tropical 30- to 60-day oscillation Part II: the prediction of equatorial velocity potential and its skill, *Climate Dyn.* **6**, 1-12.

35. Wallace, J.M. and Dickinson, R.E. (1972) Empirical orthogonal representation of time series in the frequency domain. Part I: theoretical considerations, *J. Appl. Meteor.* **11**, 887-892.

36. Wallace, J.M. and Gutzler, D.S. (1981) Teleconnections in the geopotential height field during the northern hemisphere winter, *Mon. Wea. Rev.* **109**, 784-812 .

37. Wallace, J.M., Smith, C. and Bretherton, C.S. (1992) Singular Value Decomposition of wintertime sea surface temperature and 500 mb height anomalies, *J. Climate* **5**, 561-576.

38. Weare, B.C. and Nasstrom, J.S. (1982) Examples of Extended Empirical Orthogonal Function Analyses, *Mon. Wea. Rev.* **110**, 481-485.

39. Zwiers, F. and von Storch, H. (1990) Regime-dependent autoregressive time series modeling of the Southern Oscillation, *J. Climate* **3**, 1347-1363.

USE OF SIMPLIFIED ATMOSPHERIC MODELS

J. THUBURN
Department of Meteorology
University of Reading
Earley Gate
Reading RG6 6BB
UK

1. Introduction

The NATO Advanced Study Institute was principally about the use of full global circulation models (GCMs) with comprehensive suites of parametrization schemes to represent the effects of unresolved processes. However, because there are complex interactions among the many different processes represented, it may be difficult to design and set up the ideal GCM experimental configurations needed to address a particular scientific question, while the results from GCM experiments can sometimes be difficult to interpret. In addition, GCMs are expensive to run. For these reasons, research is often carried out using simplified models.

There are many kinds of model that are simpler than a full GCM but that still retain some of the essential physics. These range in complexity from highly idealized one-dimensional models to models with almost the full complexity of a GCM. Examples include: models with modified equations of motion, such as quasigeostrophic models, steady-state models like linear or nonlinear stationary wave models, and chemical transport and trajectory models that predict chemical evolution, but need to be supplied with wind and temperature data; models with a reduced number of spatial dimensions, such as energy balance models, single column radiative-convective models, single layer models, or zonally averaged models; highly truncated models, for example, retaining only a subset of zonal wavenumbers or two or three vertical layers; models with reduced domain, such as stratosphere-mesosphere models; models with simplified physical parametrizations, such as dry adiabatic models; and models with simplified forcing or boundary conditions, such as aquaplanet models or GCMs run with sea surface temperature and solar forcing fixed at their January values.

P. Mote and A. O'Neill (eds.), Numerical Modeling of the Global Atmosphere in the Climate System, 105–118.
© 2000 *Kluwer Academic Publishers. Printed in the Netherlands.*

Simplified models have two main advantages over GCMs. First, they are usually cheaper to run, often one or two orders of magnitude cheaper for a given resolution and length of integration. This means that simplified models can be used for very long integrations, or for very high resolution integrations, or to perform a large number of integrations in order to map the sensitivity of the solution to one or more parameters.

Secondly, simplified models are, of course, simpler than full GCMs, both in the range of physical processes they represent and in their numerical design and operation. Thus it is usually easier to design and set up clean experiments for simplified models, and it is often more straightforward to interpret their results. In addition, simplified models can be a valuable test bed for new numerical schemes or new parametrization schemes before they are implemented in a full GCM.

However, simplified models, by their nature, simplify or omit certain physical processes, which may turn out to play a role in the phenomenon under investigation. Therefore, simplified models should not be considered as an alternative to GCMs, but as complementary to them. For example, simplified model results might help to interpret GCM results by suggesting new experiments or diagnostics, while GCM results might suggest a hypothesis that can be tested most cleanly in a simplified model. Some examples are given below. An ideal situation for a researcher is to have access to a *hierarchy of models* of varying complexity and completeness.

Finally, while GCMs are the most complete and sophisticated global models that we can run in the current state of knowledge and computer resources, they too contain many simplifications of the physics of the real atmosphere, particularly in their parametrization schemes. This limitation should be borne in mind when designing experiments for GCMs or interpreting their results.

In this Chapter we will look at just four examples of simplified models: shallow water models (Section 2); zonally averaged models (Section 3); models with highly simplified parametrized physics (Section 4); and stratosphere-mesosphere models (Section 5).

2. Shallow water models

The shallow water equations describe the evolution of a single layer of fluid of constant density but variable depth. On the surface of a rotating sphere the equations can be written in several ways (e.g., [21]). One of the most intuitive ways is in terms of the vector (horizontal) velocity and the depth:

$$\frac{D\mathbf{v}}{Dt} + f\hat{\mathbf{k}} \times \mathbf{v} = -g\nabla(h + h_*) + \mathbf{X}, \tag{1}$$

$$\frac{\partial h}{\partial t} + \nabla \cdot (h\mathbf{v}) = H. \tag{2}$$

Here, f is the Coriolis parameter, $\hat{\mathbf{k}}$ is the unit vertical vector, h_* is the height of the bottom boundary, and D/Dt means a material derivative. $\mathbf{X}$ represents momentum forcing and H represents a mass source. An alternative is to rewrite (1) in terms of the relative vorticity ζ and the divergence D:

$$\frac{\partial \zeta}{\partial t} + \nabla \cdot [\mathbf{v}(f + \zeta)] = \hat{\mathbf{k}} \cdot \nabla \times \mathbf{X} \tag{3}$$

$$\frac{\partial D}{\partial t} = \nabla \cdot \{\mathbf{X} - (f + \zeta)\hat{\mathbf{k}} \times \mathbf{v} - \nabla[g(h + h_*) + \mathbf{v}^2/2]\} \tag{4}$$

This vorticity-divergence form, in conjunction with (2), is often used as the basis for numerical shallow water models. Yet another alternative is to replace the vorticity equation (3) by an equation for the potential vorticity (PV) $Q = (f + \zeta)/h$:

$$\frac{DQ}{Dt} = \frac{1}{h}[-\nabla \cdot (\hat{\mathbf{k}} \times \mathbf{X}) - HQ]. \tag{5}$$

PV is a valuable diagnostic of atmospheric flow because it is conserved following fluid parcels, to the extent that $\mathbf{X}$ and H are negligible, and because it satisfies an invertibility principle that allows the mass and wind fields to be deduced from the PV field when the flow is balanced in an appropriate sense (e.g., [7]). A tracer transport equation

$$\frac{D\chi}{Dt} = S \tag{6}$$

can also be included, where S represents chemical sources and sinks.

Despite representing only a single layer of fluid, the shallow water equations support many of the phenomena and kinds of fluid motion supported by the full three-dimensional primitive equations (see, e.g., Chapter 3 of this volume), including Rossby waves, gravity waves, balanced motions and adjustment towards balance, and barotropic (but not baroclinic) instability. Layerwise two-dimensional eddies are explicitly represented, allowing mixing of PV and tracers and cascades to small scales. Forcing of eddies can be included through the lower boundary height term h_*. The equations have a consistent set of conservation laws for mass, energy, angular momentum, PV, potential enstrophy (hQ^2), and more exotic quantities like wave activity (pseudomomentum). It is often useful to draw an analogy between a shallow water model and a thin layer of a three-dimensional stratified atmosphere bounded by a pair of isentropes. Under this analogy, the mass source term H is interpreted as a diabatic (e.g., radiative) heating term at the lower bounding isentrope. A positive mass source corresponds to positive diabatic heating and ascent. Thus the shallow water equations allow the study of a wide range of fluid dynamical phenomena of relevance to the real atmosphere.

The most well-known application of shallow water models is to the study of planetary wave breaking in the winter stratosphere. The pioneering studies by

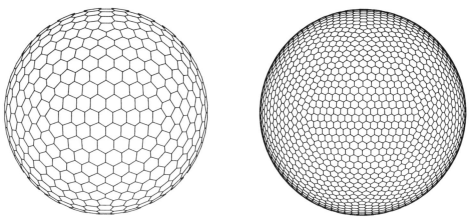

Figure 1. The hexagonal-icosahedral grid at two different resolutions.

Juckes and McIntyre [11] and Juckes [10] used single layer but high horizontal resolution models to reveal details of the wave breaking process previously only hinted at by coarse resolution satellite observations. Other applications include idealized turbulence studies (e.g., [20]), studies of flows on other planets, studies of planetary wave critical layers in the subtropical upper troposphere [3], and tests of new numerical methods (e.g., [21]).

Here we give one example of a simulation of planetary wave breaking in the winter stratosphere using a shallow water model. The particular model used happens to be that of Thuburn [18], which integrates the equations in PV-divergence form (2), (4), (5), on a hexagonal-icosahedral grid (Figure 1) with a resolution of about 120 km. The hexagonal-icosahedral grid avoids the pole problems that can arise with a regular longitude-latitude grid (see, e.g., Chapter 7 of this volume). The experiment begins with a steady, zonally symmetric, westerly vortex in balance with a zonally symmetric lower boundary height. This initial state is obtained by spinning up the model from rest with a strong relaxation of the depth towards a 'radiative equilibrium' value of 4000 m. For the experiment itself, the radiative relaxation timescale is set to 20 days, representative of the lower stratosphere. The flow is disturbed from zonal symmetry by a strong asymmetry in the lower boundary height of maximum amplitude ±720m. For more details of this particular experiment see [19].

Figure 2 shows PV maps from day 16 and day 46 of the simulation. At day 16 two anticyclones have formed outside the main cyclonic vortex, and they are stripping tongues of high PV air from the edge of the main vortex. One of the anticyclones is clearly being fed by a streamer of low PV air from low latitudes. By day 46 the continual stripping of air from the main vortex edge has resulted in a somewhat shrunken main vortex, with a sharp PV gradient at its edge. Outside the main vortex, mixing of air from different latitudes is leading to a region of

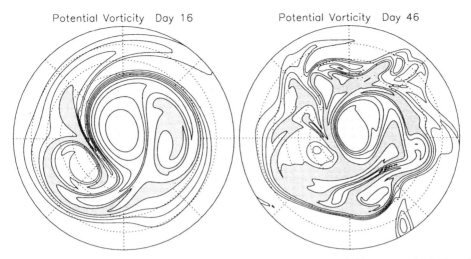

Figure 2. Potential vorticity at day 16 (left) and day 46 (right) from a shallow water simulation of stratospheric wave breaking. Contours range from 5.23×10^{-9} m^{-1}s^{-1} to 40.27×10^{-9} m^{-1}s^{-1} in steps of 4.38×10^{-9} m^{-1}s^{-1}. Light shading indicates high PV values and dark shading indicates intermediate PV values.

increasingly homogenized PV, often called the 'surf zone', as indicated by the increased area of dark shading.

Figure 3 shows the evolution of the zonal mean zonal wind $\bar{u}$ and zonal mean depth $\bar{h}$ for the simulation. There is a tendency for the jet strength and the high-latitude depth to increase as the wave breaking erodes and sharpens the vortex edge. However, this general trend is apparently punctuated by dramatic decelerations and deepenings, e.g., around day 30, again around day 40, and particularly from day 55 onwards. Actually, these apparent deceleration events are largely an artefact of the zonal mean diagnostics; in fact the main vortex remains intact and maintains its strength, but is temporarily displaced from the pole.

3. Zonally averaged models

Zonally averaged models reduce the spatial dimension of the equations describing the atmosphere by averaging in the east-west direction. There are several ways to do this, depending on the vertical coordinate used. A straightforward Eulerian zonal mean in pressure coordinates (e.g., [5]) is an obvious choice, but turns out to be a poor choice because the mean mass transport has large and often cancelling contributions from the Eulerian mean meridional circulation ($\bar{v}, \bar{w}$) and the eddy terms in the budgets. Use of the transformed Eulerian mean (TEM) equations (e.g., [1]) greatly reduces this problem.

An alternative is to use density-weighted zonal means in isentropic coordinates (e.g., [13]). Assuming thermal wind balance, the resulting equations are

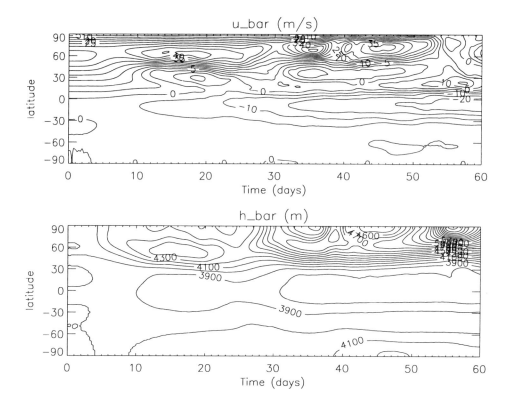

Figure 3. Zonal mean zonal wind (top) and depth (bottom) as functions of time and latitude during the 60 days of a shallow water simulation of stratospheric wave breaking.

$$\overline{\sigma}_t + \frac{(\overline{\sigma}\,\overline{v}^* \cos\varphi)_\varphi}{a\cos\varphi} = -(\overline{\sigma H})_\theta, \tag{7}$$

$$\overline{u}_t - \overline{v}^*\overline{\zeta_a} = \overline{\sigma \hat{v}\hat{Q}}^* + \overline{X} - \overline{Hu_\theta}, \tag{8}$$

$$\left(f + \frac{\overline{u}}{a}\tan\varphi\right)\overline{u} + \frac{\overline{M}_\varphi}{a} = 0, \tag{9}$$

$$\overline{M}_\theta = C_p\left(\frac{p}{p_s}\right)^\kappa, \tag{10}$$

where $\sigma = -p_\theta/g$ is the analogue of density in isentropic coordinates, $H = D\theta/Dt$ is the diabatic heating rate, $M = C_pT + \Phi$ is the Montgomery potential, and Q is Ertel's potential vorticity. Here overbar-asterisk indicates a density-weighted zonal mean and a circumflex indicates a departure from a density-weighted zonal mean. Subscripts ϕ and θ indicate derivatives with respect to latitude and potential temperature respectively. The corresponding equation for a chemical tracer with mixing ratio χ is

$$(\overline{\sigma}\,\overline{\chi})_t + \frac{(\overline{\sigma}\,\overline{v}^*\overline{\chi}\cos\varphi)_\varphi}{a\cos\varphi} + (\overline{\sigma H\chi})_\theta \approx \overline{\sigma S} - \frac{(\overline{(\sigma v)'\chi'}\cos\varphi)_\varphi}{a\cos\varphi} - (\overline{(\sigma H)'\chi'})_\theta. \quad (11)$$

A major advantage of using isentropic coordinates is that the vertical velocity is directly related to the diabatic heating. Other advantages will be mentioned below. However, isentropic coordinates are less advantageous for modelling the troposphere because the ground is not a coordinate surface and so requires special treatment, and because in the boundary layer, for example, the atmosphere can become unstably stratified so that it no longer increases monotonically with height.

The major difficulty with all zonally averaged models is that the effects of departures from zonal symmetry, i.e. eddy heat, momentum, and tracer fluxes, are important for the zonal mean budgets, but are not predicted directly by the zonal mean equations. They must therefore be parametrized in some way. The problem of parametrizing momentum fluxes due to gravity waves is similar to that in a full GCM (see, e.g., Chapter 12 of this volume). However, in a zonally averaged model the effects of planetary waves must also be parametrized. A common way to parametrize eddy tracer transport is in terms of an eddy diffusivity tensor **K**. The second major advantage of isentropic coordinates is that vertical eddy transports can justifiably be neglected so that only the K_{yy} component of **K** is nonzero. Values of K_{yy} may be taken from observations or a full GCM (e.g., [16]), or may be estimated from a simplified planetary wave model built into the zonally averaged model (e.g., [4], [12]). Such a built-in planetary wave model also allows the eddy PV flux term $\overline{\sigma \hat{v} \hat{Q}}^*$ to be estimated. Note that a third advantage of using isentropic coordinates is that the effects of eddy heat and momentum fluxes are combined in this single eddy PV flux term, which is closely related to the divergence of the Eliassen-Palm flux (e.g., [1]).

Zonally averaged models have been applied particularly to studies of chemistry in the stratosphere and mesosphere (e.g., [17], [14]). Long integrations with several dozens of chemical species are still not routinely feasible with full GCMs. Another application of such models is to the development and testing of gravity wave drag parametrization schemes.

Figure 4, taken from [12], gives an example of the quite realistic mean winds and temperatures that can be obtained with zonally averaged models these days with suitably calibrated gravity wave drag and planetary wave parametrizations. Note in particular the upward-equatorward slope of the winter stratospheric jet, a realistic feature which many full middle atmosphere GCMs fail to reproduce.

112

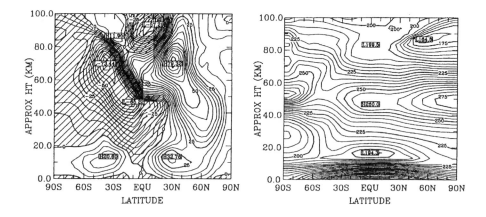

Figure 4. Climatological zonal mean zonal wind for January and temperature for July, from the 'THIN AIR' model of Kinnersley [12]. Reproduced with the permission of Dr J. S. Kinnersley and the Royal Meteorological Society.

4. 3D models with highly simplified parametrized physics

One class of simplified models uses the full hydrostatic primitive equations on the sphere, i.e. the 'dynamical core' of a full GCM, but with greatly simplified or nonexistent parametrized physical processes. (Note that when a numerical model requires explicit scale-selective dissipation terms these are best considered to be part of the dynamical core formulation rather than a physical parametrization and so they should always be included even when all other physical parametrizations are switched off.) Dynamical core models can represent all of the fluid dynamical phenomena resolved in a full GCM, including Rossby and gravity waves, barotropic and baroclinic instability, orographic effects, and nonlinear eddy effects including heat and momentum fluxes and mixing of tracers and potential vorticity. In practice, a dynamical core model can often be created by stripping down a full GCM.

Dynamical core models can be used to study basic dynamical phenomena in isolation from the complicating effects of realistic parametrized processes. Examples of applications include baroclinic instability life cycles (e.g., [8]), ultra-low frequency variability (e.g., [9]), midlatitude tropopause formation and sharpening, and the dynamical response to a localized tropical heat source. Experiments with minimal parametrized physics have also been used to assess numerical methods for integrating the dynamical core components of GCMs (e.g., [6], [2]).

We now consider an example of a baroclinic instability life cycle simulated by a dynamical core model. There are no parametrized physical processes at all in this example. The initial zonal mean state (Figure 5a) was obtained by specifying the zonal mean zonal wind $\bar{u}$, then using an initialization procedure to find a zonal

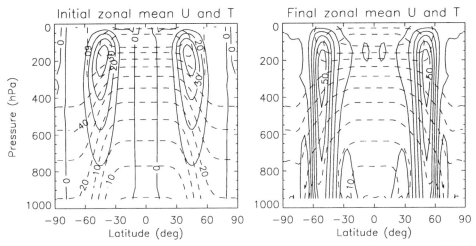

Figure 5. (a) Initial and (b) final zonal mean zonal wind (solid contours) and temperature (dashed contours) for a baroclinic instability life cycle simulation.

mean temperature $\overline{T}$ in balance with it. Superimposed on this is a small-amplitude perturbation with the structure of the most unstable normal mode, which has zonal wavenumber 6.

For the first few days, the dynamics is essentially linear and the wave grows exponentially. However, nonlinearity soon becomes important as the low level temperature contours begin to wrap up irreversibly, forming fronts (Figure 6). At later times, nonlinearity becomes important at higher levels too and the wave amplitude saturates and then eventually decays. Figure 7 shows the zonal mean poleward heat and momentum fluxes averaged over 15 days of the life cycle. The strong poleward heat flux at low levels is associated with the early baroclinic growth of the wave and upward propagation of wave activity. The poleward momentum flux at upper levels is associated with the decaying phase of the life cycle during which wave activity propagates equatorward. Comparison of Figure 5b with Figure 5a shows that the effect of the eddies has been to weaken the north-south temperature gradient and to make the jets stronger and more barotropic, in particular to spin up a strong surface wind.

Similar patterns of poleward heat and momentum flux are seen in full GCMs. Results from dynamical core models such as these help to relate these patterns to the growth and decay phases of baroclinic eddies, and so help to interpret the GCM results. (However, some time filtering is usually necessary to distinguish between contributions from baroclinic eddies and larger scale quasi-stationary waves.)

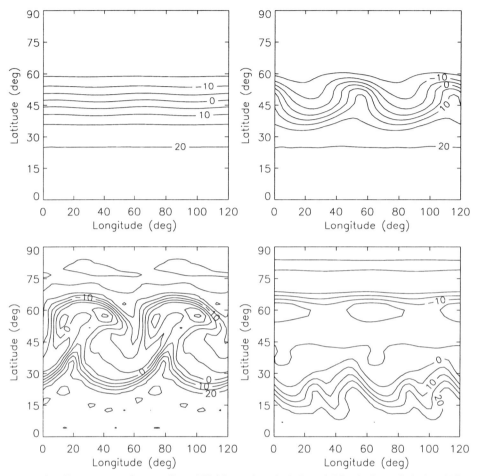

Figure 6. Temperature (degrees C) at 950 hPa at days 0, 4, 7, and 15, of a baroclinic instability life cycle simulation.

5. Stratosphere-mesosphere models

Stratosphere-mesosphere models integrate the full hydrostatic primitive equations on the sphere, and may include quite sophisticated parametrizations of radiation and gravity wave drag. Nevertheless, they are simpler than full GCMs because they include only that region of the atmosphere above some level near the tropopause, for example, 100 hPa. Thus parametrizations specifically related to the troposphere, such as boundary layer, cloud, and convection schemes, can be omitted.

Stratosphere-mesosphere models are forced by radiation, usually acting on a specified climatological ozone field, plus a specified lower boundary, e.g., specified geopotential if the lower boundary is a pressure surface, or specified Mont-

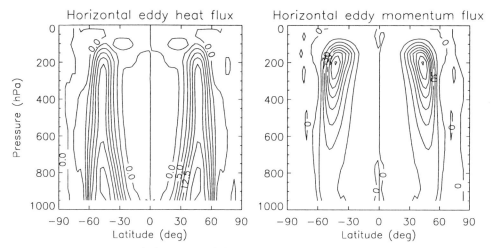

Figure 7. Zonal mean of $v'T'$ (left) and $u'v'$ (right) time averaged over a baroclinic instability life cycle.

gomery potential if the lower boundary is an isentropic surface. This specified lower boundary is both the strength and the weakness of stratosphere-mesosphere models. It allows clean experiments to be performed with attention focussed on the middle atmosphere, avoiding the complications of tropospheric variability. However, it precludes certain kinds of interactions that may be important for the real atmosphere, such as resonant wave modes or other deep modes whose structure extends over a greater vertical range.

Stratosphere-mesosphere models have been applied to idealized studies of stratospheric dynamics in which an idealized lower boundary is specified, perhaps in conjunction with idealized (e.g., 'perpetual January') radiative forcing (e.g., [15]), and to case studies using observed data for the lower boundary condition, and perhaps the initial condition. They have also been used to investigate chemical-radiative-dynamical feedbacks in the middle atmosphere, and as a test bed for new gravity wave drag parametrizations.

Figures 8 and 9 show example results from one particular stratosphere-mesosphere model. This model was developed and run by Andrew Gregory at the University of Reading. It integrates the hydrostatic primitive equations, rewritten so that PV is one of the predicted variables, using an isentropic vertical coordinate on the hexagonal-icosahedral grid mentioned in Section 2. The model is forced with a realistic seasonal cycle of radiation but an idealized lower boundary at the 414K potential temperature surface (close to 15 km altitude) consisting of a zonal wave 1 undulation with a fixed amplitude of 300 m. Figure 8 shows the zonal mean zonal wind as a function of time and latitude on the potential temperature surface (close to 35 km altitude). It shows a gradual strengthening and poleward shift of the jet at the onset of winter. Later the mean wind becomes much weaker and undergoes vacillations, with a recovery of the jet strength around early January.

116

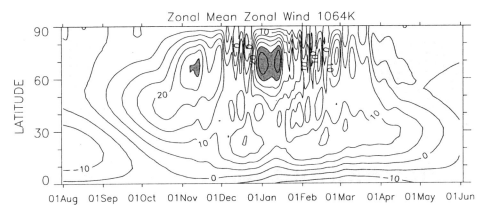

Figure 8. Zonal mean zonal wind (ms^{-1}) as a function of time and latitude on the 1064K potential temperature surface from a stratosphere-mesosphere model.

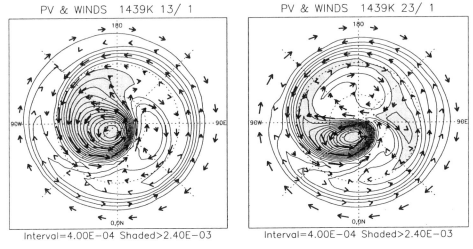

Figure 9. Potential vorticity and wind vectors on the 1439K potential temperature surface on 13 January (left) and 23 January (right).

Figure 9 shows a pair of PV maps at the 1439K level (close to 45 km altitude) on 13 January and 23 January. They show how the main cyclonic vortex is displaced from the pole as a large anticyclone is fed by low-latitude low-PV air; the anticyclone then migrates eastwards around the main vortex, stripping off a long tongue of high PV air.

Finally, note how the results from this model complement those from the shallow water model presented in Section 2. The shallow water model can be run at high horizontal resolution and so can capture many details of the wave breaking and surf zone mixing. The stratosphere-mesosphere model, on the other hand, can capture the vertical propagation of signals from the lower boundary and the three-dimensional structure of the interacting vortices.

6. Conclusion

GCMs are complicated and computationally expensive research tools. Used in isolation, their results can sometimes be difficult to interpret. Simplified models are often simpler to operate and interpret, as well as computationally cheaper, and so can complement GCMs in a valuable way. An ideal research toolkit would include a hierarchy of models of varying complexity. There are very many kinds of simplified model; I have illustrated the above ideas using just four examples of simplified models.

Acknowledgements

Figure 4 is reproduced with the kind permission of Dr J. S. Kinnersley and the Royal Meteorological Society. I am also grateful to Andrew Gregory for allowing me to show his unpublished results (Figures 8 and 9). Finally, I thank Euain Drysdale, Alan Iwi, Rawlings Miller, and Toni Pujol for their valuable comments on a draft version of this Chapter.

References

1. Andrews, D.G., Holton, J.R. and Leovy, C.B. (1987) *Middle Atmosphere Dynamics*, Academic Press, New York.
2. Boer, G.J. and Denis, B. (1997) Numerical convergence of the dynamics of a GCM, *Clim. Dyn.* **13**, 359-374.
3. Brunet, G. and Haynes, P.H. (1996) Low-latitude reflection of Rossby wave trains *J. Atmos. Sci.* **53**, 482-496.
4. Garcia, R.R., Stordal, F., Solomon, S. and J. Kiehl (1992) A new numerical model of the middle atmosphere 1. Dynamics and transport of tropospheric source gases, *J. Geophys. Res.* **97**, 12,967-12,991.
5. Harwood, R.S., and Pyle, J.A. (1975) A two-dimensional mean circulation model for the atmosphere below 80 km, *Q. J. R. Meteorol. Soc.* **101**, 723-747.
6. Held, I.M. and Suarez, M.J. (1994) A proposal for the intercomparison of the dynamical cores of atmospheric general-circulation models, *Bull. American Met. Soc.* **75**, 1825-1830.
7. Hoskins, B.J., McIntyre, M.E. and Robertson, A.W. (1985) On the use and significance of isentropic potential vorticity maps, *Q. J. R. Meteorol. Soc.* **111**, 877-946.
8. Hoskins, B.J. and Simmons, A.J. (1975) A multi-layer spectral model and the semi-implicit method, *Q. J. R. Meteorol. Soc.* **101**, 637-655.
9. James, I.N. and James, P.M. (1992): Spatial structure of ultra-low frequency variability of the flow in a simple atmospheric circulation model, *Q. J. R. Meteorol. Soc.* **118**, 1211-1233.
10. Juckes, M.N. (1989) A shallow water model of the winter stratosphere, *J. Atmos. Sci.* **46**, 2934-2955.
11. Juckes, M.N. and McIntyre, M.E. (1987) A high resolution one-layer model of breaking planetary waves in the stratosphere, *Nature* **328**, 590-596.
12. Kinnersley, J.S. (1996) The climatology of the stratospheric 'THIN AIR' model, *Q. J. R. Meteorol. Soc.* **122**, 219-252.

13. Kinnersley, J.S. and Harwood, R.S. (1993) An isentropic two-dimensional model with an interactive parametrization of dynamical and chemical planetary wave fluxes, *Q. J. R. Meteorol. Soc.* **119**, 1167-1193.

14. Nevison, C.D., Solomon, S., Garcia, R.R., Fahey, D.W., Keim, E.R., Loewenstein, M., Podolske, J.R., Gao, R.S., Wamsley, R.C., Donnelly, S.G. and DelNegro, L.A. (1997) Influence of Antarctic denitrification on two-dimensional model NO_y/N_2O correlations in the lower stratosphere, *J. Geophys. Res.* **102**, 13,183-13,192.

15. O'Neill, A. and Pope, V.D. (1988) Simulations of linear and nonlinear disturbances in the stratosphere, *Q. J. R. Meteorol. Soc.* **114**, 1063-1110.

16. Plumb, R.A. and Mahlman, J.D. (1987) The zonally averaged transport characteristics of the GFDL general circulation/transport model, *J. Atmos. Sci.* **44,** 298-327.

17. Portmann, R.W., Solomon, S., Garcia, R.R., Thomason, L.W., Poole, L.R. and McCormick, M.P. Role of aerosol variations in anthropogenic ozone depletion in the polar regions (1996) *J. Geophys. Res.* **101**, 22,991-23,006.

18. Thuburn, J. (1997) A PV-based shallow water model on a hexagonal-icosahedral grid, *Mon. Wea. Rev.* **125**, 2328-2347.

19. Thuburn, J. and Lagneau, V. (1999) Eulerian mean, contour integral, and finite-amplitude wave activity diagnostics applied to a single layer model of the winter stratosphere, *J. Atmos. Sci.* **56**, 689-710.

20. Vallis, G.K. (1996) Potential vorticity inversion and balanced equations of motion for rotating and stratified flows, *Q. J. R. Meteorol. Soc.* **122**, 291-322.

21. Williamson, D.L., Drake, J.B., Hack, J.J., Jakob, R. and Swarztrauber, P.N. (1992) A standard test set for numerical approximations to the shallow water equations in spherical geometry, *J. Comput. Phys.* **102**, 211-224.

DESIGNING A GCM EXPERIMENT

Fundamentals of the Planning Process

P.W. MOTE
JISAO Climate Impacts Group
University of Washington
Box 354235
Seattle, WA 98195 USA

1. Introduction

This chapter synthesizes and expands upon a panel discussion held at the NATO ASI on "Numerical modelling of the global atmosphere," and augments chapters 2–5. Participants in the panel were Hans Feichter, John Thuburn, Paul Valdes, and David Williamson. The panel discussion was intended to unearth some fundamental principles and techniques that experienced modelers bring to bear when changing GCMs and designing experiments. To our knowledge, these principles have not been collected and published; modelers (at best) learn them slowly and haphazardly from experienced colleagues, or rediscover them by trial and error, or (at worst) never discover them at all.

2. Control Runs

Without a "good" control climate, an experiment may be utterly meaningless. Mature models have been laboriously improved year by year; a newer model should be rigorously tested to ensure that important features of its climate adequately resemble observations, before being used in a series of experiments. Of course, there is no consensus on what constitutes a "good" climate; the burden lies on the scientist to report the model's biases quantitatively.

Furthermore, the control run should be good for the right reasons; that is, it should not be the result of large, cancelling errors in various physical parameterizations. It should therefore be subjected to a variety of sensitivity tests as outlined

P. Mote and A. O'Neill (eds.), Numerical Modeling of the Global Atmosphere in the Climate System, 119–125.
© *2000 Kluwer Academic Publishers. Printed in the Netherlands.*

in Chapter 2. The reason modelers should determine whether there are large cancelling errors is that perturbation experiments (doubled CO_2, for example) may change the error in one component but not others, leading to erroneous results.

It is usually better in the long run to avoid unphysical "quick fixes" that may appear to improve the model climate but in fact misrepresent the nature of the process. Examples include the use of Rayleigh friction (linear drag) instead of gravity wave drag in models of the stratosphere, and the practice of relaxing soil temperatures to climatology.

Finally, as is explained in section 4.2, the control run ought to be of the same length (or the same multiplicity in the case of ensembles) as the experiment. One year of a control integration should not be compared with 10 years of a perturbed model run, or vice versa.

3. Changing a Model

When possible, changes in the formulation of a model should be made one at a time so that the effects of each change can be evaluated independently (though the effects may not be linearly additive). For example, changing the horizontal resolution requires changing the time step (in order to maintain numerical stability; see Chapter 7, section 2), which could affect the parameterizations and also requires retuning the horizontal diffusion (Chapter 17, section 2). These changes can be made one at a time by first reducing the time step and then changing the resolution.

Before a new parameterization is introduced, it should be tested by running it diagnostically in parallel with the original parameterization. In such a diagnostic test, both parameterizations are used to calculate forcing at every time step, but only the original parameterization influences the simulation. This process helps ensure that the new simulation was introduced correctly (since the simulation should be identical to one without the new parameterization). Also, comparing the two forcings provides the first-order difference in the forcing, which may be useful in diagnosing the cause of differences in the simulation when the new parameterization is active.

David Williamson described a procedure that his group at NCAR (the National Center for Atmospheric Research, USA) used for comparing two approaches to formulating the dynamics: Eulerian and semi-Lagrangian (see Chapter 7 for details about these dynamical schemes). The two schemes gave different results, but the causes of these differences were difficult to pinpoint because the dynamical scheme is such a fundamental part of a GCM and because of nonlinear feedbacks. In order to understand the differences, the NCAR group performed simulations in which the tendencies were calculated separately at each time step using both schemes. If the prognostic equation for a state variable χ is summarized as

$$(\frac{\partial \chi_1}{\partial t})_{S-L} = -(\text{dynamics})_{S-L}(\chi_1) + (\text{parameterizations})(\chi_1) \qquad (1)$$

with semi-Lagrangian dynamics and

$$(\frac{\partial \chi_2}{\partial t})_{E} = -(\text{dynamics})_{E}(\chi_2) + (\text{parameterizations})(\chi_2) \qquad (2)$$

with Eulerian dynamics, then one can also calculate the tendencies with one scheme of the fields from the other scheme: $(\text{dynamics})_E(\chi_1)$ and $(\text{dynamics})_{S-L}(\chi_2)$. This calculation is performed at each time step and summed over all time steps to find the net effect of dynamics alone without the complex nonlinear feedback of the complete model. Using this technique, the NCAR group traced the difference to a vertical truncation error in the Eulerian approximations when applied to the particular temperature profile and forced by the boundary-layer parameterization over polar land areas [6].

4. Designing an Experiment

A complete design of a GCM experiment should be in place before the relevant model runs are begun. Designing an experiment includes not just planning a sequence of runs but also—and equally important—considering what diagnostics will be needed (see Chapters 3 and 4), anticipating possible outcomes, and considering how to interpret them. One goal of anticipating outcomes is to determine not only what happened but, as much as possible, why it happened. In some cases the appropriate diagnostics will require additional variables, or variables saved at more frequent intervals than the default, and the modeler is advised to anticipate these needs to avoid rerunning the model. Because some runs might depend on the outcome of other runs, the experiment design could be envisioned as an experiment "tree".

4.1. CAUSE AND EFFECT: CHOOSING THE RIGHT TOOL

GCMs are sometimes used to identify causes of a phenomenon in the climate system, like the Asian monsoon. Modelers pursuing such endeavors should think through a sequence of controlled experiments designed to test ideas of cause and effect. This well-planned approach is most effective if the modeler has a hypothesis to test, rather than simply trying one thing after another. Disentangling a GCM's complex response to a change in forcing can be a challenge. For example, changing radiative properties of tropical cirrus clouds can be accompanied by large (but not necessarily statistically significant–see section 4.2 or Chapter 4) changes in the polar stratospheric jet. One must evaluate whether the radiative change caused the remote change in the wind.

Sometimes, a GCM may not be the best, or should not be the only, tool to use for evaluating such cause-and-effect questions. Thuburn (Chapter 5) describes several examples of simplified models that may be appropriate for guiding a GCM experiment. We discuss here two additional approaches.

The first is the fixed dynamical heating (FDH) model of Fels *et al.* [3]. The simplest way to study a radiative change is using a radiative-convective (RC) model, but neglecting heat advection leads to enormous errors in many parts of the atmosphere where the temperature is far from RC equilibrium. In the FDH model, the radiative heating rates are calculated throughout the model atmosphere for a realistic time-mean basic state. That is, one assumes that

$$\frac{\partial T}{\partial t} = Q_{\text{rad}}(T) + Q_{\text{dyn}} = 0 \tag{3}$$

where T is the temperature and Q_{dyn} obtained by solving (3) is the "dynamical heating" rate. The time tendencies due to radiation can then be calculated for various perturbed cases $(Q_{\text{rad}}(T + T'))$. The FDH approach has two main advantages over a GCM. First, complicated feedbacks between the radiation and dynamics are avoided, allowing one to evaluate the purely radiative response to a change in forcing. Second, FDH is vastly cheaper than a GCM, since only the RC code is needed. In effect, FDH allows a detailed one-dimensional RC model to be run globally, with simplified advection redistributing heat.

In some cases, adjoint methods [4] may be appropriate, but so far they have not been widely used by climate modelers. The usual approach in linking a cause to a given effect is to hypothesize a cause, introduce an appropriate perturbation, and see what effect the perturbation has. Adjoint methods allow the modeler to reverse the direction of causality, choosing an effect and letting the adjoint model identify the cause. To construct the adjoint of a full GCM, one first constructs a tangent-linear version of the GCM, then finds its adjoint. Adjoint methods presently have several applications: (1) to generate optimal perturbations to initial conditions for an ensemble of forecasts [5]; (2) in a research mode in numerical weather prediction (for example, to trace a bad forecast of a storm over Ireland to poor initialization upstream over the western Atlantic a few days earlier [1]); and (3) for data assimilation (see Chapter 15). Adjoint methods are potentially powerful for climate research, but efforts are needed to develop tangent-linear and adjoint models.

4.2.   LONG RUNS AND ENSEMBLES

As is emphasized in Chapter 4, calculating statistical significance requires either long model runs or multiple runs. Calculating significance separates the robust response to a change from the model's inherent variability, and with a single short model run the two may be difficult to distinguish. To put it another way, it is pointless to devote all of one's computer resources to a single pair of runs if the results are statistically meaningless.

The approach to obtaining significant results depends on whether one is interested in a transient response to a change in forcing, like removing the snow cover, or a change in the model climate, for example when introducing a new convective parameterization. For transient runs, the best approach to obtaining significant results is to perform not one run but several runs (an ensemble) with the runs differing in a way that should have no bearing on the response to a change; usually the runs use slightly different initial conditions. For runs with climatological forcing, used to evaluate model "climate", it is often sufficient to run the model once for a long time.

How many ensemble members are needed? How long is long enough? For reasons explained in Chapter 4, it would be preferable to have on the order of ten realizations in order to calculate statistical significance. For transient runs, this means ten members in each of the categories (control, experiment 1, experiment 2, etc.), while for runs with climatological forcing it means running the model ten times longer than the timescale of interest. In the latter case, to use the example of evaluating the model's climate (where the timescale of interest is at least a year), a 10-year run would be needed. An initial adjustment period, whose length depends on the time for the model to reach "equilibrium" with any new forcing, is often discarded before averages are calculated.

Of course, whenever two runs are being compared, statistical evaluation is best done when both have the same number of degrees of freedom (section 2 and Chapter 4), since extra degrees of freedom in one run have little value in reducing uncertainty. In the case of long runs, the period evaluated from each run should therefore be of the same length; in the case of ensembles, the same number of ensemble members should be used from each ensemble.

Since no model is perfect, it is good to perform the same experiment using more than one model. Few groups are capable of running more than one model, but cooperative intercomparisons (see Chapter 19) offer multimodel integrations for a very limited set of experiments. Parameterizations ought to be coded so that they can be exchanged among modeling groups and run either in stand-alone mode or in other GCMs.

4.3. THE "RIGHT KICK"

When evaluating the impact of perturbations in the boundary conditions (or, for short-term transient experiments, changes in initial conditions), it is important to decide how big to make the perturbation, or "kick". For example, changing midlatitude sea surface temperature by a degree or two does not result in robust differences in the time-mean state of the atmosphere. To ensure statistically significant results, one might apply a big kick. It may be useful to try a small kick too, in part to evaluate whether the response is linear; but the small kick is pointless if the big kick yielded an ambiguous response. Unless the modeler's goal is to explore climate states very different from the present one (like the group at the ASI that ran a climate model with the Sun turned off), then the "right kick" is one that is realistically big, since a kick that is unrealistically big might push the model into an entirely different climate state.

4.4. THE INDIRECT APPROACH

While the topic did not come up in the panel discussion, it is worth mentioning that experienced modelers often come up with clever, indirect ways of manipulating the model. The most obvious approach may not yield the best results, while a good understanding of the physical mechanisms and of the model's parameterizations may expand the possible approaches. For example, one of the groups at the ASI wanted to study the North Atlantic Oscillation (NAO), a transient, low-frequency phenomenon. Rather than waiting for it to appear in the model simulation, Byron Boville suggested moving the sea ice line in the north Atlantic in order to nudge the surface pressure distribution toward a pattern characteristic of the NAO.

Another example was the ozone-hole study of Cariolle *et al.* (1990) [2]. Because of its strong radiative-chemical feedbacks, ozone is a very difficult atmospheric trace constituent to model, and in most GCMs until recently its radiative forcing was calculated using a climatological distribution of ozone. Furthermore, the mechanisms for catalytic ozone destruction in the ozone hole over Antarctica were, when Cariolle *et al.* were performing their study, far from being sufficiently well understood and parameterized to be included in a GCM. Instead, Cariolle *et al.* used a linear loss term to create an Antarctic ozone hole and study its effects on midlatitude ozone.

5. Checking for Errors

This may sound obvious, but too often modelers discover errors only after consuming large amounts of computer time. After changing the model code, initial conditions, or boundary conditions, make very sure that everything has been done correctly. Run the model for the shortest time over which the change will manifest itself; this may be as little as one time step. Are the differences consistent with what was changed? If not, double check that the correct subroutines were compiled, that the correct data sets were read, that all model parameters were set correctly, and that the change was made in the correct places in the code. It is far less frustrating to discover errors after running the model for one time step than after running it for several years.

6. Summary

We (though not divinely inspired) drew up a list of ten "commandments" for climate modeling:

1. Back up the quality of the control climate with technical reports or (far better) with published sources.
2. In developing a GCM, tackle biases by improving the physics, not by using unphysical "quick fixes."
3. Introduce changes one at a time.
4. Plan the experiment "tree" before you begin.
5. Choose the right tool for the job. It may not be a GCM.
6. Anticipate the diagnostics you will need.
7. Make the perturbation big enough to get results, but within a realistic range.
8. Look for indirect ways to generate the perturbation you want.
9. Check carefully for errors before running the model a long time, and keep checking while it is running. Never stop looking for inconsistencies that might imply an error.
10. Do not let anyone mistake your model's noise for a signal. Run the model enough times, or long enough, to ensure statistical significance.

Acknowledgements

The author wishes to thank rapporteurs John Knox and Evi Schuepbach, the panelists, and members of the audience who contributed. David Williamson suggested numerous valuable improvements to the chapter, and further improvements were suggested by John Knox and Jeffrey Yin. This is JISAO contribution number 584.

References

1. Burridge, D. (1997) presentation at the Royal Meteorological Society, Edinburgh.
2. Cariolle, D., Lasserre-Bigorry, A., and Royer, J.F. (1990) A general circulation model simulation of the springtime Antarctic ozone decrease and its impact on mid-latitudes, *J. Geophys. Res.* **95**, 1883–1898.
3. Fels, S.B., Mahlman, J.D., Schwarzkopf, M.D., and Sinclair, R.W. (1980) Stratospheric sensitivity to perturbations in ozone and carbon dioxide: Radiative and dynamical response, *J. Atmos. Sci.* **37**, 2265–2297.
4. Hall, M.C.G. (1986) Application of adjoint sensitivity theory to an atmospheric general circulation model, *J. Atmos. Sci.* **43**, 2644–2651.
5. Molteni, F., Buizza, R., Palmer, T.N., and Petroliagis, T. (1996) The ECMWF Ensemble Prediction System: Methodology and validation, *Q. J. R. Meteorol. Soc.* **122**, 73–119.
6. Williamson, D.L., and Olson, J.G. (1998) A comparison of semi-Lagrangian and Eulerian polar climate simulations, *Mon. Weather Rev.* **126**, 991–1000.

NUMERICAL APPROXIMATIONS FOR GLOBAL ATMOSPHERIC GENERAL CIRCULATION MODELS

DAVID L. WILLIAMSON
National Center for Atmospheric Research
Box 3000
Boulder, Colorado 80307
U.S.A.

RENÉ LAPRISE
Université du Québec à Montréal
B.P. 8888, Succ. Centre-ville
Montréal, QC
Canada H3C 3P8

1. Introduction

In this chapter we provide an overview of numerical approximations that are used in global atmospheric general circulation models (AGCMs) today, and concentrate on those aspects of the schemes that are most important with respect to the spherical geometry of the earth or are most influenced by the spherical geometry. The global geometry adds a degree of complexity not found in many disciplines involving numerical approximations. Many of the major AGCMs today are based on numerical methods whose fundamental aspects were developed two decades ago with what might be considered refinements since then. Newer, fundamentally different methods with very attractive properties have been developed in other branches of computational fluid dynamics in the last two decades for Cartesian geometry. However, difficulties in adapting them to spherical geometry in an economical way has prevented their application to AGCMs. This picture is beginning to change and we should expect to see models based on such schemes in the future. Nevertheless, in keeping with the objectives of this NATO Advanced Study Institute to "explain the main ideas behind general circulation modelling of the atmosphere and their practical implementations", we restrict ourselves here to schemes in use today in application models.

P. Mote and A. O'Neill (eds.), Numerical Modeling of the Global Atmosphere in the Climate System, 127–219.
© 2000 *Kluwer Academic Publishers. Printed in the Netherlands.*

As implied above, current application models are the result of an evolutionary process over the last two and a half decades as the fundamental numerical approximations have been refined and the resolution of typical applications modestly increased as computer power has increased. Although spatial resolution has not increased dramatically, the length of simulations has, with century-long simulations not uncommon. A more fundamental change has occurred in the completeness and complexity of physical parameterizations, which are discussed in other chapters. The earliest atmospheric models were based on finite difference approximations. The spectral transform method became popular by the late 1970s for what would be considered today as very coarse resolution (R15) approximations to large scale global flow. At that time, parameterizations were not strongly interactive, for example, clouds were specified as zonal averages, and there was much less emphasis on moist processes in the models than today. As we will see, the spectral transform method provides a natural solution to the problems introduced by spherical geometry and is a very accurate method for smoothly varying fields without sharp gradients and discontinuities. Another supposed advantage espoused at that time was that the spectral method did not appear to have the many arbitrary choices offered by finite difference approximations. This has been shown more clearly recently by Temperton [69]. Also, spectral transform models can easily avoid aliasing, which in (some) grid point schemes results in nonlinear instability requiring the use of heavier diffusion in grid point models than in spectral models. Finite difference models were never completely abandoned. However, a completely satisfactory solution to the pole problem has not been implemented in those models, although several engineering fixes have been applied. More recently, semi-Lagrangian time approximations have been implemented in both grid-point and spectral models. The semi-Lagrangian method when coupled with semi-implicit approximations also provides a natural solution to the spherical geometry problems.

The state of AGCMs in the early 1990s is summarized by the Atmospheric Model Intercomparison Project (AMIP) which included almost all the AGCMs in the world (Gates [26]). The AMIP summary documentation (Phillips [48]) indicates that of the 30 models participating, 19 were spectral transform, one of which was semi-Lagrangian, the others being Eulerian, and 11 models were finite difference. The horizontal resolution of a little over half the spectral models in AMIP was around T42, the remaining were around T30 or lower. Truncation at T42 is common in climate applications, but considered very coarse by Numerical Weather Prediction (NWP) standards. Most grid point models in AMIP were $4° \times 5°$ with only a few exceptions.

In this chapter we provide a general review of some basic properties of finite difference schemes, especially those that influence choices made in designing models and those related to the spherical geometry complications. We concentrate on simple examples in one space dimension that illustrate the numerical issues involved in the more complex spherical geometry. We will not provide details of the more complex schemes actually used in any particular model nor discuss issues associated with formal accuracy and convergence rates, nor many of the various options in designing *a priori* conservation. Many important concepts in numerical approximation theory are not covered here and the interested reader should see the references listed below. We review the basic properties of the spectral transform and semi-Lagrangian methods, and provide more details about the practical implementation of these methods in the NCAR Community Climate Model.

For more details about basic numerical methods used in atmospheric models and the concepts mentioned in this chapter, the reader is referred to the textbook by Haltiner and Williams [27] and to the less accessible, but still very valuable, GARP Publications Series No 17 (Mesinger and Arakawa [44], and Kasahara [36]) on Numerical Methods Used in Atmospheric Models. For a summary of numerical methods that have been used for global models and additional references see Williamson [71]. An excellent book by Durran [22] covers the fundamentals of numerical methods for atmospheric problems and fluid flow in general.

All global atmospheric models in application use today are based on spherical coordinates (latitude and longitude) or a transformation of these coordinates such as by Schmidt [62], and numerical approximations that involve an underlying latitude-longitude grid structure based on these coordinates. One such grid is illustrated in Figure 1.1a. The lines indicate the edges of grid boxes and the grid points can be thought of as being in the centers of the boxes. The type of grid in Figure 1.1a is referred to as a uniform latitude-longitude grid. The grid boxes have the same width in longitude everywhere, and (almost) the same height in latitude everywhere, inequality being present in the Gaussian grid associated with the spectral transform method to be discussed later. The spherical coordinate system has pole points to which the meridians converge, and at which horizontal (relative to the Earth's surface) vector components become undefined. These singularities introduce a variety of difficulties into numerical modeling. Collectively they are referred to as the "pole problem". The singularity itself is an irritant rather than a major problem and has been successfully dealt with. The so-called "pole problem" is more of an economical problem than a technical problem as will be explained later. Global grids are not required to be uniform in latitude and longitude. Figure 1.1b shows a grid in which the longitudinal grid interval is reasonably constant in terms of physical distance rather than angular distance. Nevertheless, the boxes (and points) are aligned in longitude. Such grids, first introduced by Kurihara [39] and reintroduced

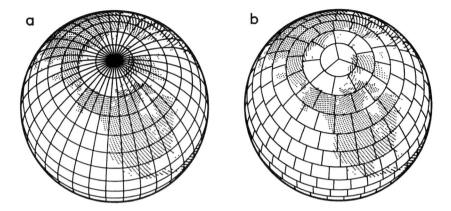

Figure 1.1. (a) Example of a uniform latitude-longitude grid in which all grid boxes have the same width in longitude and the same height in latitude. (b) Example of a reduced grid in which the longitudinal grid interval is reasonably constant in terms of physical distance rather than angular distance.

more recently in a different context by Hortal and Simmons [32], are being successfully used in atmospheric models today with some types of approximations as will also be described below. However, these grids still present difficulties for other types of approximations.

In the following sections we determine the amplitude and phase errors of various types of numerical approximations. We also consider the stability of these approximations. Both are accomplished by considering standard wave form solutions. We will see that the most intuitively obvious schemes can be unsuitable for modeling. We will then show how these various concepts are combined to create a complete atmospheric model by presenting the details of the NCAR CCM.

2. Finite Difference Approximations

We first introduce some fundamental aspects of finite difference approximations, especially those that influence their application on spherical grids such as in Figure 1.1, and those that remain relevant for spectral and semi-Lagrangian approaches. We will introduce the concept of stability of approximations and illustrate phase and amplitude errors of various schemes.

Consider the one-dimensional advection equation of a quantity $q(x, t)$:

$$\frac{\partial q}{\partial t} + U \frac{\partial q}{\partial x} = S, \qquad (2.1)$$

over the interval $0 \leq x \leq 2\pi$ with U constant and periodic boundary conditions $q(0, t) = q(2\pi, t)$. The term S represents possible sources or sinks of the quantity

q; for now we assume $S = 0$. Assume wavelike solutions of the form

$$q(x, t) = Q e^{i(kx + \nu t)} \tag{2.2}$$

for integer $k \geq 1$, where Q is the amplitude. By substitution of (2.2) into (2.1) the frequency ν is related to the wavenumber k by

$$\nu = -kU , \tag{2.3}$$

which says that the wave structure e^{ikx} moves with uniform speed U with no change of shape or amplitude.

2.1 EXPLICIT ADVECTION

Now calculate an approximation to the solution of (2.1) on a discrete set of points in space and time

$$x_j = (j - 1)\Delta x , \tag{2.4}$$

$$\Delta x = \frac{2\pi}{J} , \tag{2.5}$$

where $j = 1, \ldots, J$ and the discrete times are denoted $t^n = n\Delta t$, $n \geq 0$. A common approximation to (2.1) is obtained by replacing the differentials with differences:

$$\frac{q_j^{n+1} - q_j^{n-1}}{2\Delta t} + U \frac{q_{j+1}^n - q_{j-1}^n}{2\Delta x} = 0 . \tag{2.6}$$

The shorthand notation q_j^n stands for $q(x_j, t^n)$. The approximation (2.6) is referred to as centered differences or leapfrog and it can be solved in closed form in the linear case with U constant. Assume solutions of the form

$$q(x_j, n\Delta t) = Q \rho^n e^{ikx_j} . \tag{2.7}$$

That is, the solution at grid point x_j and time $n\Delta t$ is written as the initial amplitude Q times a factor ρ, possibly complex, raised to the power n, times a wave–like structure in x. Note that Q and ρ are specific to the wavenumber k, but otherwise constant.

Because the solutions are defined at a discrete set of points x_j, the wavenumber k is restricted to the range $k \leq J/2$. On the discrete set of points (2.4), a wave with $k > J/2$ is indistinguishable from one with $k < J/2$ and thus need not be considered. To see this, consider a wavenumber in the range $J/2 < k \leq J$ and define $k' = J - k$. Substitution of (2.4) for x_j and of (2.5) for Δx into (2.7) gives

$$e^{ikx_j} = e^{ik(j-1)\frac{2\pi}{J}} = \left(e^{i2\pi}\right)^{j-1} e^{-ik'(j-1)\frac{2\pi}{J}} = e^{-ik'x_j} , \tag{2.8}$$

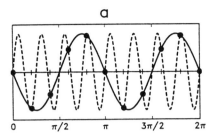
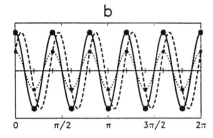

Figure 2.1. (a) Illustration of a short wave (dashed line) observed on a course grid (dots) misrepresented or aliased as a longer one (solid line.) (b) Illustration of shortest resolvable $2\Delta x$ wave with zero phase (solid line), and of a $2\Delta x$ wave with nonzero phase (dashed line) misrepresented as a smaller amplitude wave with zero phase (dotted line).

which is equivalent to one of the original waves in the range $0 \le k' < J/2$ with a 180 degree phase shift. The cases for $k > J$ likewise easily map onto the cases above. The wave expression can also be written in terms of a wavelength L,

$$e^{ikx_j} = e^{i\frac{2\pi}{L}x_j}\,, \tag{2.9}$$

and the wavelength $L = 2\pi/k$ is bounded by $L \ge 2\Delta x$. The shortest resolvable wave on the discrete grid is two grid intervals long.

The misrepresentation or aliasing of a short wave onto a longer one as in (2.8) because it is observed only at a discrete set of points as described above is illustrated in Figure 2.1a. If the short wave (dashed line) is observed only at the points indicated with the dots, it cannot be distinguished from the long wave (solid line). Figure 2.1b illustrates the shortest $2\Delta x$ wave that is resolvable on the grid (solid line). In addition, the phase of this wave is fixed such that the peaks are at the grid points. We will refer to this as a zero phase angle. If the true $2\Delta x$ wave has a different phase (dashed line), it is misrepresented as one with a smaller amplitude and zero phase angle (dotted line).

2.1.1 *Stability Condition*
Substitution of (2.7) into (2.6) and elimination of $Q\rho^{n-1}e^{ik(j-1)\Delta x}$ gives

$$\rho^2 + \rho\left[2iU\frac{\Delta t}{\Delta x}\sin k\Delta x\right] - 1 = 0\,, \tag{2.10}$$

which has two solutions

$$\rho = -iU\frac{\Delta t}{\Delta x}\sin k\Delta x \pm \left[1 - \left(U\frac{\Delta t}{\Delta x}\right)^2 (\sin k\Delta x)^2\right]^{\frac{1}{2}}. \tag{2.11}$$

If $|U (\Delta t/\Delta x) \sin k\Delta x| \leq 1$ for all admissible k, i.e.,

$$\left| U \frac{\Delta t}{\Delta x} \right| \leq 1, \tag{2.12}$$

then the term in brackets $[\ ]^{1/2}$ in (2.11) is real, and $|\rho| = 1$ for all k. In this case the amplitude of the solution is Q as in the analytic case (2.2) and neither grows nor decays. The condition $|\rho| = 1$ means that ρ is of the form $\rho = e^{i\omega t}$ where ω is a real number giving the frequency of the numerical solution. Recall that the exact value of the frequency is $\nu = -kU$. We will use the exponential form for the temporal aspects of the solution in the next section and solve for the frequency of the approximate solution. If $|U \frac{\Delta t}{\Delta x}| > 1$, one of the two solutions of (2.11) will have amplitude $|\rho| > 1$ and that solution will amplify exponentially with time. Such erroneous behavior of the solution is referred to as computational instability because in a short time the amplitude becomes so large that it cannot be represented on a computer and the computation halts. The condition (2.12) is referred to as the von Neumann stability condition and the analysis leading to it as von Neumann's method for determining stability. A related but less rigorous condition was identified by Courant, Friedrichs, and Lewy [20] and is referred to as the CFL stability condition. The combination $|U \frac{\Delta t}{\Delta x}|$ is also commonly referred to as the Courant number and is denoted by

$$C = \left| U \frac{\Delta t}{\Delta x} \right|. \tag{2.13}$$

In numerically solving an equation such as (2.1) with approximations (2.6) one first defines the grid interval Δx and grid x_j on which the solution will be calculated. This defines the smallest scale that will be present in the numerical solution since through (2.8) the smallest wave in the solution will have a wavelength $L = 2\Delta x$. Or put another way, the grid must be chosen fine enough (Δx small enough) so the scales of the phenomena of interest L_{ph} satisfy $L_{ph} \geq 2\Delta x$. The velocity U is determined by the problem and is not under our control. Thus the only remaining parameter Δt must be chosen to satisfy (2.12),

$$\Delta t \leq \text{Min} \frac{\Delta x}{|U|}. \tag{2.14}$$

On a global grid as in Figure 1.1a, the grid interval in the longitudinal (east-west) direction is $\Delta x = \Delta\lambda cos\varphi$ and $\Delta\lambda$ is uniform, where λ denotes longitude and φ denotes latitude. The convergence of meridians in a uniform latitude-longitude grid implies that Δx becomes very small near the poles. The zonal wind com-

134

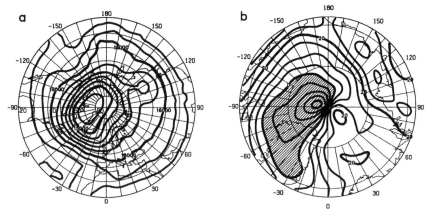

Figure 2.2. Northern Hemisphere 100 hPa(a) height field and (b) zonal wind field for 00 UCT
23 Jan 1984. The $10°$ grid boxes indicated by dashed lines are the same in both
panels, (a) covers the region poleward of $20°$N and (b) poleward of $68°$N. Contour
intervals are 200 m for height field and 5 m sec^{-1} for zonal wind component.

ponent u, on the other hand, which serves as U in the previous equations, is
not necessarily small there. For example, Figure 2.2 shows an asymmetric polar vortex, i.e., one not centered on the pole. Such structures commonly occur
in the atmosphere in the northern hemisphere in winter. Figure 2.2a shows
the Northern Hemisphere 100 hPa height field for 00 UCT 23 Jan 1984. The
geostrophic relationship implies that the wind is parallel to the height contours
with its strength proportional to the distance between contours. Because the
center of the vortex is shifted away from the pole the wind is blowing across
the pole and between $20°$E and $70°$E is parallel to the latitude circles resulting
in a large zonal wind component there. Figure 2.2b shows the analyzed zonal
wind u in an enlargement of the polar region (the grid lines are the same in
both panels). The zonal wind u exceeds 20 m s^{-1} poleward of $85°$ latitude. The
effect of such winds near the poles on the maximum allowed time step is illustrated in Figure 2.3 which shows the maximum (over height and longitude) zonal
wind speed as a function of latitude for the same case as in Figure 2.2. In both
hemispheres the maximum wind is found in mid-latitudes where the jet stream
occurs. However, the bound for Δt in the stability condition (2.14), $(\Delta x)/U$, becomes $(\Delta \lambda \cos \varphi)/u$ on the global grid. Although u is not excessively large, $\cos \varphi$
becomes so small near the poles that $u/(\cos \varphi)$ is significantly larger than u. For
this particular day, the time step near the poles would have to be over 7 times
smaller than the time step determined for the mid-latitudes where the winds are
strongest. Such a time step is too small to be economically viable. This is why we
referred earlier to the "pole problem" as being an economical one. Much effort
has been expended to find schemes that do not require such a small time step.

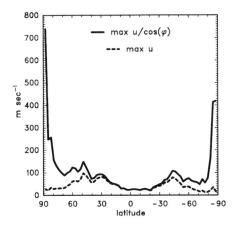

Figure 2.3. Maximum (over height and longitude) zonal wind speed and maximum zonal wind speed divided by cosine of latitude for 00 UCT 23 Jan 1984.

2.1.2 Response Function

Let us return to (2.6) and rewrite it using a discrete operator notation:

$$\delta_{2t}q_j^n + U\delta_{2x}q_j^n = 0\,, \tag{2.15}$$

where the centered difference operators are defined by

$$\delta_{2t}q_j^n = \frac{q_j^{n+1} - q_j^{n-1}}{2\Delta t}\,, \tag{2.16}$$

and

$$\delta_{2x}q_j^n = \frac{q_{j+1}^n - q_{j-1}^n}{2\Delta x}\,. \tag{2.17}$$

Assume wavelike solutions in both space and time of the form (2.2). The discrete operators (2.16) and (2.17) applied to such waves give

$$\delta_{2t}Qe^{i(kx+\nu t)} = i\nu\mathcal{D}_{2t}Qe^{i(kx+\nu t)}\,, \tag{2.18}$$

and

$$\delta_{2x}Qe^{i(kx+\nu t)} = ik\mathcal{D}_{2x}Qe^{i(kx+\nu t)}\,, \tag{2.19}$$

where

$$\mathcal{D}_{2t} = \frac{\sin\nu\Delta t}{\nu\Delta t}\,, \tag{2.20}$$

and

$$\mathcal{D}_{2x} = \frac{\sin k\Delta x}{k\Delta x} \tag{2.21}$$

136

are referred to as the response functions of the centered difference operators. They are just the ratio of the result of the discrete operator to the result of the analytic operator and represent a relative error introduced by the approximation. Compared to the differential operators $\partial/\partial t$ and $\partial/\partial x$ the difference operators contain the extra terms $\mathcal{D}_{2t}$ and $\mathcal{D}_{2x}$. Substitution of (2.18) and (2.19) into (2.15) gives

$$i\nu\mathcal{D}_{2t} + ikU\mathcal{D}_{2x} = 0 . \tag{2.22}$$

This equation has solution

$$\nu\frac{\mathcal{D}_{2t}}{\mathcal{D}_{2x}} = -kU , \tag{2.23}$$

which may be compared to the analytical solution (2.3). Alternatively, (2.22) may be written as

$$\sin\nu\Delta t + U\frac{\Delta t}{\Delta x}\sin k\Delta x = 0 , \tag{2.24}$$

which can be solved for the frequency ν,

$$\nu = -\frac{1}{\Delta t}\arcsin\left(U\frac{\Delta t}{\Delta x}\sin k\Delta x\right) . \tag{2.25}$$

Once again, in order to have a real solution for ν and thus a propagating wave,

$$\left|U\frac{\Delta t}{\Delta x}\sin k\Delta x\right| \leq 1 \tag{2.26}$$

for all $k \leq J/2$ which is just a restatement of the stability condition (2.12) found earlier.

Figure 2.4a shows the solution of (2.25) for various values of $C = |U\Delta t/\Delta x|$, as the ratio of the frequency of the approximate solution to the true frequency, $\nu/(kU)$. The abscissa covers the range of possible wavenumbers $0 \leq k\Delta x \leq \pi$. For $C = 1$, all waves with wavenumber $k\Delta x \leq \pi/2$ (wavelength $L \geq 4\Delta x$) move with the correct frequency, but shorter waves with $\pi/2 < k\Delta x \leq \pi$ ($2\Delta x \leq L < 4\Delta x$) are retarded. The shortest $2\Delta x$ wave ($k\Delta x = \pi$) is stationary. Figure 2.4a was created by using the principal values of the arcsin, and thus $-\pi/2 \leq \nu\Delta t \leq \pi/2$. For the case $C = 1$, when $0 \leq k\Delta x \leq \pi/2$, $\nu\Delta t = -k\Delta x = -kU\Delta t$, the true value, but when $\pi/2 \leq k\Delta x \leq \pi$, $\nu\Delta t = -\pi + k\Delta x$, and thus the frequency goes to 0 as $k\Delta x$ goes to π.

In practice one cannot choose Δt so that $C = 1$ because U is unknown beforehand and generally varies over the domain. Thus Δt must be chosen so that $C < 1$ everywhere. The wind speeds in the atmosphere are 100 m s^{-1}

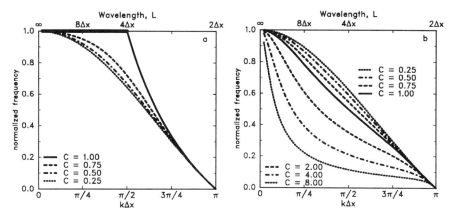

Figure 2.4. Normalized frequency (approximate divided by analytic) for (a) explicit centered difference approximations and (b) implicit centered difference approximations as a function of wavelength and wavenumber for various Courant numbers.

or more in the regions of the jet, but 25 m s^{-1} or less over much of the domain. Thus a local value of Courant number of 0.25 or less is not uncommon in atmospheric models. In the case $C < 1$, all waves are slowed down in the approximate numerical solution. As $C \to 0$, the frequencies asymptote to a curve slightly below the $C = 0.25$ line in Figure 2.4a. Thus the $C = 0.25$ provides a good indication of the worst phase errors of centered differences. The longer the waves are relative to the grid interval, the closer the phase speeds are to the correct values. Therefore in order to obtain accurate solutions with approximations (2.6) we must choose Δx small enough compared to the scale of the phenomena of interest keeping in mind the curve for $C = 0.25$ in Figure 2.4a.

2.1.3 Physical versus Computational Solutions

There is a second set of solutions to (2.25) in addition to that in Figure 2.4a. Just as the wavenumber k is limited by the spatial grid interval, $k \leq 2\pi/2\Delta x$, the frequency has an upper limit related to the time step, $|\nu\Delta t| \leq \pi$. Thus we must consider all solutions to (2.25) with $|\nu\Delta t| \leq \pi$, not just the principal branch $|\nu\Delta t| \leq \pi/2$. For a given wavenumber, if $\nu_k\Delta t$ is the principal branch solution of (2.25) presented in Figure 2.4a, then there is a second solution $[\text{sgn}(\nu_k)\pi - \nu_k\Delta t]$ where $\text{sgn}(\nu_k)$ is $+1$ if $\nu_k > 0$ and -1 if $\nu_k < 0$. This general form of the second solution is obtained by recognizing that the sine function is symmetric about $\pi/2$ and also about $-\pi/2$. The existence of two solutions is also seen in the preceding analysis where we solved for the amplitude (2.11) of the solution, and the radical enters with plus and minus signs. In the limit of long waves, $k\Delta x \to 0$, i.e. for very good spatial resolution, ρ in (2.11) goes to $+1$ and -1. The $\rho = +1$ solution corresponds to the true solution and is generally referred to as the "physical" mode. The $\rho = -1$ solution changes sign every time step giving a solution with a $2\Delta t$ oscillation. This solution, referred to as the computational mode, is an artifact of the approximations and is not related to the true solution. As discussed above, for large wavelengths the principal branch solution of (2.25) gives the physical mode $\nu_k \Delta t$, the second solution, $[\text{sgn}(\nu_k)\pi - \nu_k \Delta t]$, is the computational mode. For shorter waves, $\pi/2 < |k\Delta x| \leq \pi$, the separation is less clear as the poorly resolved space and time approximations interact and lead to cancellation of errors in some cases. In fact in the case of $C = 1$ discussed above, the so-called computational solution just happens to give the true solution. However, this serendipitous behavior does not occur in a real application with non–constant velocity. The computational mode is caused by the centered time differences as will be illustrated below.

2.1.4 *Spatial versus Temporal Discrete Response*
We can examine the time and space approximations of (2.15) separately by considering the differential equations

$$\frac{dq}{dt} + ikUq = 0 , \tag{2.27}$$

and

$$i\nu q + U\frac{dq}{dx} = 0 . \tag{2.28}$$

First consider the temporal centered difference approximation to (2.27)

$$\delta_{2t}q^n + ikUq^n = 0 , \tag{2.29}$$

and assume wavelike solutions as in (2.2). The equation for ν is the same as (2.23) with $\mathcal{D}_{2x}$ replaced by 1 and has solution

$$\nu = -\frac{1}{\Delta t} \arcsin\left(kU\Delta t\right) , \tag{2.30}$$

which implies $|kU\Delta t| \leq 1$ for real ν and those waves propagate faster than the true solution (2.3). Eqn. (2.30) also has a computational mode as discussed above. For a given wave with $|kU\Delta t| \leq 1$, (2.30) has a solution $\nu_k \Delta t$ from the principal branch and a higher frequency solution $[\text{sgn}(\nu_k)\pi - \nu_k \Delta t]$ in the range $\pi/2 < |\nu_k \Delta t| \leq \pi$. This illustrates that the centered time differencing is responsible for the introduction of the computational mode.

Now consider the spatial centered difference approximations to (2.28):

$$i\nu q_j + U\delta_{2x}q_j = 0 . \tag{2.31}$$

The equation for ν is the same as (2.23) with $\mathcal{D}_{2t}$ replaced by 1 and has solution

$$\nu = -kU\frac{\sin k\Delta x}{k\Delta x}\,. \tag{2.32}$$

Those waves that can be represented on the grid, i.e., $k\Delta x \leq \pi$, propagate slower than the true solution (2.3). As seen above, when the space and time approximations are combined, the retardation from the spatial differencing dominates the acceleration from the time differencing, and the net effect is to retard all waves except when $|U\Delta t/\Delta x| = 1$ and the two effects exactly cancel. In fact, the retardation from the spacial differencing stabilizes those waves with $1 < |kU\Delta t| \leq \pi$ that were unstable with the time differencing alone.

2.1.5 *Forward Approximations*
Eqn. (2.15) and its equivalent (2.6) look overly complicated in that the predicted values at time $n+1$ depend on those at both n and n–1. A simpler approximation can be defined involving only time $n+1$ and n as

$$\delta_t q_j^{n+1/2} + U\delta_{2x}q_j^n = 0\,, \tag{2.33}$$

where the temporal difference operator is defined by

$$\delta_t q_j^{n+1/2} = \frac{q_j^{n+1} - q_j^n}{\Delta t}\,. \tag{2.34}$$

Substitution of wavelike solutions (2.2) into the difference equation (2.33) gives an equation for the frequency:

$$e^{i\nu\Delta t} = 1 - iU\frac{\Delta t}{\Delta x}\sin k\Delta x\,. \tag{2.35}$$

For pure wave solutions (ν real), $\left|e^{i\nu\Delta t}\right| = 1$. This is only satisfied for all resolvable waves if $U\Delta t/\Delta x = 0$ and thus with the forward scheme (2.33) ν must have an imaginary component and the solution either damps or amplifies with time. In fact $\left|e^{i\nu\Delta t}\right| > 1$ for non-zero U and the solution grows with time. To see this more clearly write ν in its complex form,

$$\nu = \nu_r + i\nu_i\,. \tag{2.36}$$

Substitution into (2.35) gives

$$e^{-\nu_i\Delta t}(\cos\nu_r\Delta t + i\sin\nu_r\Delta t) = 1 - iU\frac{\Delta t}{\Delta x}\sin k\Delta x\,. \tag{2.37}$$

140

Equating the real and imaginary parts of (2.37) and taking the ratio of the resulting equations gives

$$\tan \nu_r \Delta t = -U \frac{\Delta t}{\Delta x} \sin k \Delta x . \tag{2.38}$$

Solving the real part of (2.37) for ν_i gives

$$\nu_i = -\frac{1}{\Delta t} \ln \left(\frac{1}{\cos \nu_r \Delta t} \right) , \tag{2.39}$$

since $|\nu_r \Delta t| \leq \pi/2$ from (2.38), $0 \leq \cos \nu_r \Delta t \leq 1$ and $1 \leq (1/\cos \nu_r \Delta t) \leq \infty$. It follows that ν_i is negative and the solution $e^{i \nu n \Delta t} = e^{-\nu_i n \Delta t} e^{i \nu_r n \Delta t}$ grows with time. The forward scheme (2.33) is unstable.

2.2 IMPLICIT ADVECTION

The approximations (2.6) and (2.33) are members of a class referred to as explicit. The characteristic defining explicit schemes is that the forecast at a grid point q_j^{n+1} depends only on the neighboring grid-point values at past times and not on future predicted values. Another class of approximations is referred to as implicit. The characteristic of this class is that the forecast at a grid point q_j^{n+1} depends on the forecast at neighboring grid points, e.g., q_{j+1}^{n+1} and q_{j-1}^{n+1}, as well as on the data from the past times. An example of such a scheme applied to (2.1) is

$$\frac{q_j^{n+1} - q_j^{n-1}}{2\Delta t} + U \left[\left(\frac{1+\alpha}{2} \right) \frac{q_{j+1}^{n+1} - q_{j-1}^{n+1}}{2\Delta x} + \left(\frac{1-\alpha}{2} \right) \frac{q_{j+1}^{n-1} - q_{j-1}^{n-1}}{2\Delta x} \right] = 0 \tag{2.40}$$

with $-1 \leq \alpha \leq 1$. Rewriting (2.40) with the operator notation (2.16) and (2.17) gives

$$\delta_{2t} q_j^n + U \delta_{2x} \overline{(q_j^n)}_\alpha^{2t} = 0 , \tag{2.41}$$

where the overbar denotes the time average and the subscript α indicates the weight

$$\overline{(q_j^n)}_\alpha^{2t} = \left[\left(\frac{1+\alpha}{2} \right) q_j^{n+1} + \left(\frac{1-\alpha}{2} \right) q_j^{n-1} \right] . \tag{2.42}$$

The averaging operator applied to the wave structure (2.2) can also be written in terms of a response function $\mathcal{A}_\alpha^{2t}$ as

$$\overline{\left(Q e^{i(kx+\nu t)} \right)}_\alpha^{2t} = \mathcal{A}_\alpha^{2t} Q e^{i(kx+\nu t)} , \tag{2.43}$$

where

$$\mathcal{A}_\alpha^{2t} = \cos \nu \Delta t + i\alpha \sin \nu \Delta t . \tag{2.44}$$

Eqn. (2.40) is a little more complicated to solve than (2.6) because each q_j^{n+1} is related to its unknown neighbors $(q_{j+1}^{n+1}, q_{j-1}^{n+1})$ and all values must be determined simultaneously. But it is not extremely complicated and in this case simply results in a tridiagonal matrix that must be inverted. There are standard methods for doing so efficiently. Note that (2.40) could be written to predict the values at time $n+1$ from those at n instead of $n-1$ since values at time n do not enter (2.40) as written. In fact (2.40) results in two independent interlaced sequences of values $(q^{n-1}, q^{n+1}, q^{n+3}\ldots)$ and $(q^n, q^{n+2}, q^{n+4}\ldots)$. We chose the form (2.40) here to be consistent with the centered differences (2.6) and (2.15) and with the following section. We will return to the issue of two interlaced solutions later.

Assuming solutions to (2.40) of the form (2.2) leads to an equation like (2.22) but with the time average response function included:

$$i\nu \mathcal{D}_{2t} + ikU\mathcal{A}_\alpha^{2t}\mathcal{D}_{2x} = 0, \tag{2.45}$$

or

$$\sin \nu \Delta t + U\frac{\Delta t}{\Delta x}(\cos \nu \Delta t + i\alpha \sin \nu \Delta t)\sin k\Delta x = 0. \tag{2.46}$$

Consider first the case $\alpha = -1$. This is just the same as the forward difference approximations (2.33) except over the interval $2\Delta t$ instead of Δt. Recognizing that for $\alpha = -1$, $(\cos \nu \Delta t + i\alpha \sin \nu \Delta t) = e^{-i\nu \Delta t}$ and that $\sin \nu \Delta t = -(i/2)\left(e^{i\nu \Delta t} - e^{-i\nu \Delta t}\right)$ and multiplying (2.46) by $e^{i\nu \Delta t}$ recovers (2.35) with Δt replaced by $2\Delta t$.

Now consider the case $\alpha = +1$. This scheme is referred to as backward implicit. Following algebra similar to that described above for the $\alpha = -1$ case yields

$$e^{-i\nu 2\Delta t} = 1 + iU\frac{2\Delta t}{\Delta x}\sin k\Delta x. \tag{2.47}$$

Since $\left|e^{-i\nu 2\Delta t}\right| > 1$ for non-zero U, $\left|e^{i\nu 2\Delta t}\right| < 1$ and the solution damps with time. Solving for the real and imaginary parts of ν as was done following (2.35) gives

$$\tan \nu_r 2\Delta t = -U\frac{2\Delta t}{\Delta x}\sin k\Delta x, \tag{2.48}$$

and

$$\nu_i = \frac{1}{2\Delta t}\ln\left(\frac{1}{\cos \nu_r 2\Delta t}\right). \tag{2.49}$$

In this case ν_i is positive and the solution is damped with time. Thus the scheme is stable, but an amplitude error is introduced by the damping.

Finally, consider the symmetric centered case $\alpha = 0$, commonly referred to as Crank-Nicholson. The solution of (2.46) is then

$$\nu = -\frac{1}{\Delta t}\arctan\left[U\frac{\Delta t}{\Delta x}\sin k\Delta x\right]. \tag{2.50}$$

Unlike the leapfrog scheme (2.15), which required a bound on $U\Delta t/\Delta x$ in order to have a suitable argument for the arcsin (2.26), there is no limit on $U\Delta t/\Delta x$ in (2.50) for the arctan and the scheme is stable for all choices of Δt. Note also that since ν is real, the solution is neutral, neither damping nor amplifying with time. Because of the lack of a stability restriction on Δt, (2.40) with appropriately chosen α seems to be an ideal solution of the pole problem at first glance. Consideration of the actual solution of (2.50) and thus the phase speed of the approximate solution shows this not to be the case. Figure 2.4b shows the normalized phase speed from (2.50) for a variety of Courant numbers C. As $C \to 0$ the asymptote of the frequencies is the same as that for the explicit differences, and as in that case, the $C = 0.25$ line provides a good indication of it. With the implicit scheme in Figure 2.4b, the frequencies decrease with increasing time step, away from the correct value. From Figure 2.4b, we see that the stability of the implicit approximation (2.40) is obtained at a price: the phase error is very large even for the longest waves as the Courant number exceeds 1. In Figure 2.3 we see that the Courant number near the poles would be between 4 and 8 if the mid-latitude Courant number were between 0.5 and 1.0. Figure 2.4b shows that in such a case the signal would be significantly slowed down compared to the true solution. The stability of the implicit scheme is obtained by slowing the waves down significantly, such that the scheme is not really useful (for large C) for predicting atmospheric flows in which advection is a dominant process. Thus we have not identified a satisfactory solution to the pole problem, to which we will return shortly.

2.3 TIME FILTER

Since the general form (2.40) is unstable as a forward approximation ($\alpha = -1$), damps as a backward scheme ($\alpha = +1$), and in its centered form ($\alpha = 0$) offers no time step advantage over centered approximations (2.15) for accurate solutions but is more complex to solve, it is not commonly used and many models use explicit centered time approximations (Section 2.1) for the advection. With these approximations the time computational mode can be a problem. In the linear problem used for illustration above, the computational and physical modes evolve in time independently of each other and their amplitudes depend only on the initial conditions q_j^0 and q_j^1. In more complex nonlinear problems, the nonlinear terms introduce a coupling between the computational and physical modes which often leads to a growth of the computational mode until it dominates the solution. The spurious growth of the mode is often controlled by the application of a time filter as part of the time marching process. The most common filter was first used by Robert [59] and later analyzed by Asselin [6].

Let a tilde denote the filtered value. Then the filter is defined by

$$\tilde{q}^n = q^n + \gamma \left(\tilde{q}^{n-1} - 2q^n + q^{n+1} \right) . \tag{2.51}$$

Note that this filter is recursive, in that the filtered value $\tilde{q}^n$ depends on a previously filtered value $\tilde{q}^{n-1}$. The term in parentheses looks like a discrete approximation to $\partial^2 q/\partial t^2$ and as such (2.51) resembles some sort of temporal diffusive smoothing. Approximations to the true diffusion equation will be considered in Section 2.6.

Assume the unfiltered solution is oscillatory in time

$$q^n = e^{i\nu n\Delta t} . \tag{2.52}$$

We look for solutions of the form

$$\tilde{q}^n = Q(\nu, \gamma)q^n , \tag{2.53}$$

where Q gives the modulation of the solution by the filter, and in general depends on the damping parameter γ and the frequency of the solution ν. Substitution of (2.53) into (2.51) and solving for Q gives

$$Q(\nu, \gamma) = \frac{(1-\gamma)^2 + 2\gamma^2(1-\cos\nu\Delta t)e^{i\nu\Delta t}}{(1-\gamma)^2 + 2\gamma(1-\cos\nu\Delta t)} . \tag{2.54}$$

Recall that with the discrete time grid the resolvable frequencies are bounded by $|\nu\Delta t| \leq \pi$. Consider first the shortest resolvable waves $\nu\Delta t = \pm\pi$. The response function simplifies to

$$Q(\nu, \gamma) = \frac{1-3\gamma}{1+\gamma} , \tag{2.55}$$

and these $2\Delta t$ waves are damped rapidly for γ even as small as 0.05 where Q is about 0.81. Figure 2.5 shows the damping as a function of frequency for selected values of γ. The NCAR CCM typically uses a value of $\gamma = 0.06$, thus the low–frequency waves are only marginally affected by the filter, while the high–frequency waves are well controlled. The choice is arbitrary and is generally made by observing the temporal behavior of the solution and ensuring that $2\Delta t$ waves are not noticeable in time traces of grid point values. Since Q is complex the filter also introduces a phase shift, but it is negligible for typical values of γ. The phase of Q normalized by ν is less than 1×10^{-5} for $\gamma = 0.01$ and less than 2.5×10^{-4} for $\gamma = 0.05$.

The filtered value is used in the time differencing. Thus in the centered difference operator (2.16), q_j^{n-1} is replaced by $\tilde{q}_j^{n-1}$. The recursive form (2.51) is chosen over the seemingly more straightforward form

$$\tilde{q}^n = q^n + \gamma \left(q^{n-1} - 2q^n + q^{n+1} \right) \tag{2.56}$$

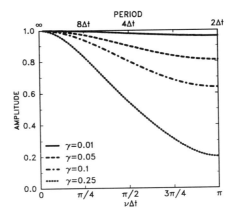

Figure 2.5. Damping by the time filter (2.51) as a function of frequency for selected values of coefficient γ.

to minimize storage. With (2.56) q^n would have to be stored along with $\tilde{q}^n$ for use the next time step to provide q^{n-1} for the filter, and the filtered value $\tilde{q}^{n-1}$ for use in the time differencing. A more complete analysis of the filter combined with centered differencing is provided by Asselin [6].

2.4 FINITE DIFFERENCE POLE PROBLEM

We return now to the basic advective stability condition

$$\left| U \frac{\Delta t}{\Delta x} \sin k\Delta x \right| \leq 1, \tag{2.57}$$

which becomes excessively restrictive near the poles where Δx becomes small. Rather than restricting Δt, a second possible way to satisfy the condition is to limit the wavenumbers k included in the solution in the polar regions so that $\sin k\Delta x$ is small enough for (2.57) to be satisfied. Such an approach requires filtering in longitude using a Fourier transform and resynthesis. The filter can be applied to the predicted variables as in the grid point model of Kalnay–Rivas et al. [34] or just to those terms in the equations responsible for the instabilities as in the model of Arakawa and Lamb [2]. The Kalnay–Rivas et al. [34] model was based on approximations that are more accurate than those presented in Section 2. They involve additional neighboring grid points and are referred to as 4th order accurate while those in Section 2 and in the Arakawa and Lamb [2] model are 2nd order accurate. See Durran [22] for a discussion of these concepts.

A stabilizing procedure related to filtering is to damp those waves which would be unstable for a chosen time step at a rate that exceeds the exponential growth. This involves application of a somewhat arbitrary longitudinal grid point filter of sufficient strength and width to overcome the exponential growth. Of course, these damped waves no longer provide an accurate solution even though they are included in the solution. Both of these approaches are somewhat arbitrary, introduce distortions into the flow, and significantly degrade the accuracy of the retained two dimensional structures (Purser [49]). The errors are introduced because the longitudinal structures are manipulated independently of the adjacent latitudinal structures. No account is taken of the appropriate latitudinal structures that should be maintained to be consistent with the overall two dimensional structures near the poles. We will discuss this further after the spectral transform method has been discussed.

A third method that has been attempted to satisfy (2.57) is to increase $\Delta\lambda$ approaching the poles so that Δx remains constant, i.e., instead of keeping $\Delta\lambda$ constant as in Figure 1.1a, $\Delta x = \Delta\lambda\cos\varphi$ is held constant. This type of grid is shown in Figure 1.1b. Unfortunately, the atmosphere also contains structures, such as Rossby waves, that have a longitudinal wave-like structure $A(\varphi)e^{i(k\lambda+\omega t)}$ and propagate with the same longitudinal phase speed at all latitudes. In this case the angular propagation velocity is the same at each latitude, and an effective Courant number is more like $C = \omega\Delta t/\Delta\lambda$ where ω is a constant propagation velocity in degrees longitude per second, rather than meters per second as described earlier. These coherent waves are important in the atmosphere and must be solved for accurately. They tend to have rather long wavelengths, so the longitudinal filtering described above does not affect them. That filtering is applied to shorter waves. In addition, they are representable with the longer $\Delta\lambda$ associated with the type of grid in Figure 1.1b. However, if the longitudinal grid interval $\Delta\lambda$ increases approaching the poles so that the distance $\Delta x = \Delta\lambda\cos\varphi$ is reasonably constant, then $C = \omega\Delta t/\Delta\lambda$ decreases with increasing latitude. From Figure 2.4a we see that this will introduce a differential phase error as a function of latitude that leads to a latitudinal tilt in the wave structures (Kreiss and Oliger [38]). Such a tilt will produce spurious poleward transport of heat and momentum. The Rossby waves in the example above can be represented and predicted in terms of scalar variables only using the barotropic vorticity equation. Additional errors are introduced with the reduced grid when the equations include vector variables and are expressed in terms of spherical vector components. These errors arise from the large rotational variation in the spherical unit vectors across the long longitudinal grid interval of the reduced grid near the poles. This will also be discussed further after the spectral transform method is discussed.

2.5 SEMI–IMPLICIT APPROXIMATIONS

Although implicit schemes are not useful for solving the advection pole problem, they are useful for some components of the complete equations of atmospheric models, and one component consists of gravity waves. Consider the linearized system of equations in a rotating framework, and assume the independent variables are taken as uniform in the y direction:

$$\frac{\partial u}{\partial t} + U\frac{\partial u}{\partial x} - fv + g\frac{\partial h}{\partial x} = 0, \qquad (2.58)$$

$$\frac{\partial v}{\partial t} + U\frac{\partial v}{\partial x} + fu = 0, \qquad (2.59)$$

$$\frac{\partial h}{\partial t} + U\frac{\partial h}{\partial x} + H\frac{\partial u}{\partial x} = 0. \qquad (2.60)$$

In these equations, referred to as the shallow water equations (illustrated in Chapter 5), (u, v) denote the velocity components in the (x, y) directions, h is the height of the surface of the water above some mean height H, and U is a constant advecting velocity in the x direction. The Coriolis parameter f associated with the rotating system is taken as constant here, thus making the intrinsic phase speed of Rossby waves zero. Assume solutions of the form

$$\begin{pmatrix} u \\ v \\ h \end{pmatrix} = \begin{pmatrix} u_0 \\ v_0 \\ h_0 \end{pmatrix} e^{i(kx + \nu t)}. \qquad (2.61)$$

Substitution of (2.61) into (2.58)–(2.60) gives

$$(\nu + kU)\left[(\nu + kU)^2 - (gHk^2 + f^2)\right] = 0, \qquad (2.62)$$

which has three solutions

$$\nu = -kU, \qquad (2.63)$$

$$\nu = -k\left[U \pm \sqrt{gH + \frac{f^2}{k^2}}\right]. \qquad (2.64)$$

The first solution (2.63) is just pure advection (corresponding to Rossby waves on an f–plane) as seen earlier in (2.3). The other two solutions (2.64) are inertial-gravity waves.

If we apply the centered difference operators (2.16) and (2.17) to the terms in (2.58)–(2.60),

$$\delta_{2t}u_j^n + U\delta_{2x}u_j^n - fv_j^n + g\delta_{2x}h_j^n = 0, \qquad (2.65)$$

$$\delta_{2t}v_j^n + U\delta_{2x}v_j^n + fu_j^n = 0, \qquad (2.66)$$

$$\delta_{2t}h_j^n + U\delta_{2x}h_j^n + H\delta_{2x}u_j^n = 0,\tag{2.67}$$

the equation for the frequencies of the approximate solutions has the same form
as that for the continuous solutions (2.62)

$$\left(\sin\nu\Delta t + U\frac{\Delta t}{\Delta x}\sin k\Delta x\right)\times$$

$$\left[\left(\sin\nu\Delta t + U\frac{\Delta t}{\Delta x}\sin k\Delta x\right)^2 - (\Delta t)^2\left(gH\left(\frac{\sin k\Delta x}{\Delta x}\right)^2 + f^2\right)\right] = 0.\tag{2.68}$$

Hence the solutions also have the same form (2.63) and (2.64) but with the
response functions included

$$\sin\nu\Delta t = -U\frac{\Delta t}{\Delta x}\sin k\Delta x,\tag{2.69}$$

$$\sin\nu\Delta t = -\left[U \pm \sqrt{gH + f^2\frac{\Delta x^2}{(\sin k\Delta x)^2}}\right]\frac{\Delta t}{\Delta x}\sin k\Delta x.\tag{2.70}$$

Equation (2.69) is the same as (2.24) discussed above and gives the stability
condition (2.26), while (2.70) is very similar to (2.69), but with U replaced by
$U \pm \sqrt{gH + f^2\Delta x^2/(\sin k\Delta x)^2}$ and will thus add another stability condition for
the explicit centered system:

$$\left|U \pm \sqrt{gH + f^2\frac{\Delta x^2}{(\sin k\Delta x)^2}}\right|\frac{\Delta t}{\Delta x}\sin k\Delta x \le 1.\tag{2.71}$$

For atmospheric-type flows the advecting velocity U is the order of 10 to 100
m s^{-1}. The size of H is determined by the nature of the fluid or gas in the
problem. For the atmosphere H is around 10 km and $\sqrt{gH}$ is around 300 m
s^{-1}, over three times as large as the maximum U. The magnitude of $\sqrt{gH}$
introduces a severe burden to the calculation for this more complex case. Since
in (2.71) the maximum propagation speed is 400 m s^{-1}, Δt must be at least
4 times smaller than for the simple advection case (2.12), where the maximum
propagation speed is 100 m/sec. Because of this time step alone a forecast with
the complete equations would cost four times as much as with just the advection
equations.

In the large scale global atmospheric models the high-frequency motions associated with inertial-gravity waves and thus the terms (2.64) and (2.71) with
frequency $-k\left[U \pm \sqrt{gH + f^2/k^2}\right]$ are of very small amplitude compared to the
advection component and do not influence the forecast in any significant way.
Even though they are unimportant, they are present in the system of equations

148

and any small perturbations would grow unboundedly and spoil the solution if the time step did not satisfy (2.71). However, because they are unimportant, they can be treated implicitly to decrease their frequencies and allow longer time steps without ruining the forecast. Such an approach, introduced by Robert [60], is referred to as semi-implicit because only some of the terms, rather than all terms in the equations, are approximated implicitly. The semi-implicit approximations for (2.58)–(2.60) are

$$\delta_{2t} u_j^n + U\delta_{2x} u_j^n - f v_j^n + g\delta_{2x} \overline{(h_j^n)}_\alpha^{2t} = 0,\qquad(2.72)$$

$$\delta_{2t} v_j^n + U\delta_{2x} v_j^n + f u_j^n = 0,\qquad(2.73)$$

$$\delta_{2t} h_j^n + U\delta_{2x} h_j^n + H\delta_{2x} \overline{(u_j^n)}_\alpha^{2t} = 0.\qquad(2.74)$$

Note that we have included the general off-centering in (2.72) and (2.74). Only the terms that contribute to the high-frequency components are treated implicitly. The advection terms remain explicit. Again, assuming solutions of the form (2.61) gives an equation like (2.68) but with the time average response function (2.44) included in the terms responsible for the gravity waves:

$$\left(\sin\nu\Delta t + U\frac{\Delta t}{\Delta x}\sin k\Delta x\right) \times$$

$$\left[\left(\sin\nu\Delta t + U\frac{\Delta t}{\Delta x}\sin k\Delta x\right)^2\right.\qquad(2.75)$$

$$\left. - (\Delta t)^2\left(gH\left(\cos\nu\Delta t + i\alpha\sin\nu\Delta t\right)^2\left(\frac{\sin k\Delta x}{\Delta x}\right)^2 + f^2\right)\right] = 0.$$

The solution for the advected wave is

$$\sin\nu\Delta t = -U\frac{\Delta t}{\Delta x}\sin k\Delta x,\qquad(2.76)$$

which is unaffected by the off-centered time averaging and identical to that from the explicit approximations (2.69). The inertial-gravity waves are affected by the time averaging. In the case $\alpha = 0$,

$$\sin\nu\Delta t = -\left[U \pm \sqrt{gH\left(\cos\nu\Delta t\right)^2 + f^2\frac{\Delta x^2}{(\sin k\Delta x)^2}}\right]\frac{\Delta t}{\Delta x}\sin k\Delta x.\qquad(2.77)$$

With the centered averaging, compared to (2.70) the equation for the inertial-gravity wave frequencies (2.77) has an additional $(\cos\nu\Delta t)^2$ factor multiplying gH, which effectively reduces its value. If we consider the case of no rotation

$(f = 0)$ and zero mean wind $(U = 0)$, this factor just acts to slow the frequency exactly as in the case with implicit advection in (2.46). When $0 < \alpha < 1$, the inertial-gravity waves are damped as well as retarded. Experience has shown that semi-implicit approximations such as (2.72)–(2.74) with time steps bounded by (2.76) alone produce large-scale forecasts that are as accurate as those from an explicit, centered approximation with a time step at least four times shorter. Thus semi-implicit approximations are commonly adopted in atmospheric models. Although the gravity wave components of semi-implicit schemes are stable for long time steps, the advective component continues to impose a restriction on Δt through (2.76). Thus the "pole problem" remains for these approximations.

The system (2.72)–(2.74) is solved by substituting $h_{j\pm1}^{n+1}$ from (2.74) into (2.72). First, isolate the unknowns in (2.72) and (2.74):

$$u_j^{n+1} + \left(\frac{1+\alpha}{2}\right) g\Delta t \left(h_{j+1}^{n+1} - h_{j-1}^{n+1}\right) = RU_j\,, \tag{2.78}$$

$$h_j^{n+1} + \left(\frac{1+\alpha}{2}\right) H\Delta t \left(u_{j+1}^{n+1} - u_{j-1}^{n+1}\right) = RH_j\,, \tag{2.79}$$

where the terms on the right hand side, RU_j and RH_j involve known values at time n and n–1,

$$RU_j = u_j^{n-1} - 2\Delta t U \delta_{2x} u_j^n + 2\Delta t f v_j^n - g\Delta t (1-\alpha)\,\delta_{2x}\left(h_j^{n-1}\right)\,, \tag{2.80}$$

$$RH_j = h_j^{n-1} - 2\Delta t U \delta_{2x} h_j^n - H\Delta t (1-\alpha)\,\delta_{2x}\left(u_j^{n-1}\right)\,. \tag{2.81}$$

After combining (2.78) and (2.79) we get

$$u_j^{n+1} - \left(\frac{1+\alpha}{2}\right)^2 gH\Delta t^2 \left(u_{j+2}^{n+1} + u_{j-2}^{n+1}\right) =$$
$$RU_j - \left(\frac{1+\alpha}{2}\right) g\Delta t(RH_{j+1} - RH_{j-1})\,, \tag{2.82}$$

which, since u_j^{n+1} depends on u_{j+2}^{n+1} and u_{j-2}^{n+1} and not on u_{j+1}^{n+1} and u_{j-1}^{n+1}, is just two interlaced systems of equations for even and odd j, or two sets of tridiagonal systems. Once u_j^{n+1} are known, h_j^{n+1} can be determined from (2.74) and v_j^{n+1} from (2.73). When more than one space dimension is involved, this approach leads to a Helmholtz elliptic problem that is more difficult to solve than the tridiagonal system arising in the one-dimensional system above.

150

2.6 DIFFUSION APPROXIMATIONS

In addition to advection and wave-like motions, AGCMs include components of a different nature that behave like damping and diffusive processes rather than like propagating waves. These include horizontal diffusion, which is often included in models to mimic the cascade of energy or enstrophy to smaller unresolved scales or simply to control small scale noise, vertical diffusion as a component in the planetary boundary layer parameterization, linear drag as part of the surface exchange, and although less obvious, even the radiative heating and cooling. While most of these processes are included to directly model physical processes that are not resolved by the model grid and result in diffusive-like equations, the horizontal diffusion is much more arbitrary, sometimes being included to suppress spatial grid-scale noise and instabilities introduced by aliasing. Although, as will be seen, spectral transform models do not suffer from the aliasing problem, they still require a process to prevent the accumulation of energy in the smallest resolved scales. The actual form of diffusion in models is often arbitrarily taken as a ∇^2, ∇^4 or even ∇^8 term with the coefficient chosen empirically such that the kinetic energy spectrum does not change shape near the truncation limit, e.g. Boville [12]. There are also instances where turbulence concepts are used to design the form of the horizontal diffusion, e.g. Boer *et al.* [10].

The linear drag equation is

$$\frac{\partial q}{\partial t} = -Kq\,, \tag{2.83}$$

and second order or ∇^2 type diffusion in one spatial dimension is

$$\frac{\partial q}{\partial t} = K\frac{\partial^2 q}{\partial x^2}\,. \tag{2.84}$$

Alternatively, higher order diffusions such as ∇^4 or ∇^8 are often included in models. We can consider all cases by treating the following equation

$$\frac{\partial q}{\partial t} = -(-1)^\ell K\frac{\partial^{2\ell} q}{\partial x^{2\ell}}\,, \tag{2.85}$$

with $\ell = 0$ for (2.83) and $\ell = 1$ for (2.84). In all cases we take $K > 0$. Note, the $(-1)^\ell$ is included in (2.85) to counteract the $(i^2)^\ell = (-1)^\ell$ introduced by the "2ℓ"th derivative of wavelike structures e^{ikx} and provides a damping of such structures. Since we expect solutions that damp with time, we assume solutions of the form

$$q(x,t) = Qe^{-\mu t + ikx}\,, \tag{2.86}$$

rather than the propagating wave form (2.2). Substitution into (2.85) gives

$$\mu = k^{2\ell}K \tag{2.87}$$

for the analytic solution, which is therefore always damped.

To be consistent with the stable centered difference advection approximations (2.6) and (2.15), we first consider centered approximations to (2.85)

$$\delta_{2t}q_j^n = -(-1)^\ell K \delta_x^{2l} q_j^n , \qquad (2.88)$$

where the discrete diffusion operators are given by

$$\delta_x^0 q_j^n = q_j^n \qquad (2.89)$$

for $\ell = 0$,

$$\delta_x^2 q_j^n = \frac{q_{j+1}^n - 2q_j^n + q_{j-1}^n}{\Delta x^2} \qquad (2.90)$$

for $\ell = 1$, and so on. Assume solutions of the form

$$q(x_j, n\Delta t) = Qe^{-\mu n\Delta t} e^{ik(j-1)\Delta x} , \qquad (2.91)$$

which is just the form (2.7) with $\rho = e^{-\mu\Delta t}$, or (2.2) with $\mu = -i\nu$. Rewriting the centered time difference operator (2.18) and its response function (2.20) in terms of μ instead of ν gives

$$\delta_{2t}Qe^{-\mu t} = -\mu \mathcal{D}_{2t}Qe^{-\mu t} , \qquad (2.92)$$

and

$$\mathcal{D}_{2t} = -\left(\frac{e^{-\mu\Delta t} - e^{\mu\Delta t}}{2\mu\Delta t} \right) . \qquad (2.93)$$

This form of $\mathcal{D}_{2t}$ will be more convenient then (2.20) for the analysis that follows since we are dealing here with damped solutions rather than propagating waves. The response function of the discrete diffusion operator is

$$\delta_x^{2\ell} Qe^{ikx} = (-1)^\ell k^{2\ell} \mathcal{D}_x^{2\ell} Qe^{ikx} , \qquad (2.94)$$

where

$$\mathcal{D}_x^0 = 1 , \qquad (2.95)$$

and

$$\mathcal{D}_x^2 = \frac{2(1 - \cos k\Delta x)}{k^2 \Delta x^2} = \frac{\sin^2 (k\Delta x/2)}{(k\Delta x/2)^2} . \qquad (2.96)$$

Thus the centered difference approximations (2.88) give

$$\mu \mathcal{D}_{2t} = K k^{2\ell} \mathcal{D}_x^{2\ell} . \qquad (2.97)$$

Substituting the time difference response function but retaining the general space form gives

$$e^{-2\mu\Delta t} + e^{-\mu\Delta t} K 2\Delta t k^{2\ell} \mathcal{D}_x^{2\ell} - 1 = 0, \tag{2.98}$$

which has solutions

$$e^{-\mu\Delta t} = -K\Delta t k^{2\ell} \mathcal{D}_x^{2\ell} \pm \sqrt{\left(K\Delta t k^{2\ell} \mathcal{D}_x^{2\ell}\right)^2 + 1}. \tag{2.99}$$

Since the response functions $\mathcal{D}_x^{2\ell}$ are real and greater than zero for all resolvable wavenumbers $1 \leq k \leq J/2$ (at least for the two operators explicitly defined above, (2.95) and (2.96)), the amplitude $\left|e^{-\mu\Delta t}\right| > 1$ for the negative root and the centered approximations give solutions which grow with time and are unstable.

Now consider the general implicit form using times $n-1$ and $n+1$. Note, we consider these time levels instead of the more natural n and $n+1$ as we anticipate that the diffusion approximations are likely to be coupled with centered advection approximations. However, the analysis is essentially the same for both cases with the difference being the appearance of Δt or $2\Delta t$ in the formulas. Instead of (2.88) we have

$$\delta_{2t} q_j^n = -(-1)^{\ell} K \delta_x^{2l} \overline{(q_j^n)}_\alpha^{2t}, \tag{2.100}$$

where the time averaging was defined in (2.42). In terms of the response functions (2.100) is

$$\mu \mathcal{D}_{2t} = K k^{2\ell} \mathcal{A}_\alpha^{2t} \mathcal{D}_x^{2\ell}, \tag{2.101}$$

where $\mathcal{A}_\alpha^{2t}$ is as (2.44). Rather than (2.44), however, it is more convenient here to write the response function as

$$\mathcal{A}_\alpha^{2t} = \left[\left(\frac{1+\alpha}{2}\right) e^{-\mu\Delta t} + \left(\frac{1-\alpha}{2}\right) e^{+\mu\Delta t} \right], \tag{2.102}$$

in which case (2.101) becomes

$$e^{-\mu 2\Delta t} = \frac{1 - (1-\alpha) K\Delta t k^{2\ell} \mathcal{D}_x^{2\ell}}{1 + (1+\alpha) K\Delta t k^{2\ell} \mathcal{D}_x^{2\ell}}. \tag{2.103}$$

We consider three cases, forward explicit case ($\alpha = -1$)

$$e^{-\mu 2\Delta t} = 1 - K 2\Delta t k^{2\ell} \mathcal{D}_x^{2\ell}, \tag{2.104}$$

Crank-Nicholson ($\alpha = 0$)

$$e^{-\mu 2\Delta t} = \frac{1 - K\Delta t k^{2\ell} \mathcal{D}_x^{2\ell}}{1 + K\Delta t k^{2\ell} \mathcal{D}_x^{2\ell}}, \tag{2.105}$$

and backward implicit case ($\alpha = +1$)

$$e^{-\mu 2\Delta t} = \frac{1}{1 + K2\Delta t k^{2\ell} \mathcal{D}_x^{2\ell}} . \tag{2.106}$$

For the schemes to be stable,

$$\left| e^{-\mu 2\Delta t} \right| \leq 1 . \tag{2.107}$$

For the Crank-Nicholson (2.105) and implicit (2.106) cases, since the $\mathcal{D}_x^{2\ell}$ are real and greater than zero for all resolvable wavenumbers, (2.107) is satisfied for both cases and the methods are stable for all (positive) Δt. For the explicit case (2.104), (2.107) implies

$$\left| 1 - K2\Delta t k^{2\ell} \mathcal{D}_x^{2\ell} \right| \leq 1 \tag{2.108}$$

and the method is stable for a limited range of Δt. Consider first the linear drag case ($\ell = 0$) for which (2.108), from the forward approximations, becomes

$$K\Delta t \leq 1 . \tag{2.109}$$

For the ∇^2 diffusion case ($\ell = 1$), (2.108) from the forward approximations becomes

$$0 \leq K \frac{2\Delta t}{\Delta x^2} (1 - \cos k\Delta x) \leq 1 , \tag{2.110}$$

or equivalently

$$0 \leq K k^2 \Delta t \frac{\sin^2 (k\Delta x/2)}{(k\Delta x/2)^2} \leq 1 . \tag{2.111}$$

Since $\cos k\Delta x$ runs from 1 to -1 for the resolvable range of k, i.e. $|k\Delta x| \leq \pi$, the condition that all waves be stable is

$$K \frac{2\Delta t}{\Delta x^2} \leq \frac{1}{2} . \tag{2.112}$$

Although conditions (2.109) and (2.112) ensure that the linear drag and ∇^2 diffusion cases are stable and do not grow exponentially with the forward approximations, they do not ensure that the schemes provide an accurate solution or even a qualitatively reasonable solution. For example, in the forward difference linear drag case

$$q(x,t) = Q e^{-\mu t + ikx} = Q \left(e^{-\mu \Delta t} \right)^{\frac{n}{2}} e^{ikx} = Q e^{ikx} (1 - 2\Delta t K)^{\frac{n}{2}} . \tag{2.113}$$

If $\frac{1}{2} < K\Delta t \leq 1$, then $-1 \leq (1 - 2\Delta t K) < 0$ and although the solution damps, it changes sign every other step in the cycle $n = 0, 2, 4 \ldots$ whereas the true solution damps without changing sign from (2.86) and (2.87). Note that the cycle involves every other n because we chose to write the approximations (2.100) involving $n-1$ and $n+1$ rather than n and $n+1$. Similarly, for the diffusion case:

$$q(x,t) = Qe^{ikx}\left[1 - K\frac{2\Delta t}{\Delta x^2}2(1 - \cos k\Delta x)\right]^{\frac{n}{2}}. \tag{2.114}$$

If the largest permissible time step, $K(2\Delta t/\Delta x^2) = 1/2$ is used,

$$q(x,t) = Qe^{ikx}\left(\cos k\Delta x\right)^{\frac{n}{2}} \tag{2.115}$$

and the waves with $\pi/2 < k\Delta x \le \pi$ oscillate in amplitude.

Oscillatory solutions can also arise from the Crank-Nicholson approximations. The solution for the frequency from the forward approximations (2.104) is

$$\mu = -\frac{1}{2\Delta t}\ln\left(1 - K2\Delta t k^{2\ell}\mathcal{D}_x^{2\ell}\right) \tag{2.116}$$

and that for the Crank-Nicholson approximations (2.105) is

$$\mu = -\frac{1}{2\Delta t}\ln\left(\frac{1 - K\Delta t k^{2\ell}\mathcal{D}_x^{2\ell}}{1 + K\Delta t k^{2\ell}\mathcal{D}_x^{2\ell}}\right). \tag{2.117}$$

In both cases, if Δt is large enough, the argument of the logarithm becomes negative and there are no real solutions. Again we must consider complex frequencies $\mu = \mu_r + i\mu_i$. In the following we detail the forward approximations (2.104). The corresponding analysis for the Crank-Nicholson (2.105) is then obvious. Equation (2.104) can be written

$$e^{-(\mu_r+i\mu_i)2\Delta t} = 1 - K2\Delta t k^{2\ell}\mathcal{D}_x^{2\ell} \tag{2.118}$$

or

$$e^{-\mu_r 2\Delta t}\left(\cos \mu_i 2\Delta t - i\sin \mu_i 2\Delta t\right) = 1 - K2\Delta t k^{2\ell}\mathcal{D}_x^{2\ell}. \tag{2.119}$$

The ratio of the imaginary and real components of (2.119) gives

$$\tan \mu_i 2\Delta t = 0 \tag{2.120}$$

and the real part of (2.119) gives

$$\mu_r 2\Delta t = -\ln\left(\frac{1 - K2\Delta t k^{2\ell}\mathcal{D}_x^{2\ell}}{\cos \mu_i 2\Delta t}\right). \tag{2.121}$$

The solutions of (2.120) for $\mu_i 2\Delta t$ in the admissible frequency range for the discrete time grid are 0 and π. Thus if the numerator in (2.121) is greater than or equal to 0, we take $\mu_i = 0$, and if the numerator is less than 0, $\mu_i 2\Delta t = \pi$ in order to have a non-negative argument for the logarithm. In the latter case the solution oscillates in time. The same situation arises with the Crank-Nicholson approximations, except Δt is double that of the forward approximations when the numerator becomes negative. In addition, the Crank-Nicholson has an another term in the denominator $\left(1 + K\Delta t k^{2\ell} \mathcal{D}_x^{2\ell}\right)$ that provides additional damping of the oscillatory solution. Nevertheless, to have realistic solutions the time step must be significantly shorter than implied by the stability conditions for both the forward approximations and the Crank-Nicholson approximations.

Condition (2.112) also looks problematic when going to high resolutions since Δt must decrease as Δx^2 to maintain stability if K is constant. However, in practice horizontal diffusion is included in models to mimic the transfer of energy out of the system into smaller unresolved waves, rather than as a diffusive large scale process. The scales to be affected by the diffusion are smaller at higher resolution. Therefore K is usually decreased when going to higher resolution so that the damping rate of the shortest resolved wave is the same at the different resolutions.

Vertical diffusion is a different story. It is a component of the planetary boundary layer parameterization and the coefficient K depends on the physical situation rather than the grid resolution. Here the more restrictive Δt at higher vertical resolution could become a problem and the implicit form is often used to avoid a stability restriction.

2.7 COMBINED ADVECTION DIFFUSION APPROXIMATIONS

The complete model contains advection and diffusion combined, rather than as independent equations. Therefore we consider the mixed equation combining (2.1) and (2.85):

$$\frac{\partial q}{\partial t} + U\frac{\partial q}{\partial x} = -(-1)^{\ell} K \frac{\partial^{2\ell} q}{\partial x^{2\ell}}. \tag{2.122}$$

Consider the combination of the two approximations that are stable for advection alone (2.15) and for diffusion alone (2.88):

$$\delta_{2t} q_j^n + U\delta_{2x} q_j^n = -(-1)^{\ell} K \delta_x^{2\ell} \overline{(q_j^n)}_\alpha^{2t}. \tag{2.123}$$

Again, substitute wavelike structures (2.2) and use the general response function notation for the spatial differences (2.19) and (2.94) but not the time difference. Eliminating the common factor gives

$$\frac{e^{+i\nu\Delta t} - e^{-i\nu\Delta t}}{2\Delta t} + ikU\mathcal{D}_{2x}$$

$$+ Kk^{2\ell}\mathcal{D}_x^{2\ell}\left[\left(\frac{1+\alpha}{2}\right)e^{+i\nu\Delta t} + \left(\frac{1-\alpha}{2}\right)e^{-i\nu\Delta t}\right] = 0. \tag{2.124}$$

Multiplication by $e^{i\nu\Delta t}$ and combining like terms gives

$$\left(e^{i\nu\Delta t}\right)^2 \left[1 + (1+\alpha)\,K\Delta t k^{2\ell}\mathcal{D}_x^{2\ell}\right] + i2\Delta t k U \mathcal{D}_{2x}\left(e^{i\nu\Delta t}\right)$$
$$- \left[1 - (1-\alpha)\,K\Delta t k^{2\ell}\mathcal{D}_x^{2\ell}\right] = 0\,, \tag{2.125}$$

with solution

$$e^{i\nu\Delta t} = \frac{-i\Delta t k U \mathcal{D}_{2x}}{\left[1 + (1+\alpha)\,K\Delta t k^{2\ell}\mathcal{D}_x^{2\ell}\right]}$$
$$\pm \frac{\left\{-\left(\Delta t k U \mathcal{D}_{2x}\right)^2 + \left[1 + (1+\alpha)\,K\Delta t k^{2\ell}\mathcal{D}_x^{2\ell}\right]\left[1 - (1-\alpha)\,K\Delta t k^{2\ell}\mathcal{D}_x^{2\ell}\right]\right\}^{\frac{1}{2}}}{\left[1 + (1+\alpha)\,K\Delta t k^{2\ell}\mathcal{D}_x^{2\ell}\right]}\,. \tag{2.126}$$

For the quantity in braces to be greater than zero, i.e.,

$$\left(\Delta t k U \mathcal{D}_{2x}\right)^2 \le \left[1 + (1+\alpha)\,K\Delta t k^{2\ell}\mathcal{D}_x^{2\ell}\right]\left[1 - (1-\alpha)\,K\Delta t k^{2\ell}\mathcal{D}_x^{2\ell}\right]\,, \tag{2.127}$$

then

$$\left|e^{i\nu\Delta t}\right|^2 = \frac{1 - (1-\alpha)\,K\Delta t k^{2\ell}\mathcal{D}_x^{2\ell}}{1 + (1+\alpha)\,K\Delta t k^{2\ell}\mathcal{D}_x^{2\ell}}\,, \tag{2.128}$$

which is just the condition we had earlier for diffusion alone (2.103). Thus the time step has one limiting condition which is the same as from the pure diffusion approximations. In the case of no diffusion $K = 0$, (2.127) is just the stability condition for centered differences (2.26).

Consider the case with the forward explicit approximation for the diffusion $(\alpha = -1)$. Eqn. (2.127) becomes

$$\left(\Delta t k U \mathcal{D}_{2x}\right)^2 \le 1 - 2K\Delta t k^{2\ell}\mathcal{D}_x^{2\ell}\,. \tag{2.129}$$

Since $K > 0$, this imposes a more severe restriction on the time step than with pure advection (2.26). For the Crank-Nicholson $(\alpha = 0)$,

$$\left(\Delta t k U \mathcal{D}_{2x}\right)^2 \le \left[1 - K\Delta t k^{2\ell}\mathcal{D}_x^{2\ell}\right]\left[1 + K\Delta t k^{2\ell}\mathcal{D}_x^{2\ell}\right] = 1 - \left(K\Delta t k^{2\ell}\mathcal{D}_x^{2\ell}\right)^2\,, \tag{2.130}$$

which is not as restrictive as (2.129) but still more restrictive than advection alone (2.26). For example the right side of (2.129) goes to zero for $K\Delta t k^{2\ell}\mathcal{D}_x^{2\ell} = \frac{1}{2}$ while that of (2.130) goes to zero for $K\Delta t k^{2\ell}\mathcal{D}_x^{2\ell} = 1$. Finally consider the backward implicit case $(\alpha = +1)$:

$$\left(\Delta t k U \mathcal{D}_{2x}\right)^2 \le 1 + 2K\Delta t k^{2\ell}\mathcal{D}_x^{2\ell}\,. \tag{2.131}$$

This in fact expands the range of stable Δt. The damping from the implicit diffusion helps control the amplification from the advection when $U\Delta t/\Delta x > 1$.

2.8 TIME SPLIT APPROXIMATIONS

Although the implicit diffusion coupled with centered advection is no more restrictive than either alone, when coupled with semi-implicit approximations as well the implicit system becomes more complex and expensive to solve, especially in multiple dimensions. An additional complication is that the parameterizations involving diffusion are awkard to couple directly with the advection. Therefore another approach is generally adopted to separate advective from diffusive processes in models. This separation is referred to as time splitting and is illustrated in this section.

A time split approximation to (2.122) is written by dividing (2.123) into two steps with an intermediate value q_j^*

$$\frac{q_j^* - q_j^{n-1}}{2\Delta t} = -U\delta_{2x}q_j^n , \tag{2.132}$$

$$\frac{q_j^{n+1} - q_j^*}{2\Delta t} = -(-1)^\ell K\delta_x^{2\ell}\left[\left(\frac{1+\alpha}{2}\right)q_j^{n+1} + \left(\frac{1-\alpha}{2}\right)q_j^*\right] . \tag{2.133}$$

In (2.132) the intermediate value q_j^* takes the place of q_j^{n+1} in (2.15) and in (2.133) of q_j^{n-1} in (2.100). Substituting wavelike structures (2.2) into (2.132) and (2.133) for q_j^{n+1} and q_j^{n-1}, but not for q_j^* whose time value is not well defined, continuing to use the response functions (2.21) and (2.94) and defining

$$Q_j^* = q_j^* Q^{-1}e^{-i[k(j-1)\Delta x+\nu n\Delta t]} \tag{2.134}$$

gives from (2.132)

$$Q_j^* = e^{-i\nu\Delta t} - i2\Delta t U k \mathcal{D}_{2x} \tag{2.135}$$

and from (2.133)

$$e^{i\nu\Delta t}\left[1 + (1+\alpha)K\Delta t k^{2\ell}\mathcal{D}_x^{2\ell}\right] = \left[1 - (1-\alpha)K\Delta t k^{2\ell}\mathcal{D}_x^{2\ell}\right]Q_j^* . \tag{2.136}$$

Substituting (2.135) for Q_j^* into (2.136) and multiplying by $e^{i\nu\Delta t}$ gives

$$\begin{aligned}
&\left(e^{i\nu\Delta t}\right)^2\left[1 + (1+\alpha)K\Delta t k^{2\ell}\mathcal{D}_x^{2\ell}\right] \\
&\quad - \left[1 - (1-\alpha)K\Delta t k^{2\ell}\mathcal{D}_x^{2\ell}\right]\left[1 - i2\Delta t k U \mathcal{D}_{2x}\left(e^{i\nu\Delta t}\right)\right] = 0 ,
\end{aligned} \tag{2.137}$$

with solution

158

$$e^{i\nu\Delta t}\left[1+(1+\alpha)\,K\Delta tk^{2\ell}\mathcal{D}_x^{2\ell}\right]=-i\Delta tkU\mathcal{D}_{2x}\left[1-(1-\alpha)\,K\Delta tk^{2\ell}\mathcal{D}_x^{2\ell}\right]$$

$$\pm\left\{-(\Delta tkU\mathcal{D}_{2x})^2\left[1-(1-\alpha)\,K\Delta tk^{2\ell}\mathcal{D}_x^{2\ell}\right]^2\right.$$
$$\left.+\left[1+(1+\alpha)\,K\Delta tk^{2\ell}\mathcal{D}_x^{2\ell}\right]\left[1-(1-\alpha)\,K\Delta tk^{2\ell}\mathcal{D}_x^{2\ell}\right]\right\}^{\frac{1}{2}}. \qquad (2.138)$$

For the quantity in braces to be greater than zero,

$$(\Delta tkU\mathcal{D}_{2x})^2\leq\frac{1+(1+\alpha)\,K\Delta tk^{2\ell}\mathcal{D}_x^{2\ell}}{1-(1-\alpha)\,K\Delta tk^{2\ell}\mathcal{D}_x^{2\ell}}, \qquad (2.139)$$

in which case

$$\left|e^{i\nu\Delta t}\right|^2=\frac{1-(1-\alpha)\,K\Delta tk^{2\ell}\mathcal{D}_x^{2\ell}}{1+(1+\alpha)\,K\Delta tk^{2\ell}\mathcal{D}_x^{2\ell}}, \qquad (2.140)$$

which is just the same condition as (2.103) for the diffusion alone. The condition associated with advection (2.139) is less restrictive than any of the mixed cases (2.129)–(2.131) and even less restrictive than for advection alone (2.26), again because the diffusion provides some damping. Of course, as discussed above, using a time step near the limit of stability does not necessarily provide accurate solutions.

3. Spectral Method

We have just seen that finite difference approximations to spatial derivatives introduce errors that can contaminate numerical solutions if the grid is not fine enough. It is in fact possible to avoid the truncation errors associated with spatial finite differences by approximating fields analytically. This is possible when fields are expressed as Fourier series rather than as grid point values. Consider the one-dimensional case where the series can be written as

$$q(x_j,t)=\sum_{k=-K}^{K}q_k(t)e^{ikx_j}, \qquad (3.1)$$

in which, to ensure real solutions, $q_{-k}=q_k^*$, where the $*$ denotes the complex conjugate. This transformation is exact for continuous functions in the limit that K tends to infinity. In practice however the spectral truncation K will be finite and a finite number of grid points are used.

In this one-dimensional discrete case the spectral representation $q_k(t)$ $k=-K\ldots K$ contains the same information as the grid point representation $q(x_j,t)$, $j=1\ldots J$ provided $K=J/2$ for even J or $K=(J-1)/2$ for odd J.

Data on the grid can be transformed to spectral coefficients and back to grid point data without loss of information. The transform method takes advantage of the best of both representations. Since the basis functions are orthogonal over the discrete grid, i.e.,

$$\frac{1}{J} \sum_{j=1}^{J} e^{ikx_j} e^{-ilx_j} = \delta_{kl} , \qquad (3.2)$$

where the Kronecker delta δ_{kl} is 1 if $k = l$, and 0 if $k \neq l$, the spectral coefficients q_k are obtained by multiplying (3.1) by e^{-ikx_j} and summing over the J grid points:

$$\frac{1}{J} \sum_{j=1}^{J} q(x_j, t) e^{-ikx_j} = q_k(t) . \qquad (3.3)$$

As seen in (3.3), q_0 is just the average of the grid point values. Spatial derivatives can be calculated as analytic derivatives of the individual terms of the series and thus are extremely accurate with the spectral method. The solution is determined by the time evolution of the amplitudes q_k of the terms of the series. Substituting the series (3.1) into the advective equation (2.1) with $S = 0$ and assuming centered time differences as in (2.6) or (2.40), but analytically differentiating the terms of the series for the x derivatives, gives

$$\frac{1}{2\Delta t} \sum_{k=-K}^{K} (q_k^{n+1} - q_k^{n-1}) e^{ikx_j} + U \sum_{k=-K}^{K} ikq_k^{n} e^{ikx_j} = 0 . \qquad (3.4)$$

The solution for the coefficients q_k^{n+1}, $k = -K, \ldots, K$, must be found in terms of the coefficients at time n and $n - 1$, q_k^{n} and q_k^{n-1}. The solution is obtained for each wavenumber by multiplying (3.4) by e^{-ikx_j} and summing over the x_j from j equal 1 to J. In the linear case (3.4) with constant velocity U this operation is trivial and each wavenumber is independent of all the others:

$$\frac{1}{2\Delta t} (q_k^{n+1} - q_k^{n-1}) + Uikq_k^{n} = 0 . \qquad (3.5)$$

This prognostic equation can be analyzed as was done in the previous section by again assuming a wave-like solution in time, $q_k^{n} = e^{i\nu_k n \Delta t}$, giving

$$\frac{i \sin \nu_k \Delta t}{\Delta t} + ikU = 0 \qquad (3.6)$$

or

$$\nu_k = -\frac{1}{\Delta t} \arcsin(kU\Delta t) , \qquad (3.7)$$

160

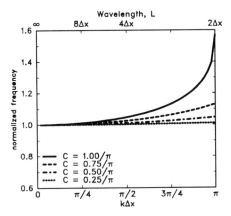

Figure 3.1. Normalized frequency (approximate divided by analytic) for spectral transform
method as a function of wavelength and wavenumber for various Courant numbers.

which is similar to the centered difference case (2.25) but with k in place of
$(\sin k\Delta x)/\Delta x$. In order to have wave-like solutions

$$|kU\Delta t| \leq 1 .$$ (3.8)

Relating the maximum k to the grid on which the calculation is done gives
$k \leq J/2$ with a nominal grid interval $\Delta x = 2\pi/J$. Thus

$$\left| U\frac{\Delta t}{\Delta x} \right| \leq \frac{1}{\pi} ,$$ (3.9)

which is a factor of π more restrictive than that for the centered differences
(2.12).

Although the time step is more restricted with the spectral method than with
centered differences, the solution is more accurate for a given wavenumber and
fewer waves need to be retained in the solution for comparable accuracy, leading
to a coarser grid and larger Δx. Figure 3.1 shows the frequency ν_k from (3.7).
In the spectral case the waves actually move faster than the true speed. This
is an effect of the centered time differences as was discussed in Section 2.1.4.
The acceleration effect of the centered time differences is also present in the
grid point scheme (2.6) but it is overwhelmed by the retardation introduced by
the spatial differences as was discussed following (2.32). The acceleration seen in
Figure 3.1 with the spectral method is significantly smaller than the deceleration
in Figure 2.4a from the centered differences. This also explains why the phase
error is smaller with larger time steps in the centered difference scheme.

In the previous example the advecting wind was assumed constant. Let us
now turn to the more realistic case of a nonlinear equation such as (2.1) with

$S = 0$, for the case where U is also a predicted quantity or linear equations with variable coefficients where U varies with x. In both cases U is also represented by a Fourier series and the problem becomes

$$\frac{1}{2\Delta t} \sum_{k=-K}^{K} (q_k^{n+1} - q_k^{n-1})e^{ikx_j} + \sum_{l=-K}^{K} U_l^n e^{ilx_j} \sum_{m=-K}^{K} imq_m^n e^{imx_j} = 0. \quad (3.10)$$

Again, the solution for one coefficient q_k^{n+1} is found by multiplying (3.10) by e^{-ikx_j} and summing over the x_j

$$\frac{1}{2\Delta t}(q_k^{n+1} - q_k^{n-1}) + \sum_{l+m=k}^{K} U_l^n imq_m^n = 0. \quad (3.11)$$

The orthogonality of the functions (3.2) yields series of quadratic terms for the right-hand term in each of which $l + m = k$. The form (3.11) provides the forecast of a spectral coefficient in terms of all spectral coefficients.

3.1 SPECTRAL TRANSFORM METHOD

Eqn. (3.11) is relatively straightforward because of the quadratic nature of the nonlinear term being approximated. For more complex nonlinear equations, especially the parameterizations in atmospheric models, such a closed form is next to impossible to write down. The spectral transform method provides an alternative approach in which the nonlinear terms are calculated in grid point space and transformed back to spectral space. The most common numerical method employed in global atmospheric models today is the spectral transform method. Machenhauer [41] provides a thorough review of this approach. In this method, at various phases of a forecast time step the predicted variables are represented by grid point values or by a series of linearly independent spectral functions. The method proceeds as follows: given spectral coefficients of dependent variables q_k and U_k, grid point values of U are synthesized by (3.1), and grid point values of the derivative $\partial q/\partial x$ are obtained from

$$\frac{\partial q(x_j, n\Delta t)}{\partial x} = \sum_{k=-K}^{K} ikq_k^n e^{ikx_j}, \quad (3.12)$$

which is obtained by analytically differentiating (3.1). The product of $U(x_j, n\Delta t)$ and $\partial q(x_j, n\Delta t)/\partial x$ is calculated on the grid and then transformed back to spectral space by (3.3). Thus the right-hand term of (3.11) is replaced by

$$\frac{\partial q_k^n}{\partial t} = -\left(U\frac{\partial q}{\partial x}\right)_k^n = -\frac{1}{J}\sum_{j=1}^{J}\left[U(x_j, n\Delta t)\frac{\partial q(x_j, n\Delta t)}{\partial x}\right]e^{-ikx_j}. \quad (3.13)$$

Since this representation is also associated with an underlying grid such as (2.4) and (2.5), the waves included are limited to be consistent with the grid and $K \leq J/2$.

3.2 ALIASING

The advantages of the spectral transform method become more apparent when it is applied to nonlinear problems in which U in (3.5) is also a function of x and is a predicted variable. Aliasing is the effect of a small scale wave being misrepresented as a large scale wave because it is sampled on a discrete set of points. This is the effect described earlier in (2.8). Aliasing of quadratic terms can be eliminated in a straightforward manner with the spectral transform method. Thus, in (3.10) and (3.11), if in the right-hand term $(l + m) > J/2$, that combination will appear erroneously as a wave with $k < J/2$. Aside from being inaccurate, such aliasing can also lead to a blow up of the solution through a computational phenomenon known as nonlinear instability.

In the spectral method aliasing can be easily avoided. The truncation limit K in (3.1) is chosen smaller than the maximum allowed by the underlying grid, $J/2$. The limit K is chosen so that the waves arising from the product of two waves which would otherwise be aliased fall in the range from K to $J/2$ and are truncated out of the system. This restricted K also allows a longer stable time step from (3.8) than determined in (3.9). In practice quadratic terms constitute the dominant nonlinearity of dynamical equations and the truncation limit K to prevent aliasing of quadratic terms in spectral transform models is $(J-1)/3$.

Finite difference schemes also suffer from aliasing. In that case there is no simple solution such as that described above for the spectral transform method. Finite difference approximations require very careful and complex formulation to avoid the problem (Arakawa [1]), or strong diffusion to damp the aliased waves as mentioned in Section 2.6.

3.3 SPHERICAL HARMONICS

Another advantage of the spectral approach is that it provides a natural solution to the spherical "pole problem". In the two dimensional spherical case the natural representation is in terms of spherical harmonics. These functions are obtained by the product of Fourier series in longitude, and associated Legendre functions P_n^m in latitude:

$$q(\lambda, \varphi, t) = \sum_{n=0}^{N} \sum_{m=-n}^{n} q_n^m(t) Y_n^m(\lambda, \mu), \qquad (3.14)$$

$$Y_n^m(\lambda, \mu) = P_n^m(\mu) e^{im\lambda}, \qquad (3.15)$$

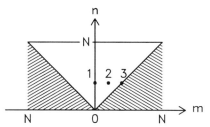

Figure 3.2. Indication of triangular truncation in the (m,n) plane. Points labeled 1, 2 and 3 indicate spherical harmonics illustrated in Figure 3.3.

where

$$\mu = \sin\varphi. \qquad (3.16)$$

It has become customary to use m for the longitudinal wavenumber when dealing with spherical harmonics rather than the k usually used with Fourier series. We will adopt that notation here as well. In three dimensional atmospheric models k is often used for the vertical grid index, a convention we will also adopt in later sections. Note also that n is customarily used to indicate the degree of the associated Legendre function. The index n is also often used to denote the time step as we have done earlier. We will continue to use n for both. It should not lead to excessive confusion as to which is meant in any particular context since the time step index appears as a superscript, and the index for the degree of the associated Legendre function appears as a subscript.

The truncation in the (m,n) plane implied by (3.14) is illustrated in Figure 3.2. The coefficients retained in the expansion fall within the white triangle, leading to this type of truncation being referred to as *triangular*. Triangular truncation provides uniform resolution over the sphere and hence has become the most common truncation in use today. Spherical harmonics with the same n have the same two-dimensional wavenumber as will be seen later in the response function of the spherical Laplacian operator (3.26). Thus n can be considered as representing the scale of the spherical harmonic. Three spherical harmonics with the same n are shown in Figure 3.3. They are (left to right) $(m,n) = (0,6)$, $(3,6)$ and $(6,6)$ and their locations in the triangular truncation are indicated in Figure 3.2 by 1,2 and 3, respectively. The harmonics with $|m| = n$ (3 in Figure 3.2) take the shape of orange wedges, those with $m = 0$ (1 in Figure 3.2) are more like pineapple slices, and those in between with $n > |m|$ have more of a checkerboard pattern. The harmonic $(m,n) = (0,0)$ is a constant and represents the mean over the sphere.

The latitudinal structures $P_n^m(\mu)$ have the useful property that, for large longitudinal wavenumber m (small zonal wavelength), their amplitude is essentially zero near the poles. A condition such as (3.9) or (2.12) for individual waves does not have to be satisfied in longitude for the larger m's near the poles since these

Figure 3.3. Spherical harmonics (left to right) $(m,n) = (0,6)$, $(3,6)$ and $(6,6)$, indicated by 1, 2 and 3 in Figure 3.2.

shorter longitude structures are not present in the spectral representation of the solution. The time step is not excessively restricted by the small longitudinal grid distance near the poles; instead the more natural limit associated with mid-latitude flow holds. Figure 3.4 shows the latitudinal structure of selected $P_n^m(\mu)$ for a few wavenumbers, m, to illustrate that the amplitudes are indeed zero near the poles for the larger wavenumbers m.

The coefficients of the spectral representation (3.14) are given by

$$q_n^m = \int_{-1}^{1} \frac{1}{2\pi} \int_0^{2\pi} q(\lambda,\mu) e^{-im\lambda} d\lambda P_n^m(\mu) d\mu \,. \tag{3.17}$$

The inner integral represents a Fourier transform,

$$q^m(\mu) = \frac{1}{2\pi} \int_0^{2\pi} q(\lambda,\mu) e^{-im\lambda} d\lambda \,, \tag{3.18}$$

which can be calculated without error as described earlier with (3.3) if the grid and truncation are chosen consistently. In practical models the transform is usually performed by a Fast Fourier Transform (FFT) subroutine. The outer integral is performed via Gaussian quadrature,

$$q_n^m = \sum_{j=1}^{J} q^m(\mu_j) P_n^m(\mu_j) w_j \,, \tag{3.19}$$

where μ_j denotes the Gaussian grid points in the meridional direction, w_j the Gaussian weight at latitude μ_j, and J the number of Gaussian grid points from pole to pole. Associated Legendre functions enjoy a property similar to Fourier

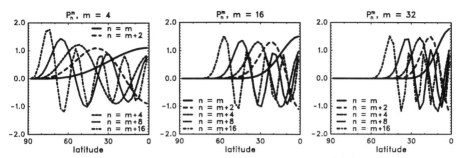

Figure 3.4. Latitudinal structure of selected $P_n^k(\mu)$.

functions in that a grid and associated integration weights can be found with which the outer integral of (3.17) is approximated without error by (3.19) for polynomial integrands less than a certain degree, and a quadratically unaliased transform may be calculated as was described for Fourier transforms above (see Machenhauer [41] for details.) The Gaussian grid points (μ_j) are given by the J roots of the Legendre polynomial $P_J(\mu)$, and the corresponding weights are given by

$$ w_j = \frac{2(1 - \mu_j^2)}{\left[J \, P_{J-1}(\mu_j) \right]^2} \, . \tag{3.20} $$

The weights themselves satisfy

$$ \sum_{j=1}^{J} w_j = 2 \, . \tag{3.21} $$

The description of the spectral method on the sphere would not be complete without providing relevant equations for horizontal derivatives. In spherical coordinates the horizontal gradient of a scalar q is

$$ \nabla q = \frac{1}{a} \left(\frac{1}{\cos\varphi} \frac{\partial q}{\partial\lambda} \hat{\mathbf{i}} \, , \, \frac{\partial q}{\partial\varphi} \hat{\mathbf{j}} \right) \, , \tag{3.22} $$

where a is the radius of the earth and $\hat{\mathbf{i}}$ and $\hat{\mathbf{j}}$ are the spherical unit vectors in the λ and φ directions, respectively. The Laplacian of a scalar q is

$$ \nabla^2 q = \frac{1}{a^2} \left[\frac{1}{\cos^2\varphi} \frac{\partial^2 q}{\partial\lambda^2} + \frac{1}{\cos\varphi} \frac{\partial}{\partial\varphi} \left(\cos\varphi \frac{\partial q}{\partial\varphi} \right) \right] \, . \tag{3.23} $$

Because of the properties of Fourier and Legendre functions, the derivatives of spherical harmonics are as follows

$$ \frac{\partial Y_n^m}{\partial\lambda} = im Y_n^m \, , \tag{3.24} $$

$$\cos\varphi\frac{\partial Y_n^m}{\partial\varphi} = (n+1)\epsilon_n^m Y_{n-1}^m - n\epsilon_{n+1}^m Y_{n+1}^m, \tag{3.25}$$

$$\nabla^2 Y_n^m = \frac{-n(n+1)}{a^2} Y_n^m, \tag{3.26}$$

with

$$\epsilon_n^m = \left(\frac{n^2-m^2}{4n^4-1}\right)^{\frac{1}{2}}. \tag{3.27}$$

The fact that spherical harmonics are the eigenfunctions of the Laplacian operator on a sphere (3.26) results in a significant simplification of the solution of the Helmholtz problem associated with the use of semi-implicit marching scheme. In fact the spectral method combined with a semi-implicit linearization about a reference atmosphere converts a three-dimensional elliptic problem into a set of independent one-dimensional problems in the vertical for each spectral component. This will be seen later in Section 9.3.7 when the details of the NCAR CCM are presented. The eigenfunction property also significantly simplifies the implementation of horizontal diffusion, allowing high order diffusion ($\ell > 1$ in spherical equivalent of (2.85)) at no extra cost. It also allows the use of other formulations such as Leith's diffusion (e.g. Boer *et al.* [10].

3.4 SHALLOW WATER EXAMPLE

Most spectral transform models follow the approach devised by Bourke [11] (mentioned in Chapter 1) to minimize the number of transforms and thus the overall cost of the model. This method avoids the explicit calculation of many derivatives at grid points that would otherwise have to be done via Gaussian quadrature and the above formulas (3.24)–(3.26). Instead, the equations are written in such a way that the derivatives are effectively obtained as part of the transformation from grid point to spectral space. This approach uses the vorticity and divergence prognostic equations, which have scalars for prognostic variables as is appropriate for the spherical harmonic spectral representation (3.14) described above. In contrast, vector velocity components are multi-valued at the pole and would require the use of vector spherical harmonics. Browning *et al.* [13] compare these two approaches, and provide additional references. Temperton [69] has shown that the two approaches are algebraically equivalent.

The vorticity and divergence are related to the vector velocity components by

$$\zeta = \frac{1}{a\cos\varphi}\left[\frac{\partial v}{\partial\lambda} - \frac{\partial}{\partial\varphi}(u\cos\varphi)\right], \tag{3.28}$$

$$\delta = \frac{1}{a\cos\varphi}\left[\frac{\partial u}{\partial\lambda} + \frac{\partial}{\partial\varphi}(v\cos\varphi)\right], \tag{3.29}$$

where ζ is vorticity and δ is divergence. The shallow water equations in vorticity-divergence form are

$$\frac{\partial \zeta}{\partial t} = -\frac{1}{a\cos^2\varphi}\frac{\partial}{\partial\lambda}[(\zeta + f)U] - \frac{1}{a\cos\varphi}\frac{\partial}{\partial\varphi}[(\zeta + f)V], \qquad (3.30)$$

$$\frac{\partial \delta}{\partial t} = \frac{1}{a\cos^2\varphi}\frac{\partial}{\partial\lambda}[(\zeta + f)V] - \frac{1}{a\cos\varphi}\frac{\partial}{\partial\varphi}[(\zeta + f)U] - \nabla^2\left[gh + \frac{U^2 + V^2}{2\cos^2\varphi}\right],$$
$$(3.31)$$

$$\frac{\partial h}{\partial t} = -\frac{1}{a\cos^2\varphi}\frac{\partial}{\partial\lambda}(hU) - \frac{1}{a\cos\varphi}\frac{\partial}{\partial\varphi}(hV), \qquad (3.32)$$

(See Williamson *et al.* [74] for a summary of a wide variety of forms of the shallow water equations in spherical geometry.) The Laplacian ∇^2 is given by (3.23). The wind vector components that remain in the equations are written as

$$U = u\cos\varphi \qquad (3.33)$$
$$V = v\cos\varphi \qquad (3.34)$$

to remove the multi-valued pole value and make them suitable for representation by the spherical harmonics (3.14).

An explicit approximation to (3.30)–(3.32) is

$$\zeta^{n+1} = \zeta^{n-1} - \frac{2\Delta t}{a\cos^2\varphi}\frac{\partial}{\partial\lambda}[(\zeta^n + f)U^n] - \frac{2\Delta t}{a\cos\varphi}\frac{\partial}{\partial\varphi}[(\zeta^n + f)V^n], \quad (3.35)$$

$$\delta^{n+1} = \delta^{n-1} + \frac{2\Delta t}{a\cos^2\varphi}\frac{\partial}{\partial\lambda}[(\zeta^n + f)V^n] - \frac{2\Delta t}{a\cos\varphi}\frac{\partial}{\partial\varphi}[(\zeta^n + f)U^n]$$
$$-2\Delta t\nabla^2\left[gh^n + \frac{(U^n)^2 + (V^n)^2}{2\cos^2\varphi}\right], \qquad (3.36)$$

$$h^{n+1} = h^{n-1} - \frac{2\Delta t}{a\cos^2\varphi}\frac{\partial}{\partial\lambda}(h^nU^n) - \frac{2\Delta t}{a\cos\varphi}\frac{\partial}{\partial\varphi}(h^nV^n). \qquad (3.37)$$

The forecast starts with spectral coefficients ζ_n^m, δ_n^m, and h_n^m at time n. (Be careful to recall the two uses of n here, one as a subscript for the degree of the associated Legendre function as in (3.14) and the other as a superscript for the time index as defined originally following (2.6).) Grid values of $\zeta_{i,j}^n$, $\delta_{i,j}^n$ and $h_{i,j}^n$ are calculated at time n from the coefficients using (3.14). The grid point velocity components $U_{i,j}^n$ and $V_{i,j}^n$ can also be calculated for time n directly from the spectral coefficients ζ_n^m and δ_n^m. The required time $(n-1)$ grid point values, $\zeta_{i,j}^{n-1}$, $\delta_{i,j}^{n-1}$ and $h_{i,j}^{n-1}$, were saved from the previous time step.

The terms on the right-hand sides of (3.35)–(3.37) that are to be differentiated can be calculated at the grid points. Eqns. (3.35)–(3.37) are then transformed back to spectral space to give the spectral coefficients of ζ_n^m, δ_n^m and

h_n^m at time $n+1$. The undifferentiated terms ζ^{n-1}, δ^{n-1} and h^{n-1} are transformed directly via (3.19). The longitudinally differentiated term is handled by integration by parts, using the cyclic boundary conditions. For example, the Fourier coefficients (equivalent of 3.18) at each latitude for the longitudinally differentiated term in the divergence equation (3.36) are determined by

$$\left\{ \frac{\partial}{\partial \lambda} [(\zeta + f)V] \right\}^m = \frac{1}{2\pi} \int_o^{2\pi} \frac{\partial [(\zeta + f)V]}{\partial \lambda} e^{-im\lambda} d\lambda$$

$$= im \frac{1}{2\pi} \int_o^{2\pi} [(\zeta + f)V] e^{-im\lambda} d\lambda \,, \tag{3.38}$$

so that the Fourier transform is performed first, then the differentiation is carried out in spectral space. The transformation to spherical harmonic space then follows (3.19):

$$\left\{ \frac{\partial}{\partial \lambda} [(\zeta + f)V] \right\}_n^m = im \sum_{j=1}^J [(\zeta + f)V]_j^m \, P_n^m(\mu_j) w_j \,, \tag{3.39}$$

where $[(\zeta + f)V]_j^m$ is the Fourier coefficient of $[(\zeta + f)V]$ with wavenumber m at the Gaussian grid latitude μ_j.

The latitudinally differentiated term in the divergence equation is also handled by integration by parts using zero boundary conditions at the poles

$$\left\{ \frac{\partial}{\partial \mu} [(\zeta + f)U] \right\}_n^m = \int_{-1}^1 \frac{\partial}{\partial \mu} [(\zeta + f)U]^m \, P_n^m d\mu = -\int_{-1}^1 [(\zeta + f)U]^m \frac{dP_n^m}{d\mu} d\mu \,. \tag{3.40}$$

Defining the derivative of the associated Legendre polynomial by

$$H_n^m = (1 - \mu^2) \frac{dP_n^m}{d\mu} \,, \tag{3.41}$$

(3.40) can be written

$$\left\{ \frac{\partial}{\partial \mu} [(\zeta + f)U] \right\}_n^m = -\sum_{j=1}^J [(\zeta + f)U]_j^m \frac{H_n^m(\mu_j)}{(1 - \mu_j^2)} w_j \,. \tag{3.42}$$

The derivative H_n^m is obtained from the relationship (3.25).

Similarly, the ∇^2 operator in the divergence equation can be converted to spectral space by sequential integration by parts and then application of the relationship (3.26) to each spherical harmonic function individually so that

$$\left\{ \nabla^2 \left[gh + \frac{(U)^2 + (V)^2}{2\cos^2 \varphi} \right] \right\}_n^m = \frac{-n(n+1)}{a^2} \sum_{j=1}^J \left[gh + \frac{(U)^2 + (V)^2}{2\cos^2 \varphi} \right]_j^m P_n^m(\mu_j) w_j \tag{3.43}$$

where $[\quad]_j^m$ denotes the Fourier coefficient of $\left[gh + \frac{(U)^2+(V)^2}{2\cos^2\varphi}\right]$, the original grid variable.

3.5 FURTHER CONSIDERATIONS

Although quadratic terms constitute the dominant nonlinearity of dynamical equations, there are some higher order terms that are also present. One such term is the pressure gradient force in the horizontal momentum equations. In the terrain following sigma coordinate (and see later in (9.2) and (9.3) for the hybrid equivalent) this term contains a component of the form $(RT/p_s)\nabla p_s$. Because of the $1/p_s$ factor, this term will be aliased if it is evaluated on a transform grid that only satisfies the relationship for a quadratic unaliased grid, $K = (J-1)/3$. Increasing the grid dimension J while holding the truncation K fixed would degrade the efficiency of models. Alternatively the pressure gradient term can be rearranged as $RT\nabla \ln p_s$ which is now quadratic in the variables T and $\ln p_s$. However, a finite series in the variable $\ln p_s$ does not guarantee global mass conservation, which is a problem for very long term integrations. Experience has shown that the increased aliasing with the p_s form leads to unacceptably noisy fields and $\ln p_s$ has become the standard mass variable in spectral models. However, since conservation is not guaranteed in that system, some *ad hoc* mass correction is usually applied to force mass conservation in very long climate simulations.

Because the spectral transform method is based on consistent two dimensional structures it does not suffer from the same errors as those introduced into grid point schemes by polar longitudinal Fourier filtering, or by the increased longitudinal grid interval of the reduced grid. Although the amplitudes of the high wavenumber Fourier components included in the representation go to zero approaching the poles, adjacent latitudes are not independent and the approach to zero is consistent with two-dimensional structures. In fact, the truncation implied by the summation limits in the representation (3.14) provides uniform resolution on the sphere. This type of truncation is generally referred to as triangular because the coefficients involved in the expansion fill a triangle in Legendre (m, n) space.

Purser [49] introduced an alternate polar filter for grid point models that provides the higher wavenumber structures normally removed by the traditional filter, by interpolation from surrounding grid circles. It effectively uses the local two dimensional structure to provide the needed information. He shows that this procedure provides more accurate zonal differences near the poles and is stable. It has not, however, been implemented in global grid point models. Purser [50] suggests a similar approach of interpolating from adjacent latitudes for use with the reduced grid as in Figure 1.1b. Such interpolation provides substantial improvement in the accuracy of the zonal differences. It also improves the meridional differences at rows farther from the poles by providing more accurate interpolated values needed for the meridional finite difference operators. Longitudinal interpolation alone, even of very high order, is not adequate to provide these values and maintain overall high accuracy. Purser [50] also shows how to account for the longitudinal variation of the curvilinear coordinate unit vectors in the vicinity of the pole, making his approach applicable to spherical vector component equations as well as scalar equations. The latitudinal aspect can also be easily included.

Because it is an accurate discretization technique, the spectral method provides more effective resolution per degree of freedom than finite difference methods. This is because the response of finite difference discretization operators degrades substantially for scales approaching grid length as was seen in Section 2. But this accuracy comes at a cost. While the computational cost of finite difference schemes increases linearly with the number of grid points, the number of operations to perform spectral transforms increases quadratically, as can be seen from (3.1) and (3.3). In the case of Fourier transforms in longitude, fast Fourier transforms (FFT) are more efficient and their computational cost increases as $K \ln K$ rather than K^2 for the direct summations in (3.1) and (3.3). Until recently Legendre transforms in the meridional direction were performed by summation and hence their cost escalated rapidly with increased resolution. Recent work by Healy *et al.* [28] indicates the possibility of achieving computationally efficient and numerically reliable fast spherical harmonic expansions with cost increasing as $K^2 \ln^2 K$. However, the crossover point at which these fast transforms become more efficient than direct summation is close to the high truncations used in Numerical Weather Prediction models, far above the truncations normally used in climate models.

Although the spectral transform method has many advantages and seems ideal for spherical domains, it also has some serious disadvantages. The method is formally equivalent to a least square approximation and minimizes the mean square error over the global domain. This implies that the size of the error is likely to be the same everywhere, which is not a problem for a field such as temperature where, say, a one degree error is no more serious in the polar regions than in equatorial regions. However, it is a serious problem for a field such as water vapor, usually treated as mixing ratio in atmospheric models. At the Earth's surface the mixing ratio varies from close to 20 g/kg in the equatorial regions to 1 g/kg or less in the polar regions. An error of 2 g/kg may only be 10% in equatorial regions and have little effect, but such an error in polar regions completely changes the character of the field. It can even make the mixing ratio negative which, of course, is physically impossible. Yet just such a case commonly occurs in spectral models (Rasch and Williamson [52]). In addition, the error can lead to significantly larger values than should be present and create spurious supersaturation which in turn leads to spurious precipitation as illustrated in Williamson [70]. Some attempts have been made to replace mixing ratio in the spectral representation by nonlinear functions of mixing ratio (such as for example dew-point depression) but other problems arise such as the absence of conservation of total mass and the approach remains rather unsatisfactory (Laprise [40]). In addition, some such functions that involve a saturation value are only appropriate for water vapor but not for chemical constituents, which are now being included in global models.

A related problem with spectral methods is the so-called Gibbs phenomenon. In regions of strong gradients of the transformed field, a truncated series representation invariably leads to fictitious ripples (short scale oscillations) whose amplitudes decrease very slowly with increased resolution. For water vapor near frontal zones, such a phenomenon can result in locally negative values or fictitious supersaturation. In regions of abrupt topographic height changes, such as the west side of the Andes, Gibbs phenomenon results in non-zero sea level heights for example. This error can have important repercussions when interacting with physical parameterizations. Attempts have been made recently to modify the spectrum of topography to minimize Gibbs ripples (Holzer [31]). An "optimal" topography is obtained by adjusting the spectral coefficients to minimize a user-defined global "cost" function.

4. Semi-Lagrangian Method

Semi-Lagrangian methods were originally developed for application in numerical weather prediction because they offered economical advantages of long time steps without loss of forecast accuracy. They were also introduced into spectral transform global general circulation models for water vapor and chemical constituent transport to address the problem of spectral ringing or Gibbs phenomenon which was responsible for unrealistic constituent distributions and for spurious precipitation in the polar regions as discussed above in Section 3.5. Staniforth and Côté [66] provide an excellent review of the semi-Lagrangian. The method is based on the Lagrange form of the equations rather than the Euler form. For the one-dimensional case, instead of (2.1), the approximations are based on the pair of equations

$$\frac{dq}{dt} = S(x,t), \tag{4.1}$$

$$\frac{dx}{dt} = U(x,t). \tag{4.2}$$

Eqn. (4.1) states that the change in q along a trajectory is due solely to sources and sinks, denoted S. The trajectory is given by (4.2). Once again we seek a forecast at a specified set of points $\{x_j\}$. For each point x_j we approximate the trajectory that would arrive at that point at time $(n+1)\Delta t$ using the wind U along the trajectory. We refer to the point from which the trajectory departed at time $n\Delta t$ as the departure point and denote it with a subscript D.

The forecast at time $(n+1)$ for q is determined by an approximation to the integral form of (4.1):

$$q_j^{n+1} - q_{j_D}^n = \int_{(x_{j_D},t^n)}^{(x_j,t^{n+1})} S(x,t)ds \, , \tag{4.3}$$

where the integral is taken along the trajectory, denoted ds. The departure point itself is determined by integrating (4.2) back along the trajectory,

$$x_j - x_{j_D} = \int_{(x_{j_D},t^n)}^{(x_j,t^{n+1})} U(x,t)ds \, . \tag{4.4}$$

To determine the basic properties of the semi-Lagrangian method we first consider the one-dimensional case with constant wind and no sources or sinks ($S = 0$). Bates and McDonald [7] considered the stability properties of the semi-Lagrangian method with linear and quadratic interpolants used to determine the value at the departure point. Since we assume the source S is zero, the analytic solution is given by (2.2) and (2.3) just as in the Eulerian case (2.1). In the semi-Lagrangian case q is constant along the trajectory and the forecast value is the value at the departure point

$$q_j^{n+1} = q_{j_D}^n \, . \tag{4.5}$$

Since the wind is assumed to be uniform the trajectory is given by

$$x_{j_D} = x_j - U\Delta t \, . \tag{4.6}$$

For the purpose of analysis assume $U > 0$. The case $U < 0$ can be treated in a similar fashion. Let p denote the integer off-set to the grid interval in which the departure point falls

$$x_{j-p} \leq x_{j_D} < x_{j-p+1} \, . \tag{4.7}$$

Figure 4.1a illustrates the geometry. Denote the fraction of the grid interval between the departure point and the first grid point to its left by β

$$\beta = \frac{x_{j_D} - x_{j-p}}{\Delta x} \, . \tag{4.8}$$

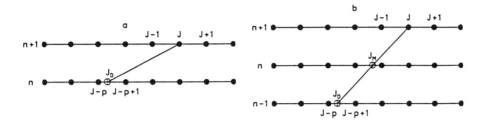

Figure 4.1. Illustration of the geometry of (a) two-time-level semi-Lagrangian trajectories and (b) three-time-level.

By this definition, $0 \leq \beta < 1$. The reference grid point is related to the arrival point by

$$x_{j-p} = x_j - (U\Delta t + \beta \Delta x) , \qquad (4.9)$$

and

$$\frac{p\Delta x}{\Delta t} = U + \beta \frac{\Delta x}{\Delta t} , \qquad (4.10)$$

which will be useful later.

4.1 LINEAR INTERPOLATION

Approximate the value at the departure point by linear interpolation between x_{j-p} and x_{j-p+1}:

$$q_j^{n+1} = q_{j_D} = (1 - \beta)q_{j-p} + \beta q_{j-p+1} . \qquad (4.11)$$

Substituting wavelike solutions (2.2), and factoring out the common factor $e^{i(\nu n\Delta t + k(j-p)\Delta x)}$ gives

$$e^{i(\nu \Delta t + kp\Delta x)} = \left[(1 - \beta) + \beta e^{ik\Delta x}\right] = 1 - \beta(1 - \cos k\Delta x) + i\beta \sin k\Delta x . \quad (4.12)$$

Consider the real and imaginary components of ν as done earlier in (2.36). The square of the amplitude of the solution is given by

$$\begin{aligned} \left|e^{-\nu_i \Delta t}\right|^2 &= \left|(1 - \beta) + \beta(\cos k\Delta x + i \sin k\Delta x)\right|^2 \\ &= 1 - 2\beta(1 - \beta)(1 - \cos k\Delta x) . \end{aligned} \qquad (4.13)$$

Since $0 \leq \beta < 1$, $\left|e^{-\nu_i \Delta t}\right| \leq 1$ and the approximation is stable for any wind speed U and Δt, but does have an amplitude error if $\beta \neq 0$, due to the damping implied by (4.13).

Now consider the phase speed of the solution. Taking the ratio of the imaginary to the real components of (4.12) gives

$$\tan(\nu_r \Delta t + kp\Delta x) = \frac{\beta \sin k\Delta x}{1 - \beta(1 - \cos k\Delta x)}, \tag{4.14}$$

or with (4.10)

$$\nu_r = -kU - k\beta \frac{\Delta x}{\Delta t} + \frac{1}{\Delta t} \arctan\left[\frac{\beta \sin k\Delta x}{1 - \beta(1 - \cos k\Delta x)}\right]. \tag{4.15}$$

Consider first the special case $\beta = 0$ where the departure point coincides with a grid point. In this case, $\nu_r = -kU$ and $\nu_i = 0$; there is no phase or amplitude error and the approximate solution recovers the analytic solution. This is also the case in the limit $\Delta x \to 0$ with Δt constant.

Like the implicit approximations (2.41) with $\alpha = 0$, the semi-Lagrangian method does not have a time step restriction for stability. However, unlike the implicit case, the semi-Lagrangian method does not suffer from excessive phase speed errors with long time steps. Eqn. (4.15) shows that the component over an integer number of grid intervals is correct, and the phase error is introduced by the component within the last grid interval.

Returning now to the amplitude (4.13), consider the worst case when the departure point falls at the center of the grid interval, $\beta = \frac{1}{2}$. Then the amplitude of the smallest wave, $2\Delta x$, is reduced to zero in one time step and that of the $4\Delta x$ is reduced to $\sqrt{1/2} = 0.707$ of its previous value in one time step. This damping is too extreme for this scheme to be useful for atmospheric modeling. The amplitude error arises from the linear interpolation and is reduced by using higher order interpolation.

4.2 QUADRATIC INTERPOLATION

Consider quadratic polynomial interpolation. In this case let p denote the integer off-set to the grid point closest to the departure point, and again let β denote the fraction of the grid interval between the departure point and x_{j-p} as in (4.8). By the definition of p,

$$-\frac{1}{2} < \beta \le \frac{1}{2}. \tag{4.16}$$

If the departure point falls exactly at the grid center we take p to be the grid point to the left.

The general form of the Lagrange interpolation polynomial of degree n that takes on the values $q(x_k)$ at the set of $n+1$ distinct points $x_k, k = 0 \ldots n$ is (e.g. Hildebrand [30], page 62)

$$q(x) = \sum_{k=0}^{n} \ell_k(x) q(x_k), \tag{4.17}$$

where

$$\ell_k(x) = \frac{(x - x_0) \cdots (x - x_{k-1})(x - x_{k+1}) \cdots (x - x_n)}{(x_k - x_0) \cdots (x_k - x_{k-1})(x_k - x_{k+1}) \cdots (x_k - x_n)} . \tag{4.18}$$

The quadratic polynomial interpolant from (4.18) for our specific situation is

$$q_j^{n+1} = q_{j_D} = -\frac{1}{2}\beta(1-\beta)q_{j-p-1} + (1-\beta)(1+\beta)q_{j-p} + \frac{1}{2}\beta(1+\beta)q_{j-p+1} . \tag{4.19}$$

Substituting wavelike solutions (2.2) gives

$$e^{i(\nu \Delta t + kp\Delta x)} = 1 - \beta^2 (1 - \cos k\Delta x) + i\beta \sin k\Delta x . \tag{4.20}$$

Write ν in terms of a real and imaginary part as in (2.36). The square of the amplitude of the solution is given by

$$\left| e^{-\nu_i \Delta t} \right|^2 = 1 - 2\beta^2(1 - \cos k\Delta x) + \beta^4(1 - \cos k\Delta x)^2 + \beta^2(1 - \cos^2 k\Delta x) . \tag{4.21}$$

Combining the β^2 terms gives

$$\left| e^{-\nu_i \Delta t} \right|^2 = 1 - \beta^2(1 - \cos k\Delta x)^2 + \beta^4(1 - \cos k\Delta x)^2 , \tag{4.22}$$

leading to

$$\left| e^{-\nu_i \Delta t} \right|^2 = 1 - \beta^2(1 - \beta^2)(1 - \cos k\Delta x)^2 . \tag{4.23}$$

Thus, by the definition of p and the restricted range of β in (4.16), the scheme with quadratic interpolation is stable. The maximum damping for this scheme is for $\beta = \pm 1/\sqrt{2}$ where the shortest wave would be reduced to zero amplitude in one time step as with linear interpolation. However, this is outside the range of β and illustrates why the choice was made earlier that p defines the grid point closest to the departure point. Although the scheme is stable as long as the departure point is within the interpolation stencil, it is more accurate if $-1/2 < \beta \le 1/2$. In the permitted range, the maximum damping is at $\beta = \pm 1/2$ where the amplitude of the shortest wave is reduced by half and the $4\Delta x$ wave is reduced by $\sqrt{13/16} = 0.90$.

4.3 CUBIC INTERPOLATION

Even better performance is obtained with higher order interpolants such as cubic polynomial which is commonly used in semi-Lagrangian models today. Again let

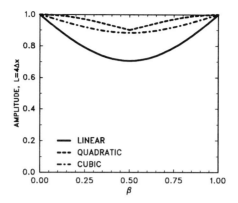

Figure 4.2. Normalized amplitude error of $4\Delta x$ wave as a function of the relative location of the departure point β for linear, quadratic and cubic polynomial interpolants.

p denote the integer off-set to the grid interval in which the departure point falls as in (4.7). Continue to define β to be the fraction of the grid interval between the departure point and x_{j-p} as in (4.8). The cubic polynomial interpolant from (4.17) and (4.18) for our specific situation is

$$q_j^{n+1} = q_{j_D} = -\frac{1}{6}\beta(1-\beta)(2-\beta)q_{j-p-1} + \frac{1}{2}(1+\beta)(1-\beta)(2-\beta)q_{j-p}$$
$$+ \frac{1}{2}(1+\beta)\beta(2-\beta)q_{j-p+1} - \frac{1}{6}(1+\beta)\beta(1-\beta)q_{j-p+2} .$$
$$(4.24)$$

Figure 4.2 shows the amplitude $\left|e^{-\nu_i \Delta t}\right|$ normalized by the correct amplitude for linear, quadratic and cubic polynomial interpolants applied to the $4\Delta x$ wave as a function of the relative location β of the departure point within a

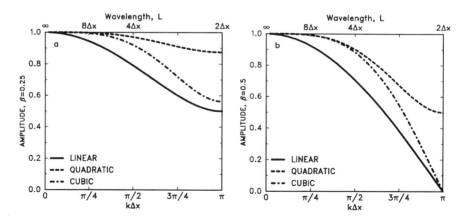

Figure 4.3. Normalized amplitude error for (a) $\beta = 0.25$ and (b) $\beta = 0.5$ as function of wavelength.

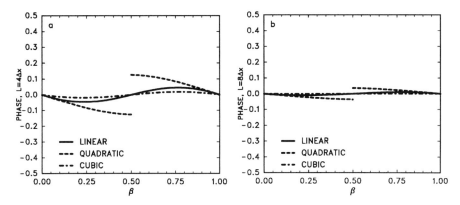

Figure 4.4. Absolute phase error expressed as a fraction of the grid interval. The abscissa is the relative location of the departure point β for (a) $L = 4\Delta x$ and (b) $L = 8\Delta x$.

grid interval. In order to be comparable, the quadratic case is plotted relative to the interval 0 to 1, rather than from -0.5 to 0.5, the way it is calculated, i.e. the values -0.5 to 0 are plotted shifted to 0.5 to 1. As pointed out above, the error is independent of the integer number of grid intervals between the departure point and arrival point. In all cases the damping is maximum for $\beta = 0.5$. Figure 4.3 shows the normalized amplitude plotted as a function of wavelength for $\beta = 0.25$ and the worst case $\beta = 0.5$. In the latter case, the amplitude of the shortest $2\Delta x$ wave is zero with the linear and cubic interpolants, while the quadratic retains half the amplitude. Nevertheless, after several timesteps the $2\Delta x$ wave will be annihilated for practical purposes by the quadratic scheme. The linear interpolant tends to damp rather strongly, making it unsuitable for atmospheric models. For all wavelengths and departure point locations, the quadratic has less amplitude error than the cubic, although for the waves longer than $6\Delta x$ the differences are not significant. Given that the quadratic has less damping than the cubic one must question the value of the extra calculations involved in the cubic. The value of the cubic is in the phase error. Figure 4.4 shows the absolute phase error expressed as a fraction of the grid interval,

$$\frac{(\nu_{approx} - \nu)\,\Delta t}{k\Delta x} = \frac{1}{k\Delta x}\arctan[\quad] - \beta, \qquad (4.25)$$

where ν_{approx} is the approximate frequency, ν is the true frequency and the $[\quad]$ contain the quantity in square brackets in (4.15) for the linear case and corresponding quantities for the quadratic and cubic. Again the error is independent of the integer number of grid intervals between the departure point and arrival point. The errors are plotted as a function of β. The error is antisymmetric about the center of the grid interval. The phase errors of the linear and cubic maximize at $\beta = 0.25$ and $\beta = 0.75$ and are zero at $\beta = 0.5$. The error of

178

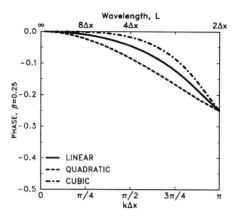

Figure 4.5. Absolute phase error expressed as a fraction of the grid interval as a function of wavelength for $\beta = 0.25$.

the quadratic interpolant maximizes at $\beta = \pm 0.5$ and has different signs in the limit of the two cases. (Recall that for the quadratic case the -0.5 to 0 values are plotted here as if from .5 to 1 to provide comparison with the linear and cubic.) Discontinuities in algorithms, such as that for the quadratic interpolant in Figure 4.4 in which a small difference – as small as the machine rounding error – can result in a large difference in the output are undesirable in atmospheric models (Rosinski and Williamson [61]). However, the discontinuity in the quadratic interpolant would be an extremely rare occurrence unlike the ones studied by Roskinski and Williamson in the physical parameterizations, which push the solution toward the discontinuities. There is no similar process in the dynamics pushing the departure points near the mid-points of grid intervals. Figure 4.5 shows the phase errors as a function of wavelength for $\beta = 0.25$, where the linear and cubic have their largest errors. The phase errors of the cubic are significantly less than those of the quadratic, which are in fact larger than those of the linear interpolant.

4.4 THREE-TIME-LEVEL APPROXIMATIONS

In the above analysis for the linearized case with constant wind, the trajectory is calculated exactly and introduces no error. In practice the wind varies along the trajectory and is only known at grid points so the integral (4.4) can only be approximated. This introduces additional errors as well as extra complications. For example, if (4.4) is approximated by the average of the wind at the ends of the trajectory,

$$x_{j_D} = x_j - \frac{1}{2}\Delta t \left[U(x_{j_D}, n\Delta t) + U(x_j, (n+1)\Delta t) \right] . \qquad (4.26)$$

However, if the semi-Lagrangian method is also being used for the momentum equation, $U(x_j, (n+1)\Delta t)$ is not known until after the departure point x_{j_D} is found, and (4.26) becomes a rather complex equation. For this reason and others to become apparent shortly, the semi-Lagrangian approximations are often applied in a three-time-level form. The geometry of this form is illustrated in Figure 4.1b. Instead of (4.5) the forecast is given by

$$q_j^{n+1} = g_{j_D}^{n-1},$$ (4.27)

and the trajectory is approximated by the wind at the mid-point in space and time

$$x_{j_D} = x_j - 2\Delta t U(x_{j_M}, n\Delta t),$$ (4.28)

where x_{j_M} denotes the mid-point of the trajectory and is given by

$$x_{j_M} = x_j - \Delta t U(x_{j_M}, n\Delta t),$$ (4.29)

which is implicit since U at x_{j_M} cannot be determined until x_{j_M} is known. Thus x_{j_M} is generally found by an iterative process

$$x_{j_M}^{(k+1)} = x_j - \Delta t u(x_{j_M}^{(k)}, n\Delta t),$$ (4.30)

with a suitable first guess for $x_{j_M}^{(0)}$, often taken as the arrival point x_j. In atmospheric applications, only a few iterations are required.

Alternatively, (4.28) can be replaced by

$$x_{j_D} = x_j - \Delta t \left[U(x_{j_D}, n\Delta t) + U(x_j, n\Delta t) \right]$$ (4.31)

using the spatial average of values at the geographic ends of the trajectory, but at the central time. The value of x_{j_D} is found iteratively in a similar fashion. This form requires interpolations only at x_{j_D} with none at x_{j_M}.

4.5 SEMI-LAGRANGIAN, SEMI-IMPLICIT

The semi-Lagrangian method provides accurate approximations for advection with no time step restriction. However, the atmospheric equations also admit gravity waves as discussed earlier. If the terms responsible for these waves were treated explicitly, they would severely restrict the time step for the system of equations even with semi-Lagrangian advection approximations. Thus to obtain maximum benefit from the semi-Lagrangian approximations, they need to be combined with semi-implicit approximations.

Define the Lagrangian time derivative operator as

$$\delta_{2t}^D q_j^n = \frac{q_j^{n+1} - q_{j_D}^{n-1}}{2\Delta t},$$ (4.32)

and the Lagrangian average

$$\overline{(q_j^n)}^{2t}_{\alpha,D} = \left[\left(\frac{1+\alpha}{2}\right) q_j^{n+1} + \left(\frac{1-\alpha}{2}\right) q_{j_D}^{n-1}\right].$$ (4.33)

The semi-Lagrangian, semi-implicit approximations to the shallow-water equations equivalent to (2.72)–(2.74) are

$$\delta_{2t}^D u_j^n - f v_{j_M}^n + g \delta_{2x} \overline{(h_j^n)}^{2t}_{\alpha,D} = 0,$$ (4.34)

$$\delta_{2t}^D v_j^n + f u_{j_M}^n = 0,$$ (4.35)

$$\delta_{2t}^D h_j^n + H \delta_{2x} \overline{(u_j^n)}^{2t}_{\alpha,D} = 0.$$ (4.36)

The implicit set of equations is the same as (2.78) and (2.79) but with different RU_j and RH_j given by

$$RU_j = u_{j_D}^{n-1} + 2\Delta t f v_{j_M}^n - g\Delta t \left(1 - \alpha\right) \left[\delta_{2x} \left(h_j^{n-1}\right)\right]_D,$$ (4.37)

$$RH_j = h_{j_D}^{n-1} - H\Delta t \left(1 - \alpha\right) \left[\delta_{2x} \left(u_j^{n-1}\right)\right]_D.$$ (4.38)

The solution proceeds as in (2.82) and following.

Although (4.37) and (4.38) each have several terms interpolated to the departure point x_{j_D}, in practice they are combined before the interpolation to reduce the number of interpolations needed and thus minimize the cost. Thus, for example, (4.37) becomes

$$RU_j = \left[u_j^{n-1} - g\Delta t \left(1 - \alpha\right) \delta_{2x} \left(h_j^{n-1}\right)\right]_D + \left(2\Delta t f v_j^n\right)_M.$$ (4.39)

For the linearized example above and with polynomial interpolation, the result is the same since the combination of terms and interpolation commute. However, with nonlinear equations, or with nonlinear, shape-preserving interpolants, the operations do not necessarily commute and an additional error is introduced. Nevertheless, in practice it appears to be satisfactory to combine all like terms before interpolation.

In fact to minimize the number of interpolations required, many groups replace the values at the midpoint of the trajectory with averages on the ends, but at the same time, i.e.

$$(\quad)_M^n = \frac{1}{2}[(\quad)_A^n + (\quad)_D^n].$$ (4.40)

The terms at the departure point are combined with the others there before interpolation.

4.6 SPHERICAL COORDINATES

The semi-Lagrangian, semi-implicit method provides a natural solution to the pole problem because it provides an accurate solution without a time step restriction. Thus, near the poles, the departure point may be many grid intervals away from the arrival point without being unstable, and the convergence of meridians does not introduce a stability restriction. Although there is no stability problem near the poles, care is needed in calculating the trajectory there in order to obtain an accurate trajectory. In spherical coordinates the equivalent of (4.2) is

$$a \cos \varphi \frac{d\lambda}{dt} = u, \tag{4.41}$$

$$a \frac{d\varphi}{dt} = v, \tag{4.42}$$

where a is the radius of the earth, φ is latitude and λ is longitude. The natural three-time-level approximation equivalent to (4.28) is

$$\lambda_D = \lambda_A - 2\Delta t \, \frac{u(\lambda_M, \varphi_M)}{a \cos \varphi_M}, \tag{4.43}$$

$$\varphi_D = \varphi_A - 2\Delta t \, \frac{v(\lambda_M, \varphi_M)}{a}, \tag{4.44}$$

where A denotes the arrival grid point (i, j) at which the forecast is being made, and as before D and M denote the trajectory departure and mid-points associated with that arrival point.

The approximations (4.43) and (4.44) lead to inaccurate trajectories near the poles. There, the spherical coordinate unit vectors vary rapidly in space and the assumption that the velocity components at the mid-point of the trajectory represent the flow along the entire length of the trajectory breaks down. This problem is easily overcome by transforming to another coordinate system to calculate the trajectories. Ritchie [53] developed a method based on a straight line trajectory in three-dimensional Cartesian geometry that does not follow the surface of the sphere. He then maps the departure point back to the surface of the sphere. McDonald and Bates [42] and Williamson and Rasch [72] applied a coordinate transform before calculating the trajectory. For the calculation associated with each grid point, they transform to a spherical coordinate system that is rotated with respect to the original system so that arrival point is on the equator of the rotated system. The formulas relating this transformed system to the original spherical coordinates are given by Williamson and Rasch [72]. The trajectory is calculated by applying (4.43) and (4.44) in the transformed system. Once the departure point locations are found in the new system they are transformed back to the original spherical system and the interpolation is done in the original spherical coordinates where the grid points are aligned on the coordinate lines and the tensor product form (to be explained in Section 4.8) can be applied. The grid points are not aligned on the coordinate lines of the transformed system, making interpolation in that system much more complicated. Since the transformations involved require additional expense, the original spherical system is used in the region sufficiently far from the poles, say equatorward of $\pm 70°$ latitude.

4.7 MONOTONIC INTERPOLANTS

As mentioned above, the semi-Lagrangian method has an additional advantage in that the interpolation used to obtain q_D^{n-1} in (4.27) can incorporate some physical constraints. One that has been used for water vapor prediction is monotonicity (Rasch and Williamson [51], Williamson and Rasch [72]). This ensures that the value of q_D will be within the range defined by the neighboring values used in the interpolation. Thus negative values cannot be introduced by the scheme if the field is positive to start with, and overshooting will not produce spurious supersaturation and its accompanying spurious precipitation.

The NCAR CCM uses cubic Hermite polynomial interpolation for semi-Lagrangian constituent transport as this interpolant can be easily modified if desired to provide a monotonic interpolant. With the usual definition of p (4.7) and β (4.8) the Hermite interpolant is

$$
\begin{aligned}
q_{j_D} = {} & [3 - 2(1 - \beta)](1 - \beta)^2 q_{j-p} && + && \left[\Delta x \beta (1 - \beta)^2\right] d_{j-p} \\
& + [3 - 2\beta]\beta^2 q_{j-p+1} && - && \left[\Delta x \beta^2 (1 - \beta)\right] d_{j-p+1},
\end{aligned}
\tag{4.45}
$$

where d_{j-p} and d_{j-p+1} are estimates of the derivatives of q at the grid points. This specific form can be derived from the general form for Hermite polynomial interpolants given in introductory Numerical Analysis text books such as Hildebrand [30], page 316. If the derivative estimates satisfy certain conditions, then the interpolant (4.45) is monotonic. Consider two arbitrary points within the interpolation grid interval such that $x_{j-p} \le x_a \le x_b \le x_{j-p+1}$. For $q(x_{j-p}) \le q(x_{j-p+1})$, the function is monotonic if $q(x_a) \le q(x_b)$, and for $q(x_{j-p}) \ge q(x_{j-p+1})$, the function is monotonic if $q(x_a) \ge q(x_b)$.

Fritsch and Carlson [24] have found a necessary condition for monotonicity of the Hermite cubic interpolant. This condition is

$$
0 \le \frac{d_{j-p}}{\Delta_{j-p}q} \le 3
\tag{4.46}
$$

and

$$
0 \le \frac{d_{j-p+1}}{\Delta_{j-p}q} \le 3,
\tag{4.47}
$$

where

$$\Delta_{j-p}q = \frac{q_{j-p+1} - q_{j-p}}{\Delta x} . \tag{4.48}$$

For our application, the derivative estimates d_{j-p} and d_{j-p+1} are first approximated by differentiating a Lagrange cubic polynomial interpolant through the four points $\{x_{j-p-1}, x_{j-p}, x_{j-p+1}, x_{j-p+2}\}$ and evaluating it at x_{j-p} and x_{j-p+1}, respectively. With these derivative estimates the Hermite cubic interpolant (4.45) is identical to the Lagrange cubic polynomial interpolant (4.24). If either (4.46) or (4.47) is not satisfied, the offending derivative estimate is modified to bring it to just within the condition. More details are given by Rasch and Williamson [51].

The cubic polynomial as a function of x is given by (4.24); here $\beta = (x - x_{j-p})/\Delta x$. By the chain rule

$$\frac{dq}{dx} = \frac{dq}{d\beta}\frac{d\beta}{dx} = \frac{1}{\Delta x}\frac{dq}{d\beta} . \tag{4.49}$$

Differentiating (4.24) and evaluating the result for $\beta = 0$ and $\beta = 1$ gives the desired derivative estimates

$$d_{j-p} = -\frac{q_{j-p-1}}{3\Delta x} - \frac{q_{j-p}}{2\Delta x} + \frac{q_{j-p+1}}{\Delta x} - \frac{q_{j-p+2}}{6\Delta x} , \tag{4.50}$$

and

$$d_{j-p+1} = +\frac{q_{j-p-1}}{6\Delta x} - \frac{q_{j-p}}{\Delta x} + \frac{q_{j-p+1}}{2\Delta x} + \frac{q_{j-p+2}}{3\Delta x} . \tag{4.51}$$

4.8 TENSOR PRODUCT INTERPOLANTS

So far we have only discussed one-dimensional polynomial interpolation. The Lagrange and Hermite polynomial interpolants can be generalized to multiple dimensions, as can the monotonicity conditions. However, the monotonicity conditions become very complicated in multiple dimensions, and the polynomial interpolants themselves become disproportionately expensive. Therefore, almost all semi-Lagrangian models use a tensor product interpolant based on one-dimensional interpolants. Consider the two-dimensional case with space coordinates (x, y) and suppose we have a grid $\{(x_i, y_j)\}$ analogous to (2.4). Similarly define the grid box in which the departure point falls by (x_{i-p}, y_{j-q}). Then, for example, for a cubic interpolant, four one-dimensional interpolations are performed in x to provide values at $\{(x_{i_D}, y_{j-q-1}), (x_{i_D}, y_{j-q}), (x_{i_D}, y_{j-q+1}), (x_{i_D}, y_{j-q+2})\}$. Then one cubic interpolation is performed in y using the values at those four points to give the value at (x_{i_D}, y_{j_D}). To save even more computer time, it has become common to use a *quasi-cubic* interpolant in which the outer interpolants for (x_{i_D}, y_{j-q-1}) and (x_{i_D}, y_{j-q+2}) use linear interpolation and the inner ones for (x_{i_D}, y_{j-q}) and (x_{i_D}, y_{j-q+1}) use cubic, while the final interpolant in y is cubic. In the monotonic Hermite case, the monotonicity conditions (4.46) and (4.47) are applied to each one-dimensional cubic interpolant. The outer, linear interpolants are automatically monotonic without further modification.

As illustrated in (2.58)–(2.60), advection is only one component making up the equations representing atmospheric motions. Semi-Lagrangian advection approximations have been successfully combined with finite difference, spectral, and finite element based approximations for the remaining components. Examples of spectral, semi-Lagrangian models are Ritchie *et al.* [57] which has been applied to high resolution numerical weather prediction and Williamson and Olson [73] applied to moderate resolution climate simulation. Chen and Bates [15] developed a finite difference, semi-Lagrangian model applied to both numerical weather prediction and climate simulation. Côté *et al.* [17], [19] developed a variable resolution, finite element, semi-Lagrangian model for the sphere which also has the capability of concentrating resolution over an area of interest, while retaining a global domain.

5. Orographic and Other Stationary Forcings

We have seen in Section 4.5 that the combination of semi-implicit (SI) and semi-Lagrangian (SL) numerical algorithms results in an efficient marching scheme to integrate the meteorological equations because the time step is determined by accuracy rather than stability considerations. With the SI scheme a model may be integrated with a time step larger than the grid length divided by the phase speed of the fastest waves supported by the equations. For the hydrostatic equations these are the Lamb mode in the horizontal and internal gravity waves in the vertical. With the fully elastic Euler equations, vertically propagating elastic modes travel at the speed of sound. With the SL scheme the time step of a model is not limited by the advection under strong wind conditions.

The aforementioned conclusions result from the analysis of free motions, that is, of linear traveling waves and/or simple advection. Forced problems constitute a separate class of problems that receive much less attention in numerical analysis. The reason is that with Eulerian methods forced, stationary problems are usually less restrictive than free, traveling problems. This is not necessarily the case however for SI-SL schemes as was first noted by Coiffier *et al.* [16] in the context of mountain-induced internal gravity waves. Forcing that is stationary in the Eulerian framework is Doppler shifted to higher frequencies in the Lagrangian framework, and may result in large truncation errors associated with the time averaging in semi-implicit schemes. The problem is illustrated in

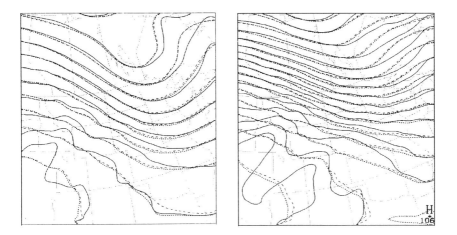

Figure 5.1. 500 hPa(left) and 250 hPa(right) geopotential obtained with CRCM after 18 h with 15 min. time step and $\alpha = 0.01$ (full), with 15 min. time step and $\alpha = 0.10$ (dashed), and with 5 min. time step and $\alpha = 0.01$ (dotted). α is the un-centering parameter as in (4.3).

Figure 5.1, from a short integration of the Canadian Regional Climate Model (CRCM, Caya and Laprise [14]) with a 45 km grid interval coupled with long (15 min) and short (5 min) time steps. It can be seen that noise develops over and in the lee of the Rocky Mountains with the longer time step and weakly un-centered approximations. The problem is present at great heights above the mountains, and in fact amplifies up to the tropopause.

Variants of the classical SI-SL algorithm have been proposed to address this problem, usually involving some un-centering of the time average operator along the trajectory, e.g. Kaas [33], Tanguay *et al.* [68]. Rivest *et al.* [58], Côté *et al.* [18], Ritchie and Tanguay [56]. Figure 5.1 shows that good results can be obtained with un-centering and a 15-min time step in this case. Héreil and Laprise [29] have shown though that none of the proposed alternatives completely eliminates the problem for mountain wave solutions.

In this section, a simplified canonical equation is used to show the origin of the numerical problem and some possible ways of alleviating it. We consider as a prototype to the meteorological equations the following linear one-dimensional forced oscillator-type equation with dependent variable q:

$$\frac{dq}{dt} = \frac{\partial q}{\partial t} + U \frac{\partial q}{\partial x} = i\omega q + S \,, \tag{5.1}$$

where dq/dt is the Lagrangian (following the flow) rate of change of q, $\partial q/\partial t$ is the Eulerian (local) rate of change of q, $U\partial q/\partial x$ is the contribution of advection by the flow U, and $i\omega q$ the oscillatory behavior of the atmosphere (as we shall see below, ω may be one of the natural frequencies of the atmosphere). These modes include the inertial-gravity waves seen earlier in the shallow water system (2.58)–(2.60) with advecting wind U zero. The source term S, also included in (2.1) and (4.1), can represent a steady-state forcing (e.g. land-sea contrast, orography). For simplicity in the following analysis we define this forcing S as a single Fourier wave,

$$S(x) = S_k e^{ikx}, \tag{5.2}$$

which is independent of time.

5.1 ANALYTICAL SOLUTION

We first find the analytical solution to (5.1). As before (e.g., 2.2), assume a solution of the form

$$q(x,t) = Q_k e^{i(kx+\nu t)}. \tag{5.3}$$

Here we explicitly indicate that Q_k is the amplitude of the wave with wavenumber k, unlike (2.2) where it was only implied. Substituting (5.3) into (5.1) gives

$$i(\nu - \omega + kU) Q_k e^{i(\nu t + kx)} = S_k e^{ikx}. \tag{5.4}$$

Equation (5.4) has three special cases that we consider separately: free waves, forced waves, and resonance.

5.1.1 *Free wave*
In the absence of forcing, $S_k = 0$ and (5.4) yields

$$\nu = -kU + \omega, \tag{5.5}$$

which is the dispersion relationship for this problem. The frequency of the free solution in the fixed frame of reference is ν, ω is the intrinsic (i.e., following the flow) frequency, and $-kU$ is the Doppler shift in frequency due to advection by the flow. This is just the same form as we found earlier for the solution of the shallow water equations (2.64). The free-wave solution is thus

$$q(x,t) = Q_k e^{i[kx-(kU-\omega)t]}, \tag{5.6}$$

and the amplitude Q_k is determined by initial conditions.

5.1.2 *Steady-state forced wave*
In this case we consider $\nu = 0$. The solution for the amplitude of the standing wave is

$$Q_k = \frac{S_k}{i(kU - \omega)}. \tag{5.7}$$

For sinusoidal orographic forcing with wavenumber k, i.e.,

$$\phi_s(x) = \hat{\phi}_s e^{ikx} , \qquad (5.8)$$

where $\phi_s = gh_s$ is the surface geopotential, h_s is the height of the topography and $\hat{\phi}_s$ is the Fourier amplitude of wavenumber k, the forcing in physical space is

$$S(x) = -U\frac{\partial \phi_s}{\partial x} = -ikU\hat{\phi}_s e^{ikx} = -ikU\phi_s(x) , \qquad (5.9)$$

and thus in Fourier space is

$$S_k = -ikU\hat{\phi}_s . \qquad (5.10)$$

The form of the forcing in (5.9) appears in the continuity equation (2.60) of the shallow water equations when a surface topography h_s is included. In that case h is replaced by $h - h_s$, the depth of the fluid, and h continues to indicate the height of the fluid above some constant reference plane. The continuity equation is for the depth of the fluid, but of course $\partial h_s/\partial t = 0$. The pressure gradient in the momentum equations still depends on the gradient of the height of the fluid h (2.58), not the depth $h - h_s$.

With the forcing (5.9), the steady state solution of (5.4) is

$$Q_k = \frac{S_k}{i(kU - \omega)} = \left(\frac{-U}{U - \omega/k}\right)\hat{\phi}_s = \left(\frac{-1}{1 - \omega/(kU)}\right)\hat{\phi}_s . \qquad (5.11)$$

Numerical solutions will be compared to this analytical solution later. Note that when the intrinsic phase speed (ω/k) equals the advecting velocity, the response is singular. This occurs because the net effect of the combined terms $-U\partial q/\partial x + i\omega q$ vanishes and the steady-state forcing S excites a "natural' frequency of the system, here $\nu = 0$. This brings us to the third case, resonance.

Before proceeding to this case however, let us note that by analogy with the shallow water equations, the atmosphere tends to be in a subcritical flow regime, $|U| \ll |\omega/k|$. In this case

$$Q_k \approx -\frac{kU}{\omega}\hat{\phi}_s . \qquad (5.12)$$

5.1.3 Stationary resonant case

This case occurs when the forcing projects onto one of the free solutions (here the steady state). In this case, (5.1) simplifies because of cancellation of advection and oscillation to give

$$\frac{\partial q}{\partial t} = S = S_k e^{ikx} . \qquad (5.13)$$

188

The solution is algebraic rather than harmonic in time

$$q = S_k t e^{ikx} + constant. \tag{5.14}$$

The *constant* is determined by the particular initial value of the problem. This case will not be pursued further in the following analysis.

5.2 RESPONSE OF NUMERICAL DISCRETIZATIONS

5.2.1 *Leap-frog scheme*
Consider the frequently used leap-frog approximation (2.6) to (5.1) with the forcing evaluated exactly. Equation (5.1) is then discretized as

$$\delta_{2t} q_j^n = -U \delta_{2x} q_j^n + i\omega q_j^n + S_j , \tag{5.15}$$

where the discrete operators are defined in (2.16) and (2.17).

As with the continuous case, we assume a solution of the form (5.3). Substituting this form into (3.13) gives

$$i \left(\nu \mathcal{D}_{2t} - \omega + kU \mathcal{D}_{2x} \right) Q_k e^{i\nu t} = S_k , \tag{5.16}$$

which is like (5.4) but includes the response functions of the discrete operators defined earlier in (2.20) and (2.21). For free waves $(S = S_k = 0)$ we recover the classical result

$$\nu = -\frac{1}{\Delta t} \arcsin \left(U \frac{\Delta t}{\Delta x} \sin k\Delta x - \omega \Delta t \right) . \tag{5.17}$$

which is like (2.25) but including the oscillatory term as with the inertial-gravity waves (2.70) in the shallow water equations. Equation (5.17) is the dispersion relationship for the discretized equation. The ratio of the frequency from the approximate equation (5.17) to the analytical frequency (5.5) is

$$\frac{\nu}{\nu_{true}} = \frac{\arcsin \left(U \frac{\Delta t}{\Delta x} \sin k\Delta x - \omega \Delta t \right)}{(kU\Delta t - \omega \Delta t)} . \tag{5.18}$$

When $\omega = 0$ this is just the ratio plotted in Figure 2.4a.

As before with (2.26) or (2.71), for the frequency in (5.17) to be real,

$$\left| U \frac{\Delta t}{\Delta x} \sin k\Delta x - \omega \Delta t \right| \le 1 \tag{5.19}$$

for all $k \le J/2$, which is the stability condition for the leap-frog approximations and is analogous to (2.71).

Let us now turn to the less studied steady-state forced wave case ($\nu = 0$). From (5.16) the amplitude of the standing wave is

$$Q_k = \frac{S_k}{i\left(kU\mathcal{D}_{2x} - \omega\right)}.$$ (5.20)

Consider the ratio of the numerical and exact (5.11) solutions

$$\frac{Q_k}{(Q_k)_{true}} = \frac{kU - \omega}{kU\mathcal{D}_{2x} - \omega}.$$ (5.21)

The atmosphere, as the shallow water equations described following (2.71) tends to be in a subcritical flow regime, i.e. $|U| \leq |\omega/k|$. In the subcritical limit $kU/\omega \to 0$, the response is perfect, and the ratio in (5.21) goes to 1. Thus as we saw earlier, the Eulerian explicit leap-frog scheme is conditionally stable and the time step must be small enough to satisfy a Courant condition for stability. On the other hand the stationary forced solution is perfect in the subcritical limit.

5.2.1 Semi-implicit semi-Lagrangian
Let us now turn to the semi-implicit (SI) semi-Lagrangian (SL) approximations to (5.1). In its simplest form the SI-SL scheme is applied as follows

$$\delta_{2t}^D q_j^n = i\omega\overline{(q_j^n)}_{\alpha,D}^{2t} + S_M,$$ (5.22)

where the discrete Lagrangian time derivative and averaging operators were defined in (4.32) and (4.33), respectively. In analyzing (5.22) we will assume that the interpolation used to obtain values at the departure and mid-points of the trajectories is exact, thereby ignoring the damping associated with the interpolation that would occur in actual practice. We will also assume that the averaging operator is centered, $\alpha = 0$. In the linear, constant wind case, the departure point is given by

$$x_{j_D} = x_j - 2U\Delta t,$$ (5.23)

and the mid-point by

$$x_{j_M} = x_j - U\Delta t.$$ (5.24)

The discrete operators applied to the harmonic representation (5.3) give

$$\delta_{2t}^D Q_k e^{i(kx+\nu t)} = i(\nu + kU)\mathcal{S}_\mathcal{L} e^{-ikU\Delta t} Q_k e^{i(kx+\nu t)}$$ (5.25)

and

$$\overline{\left(Q_k e^{i(kx+\nu t)}\right)}_{\alpha,D}^{2t} = \mathcal{C}_\mathcal{L} e^{-ikU\Delta t} Q_k e^{i(kx+\nu t)},$$ (5.26)

190

where

$$S_{\mathcal{L}} = \frac{\sin\left[(\nu + kU)\Delta t\right]}{(\nu + kU)\Delta t} \tag{5.27}$$

and

$$C_{\mathcal{L}} = \cos\left[(\nu + kU)\Delta t\right] \tag{5.28}$$

are the response functions of the Lagrangian operators. Substituting these expressions in (5.22) gives, after some simplification,

$$Q_k e^{i\nu t} i \left\{ \frac{\sin\left[(\nu + kU)\Delta t\right]}{\Delta t} - \omega \cos\left[(\nu + kU)\Delta t\right] \right\} = S_k . \tag{5.29}$$

Let us again consider the free wave case ($S_k = 0$). This gives the classical dispersion relationship of SI-SL as

$$\nu = -kU + \frac{1}{\Delta t}\arctan(\omega\Delta t), \tag{5.30}$$

where again ω is the intrinsic frequency in the problem (5.1). Equation (5.30) shows the unconditional stability of the scheme because it yields real numerical frequencies for any time step. The Doppler shifted component $(-kU)$ is exact since in this section we assumed no damping from the interpolation, unlike (4.15) which includes that damping.

Consider next the steady-state forced wave ($\nu = 0$) case. The amplitude of the steady-state SI-SL solution is

$$Q_k = \frac{S_k}{i\left[kU S_{\mathcal{L}}(0) - \omega C_{\mathcal{L}}(0)\right]}, \tag{5.31}$$

where $S_{\mathcal{L}}(0)$ and $C_{\mathcal{L}}(0)$ are the response functions (5.27) and (5.28) with $\nu = 0$. Again consider the ratio of the numerical to the exact solution:

$$\frac{Q_k}{(Q_k)_{true}} = \frac{kU - \omega}{kU S_{\mathcal{L}}(0) - \omega C_{\mathcal{L}}(0)}. \tag{5.32}$$

In the subcritical limit $|\omega| \gg |kU|$, there is amplification of SI-SL by the factor

$$\lim_{\left|\frac{kU}{\omega}\right| \to 0} \left(\frac{Q_k}{(Q_k)_{true}}\right) = \frac{1}{C_{\mathcal{L}}(0)} = \frac{1}{\cos(k\Delta x C)}, \tag{5.33}$$

where C is the Courant number $U\Delta t/\Delta x$. The amplitude is twice the correct value for

$$k\Delta x C = \frac{\pi}{3}, \tag{5.34}$$

or in terms of wavelength, for

$$L = 6\Delta x C \, . \tag{5.35}$$

For $C > 1$, even fairly long waves are amplified. For example, with $C = 3$, waves with $L < 18\Delta x$ have their amplitude more than doubled, and the wave with $L = 12\Delta x$ is singular, i.e., $\cos(k\Delta x C) = 0$.

In summary the SI-SL scheme is unconditionally stable and hence the time step may be chosen on the basis of accuracy rather than stability. There is, however, a spurious amplification of forced stationary waves that may be substantial for $C > 1$. This amplification affects even rather long waves that span many grid increments, so diffusion cannot be used effectively to counteract it. Hence, if no other modifications are made to the scheme analyzed above, the time step must be reduced to obtain an accurate forced solution.

5.3 TREATMENT OF OROGRAPHY

Thus far we have assumed that the stationary forcing term S in our prototype equation (5.1) was known exactly, so that it was not affected by the choice of discretization. In practice, depending on what this term represents in the full equations, it may also have to be discretized. The question that then arises concerns the effect of the choice of discretization on the overall response of the model. In this section, we will consider various approximations to the orographic forcing S_M in (5.22), all of which will be coupled with semi-Lagrangian, semi-implicit approximations for the other terms in the equation.

As discussed in Section 5.1.2, S is defined as

$$S = -U \frac{\partial \phi_s}{\partial x} \, , \tag{5.36}$$

where ϕ_s represents the geopotential of the surface topography. Clearly, since this term involves derivatives, there are a number of alternatives for its discretion. Let us assume a monochromatic (single Fourier component) representation of topography as in (5.8). Consider first Eulerian, centered finite difference approximations to (5.36). In this case, S is evaluated by

$$S_M = -U \left(\delta_{2x}\phi_s\right)_M = -ikU \mathcal{D}_{2x} e^{-ikU\Delta t} \hat{\phi}_s e^{ikx} \, , \tag{5.37}$$

where $\mathcal{D}_{2x}$ is the response function for the centered differences (2.21).

Consider now a semi-Lagrangian formulation for S. In this case, we interpret the definition of orographic forcing as

$$S = -U \frac{\partial \phi_s}{\partial x} = -\frac{d\phi_s}{dt} \tag{5.38}$$

because topography is fixed in time. This interpretation lends itself naturally to the following SL discretization

$$S = -\delta_{2t}^{D}\phi_s = -ikU\mathcal{S}_\mathcal{L}(0)e^{-ikU\Delta t}\hat{\phi}_s e^{ikx} , \qquad (5.39)$$

where δ_{2t}^{D} is the Lagrangian difference operator along the trajectory (4.32) and $\mathcal{S}_\mathcal{L}(0)$ is its response function (5.27) with frequency $\nu = 0$. Assuming as before that there is no interpolation error, we will consider the implications of this modification to S after analyzing one additional approximation.

A number of semi-Lagrangian models use what is referred to as a spatially averaged (SA) Eulerian treatment of mountains, e.g. Ritchie and Tanguay [56]. It consists of computing the orographic forcing by explicit Eulerian approximations, averaged at the ends of the trajectory,

$$S = -U\overline{(\delta_{2x}\phi_s)}_{\alpha,D} , \qquad (5.40)$$

where

$$\overline{(q_j^n)}_{\alpha,D} = \left[\left(\frac{1+\alpha}{2}\right)q_j^n + \left(\frac{1-\alpha}{2}\right)q_{j_D}^n\right] \qquad (5.41)$$

denotes the spatial average of the values at the geographical ends of the trajectory, but at the same time. It differs from $\overline{(q_j^n)}_{\alpha,D}^{2t}$ of (4.33) which is the Lagrangian average including the time aspects of the trajectory. Here we consider only the centered approximations with $\alpha = 0$, although un-centering could also have advantages. The response function for this averaging operator with $\alpha = 0$ is just (5.28) with $\nu = 0$, denoted $\mathcal{C}_\mathcal{L}(0)$. In terms of these response functions, (5.40) can be written

$$S = -ikU\mathcal{D}_{2x}\mathcal{C}_\mathcal{L}(0)e^{-ikU\Delta t}\hat{\phi}_s e^{ikx} . \qquad (5.42)$$

We now consider the amplitude for the steady-state forced case ($\nu = 0$). The amplitude for each of these schemes has the form of the analytic orography approximation (5.31) with S_k of the analytic form replaced by $S_k\mathcal{D}_{2x}$ for the centered Eulerian (5.37), $S_k\mathcal{S}_\mathcal{L}(0)$ for the semi-Lagrangian (5.39) and $S_k\mathcal{S}_\mathcal{L}(0)\mathcal{C}_\mathcal{L}(0)$ for the spatial averaged Eulerian (5.40). Consider the ratios of the numerical to the exact solutions in the subcritical limit $|\omega| \gg |kU|$. The analytical solution was given in (5.7). The SI-SL solution with Eulerian discretization of S gives

$$\lim_{|\frac{kU}{\omega}|\to 0}\left(\frac{Q_k}{(Q_k)_{true}}\right) = \frac{\mathcal{D}_{2x}}{\mathcal{C}_\mathcal{L}(0)} = \frac{[\sin(k\Delta x)]/(k\Delta x)}{\cos(k\Delta x C)} . \qquad (5.43)$$

The SI-SL solution with semi-Lagrangian interpretation of S gives

$$\lim_{|\frac{kU}{\omega}|\to 0}\left(\frac{Q_k}{(Q_k)_{true}}\right) = \frac{\mathcal{S}_\mathcal{L}(0)}{\mathcal{C}_\mathcal{L}(0)} = \frac{\tan(k\Delta x C)}{k\Delta x C} . \qquad (5.44)$$

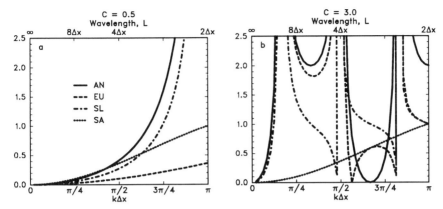

Figure 5.2. Relative error $\left|[Q_k - (Q_k)_{true}] / (Q_k)_{true}\right|$ of SI-SL scheme for (a) $C = 0.5$ and (b) $C = 3.0$ as a function of $k\Delta x$ for various treatments of orography. AN = analytic, EU = Eulerian, SL = semi-Lagrangian and SA = spatial averaged Eulerian topography.

The SI-SL solution with spatially averaged Eulerian discretization of S gives

$$\lim_{\left|\frac{kU}{\omega}\right| \to 0} \left(\frac{Q_k}{(Q_k)_{true}}\right) = \frac{\mathcal{D}_{2x}\mathcal{C}_{\mathcal{L}}(0)}{\mathcal{C}_{\mathcal{L}}(0)} = \frac{\sin(k\Delta x)}{(k\Delta x)} . \tag{5.45}$$

Figure 5.2 shows the relative absolute error in the subcritical case $|kU| \ll \omega$,

$$E = \left|\frac{Q_k - (Q_k)_{true}}{(Q_k)_{true}}\right| \tag{5.46}$$

as a function of scale. Clearly the three methods of discretization of orographic forcing are not equally accurate within the context of a SI-SL model. For $C > 1$, semi-Lagrangian treatment of orography is paradoxically the worst method, a much better one is the Eulerian treatment of orography, and the best is clearly the spatially averaged Eulerian treatment of orography at large Courant number. With this scheme, the response is never singular and hence the relative error never exceeds 100%. On the other hand, the Eulerian treatment of the term S is best at small Courant number, but none of the methods are really bad then, except for the shortest resolved scales which may easily be cured by some diffusion mechanism.

5.4 APPLICATION TO SHALLOW-WATER EQUATIONS

Let us now consider a somewhat more complex and realistic system of equations, the shallow-water equations linearized about a constant U flow and mean free surface with geopotential Φ, in one dimension, neglecting rotation effects for simplicity

$$\frac{\partial u}{\partial t} + U\frac{\partial u}{\partial x} + \frac{\partial \phi}{\partial x} = 0, \tag{5.47}$$

$$\frac{\partial \phi}{\partial t} + U\frac{\partial \phi}{\partial x} + \Phi\frac{\partial u}{\partial x} = U\frac{\partial \phi_s}{\partial x}. \tag{5.48}$$

Here we can see more clearly the source of the term used earlier to represent mountain forcing. The pressure gradient in the momentum equation involves the geopotential ϕ corresponding to the height of the free surface of the fluid above some constant reference, but the continuity equation involves the geopotential depth $\phi - \phi_s$ of fluid above topography.

Again assuming harmonic solutions for wind and geopotential

$$\begin{pmatrix} u \\ \phi \end{pmatrix} = \begin{pmatrix} \hat{u} \\ \hat{\phi} \end{pmatrix} e^{i(kx+\nu t)} \tag{5.49}$$

and (5.35) for ϕ_s gives

$$e^{i\nu t} \begin{pmatrix} \nu + kU & k \\ k\Phi & \nu + kU \end{pmatrix} \begin{pmatrix} \hat{u} \\ \hat{\phi} \end{pmatrix} = \begin{pmatrix} 0 \\ kU\hat{\phi}_s \end{pmatrix}. \tag{5.50}$$

The dispersion relationship for free waves ($\phi_s = 0$) is

$$\nu = -k\left(U \pm \sqrt{\Phi}\right), \tag{5.51}$$

and the steady-state forced solution ($\nu = 0$) is

$$\hat{\phi} = -\left(\frac{U^2}{\Phi - U^2}\right)\hat{\phi}_s. \tag{5.52}$$

In the subcritical case, $U^2 \ll \Phi$,

$$\hat{\phi} \approx -\frac{U^2}{\Phi}\hat{\phi}_s. \tag{5.53}$$

Now consider the response of numerical solutions obtained with the SI-SL scheme, first with $U\partial \phi_s/\partial x$ evaluated analytically at the mid-point of the trajectory:

$$\delta_{2t}^D u_j^n + U\overline{(\delta_{2x}\phi_j^n)}_{\alpha,D}^{2t} = 0, \tag{5.54}$$

$$\delta_{2t}^D \phi_j^n + \Phi\overline{(\delta_{2x}u_j^n)}_{\alpha,D}^{2t} = U\left(\frac{\partial \phi_s}{\partial x}\right)_M. \tag{5.55}$$

The solution in this case is

$$
e^{i\nu t} \begin{pmatrix} (\nu + kU)\mathcal{S}_{\mathcal{L}}(0) & k\mathcal{D}_{2x}\mathcal{C}_{\mathcal{L}}(0) \\ k\Phi\mathcal{D}_{2x}\mathcal{C}_{\mathcal{L}}(0) & (\nu + kU)\mathcal{S}_{\mathcal{L}}(0) \end{pmatrix} \begin{pmatrix} \hat{u} \\ \hat{\phi} \end{pmatrix} = \begin{pmatrix} 0 \\ kU\hat{\phi}_s \end{pmatrix}. \tag{5.56}
$$

The response functions $\mathcal{C}_{\mathcal{L}}(0)$ and $\mathcal{S}_{\mathcal{L}}(0)$ are defined by (5.28) and (5.27) with $\nu = 0$, and $\mathcal{D}_{2x}$ by (2.21). The dispersion relation for free waves ($\phi_s = 0$) is

$$
\nu = -k \left(U \pm \frac{\mathcal{D}_{2x}\mathcal{C}_{\mathcal{L}}(0)}{\mathcal{S}_{\mathcal{L}}(0)} \sqrt{\Phi} \right), \tag{5.57}
$$

and the steady-state forced solution ($\nu = 0$) is

$$
\hat{\phi} = - \left(\frac{\mathcal{S}_{\mathcal{L}}(0)U^2}{(\mathcal{D}_{2x})^2 \left[\mathcal{C}_{\mathcal{L}}(0)\right]^2 \Phi - \left[\mathcal{S}_{\mathcal{L}}(0)\right]^2 U^2} \right) \hat{\phi}_s. \tag{5.58}
$$

For the various approximations for the orographic forcing term considered in the previous section, the right-hand side of (5.56) is multiplied by their response functions, $\mathcal{D}_{2x}$ for the Eulerian, $\mathcal{S}_{\mathcal{L}}(0)$ for the semi-Lagrangian and $\mathcal{D}_{2x}\mathcal{C}_{\mathcal{L}}(0)$ for the spatially averaged Eulerian. In the subcritical range ($U^2 \ll \Phi$), the steady forced solutions are:
for SI-SL, exact topography,

$$
\begin{aligned}
\hat{\phi} &\approx - \frac{U^2}{\Phi} \frac{\mathcal{S}_{\mathcal{L}}(0)}{(\mathcal{D}_{2x})^2 \left[\mathcal{C}_{\mathcal{L}}(0)\right]^2} \hat{\phi}_s \\
&= - \frac{U^2}{\Phi} \hat{\phi}_s \frac{\sin(k\Delta x C)}{k\Delta x C} \left[\frac{k\Delta x}{\cos(k\Delta x C)\sin(k\Delta x)} \right]^2,
\end{aligned} \tag{5.59}
$$

for SI-SL, Eulerian topography,

$$
\begin{aligned}
\hat{\phi} &\approx - \frac{U^2}{\Phi} \frac{\mathcal{S}_{\mathcal{L}}(0)}{\mathcal{D}_{2x} \left[\mathcal{C}_{\mathcal{L}}(0)\right]^2} \hat{\phi}_s \\
&= - \frac{U^2}{\Phi} \hat{\phi}_s \frac{\sin(k\Delta x C)}{k\Delta x C} \frac{k\Delta x}{\cos^2(k\Delta x C)\sin(k\Delta x)},
\end{aligned} \tag{5.60}
$$

for SI-SL, SL topography,

$$
\begin{aligned}
\hat{\phi} &\approx - \frac{U^2}{\Phi} \frac{\left[\mathcal{S}_{\mathcal{L}}(0)\right]^2}{(\mathcal{D}_{2x})^2 \left[\mathcal{C}_{\mathcal{L}}(0)\right]^2} \hat{\phi}_s \\
&= - \frac{U^2}{\Phi} \hat{\phi}_s \left[\frac{\tan(k\Delta x C)}{k\Delta x C} \frac{k\Delta x}{\sin(k\Delta x)} \right]^2,
\end{aligned} \tag{5.61}
$$

196

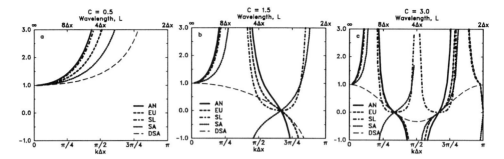

Figure 5.3. Ratios of approximate solutions to true solution for the subcritical case for various orographic forcing approximations coupled with the SI-SL approximations, for Courant numbers (a) $C = 0.5$, (b) $C = 1.5$ and (c) $C = 3.0$. AN = analytic, EU = Eulerian, SL = semi-Lagrangian, SA = spatial averaged Eulerian and DSA = double spatial averaged Eulerian topography.

and for SI-SL, SA Eulerian topography.

$$
\begin{aligned}
\hat{\phi} &\approx - \frac{U^2}{\Phi} \frac{\mathcal{S}_\mathcal{L}(0)}{\mathcal{D}_{2x}\mathcal{C}_\mathcal{L}(0)} \hat{\phi}_s \\
&= - \frac{U^2}{\Phi} \hat{\phi}_s \frac{\tan(k\Delta xC)}{k\Delta xC} \frac{k\Delta x}{\sin(k\Delta x)} .
\end{aligned}
\tag{5.62}
$$

Figure 5.3 shows the ratios of the approximate solutions (5.59)–(5.62) to the true solution (5.53) for the subcritical case. Note that since all approximations have $\sin(k\Delta x)$ in the denominator, they will be singular for the $2\Delta x$ wave except for a few fortuitous values of C in some schemes where the $\sin(k\Delta xC)$ in the numerator will cancel it. The singularity is not too harmful because it can be controlled by diffusion of shortest resolved scales. All these solutions are also singular for $C4\Delta x$ waves due to the $\cos(k\Delta xC)$ in the denominator. This singularity is very difficult to cure because for $C > 1$, rather long waves are spuriously amplified. Unlike the prediction based on our canonical equation, spatially averaging the topographic forcing term has not completely removed this singularity. It has, however, made it linear rather than quadratic in $\mathcal{C}_\mathcal{L}(0)$.

5.5 FURTHER CONSIDERATIONS

In order to eliminate the singularity of SI-SL shallow-water equations in the subcritical limit, one would have to remove the remaining $\mathcal{C}_\mathcal{L}(0)$ factor in the denominator. Consider the following "doubly spatially averaged" (DSA) Eulerian treatment of orography:

$$
S = -U \left(\overline{(\delta_{2x}\phi_s)_{\alpha,D}} \right)_{\alpha,D} .
\tag{5.63}
$$

The response function similar to (5.42) is

$$S = -ikU\mathcal{D}_{2x} \left[\mathcal{C}_{\mathcal{L}}(0) \right]^2 e^{-ikU\Delta t} \hat{\phi}_s e^{ikx} , \qquad (5.64)$$

and the forced steady-state solution with SI-SL, DSA Eulerian topography is

$$
\begin{aligned}
\hat{\phi} &\approx -\frac{U^2}{\Phi} \frac{\mathcal{S}_{\mathcal{L}}(0)}{\mathcal{D}_{2x}} \hat{\phi}_s \\
&= -\frac{U^2}{\Phi} \hat{\phi}_s \frac{\sin(k\Delta x C)}{k\Delta x C} \frac{k\Delta x}{\sin(k\Delta x)}
\end{aligned}
\qquad (5.65)
$$

in the subcritical regime; hence double averaging eliminates the singularity of long waves at large Courant number. The "soft" amplification of very short waves remains but may be counteracted by smoothing the topography field. In the simple linearized shallow water equations with a constant wind used for analysis in this section, the double averaging is reasonably straight forward. In a nonlinear model it becomes rather complex. The trajectory must be extended both backward and forward and the implications in a convoluted flow are unknown.

We have seen that stationary forcings such as orography can be seriously misrepresented at the large Courant numbers that the stability of the SI-SL scheme permits. Various methods of reducing the singular response have been proposed, but none are completely successful in eliminating it. The analysis of the response in shallow-water equations suggests an improved solution using double spatial averaging of topographic forcing, but it has not been tried in a complete atmospheric model.

The analysis in this section implies that the schemes are singular in the limit $U \to 0$. This is a little misleading since in that case $C \to 0$ as well, other aspects of the problem being held fixed. In this case we do not need to be concerned with large C. In practice, the schemes are not singular except possibly in rare pathological cases. In the analysis, $U^2 \ll \Phi$ was used to ignore the second term (involving U^2) in the denominator of (5.58). This term actually prevents the singularity by providing a non-zero value while $\sin(k\Delta x) \to 0$. Nevertheless, the actual errors from (5.58) can be comparable to those seen in Figure 5.3a, with relative error exceeding 2 for waves as long as $6\Delta x$. In addition, the amplitude can take the wrong sign when $\left| \mathcal{D}_{2x}\mathcal{C}_{\mathcal{L}}(0)\sqrt{\Phi} \right| < \left| \mathcal{S}_{\mathcal{L}}(0)U \right|$ in (5.58). These amplitude errors might be partially offset by damping introduced by uncentering and from the semi-Lagrangian interpolations, both of which were not considered in this section.

6. Finite Element Approximations.

The finite element method is based on principles similar to those of the spectral method. Both methods represent the unknown variables by a finite series of linearly independent basis functions with a prescribed spatial structure as in (3.1) for the spectral case. The coefficients of the series are the unknowns that must be determined as a function of time. They are determined by multiplying the approximate (series form of the) equations by one of the basis functions as a test function, and integrating over the domain as in going from (3.10) to (3.11) with the spectral method. This Galerkin approach results in the error being orthogonal to each basis function. The difference between the spectral and finite element approaches is that the spectral method uses global basis functions that are orthogonal as in (3.2). The finite element method uses basis functions that are zero except in limited regions, usually spanning the equivalent of only a few grid cells, where they are piecewise linear or low order polynomials. These functions are generally not orthogonal, but normally only a few adjacent elements interact. Because the elements are not orthogonal, the method results in a matrix that must be inverted. However, since only adjacent elements interact with each other, the matrix is banded and relatively inexpensive to invert. Finite element methods applied over uniform latitude-longitude grids as in Figure 1.1a suffer from the same Courant number restriction as finite difference schemes because the elements are local and do not possess the global filtering properties of the spectral basis functions of (3.14). However, coupled with semi-implicit, semi-Lagrangian approximations, finite element schemes lose this restriction and are very suitable for spherical geometry (Côté et al. [17]).

7. Reduced Grids.

In the uniform latitude-longitude grid illustrated in Figure 1.1a, the longitudinal grid points converge approaching the poles. As described earlier this convergence results in a very restrictive permissible time step for finite difference and finite element schemes. The spectral transform method avoids this restriction by excluding from its representation the shortest longitudinal waves associated with this grid near the poles. A grid more like that in Figure 1.1b, referred to today as a reduced grid, is consistent with these spectral basis functions (Hortal and Simmons [32], Naughton et al. [46]) and in fact is now used in some spectral transform models. The semi-Lagrangian method does not have a time step restriction and thus can be used with the grid in Figure 1.1a. However, that grid is computationally wasteful since the grid points are very close together in the polar region, resulting in excessive longitudinal resolution. Semi-Lagrangian schemes can also easily be based on grids like that shown in Figure 1.1b. For

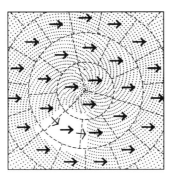

Figure 7.1. Illustration of variability of spherical coordinate unit vectors over grid boxes near the poles.

the two-dimensional interpolation required to find the value at the departure point, a series of one-dimensional longitudinal interpolations are done first at the latitudes surrounding the departure point to produce values at the longitude of the departure point as described in Section 4.8. These interpolations are readily done because the grid points are aligned along latitude circles. Then these interpolated values (at a common longitude) are interpolated in latitude to obtain the value at the departure point latitude.

As mentioned earlier, finite difference schemes are less amenable to application on grids such as that in Figure 1.1b because of the potential introduction of differential phase errors. The variability of the longitudinal and latitudinal unit vectors over the grid boxes also introduces severe difficulties. This is illustrated in Figure 7.1, which shows a polar stereographic projection of the grid in Figure 1.1b. Consider the case of a flow over the poles as indicated by the arrows in the centers of the grid boxes. Finite difference approximations often involve an underlying assumption about the variation over the grid box. The simplest is that the grid point value represents a grid box average, although modern schemes assume more complex variation. Any such assumption can lead to significant errors when applied to vector components near the poles. Consider the highlighted box in Figure 7.1 and the assumption that the grid point value represents a grid box average. The zonal wind component u illustrated on the right box wall is a reasonable approximation to the true wind in the center. But the zonal wind component illustrated on the left box wall is not, and is rotated by almost 45 degrees to the true wind. Thus an assumption that the wind over the box is representable by a single value of the vector components (u, v) is highly questionable. The variation of the unit vectors over the grid box should also be taken into account in such a way that the original assumptions of the scheme apply to the vector rather than to the individual components. Purser [50] indicates how to include the variation with respect to longitude of the unit vectors when dealing with vector components. The latitudinal aspect can also easily be included.

8. Vertical Coordinates.

So far we have considered only the horizontal aspects of numerical approximations for atmospheric models. The vertical dimension adds an additional degree of complexity out of proportion to just being a third dimension. This is in part because of the complex shape of the mountainous Earth's surface, in part because of the preferential direction of gravity toward the center of the earth coupled with the shallowness of the atmosphere relative to the radius of the earth, and in part because of the fact that the water vapor mixing ratio varies by several orders of magnitude between the Earth's surface and the tropopause. We continue to concentrate here on global models designed to forecast or simulate synoptic-scale and large-scale atmospheric motions for which hydrostatic equilibrium is a valid assumption.

Mathematically, the most natural vertical coordinate for global models is the radial distance from the center of the earth or its linear transformation, height above mean sea level, denoted z. The hydrostatic approximation

$$\frac{\partial p}{\partial z} = -\frac{gp}{RT},$$
(8.1)

where g is gravity, allows us to choose variables other than geometric height for the vertical coordinate. In all coordinate systems, the vertical terms involve integrals, which can be approximated by well known methods such as the trapezoidal rule, and differentials, which can be approximated by the methods described in the previous sections. Finite difference, finite element and semi-Lagrangian approximations are all being used in the vertical in models today. Because of the nature of the top and bottom boundary conditions, the spectral method has been less successful in the vertical. Constraints are often applied to the vertical approximations to ensure that various energy and angular momentum conservation properties of the continuous equations are satisfied by the approximate equations (Simmons and Burridge [63], Arakawa and Suarez [3], Arakawa and Konor [4]). The issues of accuracy and stability described in the previous sections with regard to the horizontal approximations are equally valid for the vertical coordinate. We will not repeat them here but rather concentrate on the choice of vertical coordinate. For more details on vertical coordinates see Haltiner and Williams [27] and Sundqvist [67].

In the z vertical coordinate system, the equation for the horizontal velocity $\mathbf{V}$ (assuming no forcing terms) is

$$\frac{d\mathbf{V}}{dt} + f\hat{\mathbf{k}} \times \mathbf{V} = -\frac{RT}{p}\nabla p,$$
(8.2)

where p is pressure, T is temperature, and R is the gas constant. Application of these equations is made difficult by the fact that the mountains stick up into the domain and thus cut holes in the horizontal coordinate surfaces (levels of constant z). Lateral boundary conditions must be applied at these irregular boundaries, and although the pressure gradient term is a simple gradient operator, its approximation at these boundaries is difficult. In addition, because the horizontal domain is incomplete, it is very difficult to apply the spectral transform method in the horizontal dimensions. Because of these and other difficulties the z system has not proven popular for global hydrostatic atmospheric models.

Pressure at first appears to be an attractive choice for a vertical coordinate because of its simple hydrostatic relation to z (8.1), because the continuity equation is particularly simple in that system, and because of the tradition of analyzing meteorological variables on pressure surfaces. However, as in the z system, mountains stick up into the domain and intersect surfaces of constant pressure. In addition, in the p system the location of these lateral boundaries is a function of time since the pressure in the atmosphere varies with time.

The most common vertical coordinate in use in global models today is a normalized pressure coordinate first introduced by Phillips [47]. This is usually referred to as the σ system after Phillips' original definition

$$\sigma = \frac{p}{p_s} \, , \tag{8.3}$$

where p_s denotes the surface pressure. In this system the Earth's surface is a coordinate surface and the difficulties of the z and p systems associated with mountains intersecting "horizontal" coordinate surfaces do not arise. Nevertheless, this advantage does not come without cost. In the σ system the equation for the horizontal velocity is

$$\frac{d\mathbf{V}}{dt} + f\hat{\mathbf{k}} \times \mathbf{V} = -\nabla\phi - RT\nabla \ln p_s \, , \tag{8.4}$$

where $\phi = gz$ is the geopotential. The pressure gradient is the sum of the two terms on the right-hand-side of (8.4). In the vicinity of mountains these two terms are of opposite sign and are significantly larger than the pressure gradient itself. Thus any errors in either term are amplified in a relative sense when they are combined. A common problem observed in atmospheric models is excessive precipitation locked to the orography. For example, almost all models show excessive precipitation over the Andes in January, probably because a pressure gradient error drives spurious moisture convergence over the mountains, leading to condensation. The error is not necessarily reduced by increasing vertical resolution as it may be due to the last term in (8.4) alone. The release of latent heat associated with the spurious condensation forces additional convergence, exacerbating the problem. The problem is amplified due to the fact that the last term in (8.4) contains a product of cross-differentiated terms. Indeed from (8.1) it can be seen that T is proportional to the vertical derivative of ϕ, and it is multiplied by the horizontal derivative of $\ln p_s$. In numerically discretizing (8.4) it is difficult to achieve a good degree of compensation between the last two terms due to this cross-differentiation effect. As a result, numerical truncation errors appear to converge very slowly with increasing spatial resolutions.

Another problem of the σ system is that the coordinate surfaces are never flat above sloping terrain. Indeed in an isothermal atmosphere, the σ surfaces all have exactly the same deviation as the underlying topography, no matter how high in the atmosphere, meaning that the problems associated with sloping surfaces are not reduced at higher levels in the σ system. A hybrid system introduced by Simmons and Burridge [63] is currently gaining favor in global models. This system looks like the σ system at the Earth's surface, but becomes a pressure system around the tropopause. Thus the pressure gradient error discussed above does not occur in the upper troposphere or stratosphere, but it does remain in the lower troposphere. Since most of the water vapor is concentrated in the lower troposphere, the orographically locked precipitation problem remains in this hybrid system. The equations in this system will be discussed in more detail in the following section.

To address this pressure gradient force problem in the region of steep terrain, Mesinger *et al.* [45] introduced a step-mountain coordinate that achieves approximately horizontal coordinate surfaces. Conceptually, the mountains are constructed from three-dimensional blocks (grid boxes), and above these blocks a σ–type system is used. This system also introduces horizontal boundaries into the problem while reducing the pressure gradient error. It is proving successful for regional models (Mesinger *et al.* [45]) but has not yet been widely adopted in global models.

Constant potential temperature (θ) or isentropic surfaces have always been appealing as a model vertical coordinate in spite of the fact that isentropes intersect the ground and the fact that isentropes can become locally nearly vertical (thus reducing vertical resolution there) or may even overturn in convectively unstable situations. Some of the potential advantages are that for adiabatic flow, isentropic surfaces are material surfaces and motions are "horizontal" in such a coordinate system; the pressure gradient force is a single term avoiding the need to difference two large terms as in the σ system; and fronts tend to be parallel to isentropes and thus the large gradients associated with them in other coordinate systems are avoided in the θ system and coarser grids can be used to provide equivalent accuracy. Although isentropic coordinates have been used for various applications almost since the beginning of atmospheric modeling, they never became common in global models because of their inherent difficulties. However, currently there is a resurgence of activity in the development of such approximations. Arakawa *et al.* [5] have developed a model in which artificial massless layers are introduced to simplify the lower boundary condition. They report difficulty, however, in defining the temperature at the surface because the number of degrees of freedom available to define the distribution of surface temperature is determined by the number of coordinate surfaces intersecting the ground. Zapotocny *et al.* [76] have developed a global model using a composite mesh approach with a σ system near the surface and θ above. Variables are interpolated between the two systems in a region of overlap. Zhu *et al.* [77] have derived a vertical discretization based on a general coordinate that changes smoothly from a terrain-following σ system at the ground to an isentropic system at higher levels. For a global model, however, the pure θ system cannot be attained much below the tropopause, so this system appears more useful for models that include a detailed stratosphere. Because of the problems with normalized vertical coordinates such as σ, because θ appears to offer many advantages, and because significant progress has been made in the development of local approximations with nonlinear constraints for transport and nonlinear fluid flow, there is a resurgence of developmental activity devoted to overcoming the remaining problems with the θ system in global models. Undoubtedly, coordinates based on θ will be much more common in global models in the future.

Other hybrid coordinate systems have been proposed. It is possible to conceive of coordinate systems that would follow topography at the surface and have a surface following, for example, the top of the boundary layer or the tropopause. Such coordinates may have advantages for special applications. The numerical difficulties associated with the implementation of semi-implicit time marching, however, may compromise the numerical stability of the scheme.

The vertical coordinate transformations that have been discussed above follow the traditional large scale meteorological approach. They simply correspond to a substitution of vertical placement of the dependent variables. Horizontal wind components are, however, still perfectly horizontal unlike "horizontal" gradients that are taken tangent to the coordinate surfaces and hence should really be called lateral gradients. On the other hand vertical derivatives and vertical motion are aligned in the vertical direction and not perpendicular to the coordinate system. Another approach would consist of employing curvilinear tensor transformations, e.g., Gal-Chen and Somerville [25] or Dutton [23]. So far the use of this approach has been restricted mostly to some small scale non-hydrostatic models.

9. Example of an AGCM

In this section we present the approximations in the NCAR CCM3 as an example of a baroclinic model and of the choices that must be made in designing a model. Complete details are provided by Kiehl et al. [37]. After listing the continuous equations we detail the vertical and temporal discretization, and then indicate how these are solved with the spectral transform method. As mentioned earlier, the CCM3 is based on the hybrid vertical coordinate system developed by Simmons and Burridge [63] and Simmons and Strüfing [64] following the generalized vertical coordinate of Kasahara [35]. The generalized coordinate, denoted $\eta(p, p_s)$, should be a monotonic function of p and should equal 1 for $p = p_s$ and 0 for $p = 0$. In the CCM3 implementation the top of the model, $\eta_t =$ constant, is taken at a nonzero constant pressure p_t. The boundary condition at the earth's surface, $\eta(p_s, p_s) = 1$, is $\dot{\eta}(p_s, p_s) = 0$, where $\dot{\eta} = d\eta/dt$, the vertical velocity in the η system. At the model top, $\eta(p_t, p_s) = \eta_t$, the boundary condition is $\dot{\eta}(p_t, p_s) = \omega(p_t) = 0$, i.e. the pressure vertical velocity is zero at the top.

The vertical coordinate is defined implicitly through

$$p(\eta, p_s) = A(\eta)p_0 + B(\eta)p_s \,, \tag{9.1}$$

where p_0 is a constant taken to be 10^5 Pa. The model equations can be written so they do not depend on the value of η, just on pressure p and the coordinate variable B in (9.1). The model levels are set by specifying A_k and B_k where k is the vertical grid level index. The algorithm used to generate the model levels is described by Williamson et al. [75]. Although η is not needed, it is convenient and more economical to define an η_k for use in the semi-Lagrangian vertical advection. In that case $\eta_k = A_k + B_k$. For lack of space we do not derive the continuous model equations in η here, but simply list them. The interested reader is referred to Simmons and Strüfing [64] and Kiehl et al. [37].

9.1 CONTINUOUS EQUATIONS

The continuous equations in the hybrid η system are

9.1.1 *Momentum*

$$\frac{du}{dt} = uv\frac{\tan\varphi}{a} + fv - \frac{1}{a\cos\varphi}\frac{\partial\Phi}{\partial\lambda} - R\frac{T_v}{a\cos\varphi}p_s\frac{B}{p}\frac{\partial\ln p_s}{\partial\lambda} + F_{uV} + F_{uH} \,, \tag{9.2}$$

$$\frac{dv}{dt} = -u^2\frac{\tan\varphi}{a} - fu - \frac{1}{a}\frac{\partial\Phi}{\partial\varphi} - R\frac{T_v}{a}p_s\frac{B}{p}\frac{\partial\ln p_s}{\partial\varphi} + F_{vV} + F_{vH} \,, \tag{9.3}$$

where

$$\frac{d\psi}{dt} = \frac{\partial\psi}{\partial t} + \mathbf{V}\cdot\nabla\psi + \dot{\eta}\frac{\partial\psi}{\partial\eta}. \qquad (9.4)$$

T_v denotes the virtual temperature, $T_v = [1 + (R_v/R)q]$, R is the gas constant for dry air, R_v is the gas constant for water vapor, and F_{u_V} and F_{v_V} represent parameterized momentum forcing including the surface drag, turbulent vertical diffusion and gravity wave drag, and F_{u_H} and F_{v_H} represent the horizontal diffusion.

9.1.2 Thermodynamic

$$\frac{dT}{dt} = \frac{R}{c_p}T_v\frac{\omega}{p} + Q + F_{T_H} + F_{F_H}, \qquad (9.5)$$

where Q represents the parameterized heating (solar and long-wave radiation, release of latent heat, surface sensible heat flux and sensible turbulent heating), F_{T_H} represents the horizontal diffusion of temperature, and F_{F_H} represents the conversion of kinetic energy into heating by the momentum diffusion.

9.1.3 Mass Continuity

$$\frac{\partial B}{\partial\eta}\left(\frac{\partial\ln p_s}{\partial t} + \mathbf{V}\cdot\nabla\ln p_s\right) = -\left(\frac{\partial p}{\partial\eta}\right)\frac{1}{p_s}\delta - \frac{1}{p_s}\frac{\partial}{\partial\eta}\left(\dot{\eta}\frac{\partial p}{\partial\eta}\right), \qquad (9.6)$$

where $\delta = \nabla\cdot\mathbf{V}$ is the horizontal divergence.

9.1.4 Hydrostatic

$$\Phi = \Phi_s + R\int_{p(\eta)}^{p(\eta_s)} T_v d\ln p. \qquad (9.7)$$

9.1.5 Water Vapor Continuity

$$\frac{dq}{dt} = S, \qquad (9.8)$$

where q is the specific humidity and S represents the parameterized change in atmospheric moisture which includes surface evaporation, convective and large scale precipitation, and vertical redistribution by convective processes. No explicit horizontal diffusion of specific humidity is included in CCM3.

9.2 TIME DIFFERENCING

The source terms Q, F_{ψ_V}, F_{ψ_H} and S in (9.2), (9.3), (9.5) and (9.8) are treated in a time split manner. The horizontal diffusion is calculated after the transformation to spectral space (described below). The stable condensation and convective

206

parameterizations are invoked after the new grid point values are re-synthesized from the predicted spectral coefficients. The time filter (2.51) is applied at the end of the sequence to the prognostic variables. The other parameterizations are also treated in a time split manner for water vapor, but they are not truly time split in the other prognostic equations. They are calculated first as if time split, but the result is converted to a rate and applied as a source term.

Consider the generic equation

$$\frac{\partial \psi}{\partial t} = \Gamma(\psi) + F(\psi),$$ (9.9)

where Γ represents the adiabatic dynamics and F represents the parameterizations. Eqn. (9.9) is approximated by

$$\psi^* = \psi^{n-1} + 2\Delta t F\left(\psi^*, \psi^{n-1}\right),$$ (9.10)

and

$$\psi^{n+1} = \psi^{n-1} + 2\Delta t \Gamma\left(\psi^{n+1}, \psi^n, \psi^{n-1}\right) + 2\Delta t F\left(\psi^*, \psi^{n-1}\right).$$ (9.11)

In (9.11) the time n is associated with explicit centered terms in the equation and times $n+1$ and $n-1$ with semi-implicit approximations. The distinction will become apparent below when specific equations are presented. Eqn. (9.10) is solved first, and the source rate $F\left(\psi^*, \psi^{n-1}\right)$ is saved. Then (9.11) is solved applying that rate as a forcing term. Substituting (9.10) into (9.11) gives

$$\psi^{n+1} = \psi^* + 2\Delta t \Gamma\left(\psi^{n+1}, \psi^n, \psi^{n-1}\right).$$ (9.12)

At first glance, the pair (9.10) and (9.12) look like normal time splitting. The subtle difference is that Γ uses ψ^{n-1} rather than ψ^* which would be the case in a proper split approximation. The main reason for the choice is that the parameterizations are formulated in terms of u and v, but (as we will see below) the spectral transform method is formulated using vorticity and divergence. For a proper split approximation the (u, v) forcings would have to be transformed to (ζ, δ) forms, requiring relatively expensive extra spectral transforms and re-synthesis. By using the form (9.11) the (u, v) forcings are included with other terms and are effectively transformed during the forecast at no extra cost.

Since the horizontal diffusion is included primarily to remove energy and enstrophy from the smallest scales, it is applied to (ζ, δ) in spectral space without ensuring that it is equivalent to what would be calculated if it were based on (u, v). The CCM uses a $\nabla^{2\ell}$ form with $\ell = 2$ throughout the troposphere and $\ell = 1$ at a few levels at the top of the model to serve as a simple sponge to absorb vertically propagating planetary wave energy and to control the strength of the winter stratospheric jets.

As indicated in (9.11) the dynamical equations involve semi-implicit approximations. Our earlier semi-implicit examples with finite difference and semi-Lagrangian approximations involved linear equations and resulted in particularly simple elliptic equations to solve. In order to achieve that simplicity in the non-linear three-dimensional equations, the terms responsible for the gravity waves are divided into a linear component and the remainder. The linear component is defined with respect to a specified reference atmosphere, and is approximated semi-implicitly. The remaining non-linear component is approximated explicitly. This will be described more completely in the following with reference to each individual equation involved. Simmons *et al.* [65] pointed out that with the sigma vertical coordinate system care must be taken in the choice of reference temperature profile. Computational instability arises if there is a significant difference between the actual temperature profile and the reference profile. They recommended use of a relatively warm isothermal reference atmosphere to avoid the instability. The hybrid vertical coordinate requires similar care and also requires specification of a reference surface pressure. Simmons and Burridge [63] showed that a careful choice was also required for the reference surface pressure. In addition they showed that a smooth transition to the pressure coordinate enhanced stability. The CCM3 adopts a 300K isothermal reference temperature and reference surface pressure of 10^5 Pa.

9.3 DISCRETE EULERIAN VERSION

9.3.1 *Water Vapor Continuity Equation*
In the "Eulerian" version of CCM3, the Eulerian spectral transform method is not actually used for water vapor. Instead, the specific humidity is predicted with a semi-Lagrangian formulation and monotonic Hermite interpolants to prevent over- and under-shooting and the associated spectral precipitation and negative values. The Eulerian spectral transform method is used for the dynamical component of CCM. A true time-split approximation is applied to the water vapor equation for horizontal and vertical advection, and the parameterized sources. Eqn. (9.8) can be written

$$\frac{dq}{dt} = \frac{\partial q}{\partial t} + \mathbf{V} \cdot \nabla q + \dot{\eta}\frac{\partial q}{\partial \eta} = S \,. \tag{9.13}$$

A time split Eulerian form would be

$$q^* = q^{n-1} + 2\Delta t S(q^{n-1}, q^*) \,, \tag{9.14}$$

$$q' = q^* + 2\Delta t \mathbf{V}^n \cdot \nabla q^* \,, \tag{9.15}$$

$$q^{n+1} = q' + 2\Delta t \dot{\eta}^n \frac{\partial q'}{\partial n} . \tag{9.16}$$

In the semi-Lagrangian form, (9.15) and (9.16) are replaced by

$$q' = L_{\lambda\varphi}(q^*) , \tag{9.17}$$

$$q^{n+1} = L_\eta(q') . \tag{9.18}$$

Eqn. (9.17) represents the horizontal interpolation of q^* at the departure point calculated with $\mathbf{V}^n$ assuming $\dot{\eta} = 0$. Eqn. (9.18) represents the vertical interpolation of q' at the departure point calculated with $\dot{\eta}^n$ assuming $\mathbf{V} = 0$.

The vertical and horizontal advections are split to save computer time. The splitting results in one two-dimensional horizontal interpolation associated with the horizontal advection, followed by one one-dimensional vertical interpolation associated with the vertical advection. The three dimensional form would require four two-dimensional horizontal interpolations followed by one one-dimensional interpolation as alluded to in Section 4.8. Experience has shown that the two approaches produce comparable simulations when coupled with the Eulerian dynamics. This is not the case when semi-Lagrangian approximations are used for the dynamical equations as will be discussed later.

The horizontal trajectories are calculated as described earlier in (4.43) and (4.44) equatorward of $\pm70°$ latitude and with the transformed system described in Section 4.6 poleward of $\pm70°$. The vertical trajectories are calculated with the equivalent of (4.28).

9.3.2 Mass Continuity Equation
The discrete semi-implicit continuity equation is

$$\Delta B_k \frac{(\ln p_s)^{n+1} - (\ln p_s)^{n-1}}{2\Delta t} = -\mathbf{V}_k^n \cdot \nabla \ln p_s^n \Delta B_k - \Delta \left(\frac{1}{p_s} \dot{\eta} \frac{\partial p}{\partial \eta} \right)_k^n - \frac{1}{p_s^n} \delta_k^n \Delta p_k^n$$
$$- \left[\overline{\left(\frac{1}{p_s^r} \delta_k \Delta p_k^r \right)}^n - \overline{\left(\frac{1}{p_s^r} \delta_k \Delta p_k^r \right)}^n \right] , \tag{9.19}$$

where horizontal grid point indices (i, j) have been suppressed, k denotes the vertical level,

$$\Delta(\ )_k = (\ )_{k+\frac{1}{2}} - (\ )_{k-\frac{1}{2}} \tag{9.20}$$

denotes a vertical difference, and the overbar denotes the average of values at the ends of the trajectory as defined earlier in (4.33) with $\alpha = 0$

$$\overline{(\ )}_j^n = \frac{1}{2} \left[(\ )_j^{n+1} + (\ )_{j_D}^{n-1} \right] . \tag{9.21}$$

The reference atmospheric pressure p^r is calculated from (9.1) with p^r_s in place of p_s. The semi-implicit approximations have been obtained by linearizing the term involving the divergence and adding the linearized variant in a form $(\ )^n - \frac{1}{2}\left[(\ )^{n+1} + (\ )^{n-1}\right]$ which goes to zero as the time step goes to zero. If the original equation were linear as in the shallow water example (2.72)–(2.74), the term at time n would cancel leaving only the average of times $n-1$ and $n+1$. Eqn. (9.19) looks a little unusual in that the right-hand side terms depend on the level k, yet the left side is independent of k, except for the ΔB_k term which can be moved to the right-hand side. In fact, $\dot\eta$ is determined from (9.19) to ensure that (9.19) holds at all levels.

The prognostic equation for $\ln p_s$ is obtained by summing (integrating) (9.19) in the vertical, applying boundary conditions $\dot\eta = 0$ at the top and bottom, and noting that $\sum_{\ell=1}^{K} \Delta B_\ell = 1$:

$$
\begin{aligned}
(\ln p_s)^{n+1} = {} & (\ln p_s)^{n-1} - 2\Delta t \sum_{\ell=1}^{K} \mathbf{V}^n_\ell \cdot \nabla \ln p^n_s \Delta B_\ell \\
& - 2\Delta t \sum_{\ell=1}^{K} \frac{1}{p^n_s} \delta^n_\ell \Delta p^n_\ell - 2\Delta t \sum_{\ell=1}^{K} \frac{1}{p^r_s} \Delta p^r_\ell \left[\overline{(\delta_\ell)}^n - \delta^n_\ell\right] .
\end{aligned}
\tag{9.22}
$$

The vertical velocity is given by summing (from the top of the model to the level k) an explicit form of (9.19) obtained by dropping the semi-implicit term in square brackets, and substituting a corresponding explicit form of (9.22) for $\left[(\ln p_s)^{n+1} - (\ln p_s)^{n-1}\right]$:

$$
\begin{aligned}
\left(\dot\eta \frac{\partial p}{\partial \eta}\right)^n_{k+\frac{1}{2}} = {} & B_{k+\frac{1}{2}} \sum_{\ell=1}^{K} [\delta^n_\ell \Delta p^n_\ell + \mathbf{V}^n_\ell \cdot p^n_s \nabla \ln p^n_s \Delta B_\ell] \\
& - \sum_{\ell=1}^{k} [\delta^n_\ell \Delta p^n_\ell + \mathbf{V}^n_\ell \cdot p^n_s \nabla \ln p^n_s \Delta B_\ell] .
\end{aligned}
\tag{9.23}
$$

To derive (9.23) note that $\sum_{\ell=1}^{k} \Delta B_\ell = B_{k+\frac{1}{2}}$.

The vertical velocity ω is related to $\ln p_s$ and $\dot\eta$ directly from its definition,

$$
\omega \equiv \frac{dp}{dt} = \frac{\partial p}{\partial t} + \mathbf{V} \cdot \nabla p + \dot\eta \frac{\partial p}{\partial \eta} .
$$

Thus ω/p, which will be needed in the thermodynamic equation, is given by

$$
\frac{\omega}{p} = \frac{p_s}{p} \left[B(\eta) \left(\frac{\partial \ln p_s}{\partial t} + \mathbf{V} \cdot \nabla \ln p_s \right) + \frac{1}{p_s} \dot\eta \frac{\partial p}{\partial \eta} \right] .
\tag{9.24}
$$

Substitution of the discrete difference from (9.22) for $\partial \ln p_s / \partial t$ and the average of (9.23) at $(k+1/2)$ and $(k-1/2)$ for $(\dot{\eta}\partial p/\partial \eta)_k$ gives

$$
\left(\frac{\omega}{p}\right)_k^n = \frac{B_k}{p_k^n} \mathbf{V}_k^n \cdot p_s^n \nabla \ln p_s^n
$$

$$
- \sum_{\ell=1}^{k} C_{k\ell} \left[\delta_\ell^n \Delta p_\ell^n + \mathbf{V}_\ell^n \cdot p_s^n \nabla \ln p_s^n \Delta B_\ell \right],
$$

(9.25)

where

$$
C_{k\ell} = \begin{cases} \frac{1}{p_k} & \text{for } \ell < k \\ \frac{1}{2p_k} & \text{for } \ell = k, \end{cases}
$$

(9.26)

and we restrict the vertical grid such that

$$
B_k = \frac{1}{2}\left(B_{k+\frac{1}{2}} + B_{k-\frac{1}{2}}\right)
$$

(9.27)

in order to avoid a term of the form $\sum_{\ell=1}^{K}$ summing throughout the entire depth of the atmosphere which is inconsistent with a vertical integral of (9.6) through only part of the atmosphere to get $\dot{\eta}$. The restriction (9.27) also makes it easier to achieve energy consistency between the thermodynamic and momentum equations.

9.3.3 Thermodynamic Equation

For the convenience of the spectral transform method, the horizontal advection $\mathbf{V} \cdot \nabla T$ in (9.5) is written as $\nabla \cdot (\mathbf{V}T) - T\delta$. In addition, since the continuity equation provides an equation for $\dot{\eta}\frac{\partial p}{\partial \eta}$ rather than $\dot{\eta}$ alone, we convert the vertical advection to the form $\dot{\eta}\frac{\partial p}{\partial \eta}\frac{\partial T}{\partial p}$. With these substitutions the discrete form of the thermodynamic equation without the diffusion terms is

$$
(\underline{T})^{n+1} = (\underline{T})^{n-1} - 2\Delta t \left[\frac{1}{a\cos^2\phi} \frac{\partial}{\partial\lambda}(\underline{UT})^n + \frac{1}{a\cos\phi}\frac{\partial}{\partial\phi}(\underline{VT})^n - \underline{\Gamma}^n \right]
$$

$$
- 2\Delta t \mathbf{D}^r \left[(\underline{\delta})^n - \underline{\delta}^n \right],
$$

(9.28)

where

$$
\Gamma_k = T_k \delta_k + \frac{RT_{vk}}{c_p}\left(\frac{\omega}{p}\right)_k - Q
$$

$$
- \frac{1}{2\Delta p_k}\left[\left(\dot{\eta}\frac{\partial p}{\partial\eta}\right)_{k+1/2}(T_{k+1} - T_k) + \left(\dot{\eta}\frac{\partial p}{\partial\eta}\right)_{k-1/2}(T_k - T_{k-1})\right],
$$

(9.29)

and

$$D^r_{k\ell} = \frac{R}{c_p} T^r C^r_{\ell k} \Delta p^r_\ell . \tag{9.30}$$

The reference isothermal atmospheric temperature is denoted T^r. The underline in terms such as $(\underline{T})^{n+1}$ and $(\underline{UT})^n$ in (9.28) and the following denotes the column vector over the vertical levels, with components T^{n+1}_k and $(U_k T_k)^n$ respectively. Note that in (9.28), only the component of $(\omega/p)_k$ from (9.25) involving the divergence is approximated in a semi-implicit manner, and again it is linearized about a reference atmosphere.

9.3.4 Hydrostatic Equation
The general form of the discrete hydrostatic equation is

$$\Phi_k = \Phi_s + R \sum_{\ell=k}^{K} H_{k\ell} (p) T_{v\ell} . \tag{9.31}$$

The obvious choice for the integration matrix $H_{k\ell}$ gives terms like $\Delta \ln p_\ell$. However that is not used in the CCM and $\mathbf{H}$ is chosen to ensure energy conservation. This will be described in Section 9.3.6 following the momentum equations.

9.3.5 Momentum Equations
Although the momentum equations in vector component form (9.2) and (9.3) can be used for spectral transform models (Ritchie [54], [55]), it is more common to base the approximations on the vorticity and divergence form of the equations following the approach described earlier for the shallow water equations. First write (9.2) and (9.3) in the form

$$\frac{\partial u}{\partial t} = (\zeta + f)v - \dot{\eta}\frac{\partial p}{\partial \eta}\frac{\partial u}{\partial p} - \frac{1}{a \cos \varphi}\frac{\partial \Phi}{\partial \lambda} - R\frac{T_v}{a \cos \varphi}p_s\frac{B}{p}\frac{\partial \ln p_s}{\partial \lambda} + F_{uv} + F_{u_H} , \tag{9.32}$$

$$\frac{\partial v}{\partial t} = -(\zeta + f)u - \dot{\eta}\frac{\partial p}{\partial \eta}\frac{\partial v}{\partial p} - \frac{1}{a}\frac{\partial \Phi}{\partial \varphi} - R\frac{T_v}{a}p_s\frac{B}{p}\frac{\partial \ln p_s}{\partial \varphi} + F_{vv} + F_{v_H} . \tag{9.33}$$

The vorticity and divergence were defined earlier in the shallow water spectral transform example by (3.28) and (3.29). Then by taking the curl and divergence of (9.32) and (9.33) we obtain

$$\frac{\partial \zeta}{\partial t} = \mathbf{k} \cdot \nabla \times (\mathbf{n}/\cos \varphi) + F_{\zeta_H} \tag{9.34}$$

and

$$\frac{\partial \delta}{\partial t} = \nabla \cdot (\mathbf{n}/\cos \varphi) - \nabla^2 (E + \Phi) + F_{\delta_H} , \tag{9.35}$$

where

$$E = \frac{1}{2}\left(u^2 + v^2\right) \tag{9.36}$$

and the vector $\mathbf{n} = (n_U, n_V)$ is given by

$$n_U = +(\zeta + f)V - \dot{\eta}\frac{\partial p}{\partial \eta}\frac{\partial U}{\partial p} - R\frac{T_v}{a}p_s\frac{B}{p}\frac{\partial \ln p_s}{\partial \lambda} + F_{U_V} \tag{9.37}$$

and

$$n_V = -(\zeta + f)U - \dot{\eta}\frac{\partial p}{\partial \eta}\frac{\partial V}{\partial p} - R\frac{T_v\cos\varphi}{a}p_s\frac{B}{p}\frac{\partial \ln p_s}{\partial \varphi} + F_{V_V} . \tag{9.38}$$

The Laplacian ∇^2 is given in (3.23) and the wind components U and V are the usual spherical components u and v multiplied by $\cos\varphi$ as in (3.33) and (3.34). The terms F_{ζ_H} and F_{δ_H} represent the horizontal diffusion which is usually applied to vorticity and divergence to prevent the build up of energy in the smallest resolved waves. The horizontal diffusion is calculated in a time split manner in spectral space after the vorticity and divergence are predicted at time $n+1$ and thus those terms will be dropped from the equations in the subsequent development. The terms F_{U_V} and F_{V_V} represent the forcing from the physical parameterizations such as surface stress and vertical mixing in the planetary boundary layer. They are naturally formulated in terms of u and v rather than ζ and δ. Hence as described earlier it is convenient to calculate them first using a time split procedure and then include them in n_U and n_V in which terms they will be transformed to their vorticity and divergence equivalents.

The discrete forms of the components of $\mathbf{n}$ are

$$(n_U)_k = (\zeta_k + f)\,V_k - RT_{vk}B_k\frac{p_s}{p_k}\frac{1}{a}\frac{\partial \ln p_s}{\partial \lambda}$$
$$- \frac{1}{2\Delta p_k}\left[\left(\dot{\eta}\frac{\partial p}{\partial \eta}\right)_{k+1/2}(U_{k+1} - U_k) + \left(\dot{\eta}\frac{\partial p}{\partial \eta}\right)_{k-1/2}(U_k - U_{k-1})\right]$$
$$+ (F_{U_V})_k \tag{9.39}$$

and

$$(n_V)_k = -(\zeta_k + f)\,U_k - RT_{vk}B_k\frac{p_s}{p_k}\frac{\cos\varphi}{a}\frac{\partial \ln p_s}{\partial \varphi}$$
$$- \frac{1}{2\Delta p_k}\left[\left(\dot{\eta}\frac{\partial p}{\partial \eta}\right)_{k+1/2}(V_{k+1} - V_k) + \left(\dot{\eta}\frac{\partial p}{\partial \eta}\right)_{k-1/2}(V_k - V_{k-1})\right]$$
$$+ (F_{V_V})_k . \tag{9.40}$$

Dropping the horizontal diffusion terms F_{ζ_H} and F_{δ_H}, the vorticity equation is

$$\underline{\zeta}^{n+1} = \underline{\zeta}^{n-1} + 2\Delta t \mathbf{k} \cdot \nabla \times (\underline{\mathbf{n}}^n / \cos \varphi) , \qquad (9.41)$$

and the divergence equation after substitution of the hydrostatic equation (9.31) for Φ is

$$
\begin{aligned}
\underline{\delta}^{n+1} = {}&\underline{\delta}^{n-1} + 2\Delta t \left[\nabla \cdot (\underline{\mathbf{n}}^n / \cos \varphi) - \nabla^2 \left(\underline{E}^n + \Phi_s \underline{1} + R\mathbf{H}^n (\underline{T_v})^n \right) \right] \\
&- 2\Delta t R\mathbf{H}^r \nabla^2 \left[\overline{(\underline{T})}^n - \underline{T}^n \right] \\
&- 2\Delta t R\underline{T}^r \nabla^2 \left[\overline{(\ln p_s)}^n - \ln p_s^n \right] ,
\end{aligned}
\qquad (9.42)
$$

where $\underline{1}$ denotes the unit column vector consisting of all ones. In choosing the semi-implicit component, we linearize the ∇^2 term to give a term involving T and linearize the $\ln p_s$ terms in $\mathbf{n}$ to give a $\nabla^2 \ln p_s$ contribution. Again notice that the terms at time n would drop out in a linear system leaving only the time average.

9.3.6 *Energy Conservation*
In the continuous equations, the energy conversion term in the thermodynamic equation (first term on the right side of (9.5)) and the pressure gradient term in the momentum equations exactly cancel when global integrals of the total energy equation are taken. There is a relationship between the two integration matrices, $\mathbf{C}$ in the ω equation and $\mathbf{H}$ in the hydrostatic equation, so that these terms also balance in the discrete equations. The relevant terms in the thermodynamic equation (9.28) are

$$\left(\frac{\partial T}{\partial t} \right)_k = \frac{RT_{vk}}{c_p} \left(\frac{\omega}{p} \right)_k + \cdots , \qquad (9.43)$$

with (9.25) for $(\omega/p)_k$. The relevant terms in the kinetic energy equation consistent with the prognostic equations (9.41) and (9.42) for vorticity and divergence are

$$\frac{\partial}{\partial t} \left(\frac{1}{2} \mathbf{V} \cdot \mathbf{V} \right)_k = -\mathbf{V}_k \cdot \left[R\nabla \sum_{\ell=k}^{K} \mathrm{H}_{k\ell} T_{v\ell} + RT_{vk} \frac{B_k}{p_k} p_s \nabla \ln p_s \right] + \cdots . \qquad (9.44)$$

Multiplying (9.43) by $c_p \partial p / \partial \eta$ and (9.44) by $\partial p / \partial \eta$, combining each with the continuity equation multiplied by T or $\mathbf{V} \cdot \mathbf{V}$ as appropriate, and integrating them throughout the atmosphere ultimately, after a lot of algebraic manipulation summarized in Kiehl *et al.* [37], leads to the desired relation:

$$C_{\ell k} \Delta p_\ell = \mathrm{H}_{k\ell} . \qquad (9.45)$$

214

Since $C_{\ell k}$ was chosen so that ω would be consistent with the continuity equation (9.26), we define $\mathrm{H}_{k\ell}$ by (9.45).

9.3.7 Semi-Implicit System

To prepare for the spectral transform and solution of the semi-implicit equations, in each prognostic equation (9.22, 9.28, 9.41, 9.42) we combine terms involving longitudinal derivatives, terms involving latitudinal derivatives, terms involving the Laplacian and all undifferentiated terms. The vorticity equation (9.41) is

$$\underline{\zeta}^{n+1} = \underline{V} + \frac{1}{a(1-\mu^2)} \left[\frac{\partial}{\partial\lambda}(\underline{V}_\lambda) - (1-\mu^2)\frac{\partial}{\partial\mu}(\underline{V}_\mu) \right], \qquad (9.46)$$

where

$$\mu = \sin\varphi, \qquad (9.47)$$

and the terms on the right-hand side consist of known quantities. The undifferentiated term $\underline{V}$ is just $\underline{\zeta}^{n-1}$. The two differentiated terms $\underline{V}_\lambda$ and $\underline{V}_\mu$ consist of the components of $\mathbf{n}^n$.

The divergence equation (9.42) is

$$\underline{\delta}^{n+1} + \Delta t \nabla^2 (R\mathbf{H}^r \underline{T}^{n+1} + R\underline{T}^r \ln p_s^{n+1}) =$$
$$\underline{D} + \frac{1}{a(1-\mu^2)} \left[\frac{\partial}{\partial\lambda}(\underline{D}_\lambda) + (1-\mu^2)\frac{\partial}{\partial\mu}(\underline{D}_\mu) \right] - \nabla^2 \underline{D}_\nabla. \qquad (9.48)$$

The undifferentiated term $\underline{D}$ is just $\underline{\delta}^{n-1}$ plus the terms from the second and third lines of (9.42) involving $\underline{T}^n$, $\underline{T}^{n-1}$, $\overline{(\ln p_s)}^n$, and $\overline{(\ln p_s)}^{n-1}$. The two differentiated terms $\underline{D}_\lambda$ and $\underline{D}_\mu$ consist of the components of $\mathbf{n}^n$, and $\underline{D}_\nabla$ is $2\Delta t \left(\underline{E}^n + \Phi_s \underline{1} + R\mathbf{H}^n (\underline{T}_v)^n \right)$.

The thermodynamic equation (9.28) is

$$\underline{T}^{n+1} + \Delta t \mathbf{D}^r \underline{\delta}^{n+1} = \underline{T} - \frac{1}{a(1-\mu^2)} \left[\frac{\partial}{\partial\lambda}(\underline{T}_\lambda) + (1-\mu^2)\frac{\partial}{\partial\mu}(\underline{T}_\mu) \right]. \qquad (9.49)$$

The undifferentiated term $\underline{T}$ includes $\underline{T}^{n-1}$ and terms in $\underline{\Gamma}^n$, $\underline{\delta}^n$, and $\underline{\delta}^{n-1}$ multiplied by appropriate factors. The differentiated terms $\underline{T}_\lambda$ and $\underline{T}_\mu$ contain $(\underline{UT})^n$ and $(\underline{VT})^n$ again multiplied by the appropriate factors.

The surface pressure equation (9.22) is

$$\ln p_s^{n+1} + \frac{\Delta t}{p_s^r} \left(\underline{\Delta p}^r \right)^T \underline{\delta}^{n+1} = PS, \qquad (9.50)$$

where PS contains everything on the right-hand side of (9.22) except the term involving $\underline{\delta}^{n+1}$. The terms on the right-hand sides of (9.48)–(9.50) are all known.

Expanding into spectral space gives

$$\underline{\zeta}_n^m = \underline{VS}_n^m , \tag{9.51}$$

where $(\ )_n^m$ denotes the spherical harmonic coefficient and the time index $(n+1)$ is suppressed from the left-hand side of the equation. The term on the right-hand side is known and represents the transformation of the terms on the right-hand side of (9.46).. The coupled implicit equations for the spectral coefficients become

$$\underline{\delta}_n^m - \Delta t \frac{n(n+1)}{a^2} [RH^r \underline{T}_n^m + R\underline{T}^r (\ln p_s)_n^m] = \underline{DS}_n^m , \tag{9.52}$$

$$\underline{T}_n^m + \Delta t \mathbf{D}^r \underline{\delta}_n^m = \underline{TS}_n^m , \tag{9.53}$$

$$(\ln p_s)_n^m + (\underline{\Delta p}^r)^T \frac{\Delta t}{p_s^r} \underline{\delta}_n^m = \underline{PS}_n^m , \tag{9.54}$$

where again the terms on the left sides of the equations are the predicted values at time $n+1$ (though recall that the $n(n+1)$ factor in (9.52) refers to the degree of the associated Legendre function) and those on the right-hand sides are the transforms of terms involving known quantities at times $n-1$ and n. Solving the system for $\underline{\delta}_n^m$ gives

$$\mathbf{A}_n \underline{\delta}_n^m = \underline{DS}_n^m + \Delta t \frac{n(n+1)}{a^2} [RH^r (\underline{TS})_n^m + R\underline{T}^r (PS)_n^m] , \tag{9.55}$$

where

$$\mathbf{A}_n = \mathbf{I} + \Delta t^2 \frac{n(n+1)}{a^2} \left[RH^r \mathbf{D}^r + R\underline{T}^r (\underline{\Delta p}^r)^T \frac{1}{p_s^r} \right] , \tag{9.56}$$

which is simply a set of K simultaneous equations for the coefficients with given wavenumbers (m, n) at each level and is solved by inverting $\mathbf{A}_n$. Note that $\mathbf{A}_n$ depends only on n and by construction is independent of time. It can be inverted once and stored. The space required is not excessive since its order is only the number of levels K, and N matrix inversions are required.

Acknowledgments

We would like to thank Bernard Nemry, Elissaveta Peneva and Philip Mote for comments on the original draft and Paul Swarztrauber for discussions on various subtle aspects of approximations. The National Center for Atmospheric Research is sponsored by the National Science Foundation.

216

References

1. Arakawa, A. (1966) Computational design for long-term numerical integration of the equations of fluid motion: Two-dimensional incompressible flow. Part I, *J. Comput. Phys.* **1**, 110-143.
2. Arakawa, A. and Lamb, V.R. (1977) Computational design of the basic dynamical processes of the UCLA general circulation model, *Methods in Computational Physics*, **17**, Academic Press, 174-265.
3. Arakawa, A. and Suarez, M.J. (1983) Vertical differencing of the primitive equations in sigma coordinates, *Mon. Wea. Rev.* **111**, 34-45.
4. Arakawa, A. and Konor, C.S. (1996) Vertical differencing of the primitive equations based on the Charney-Phillips grid in hybrid σ-p vertical coordinates, *Mon. Wea. Rev.* **124**, 511-528.
5. Arakawa, A., Mechoso, C.R. and Konor, C.S. (1992) An isentropic vertical coordinate model: Design and application to atmospheric frontogenesis studies, *Meteorol. Atmos. Phys.* **50**, 31-45.
6. Asselin, R. (1972) Frequency filter for time integrations, *Mon. Wea. Rev.* **100**, 487-490.
7. Bates, J.R. and McDonald, A. (1982) Multiply-Upstream, semi-Lagrangian advective schemes: Analysis and application to a multi-level primitive equation model, *Mon. Wea. Rev.* **110**, 1831-1842.
8. Bates, J.R., Semazzi, F.H.M., Higgins, R.W. and Barros, S.R. (1990) Integration of the shallow water equations on the sphere using a vector semi-Lagrangian scheme with a multigrid solver, *Mon. Wea. Rev.* **118**, 1615-1627.
9. Bates, J.R., Moorthi, S. and Higgins, R.W. (1993) A global multilevel atmospheric model using a vector semi-Lagrangian finite difference scheme. Part I: Adiabatic formulation. *Mon. Wea. Rev.* **121**, 244-263.
10. Boer, G.J., McFarlane, N.A., Laprise, R., Henderson, J.D. and Blanchet, J.-P. (1984) The Canadian Climate Centre spectral atmospheric General Circulation Model. *Atmosphere-Ocean* **22**, 397-429.
11. Bourke, W. (1972) An efficient, one-level, primitive-equation spectral model, *Mon. Wea. Rev.* **100**, 683-689.
12. Boville, B.A. (1991) Sensitivity of simulated climate to model resolution, *J. Climate* **4**, 469-485.
13. Browning, G.L., Hack, J.J. and Swarztrauber, P.N. (1989) A Comparison of Three Numerical Methods for Solving Differential Equations on the Sphere, *Mon. Wea. Rev.* **117**, 1058-1075.
14. Caya, D. and Laprise, R. (1998) A semi-Lagrangian semi-implicit regional climate model: The Canadian RCM, *Mon. Wea. Rev.* **126**, in press.
15. Chen, M. and Bates, J.R. (1996) A comparison of climate simulations from a semi-Lagrangian and an Eulerian GCM, *J. Climate* **9**, 1126-1149.
16. Coiffier, J., Chapelet, P. and Marie N. (1987): Study of various quasi-Lagrangian techniques for numerical models, in *Proceedings ECMWF Workshop on Techniques for Horizontal Discretization in Numerical Weather Prediction Models*, European Centre for Medium Range Weather Forecasts, Reading, United Kingdom, pp. 19-46.
17. Côté, J., Roch, M., Staniforth, A. and Fillion, L. (1993) A variable-resolution semi-Lagrangian finite-element global model of the shallow-water equations, *Mon. Wea. Rev.* **121**, 231-243.
18. Côté, J., Gravel, S. and Staniforth, A. (1995) A generalized family of schemes that eliminate the spurious resonant response of semi-Lagrangian schemes to orographic forcing, *Mon. Wea. Rev.* **123**, 3605-3613.

19. Côté, J., Gravel, S., Méthot, A., Patoine, A., Roch, M. and Staniforth, A. (1998) The operational CMC-MRB global environmental multiscale (GEM) model. Part I: Design considerations and formulation, *Mon. Wea. Rev.* **126**, 1373-1395.

20. Courant, R., Friedrichs, K.O. and Lewy, H. (1928) Über die partiellen differenzengleichungen der mathematischen physik, *Math. Annalen* **100**, 32-74.

21. Desharnais, F. and Robert, A. (1990) Errors near the poles generated by a semi-Lagrangian integration scheme in a global spectral model, *Atmosphere-Ocean* **28**, 162-176.

22. Durran, D.R. (1999) *Numerical Methods for Wave Equations in Geophysical Fluid Dynamics*, Texts in Applied Mathematics, Vol. 32, Springer-Verlag, New York.

23. Dutton, J.A. (1986) *The ceaseless wind*, McGraw-Hill, New York.

24. Fritsch, F.N. and Carlson, R.E. (1980) Monotone piecewise cubic interpolation, *SIAM J. Numer. Anal.* **17**, 238-246.

25. Gal-Chen, T. and Somerville, R.C.J. (1975) On the use of coordinate transformation for the solution of the Navier-Stokes equations, *J. Comput. Phys.* **17**, 209-228.

26. Gates, W.L. (1992) AMIP: The Atmospheric Model Intercomparison Project. *Bull. Amer. Meteor. Soc.* **73**, 1962-1970.

27. Haltiner, G.J. and Williams, R.T. (1980) *Numerical Prediction and Dynamic Meteorology*, John Wiley & Sons, New York.

28. Healy, D.M., Jr., Rockmore, D.N., Moore, S.S.B. and Kostelec, P. (1999) FFTs for the 2-sphere - Improvements and variations, *Advances in Applied Math,* in press.

29. Héreil, P. and Laprise, R. (1996) Sensitivity of internal gravity wave solutions to the time step of a semi-implicit semi-Lagrangian non-hydrostatic model, *Mon. Wea. Rev.* **124**, 972-999.

30. Hildebrand, F.B. (1956) *Introduction to numerical analysis*, McGraw-Hill, New York.

31. Holzer, M. (1996) Optimal spectral topography and its effect on model climate, *J. Climate* **9**, 2443-2463.

32. Hortal, M. and Simmons, A.J. (1991) Use of reduced Gaussian grids in spectral models, *Mon. Wea. Rev.* **119**, 1057-1074.

33. Kaas, E. (1987) *The construction of and test with a multi-level, semi-Lagrangian and semi-implicit limited area model*, Diploma thesis. Geophysics Institute, Copenhagen University, Copenhagen, Denmark.

34. Kalnay-Rivas, E., Bayliss, A. and Storch, J. (1977) The 4th order GISS model of the global atmosphere, *Beitr. Phys. Atm.* **50**, 299-311.

35. Kasahara, A. (1974) Various vertical coordinate systems used for numerical weather prediction. *Mon. Wea. Rev.* **102**, 509-522.

36. Kasahara, A. (ed), (1979) *Numerical Methods Used in Atmospheric Models, Vol. 2*, GARP Publications Series No 17, WMO and ICSU, Geneva.

37. Kiehl, J.T., Hack, J.J., Bonan, G.B., Boville, B.A., Briegleb, B.P., Williamson, D.L. and Rasch, P.J. (1996) *Description of the NCAR Community Climate Model (CCM3).* NCAR Technical Note NCAR/TN-420+STR.

38. Kreiss, H. and Oliger, J. (1973) *Methods for the approximate solution of time dependent problems.* GARP Publications Series No 10, WMO and ICSU, Geneva.

39. Kurihara, Y. (1965) Numerical integration of the primitive equations on a spherical grid, *Mon. Wea. Rev.* **93**, 399-415.

40. Laprise, R. (1989) Representation of water vapour in a spectral GCM. In: Research activities in atmospheric and oceanic modelling, WMO/TD 263, Rep. No. 11, 3.18-3.21.

41. Machenhauer, B. (1979) The spectral method, in A. Kasahara (ed.), *Numerical Methods Used in Atmospheric Models, Vol. 2*, GARP Publications Series No 17, WMO and ICSU, Geneva, pp. 121-275.

42. McDonald, A. and Bates, J.R. (1989) Semi-Lagrangian integration of a grid point shallow water model on the sphere, *Mon. Wea. Rev.* **127**, 130-137.

43. McDonald, A. and Haugen, J. (1992) A two-time-level, three-dimensional semi-Lagrangian, semi-implicit, limited-area grid point model of the primitive equations. *Mon. Wea. Rev.* **120**, 2603-2621.

44. Mesinger, F. and Arakawa, A. (1976) *Numerical Methods Used in Atmospheric Models, Vol. 1*, GARP Publications Series No 17, WMO and ICSU, Geneva.

45. Mesinger, F., Janjić, Z.I., Ničković, S., Gavrilov, D. and Deaven, D.G. (1988) The step-mountain coordinate: Model description and performance for cases of Alpine lee cyclogenesis and for a case of an Appalachian redevelopment, *Mon. Wea. Rev.* **116**, 1494-1517.

46. Naughton, M., Courtier, P. and Bourke, W. (1996) Representation errors in various grid and spectral truncations for a symmetric feature on the sphere, *Quart. J. Roy. Meteor. Soc.* **112**, 253-265.

47. Phillips, N.A. (1957) A coordinate system having some special advantages for numerical forecasting, *J. Meteor.* **14**, 184-185.

48. Phillips, T.J. (1994) *A Summary Documentation of the AMIP Models*, PCMDI Report No. 18.

49. Purser, R.J. (1988a) Degradation of Numerical Differencing Caused by Fourier Filtering at High latitudes, *Mon. Wea. Rev.* **116**, 1057-1066.

50. Purser, R.J. (1988b) Accurate Numerical Differencing near a Polar Singularity of a Skipped Grid, *Mon. Wea. Rev.* **116**, 1067-1076.

51. Rasch, P.J. and Williamson, D.L. (1990a) On shape-preserving interpolation and semi-Lagrangian transport. *SIAM J. Sci. Stat. Comput.* **11**, 656-687.

52. Rasch, P.J. and Williamson, D.L. (1990b) Computational aspects of moisture transport in global models of the atmosphere, *Quart. J. Roy. Meteor. Soc.* **116**, 1071-1090.

53. Ritchie, H. (1987) Semi-Lagrangian advection on a Gaussian grid, *Mon. Wea. Rev.* **114**, 135-146.

54. Ritchie, H. (1988) Application of the semi-Lagrangian method to a spectral model of the shallow water equations, *Mon. Wea. Rev.* **116**, 1587-1598.

55. Ritchie, H. (1991) Application of the semi-Lagrangian method to a multilevel spectral primitive-equations model. *Quart. J. Roy. Meteor. Soc.* **117**, 91-106.

56. Ritchie, H. and Tanguay, M. (1996) A comparison of spatially-averaged Eulerian and semi-Lagrangian treatments of mountains, *Mon. Wea. Rev.* **124**, 167-181.

57. Ritchie, H., Temperton, C., Simmons, A., Hortal, M., Davies, T., Dent, D. and Hamrud, M. (1995) Implementation of the semi-Lagrangian method in a high-resolution version of the ECMWF forecast model. *Mon. Wea. Rev.* **123**, 489-514.

58. Rivest, C., Staniforth, A. and Robert, A. (1994) Spurious resonant response of semi-Lagrangian discretizations to orographic forcing: diagnosis and solution, *Mon. Wea. Rev.* **122**, 366-376.

59. Robert, A.J. (1966) The integration of a low order spectral form of the primitive meteorological equations. *J. Meteor. Soc. Japan,* **44**, 237-245.

60. Robert, A. (1969) The integration of a spectral model of the atmosphere by the implicit method, *Proc. WMO/IUGG Symposium on NWP,* Tokyo, Japan, 19-24.

61. Rosinski, J.M. and Williamson, D.L. (1997) The accumulation of rounding errors and port validation for global atmospheric models. *SIAM J. Sci. Comput.* **18**, 552-564.

62. Schmidt, F. (1977) Variable fine-mesh in spectral global model, *Beitr. Phys. Atmos.* **50**, 211-227.

63. Simmons, A.J. and Burridge, D.M. (1981) An energy and angular momentum conserving vertical finite-difference scheme and hybrid vertical coordinates, *Mon. Wea. Rev.* **109**, 758-766.

64. Simmons, A. J. and Strüfing, R. (1981) *An energy and angular-momentum conserving finite-difference scheme, hybrid coordinates and medium-range weather prediction.* ECMWF Technical Report No. 28.

65. Simmons, A.J., Hoskins, B.J. and Burridge, D.M. (1978) Stability of the semi-implicit method of time integration, *Mon. Wea. Rev.* **106**, 405-412.

66. Staniforth, A. and Côté, J. (1991) Semi-Lagrangian integration schemes for atmospheric models - A Review, *Mon. Wea. Rev.* **119**, 2206-2223.

67. Sundqvist, H. (1979) Vertical coordinates and related discretization, in A. Kasahara (ed.), *Numerical Methods Used in Atmospheric Models, Vol. 2*, GARP Publications Series No 17, WMO and ICSU, Geneva, pp. 1-50.

68. Tanguay, M., Yakimiw, E., Ritchie, H. and Robert, A. (1992) Advantages of spatial averaging in semi-implicit semi-Lagrangian schemes, *Mon. Wea. Rev.* **120**, 115-123.

69. Temperton, C. (1991) On scalar and vector transform methods for global spectral models, *Mon. Wea. Rev.* **119**, 1303-1307.

70. Williamson, D.L. (1990) Semi-Lagrangian moisture transport in the NMC spectral model, *Tellus* **42A**, 413-428.

71. Williamson, D.L. (1992) Review of numerical approaches for modeling global transport, in H. van Dop and G. Kallos, (eds.), *Air Pollution Modeling and its Application IX*, Plenum Press, NY, pp. 377-394.

72. Williamson, D.L. and Rasch, P.J. (1989) Two-dimensional semi-Lagrangian transport with shape preserving interpolation, *Mon. Wea. Rev.* **117**, 102-129.

73. Williamson, D.L. and Olson, J.G. (1994) Climate simulations with a semi-Lagrangian version of the NCAR Community Climate Model, *Mon. Wea. Rev.* **122**, 1594-1610.

74. Williamson, D.L., Drake, J.B., Hack, J.J., Jakob, R. and Swarztrauber, P.N. (1992) A standard test set for numerical approximations to the shallow water equations in spherical geometry, *J. Comput. Phys.* **102**, 211-224.

75. Williamson, D.L., Olson, J.G. and Boville, B.A. (1998) A comparison of semi-Lagrangian and Eulerian tropical climate simulations, *Mon. Wea. Rev.*, **126**, 1001-1012.

76. Zapotocny, T.H., Johnson, D.R. and Reames, F.M. (1994) Development and initial test of the University of Wisconsin global isentropic-sigma model, *Mon. Wea. Rev.* **122**, 2160-2178.

77. Zhu, Z., Thuburn, J., Hoskins, B. and Hayes, P. (1992) A vertical finite-difference scheme based on $\sigma - \theta - p$ coordinate, *Mon. Wea. Rev.* **129**, 851-862.

BOUNDARY LAYER PROCESSES

N. MCFARLANE
Canadian Centre for Climate Modelling and Analysis
Atmospheric Environment Service
University of Victoria
Victoria, B.C., V8W 2Y2, CANADA

1. Introduction

Processes occurring within the atmospheric boundary layer (ABL) are of fundamental importance in the climate system. In this region, usually a few hundred
meters in depth, the air flow is typically much more turbulent than elsewhere in
the troposphere. It is necessary to take this fact into account in order to properly
represent exchanges of heat, moisture, momentum and, in some cases, trace substances between the surface and the atmosphere.

   A characteristic feature of the atmospheric boundary layer is that it is nearly
neutrally stratified throughout most of its depth, a state which is consistent with
the turbulent nature of the flow regime found therein. The region of the lower
troposphere adjacent to the boundary layer is typically much more stably stratified. Turbulent exchanges occur between the planetary boundary layer and the
free atmosphere above it. Because the stable stratification typically found in a
narrow region directly above the top of the ABL inhibits vertical motion, these
exchanges are the result of penetration of turbulent eddies into this region. When
there is sufficient moisture in the boundary layer air, clouds may form within it
as a result of moist air parcels being lifted and cooled until the moisture within
them condenses.

   Even when the atmospheric boundary layer air is mostly cloud free it sometimes happens that "turbulent eddies" originating within the boundary layer are
sufficiently deep and vigorous that they are able to penetrate far enough into the
free atmosphere above to reach their lifting condensation levels (LCL). The
latent heat release associated with the condensation process may act to make
some of these parcels buoyant relative to their environment. These parcels have
ascended from within the boundary layer to their levels of free convection

221

P. Mote and A. O'Neill (eds.), Numerical Modeling of the Global Atmosphere in the Climate System, 221–237.
© 2000 *Kluwer Academic Publishers. Printed in the Netherlands.*

(LFC). Such occurrences give rise to convective clouds that may extend throughout the troposphere. They are relatively common in the tropics and sub-tropics at all times of the year and in mid-latitudes in summer. Thus in such convectively active regions there is a strong coupling between boundary layer turbulence and the dynamics of deep convective clouds.

2. Basic Features of the Atmospheric Boundary Layer

Within the atmospheric boundary layer, profiles of potential temperature, specific humidity, and wind have characteristic structures. In circumstances where the boundary layer is relatively free of clouds these quantities usually have relatively small vertical gradients within most of the ABL. Within the surface layer, typically a few tens of meters thick, stronger gradients prevail. Within this region the vertical stratification usually varies with the diurnal cycle over land. Over open oceans the surface layer is often unstably stratified. The boundary layer is usually capped by thin layer of enhanced stable stratification. Within this layer the specific humidity typically decreases rapidly with height. These features are illustrated schematically in Figure 1.

The wind variation within the ABL is, in general, somewhat more complicated because the effects of Coriolis and pressure-gradient forces and the vertical gradient of the wind stress may all be of equal importance. In middle latitudes and polar regions a balance between these terms gives rise to the well-known Ekman model, which describes the turning of the wind with height when the boundary layer is nearly neutrally stratified. In reality, there is often a substantial variation of the wind structure in the ABL associated with nonstationarity which may, for example, be associated with diurnal variations. In circumstances where the surface layer is unstably stratified, the vertical gradient of the wind tends to become relatively small in response to the strong vertical mixing that occurs within the ABL in those circumstances. When the surface cools, as is typical over land on clear nights, the surface layer become stable, the magnitude of the surface stress becomes small, and the flow becomes disconnected from the surface allowing winds to increase and vary more rapidly with height.

Turbulent eddies within the planetary boundary layer typically have maximum spatial scales of a few tens of kilometres in the horizontal and a few tens of meters in the vertical. Associated temporal variability occurs with periods of a few minutes to hours. Thus these eddies are much too small to be resolved by atmospheric general circulation models and so their bulk, or larger scale, effects must be parameterized.

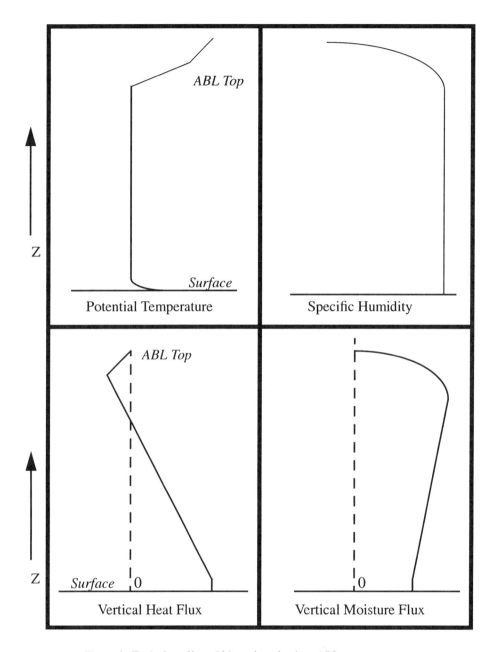

Figure 1. Typical profiles within a clear daytime ABL.

In order to accomplish this goal, it has become common to assume that it is possible to distinguish between turbulent eddies and the larger-scale flow in which they are embedded in such a way that separate sets of equations can be developed for the eddies and the larger-scale background flow. Using Cortisone coordinates for simplicity, making the hydrostatic approximation, and neglecting some generally small terms the equations for the larger-scale flow take the following form:

$$\frac{D\bar{u}}{Dt} - f\bar{v} = -\frac{\partial \Phi}{\partial x} + g\frac{\partial \tau_x}{\partial p} \tag{1}$$

$$\frac{D\bar{v}}{Dt} + f\bar{u} = -\frac{\partial \Phi}{\partial y} + g\frac{\partial \tau_y}{\partial p} \tag{2}$$

$$C_p\frac{D\bar{\theta}}{Dt} = g\frac{\partial H}{\partial p} \tag{3}$$

$$\frac{Dq_v}{Dt} = g\frac{\partial Q}{\partial p} \tag{4}$$

where u, v, θ, and q_v are respectively the eastward and northward components of the horizontal velocity, potential temperature, and specific humidity. Overbars represent averages over unresolved spatial scales. The quantities τ_x, τ_y, H, and Q are respectively the corresponding components of the vertical Reynolds stress, vertical eddy heat flux and moisture flux, defined as

$$(\tau_x, \tau_y) = \bar{\rho}(\overline{u'w'}, \overline{v'w'}) \tag{5}$$

$$H = \overline{\rho w'\theta'} \tag{6}$$

$$Q = \overline{\rho w'q_v'} \tag{7}$$

The "eddy" flux terms, represented by averages of the products of primed (unresolved) quantities in the above equations, are usually parameterized in terms of the averaged (resolved) quantities. Within the ABL, vertical gradients predominate over horizontal ones. Hence we will ignore horizontal eddy flux terms for convenience here but note that they are usually included in AGCMs in a parameterized form, often in terms of high-order horizontal diffusion operators.

Much has been written on parameterization of the effects of turbulent transfer within the ABL. There are a variety of parameterization approaches currently used in GCMs, and it is beyond the scope of this chapter to examine these in detail. Instead we will simply outline the main features of the approaches currently in use in the GCM modeling community.

2.1. THE SURFACE LAYER

The similarity theory of Monin and Obukhov [17] is commonly used as a basis for expressing the wind stress, heat flux, and moisture flux at the surface in terms of mean variables near the top of the surface layer. According to the Monin-Obukhov theory, when the turbulence is horizontally homogeneous, the winds stress, heat flux and moisture flux are nearly independent of height within the surface layer, while the profiles of the mean variables are functions of dimensionless combinations of height (z), surface roughness, wind stress, and the vertical buoyancy flux. It can be shown ([9], [19]) that in such circumstances the surface stress τ, sensible heat flux H, and moisture flux Q, are related to the mean wind, temperature, and specific humidity near the top of the surface layer through the friction velocity, temperature and humidity scales as

$$|\vec{\tau}| = \rho u_*^2 \tag{8}$$

$$H = -\rho C_p u_* \theta_* \tag{9}$$

$$Q = -\rho u_* q_* \tag{10}$$

where ρ is the density of air and C_p is the specific heat at constant pressure. The surface layer parameters are defined in terms of the mean variables through the integrated flux profile relationships [9] as

$$u_* = \frac{k U_1}{\ln\left(\dfrac{z_1}{z_0}\right) - \Psi_M\left(\dfrac{z_1}{L}\right) + \Psi_M\left(\dfrac{z_0}{L}\right)} \tag{11}$$

$$\theta_* = \frac{k(\theta_1 - \theta_s) Pr^{-1}}{\ln\left(\dfrac{z_1}{z_t}\right) - \Psi_H\left(\dfrac{z_1}{L}\right) + \Psi_H\left(\dfrac{z_t}{L}\right)} \tag{12}$$

$$q_* = \frac{k(q_1 - q_s) Pr^{-1}}{\ln\left(\dfrac{z_1}{z_q}\right) - \Psi_Q\left(\dfrac{z_1}{L}\right) + \Psi_Q\left(\dfrac{z_q}{L}\right)} \tag{13}$$

where k is the von Karmen constant, Pr is the turbulent Prandtl number (often taken to be unity), U_1, θ_1, and q_1 are respectively the wind speed, potential temperature, and specific humidity at a reference level near the top of the surface layer. The wind speed is assumed to be zero at the roughness height for momentum z_0, while θ_s and q_s are the surface values of potential temperature and specific humidity with corresponding roughness heights z_t and z_q. The Monin-Obukhov length scale is defined as

$$L = -\left(\frac{\theta_v u_*^3}{kg\overline{w'\theta'_v}}\right)_s \qquad (14)$$

where the buoyancy flux at the surface is given by

$$\overline{(w'\theta'_v)}_s = -u_*(\theta_* + 0.61\theta_{vs}q_*) \qquad (15)$$

where θ_{vs} is the virtual potential temperature evaluated near the surface.

Earlier AGCMs typically used simple representations of land surface processes. Consistent with this simplification, the roughness heights for momentum, heat, and moisture were commonly assumed to be equal to each other. More recent AGCMs include more elaborate land surface schemes, which distinguish between different surface types that may occur within a single grid cell and allow for the possibility that they may have differing roughness length scales for heat, moisture, and momentum. It is still common to assume that the roughness heights for heat and moisture are equal to each other.

Over open ocean areas the surface roughness is often taken to be a function of the friction velocity through the Charnock relationship which is of the form $z_0 = \alpha u_*^2/g$ where the empirical constant α is of the order of 10^{-2}.

The form of the flux profile functions (ψ_M, ψ_H) has been the subject of much research over the past 3 decades. For some time the forms proposed by Businger et al. [6] were commonly used, although these were somewhat controversial, among other reasons because they implied that the (supposedly universal) value of the von Karmen is 0.35, smaller than the commonly accepted value of 0.4. Later work led to refinements and extensions of these earlier studies. (See Sorbjan [21], pp 74-75, for a summary of flux profile relationships). More recently Beljaars and Holtslag [4] have proposed a new set of profile functions based on analysis of data from a more recent field experiment.

Given the flux profile functions, it is clear that equations (9) through (15) form a non-linear set of equations for the surface flux quantities that in general require an iterative procedure to solve accurately. It has been found useful to introduce the bulk Richardson number for the surface layer, defined as,

$$Ri_B = \frac{gz_1(\theta_{v1} - \theta_{vs})}{\theta_{vs} U_1^2} \tag{16}$$

In terms of this quantity and roughness parameters, surface flux quantities can be expressed in terms of bulk exchange formulae as:

$$u_*^2 = C_d F_M\left(\frac{z_1}{z_0}, Ri_B\right) U_1^2 \tag{17}$$

$$u_* \theta_* = C_h F_H\left(\frac{z_1}{z_0}, Ri_B\right) U_1(\theta_1 - \theta_s) \tag{18}$$

$$u_* q_* = C_Q F_Q\left(\frac{z_1}{z_0}, Ri_B\right) U_1(q_1 - q_s) \tag{19}$$

where C_d is the neutral drag coefficient and C_H, C_Q are the corresponding exchange coefficients for heat and moisture. A practice that is often followed in GCM applications is to use analytical approximations for the functions F_M, F_H, F_Q. In recent years a number of such simple analytical formulae have been proposed. It is also usual to assume that F_H and F_Q are equal to each other. Many of these functions are qualitatively similar to those proposed by Louis [13]. A recent example of such a scheme is that used in the second generation GCM of the Canadian Centre for Climate Modelling and Analysis [15], for which

$$(F_M, F_H) = \begin{cases} \left(1 + 10|Ri_B| \middle/ \left(1 + 10\left|\frac{Ri_B}{87 A_{M,H}^2}\right|\right)\right); \ldots\ldots Ri_B < 0 \\ (1 - 5\mu Ri_B)^2 / (1 + 10(1 - \mu)Ri_B) \ldots 0 \leq Ri_B \leq 1/(5\mu) \\ 0 \ldots\ldots\ldots\ldots\ldots\ldots\ldots\ldots\ldots\ldots Ri_B > 1/(5\mu) \end{cases} \tag{20}$$

where

$$(A_M, A_H) = \left(\frac{z_0}{z_1}\right)^{\frac{1}{2}} k^2 / (C_d, C_H) \tag{21}$$

For $\mu = 0$ this formulation is qualitatively similar to that proposed by Louis (1979) for an early version of the ECMWF forecast model. The choice $\mu = 1$ is more consistent with the traditional flux profile relationships but has the effect of allowing the surface to become thermally disconnected with the atmosphere in very stable (large positive Ri_B) conditions. This may occur during hours of darkness and permit unrealistically rapid radiative cooling of the surface, as noted by Beljaars and Holtslag [4]. These authors have proposed new flux profile relationships that do not suffer from this unrealistic behaviour. However, they also note that under very stable conditions the formulation of Louis [13] overestimates the transfer coefficients.

It could also be argued heuristically that there may be unresolved spatial variability within the relatively large surface areas of typical AGCM grid cells such that regions in which the local bulk Richardson number is small may be found even if the mean value for the grid cell is relatively large. This argument was used by Louis [13] as a justification for allowing the $F_{M,H}$ to remain non-zero for all finite values of Ri_B.

Both the Louis [13] formulation and that above suffer from the defect that they underestimate the upward heat transfer in circumstances where the mean wind is weak. As noted by Beljaars [3], the usual form of the Monin-Obukhov theory becomes singular in these circumstances. However, the singular behaviour can be removed by adding to the near-surface wind a contribution that is proportional to the convective velocity scale, defined as

$$Wc = \left(\frac{gh}{T}\overline{(w'\theta'_v)}_s\right)^{1/3} \tag{22}$$

Recently Abdella and McFarlane [1] have made use of this proposal combined with the flux profile relationships proposed by Beljaars and Holtslag [4] to derive a new parameterization of surface layer exchange coefficients.

The general structure of the bulk transfer coefficients that arise from application of the Monin-Obukov formulation is illustrated in Figure 2. Relatively

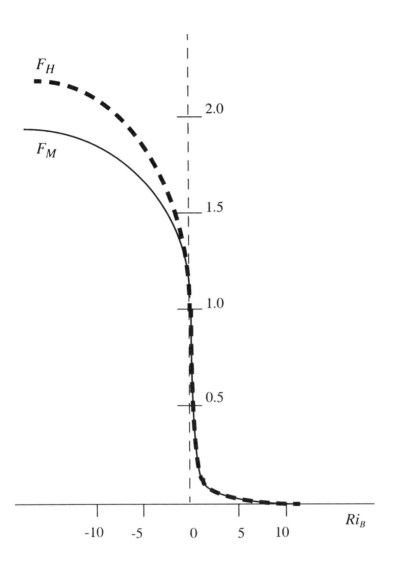

Figure 2. Typical behavior of heat and momentum flux coefficients
as a function of bulk Richardson number.

large values occur when the surface layer is unstably stratified and/or when wind
speeds near the surface are moderate to large (negative or small values of Ri_B).
Bulk transfer coefficients are small when the surface layer stratification is stable
(positive values of Ri_B).

2.2. TURBULENT TRANSFER WITHIN THE BOUNDARY LAYER

In contrast to their behaviour in the surface layer, the vertical flux quantities undergo substantial variation in the vertical within the remainder of the boundary layer, typically changing sign and/or becoming small near the top of the ABL. The associated profiles of potential temperature, specific humidity, and, to a lesser extent, mean horizontal winds often have small vertical gradients. As mentioned above, these features are associated with turbulent mixing of these quantities within the ABL.

2.2.1 Basic down-gradient representations of turbulent transfer

A commonly-used approach in AGCMs is to express vertical flux quantities in terms of eddy diffusivities as:

$$\overline{w'\vec{V}'} = -K_M \frac{\partial}{\partial z} \vec{V} \tag{23}$$

$$\overline{w'\theta'} = -K_H \frac{\partial}{\partial z} \bar{\theta} \tag{24}$$

$$\overline{w'\chi'} = -K_\chi \frac{\partial}{\partial z} \bar{\chi} \tag{25}$$

where χ is any prognostic scalar quantity such as water vapor, liquid water, or other trace substances that may be active or passive.

Here the "eddy diffusivities" (K_M, K_H, K_χ) must be specified. A variety of approaches for this are in current use. The simplest, and most commonly used, is the Prandtl mixing length formulation whereby

$$(K_M, K_H, K_\chi) = l^2 \left| \frac{\partial}{\partial z} \vec{V} \right| \{f_M(Ri), f_H(Ri), f_\chi(Ri)\} \tag{26}$$

where l is the mixing length and the gradient Richardson number is defined as

$$Ri = \left(\frac{g}{\bar{\theta}} \frac{\partial}{\partial z} \overline{\theta}_v \right) \bigg/ \left(\frac{\partial}{\partial z} \vec{V} \right)^2 \tag{27}$$

and the stability functions $(f_{M, H, \chi})$ that depend on this quantity are either chosen empirically, by analogy with the forms that are consistent with the flux profile relations in the surface layer (e.g. [13]), or through an approximate equation governing the turbulent kinetic energy. In general, the stability functions are functions of Ri defined so as to behave similarly to the bulk transfer functions $F_{M,H}$ defined above in terms of Ri_B, namely that they tend toward zero for large positive values of Ri, are close to unity for $Ri = 0$, and become larger for negative values of Ri (although negative values are almost never realized in nature in the ABL; more on this below).

The mixing length is commonly chosen, following Blackadar [5], as

$$l = \frac{kzl_\infty}{kz + l_\infty} \tag{28}$$

where k is the von Karmen constant and the outer length scale, l_∞, has to be specified empirically and is often chosen to be of the order of 10 meters for AGCM applications. It is not, however, an easily observable quantity and not strongly constrained on theoretical grounds. There is a considerable body of literature on the choice of this quantity (see for example [21], p 169). For small values of the ratio (kz/l_∞) the mixing length varies as kz and with the proper choice of the functions $f_{M,H}$ the above eddy diffusivity formulation will ensure that the vertical profiles of the mean variables are as expected from surface layer similarity theory, at least for near neutral conditions, on the assumption that vertical fluxes are nearly independent of height in that region.

2.2.2 Higher order closure and non-local effects in the ABL
The approach discussed above suffers from two basic limitations. One of these is that specification of the stability functions is often rather ad hoc and empirical in nature. The other, more significant, limitation is that the down-gradient nature of the formulation does not allow for upward heat transfer in the (commonly observed) circumstance that the ABL is either well mixed or weakly stably stratified (slightly positive $\partial \overline{\theta}_v / \partial z$) in the region above the surface layer. Turbulent transfer in these circumstances in non-local and counter-gradient.

Attempts to account for non-local effects have involved use of equations governing quadratic and higher order turbulent quantities (so called higher moments). One of the simpler approaches to generalizing the mixing length approach is to represent the eddy diffusivities in terms of the turbulent kinetic energy (TKE) as

$$(K_M, K_H) = (l, l/Pr)e^{1/2} \qquad (29)$$

where the turbulent kinetic energy ($e = \overline{(u'^2 + v'^2 + w'^2)}/2$) is determined using the following approximate equation

$$\frac{\partial e}{\partial t} = -\left(\overline{u'w'}\frac{\partial}{\partial z}\bar{u} + \overline{v'w'}\frac{\partial}{\partial z}\bar{v}\right) - \frac{\partial}{\partial z}(\overline{e'w'}) - \frac{1}{\bar{\rho}}\frac{\partial}{\partial z}(\overline{p'w'}) + \frac{g}{\bar{\theta}}\overline{w'\theta'}_v - \varepsilon \qquad (30)$$

In this equation the term involving the vertical flux of the turbulent component of the pressure field is often either ignored or assumed to be (negatively) proportional to the term involving vertical transport of turbulent kinetic energy. The vertical transport of TKE is in turn often represented in terms of a downgradient diffusion. The dissipation term ε is often taken to be proportional to $e^{3/2}/l_d$ where the dissipation length scale l_d is often taken to be proportional to the mixing length scale l.

If the time derivative of the TKE and the vertical transport terms are ignored in the TKE equation, the resulting "first order closure" system of equations gives expressions for the eddy diffusivities that are qualitatively similar to those in equation (26) above.

Retaining the prognostic nature of the TKE equation and parameterizing the vertical transport terms gives a "one-and-a half" order closure scheme that allows for some limited non-local effects. More elaborate schemes can be obtained by considering the equations for other second moment quantities as well as those for higher moments. Mellor and Yamada [14] have systematically derived a hierarchy of turbulence closure schemes in this manner; the most popular such scheme is their "level 2.5" scheme, which retains a prognostic equation for the TKE and introduces diagnostic equations for the vertical flux terms. This scheme has been used in AGCM applications (e.g., [18]); however, it still does not allow for the possibility of counter-gradient turbulent transfer effects. In the context of turbulent closure models, such effects require consideration of equations governing higher moments. For example, for a cloud-free boundary layer, an approximate equation for the vertical heat flux can be written in the form:

$$\frac{\partial}{\partial t}\overline{w'\theta'} = -\frac{1}{\bar{\rho}}\overline{\theta'\frac{\partial}{\partial z}p'} - \frac{1}{\bar{\rho}}\frac{\partial}{\partial z}(\bar{\rho}\overline{w'^2\theta'}) - \overline{w'^2}\frac{\partial}{\partial z}\bar{\theta} + \frac{g}{\bar{\theta}}\overline{\theta'\theta'}_v \qquad (31)$$

where it is assumed that the turbulence is homogeneous in the horizontal and terms involving viscous dissipation and molecular heat conduction have been ignored.

In higher order turbulence modeling it has become common to invoke the "return to isotropy" hypothesis of Rotta [20] in accordance with which the pressure correlation term is represented as a relaxational damping of the vertical heat flux of the form $\overline{w'\theta'}/\tau$ where the relaxation time scale, τ, is usually taken to be inversely proportional to the turbulent kinetic energy. Moeng and Wyngaard [16] have proposed, on the bases of high resolution simulations (Large Eddy Simulations) of turbulence in a cloud free convectively active boundary layer, that the Rotta term should also be supplemented with a term that has the effect of enhancing the magnitude of the buoyancy term ($g\overline{(\theta'\theta'_v)}/\overline{\theta}$). When these assumptions are invoked, and if the third (turbulent flux transport) term is ignored in equation 30, it is seen that

$$\overline{w'\theta'} \cong \tau \overline{w'^2}\left(\frac{\partial}{\partial z}\overline{\theta} - \gamma_c\right) \tag{32}$$

where the "counter gradient heat flux" term γ_c has the form

$$\gamma_c = \mu g\overline{\theta'\theta'}_v / (\overline{w'^2}) \tag{33}$$

where μ is a constant. This is the form proposed by Deardorff [8], after an earlier suggestion [7] that the deficiency of the traditional down-gradient heat flux formulation in the convectively active boundary layer could be alleviated substantially by introducing a constant value (of order $10^{-3}\,^\circ C m^{-1}$) of the counter-gradient term into the diffusivity formulation to give

$$\overline{w'\theta'} = -K\left(\frac{\partial}{\partial z}\overline{\theta} - \gamma_c\right) \tag{34}$$

It should be noted, however, that merely introducing the buoyancy term into the heat flux formulation does not of itself ensure that the heat transfer may be non-local or counter-gradient since, to the same order of approximation, it can be shown that this term is itself proportional to the product of the vertical heat flux and the vertical gradient of potential temperature. Non-local and counter-gradient effects arise when the third and higher order vertical transport terms are included.

Although higher order turbulent closure schemes are still not widely used in AGCMs, semi-empirical non-local schemes have been developed for use in AGCMs and found to provide a substantial improvement over earlier local down-gradient schemes when adequate vertical resolution is employed. An example of this is the scheme proposed recently by Holtslag and Boville [12] for use in the NCAR CCM2 in which a local vertical heat flux formulation is retained but supplemented by a non-local contribution of the form of equation (34). In this scheme the counter-gradient flux term is introduced for heat, moisture, and other prognostic scalar constituents. For each of these quantities it is taken to be of a form adapted from that proposed by Holtslag and Moeng [11], namely

$$\gamma_c = a W_c \overline{(w'\chi')}_s / (W_m^2 h) \tag{35}$$

where h is the height of the convective boundary layer (CABL) determined diagnostically as proposed by Troen and Mahrt [22], a is an empirical constant, and W_m is a characteristic turbulent velocity scale for the surface layer, defined as

$$W_m = \left(u^3_* + c_1 W_c^3 \right)^{1/3} \tag{36}$$

where c_1 is also an empirical constant. Note that in this formulation the counter-gradient term vanishes if the vertical flux is zero or downward.

The diffusivity for this scheme is taken to be the same for all scalar quantities and of the form

$$K_c = kz w_t \left(1 - \frac{z}{h} \right)^2 \tag{37}$$

within the CABL and vanishes above it. Here w_t is also a representative turbulent velocity scale, proportional to W_m/Pr in the outer part of the ABL (above the surface layer). It is interesting to note that, with this definition, the vertical heat flux term is qualitatively similar to equation (32) with

$$\overline{w'^2} \sim \left(\frac{z}{h} \right) \left(w_t \left(1 - \frac{z}{h} \right) \right)^2 \tag{38}$$

and

$$\tau \sim kh/w_t \tag{39}$$

Within the CABL the diffusivity for momentum is also enhanced by adding a term of the form of equation (38) above but with w_t replaced by W_m. However, no attempt is made to allow for counter-gradient effects for momentum.

More recently, Abdella and McFarlane [2] have shown, using a second order closure model that parameterization of the third order transport terms in a way which is consistent with the idea that they extend through most of the CABL, based on both observational data and results from large eddy simulations (LES), leads to a general form of non-local representation of the vertical heat flux in which earlier forms derived by Deardorff [8] and Holtslag and Moeng [11] are found to be special cases. The vertical structure of the eddy diffusivity and the magnitude of the counter-gradient term, as simulated with this model, are in good agreement with LES results and are similar to those of the empirical formulation of Holtslag and Boville [12] throughout most of the CABL.

3. Cloudy Boundary Layers

The presence of widespread stratocumulus clouds in the ABL complicates the relatively simple picture presented above to some extent, both as to the vertical structure of mean variables and the physical processes that determine this structure. In particular latent heat release associated with condensation and radiative heating/cooling due to the presence of clouds are important effects. The presence of condensed water affects buoyancy and vertical fluxes of heat and moisture in important ways. In the absence of significant precipitation generation within a cloudy ABL the "well mixed" character of it is largely preserved for quasi-conserved variables such as total water specific humidity $(\overline{q_t} = \overline{q_v} + \overline{q_w})$, equivalent potential temperature $(\theta_e \cong \theta \exp((L\overline{q_v})/(C_p T)))$, and liquid water potential temperature $(\theta_w \cong \overline{\theta} \exp(-(L\overline{q_w})/(C_p \overline{T})))$. Thus the vertical structure of these variables is qualitatively similar to that of their "dry" counterparts in a clear ABL. The structure of the associated vertical flux of these quantities is a little more complicated, however. For example, in the case where an overcast cloud layer is present in the upper part of a daytime boundary layer (with upward heat flux at the surface) the vertical flux of liquid water potential temperature typically decreases at a uniform rate below the cloud and then more rapidly within the cloud.

In some locations, notably within the sub-tropical trade wind regime, the cloud topped marine boundary layer undergoes a transformation to one with a double structure in which the well mixed ABL becomes relatively cloud-free and topped by a region in which quasi-conserved variables are less well mixed in the vertical and vertical fluxes of heat and moisture are predominantly associated with fields of shallow cumulus clouds that grow out of the ABL but do not produce significant amounts of precipitation. In the deep tropics this trade-wind regime gives way in turn to one in which the ABL is typically topped by widely separated deep convective clouds with associated heavy precipitation and flanking convective and meso-scale downdrafts that import air from the middle to lower troposphere into the ABL. In current GCM practice convective parameterizations are often coupled to the ABL parameterizations as discussed in part in the chapter on moist convection. However, although modestly successful, these parameterization schemes often fail to account for important features of marine boundary layers in the tropics and sub-tropics.

Parameterization of the important processes that determine the structure of cloudy boundary layers and transitional layers remains a challenging task for GCM modellers and is currently the topic of much active research. A more quantitative description of the approaches currently in use is beyond the scope of this chapter. The reader is referred to chapter 13 of the book by Emanuel [10] for a more complete discussion of the characteristic features of cloudy and trade-wind boundary layers.

References

1. Abdella, K. and McFarlane, N. (1996) Parameterization of the surface-layer exchange coefficients for atmospheric models, *B. Layer Met.* **80**, 223-248.
2. Abdella, K. and McFarlane, N. (1997) A new second order turbulence closure scheme for the planetary boundary Layer, *J. Atmos. Sci.* **54**, 1850-1867.
3. Beljars, A.C.M. (1995) The parameterization of surface fluxes in large-scale models under free convection, *Quart. J. Roy. Meteorol. Soc.* **121**, 255-270.
4. Beljaars, A.C.M. and Holtslag, A.A. (1991) Flux parameterization over land surfaces for atmospheric models, *J. Appl. Meteorol.* **30**, 327-341.
5. Blackadar, A.K. (1962) The vertical distribution of wind and turbulent exchange in a neutral atmosphere, *J. Geophys. Res.* **67**, 3095-3103.
6. Businger, J.A., Wyngaad, J.C., Izumi, Y., Bradley, E.F. (1971) Flux profile relationships in the atmospheric surface layer, *J. Atmos. Sci.* **28**, 181-189.
7. Deardorf, J.W. (1966) The countergradient heat flux in the lower atmosphere and in the laboratory, *J. Atmos. Sci.* **23**, 503-506.
8. Deardorff, J.W. (1972) Theoretical expression for the counter-gradient vertical heat flux, *J. Geophys. Res.* **77**, 5900-5904.
9. Dyer, A.J. (1974) A review of flux-profile relationships, *Boundary-Layer Meteorol.* **20**, 35-49.
10. Emanuel, K.A. (1994) *Atmospheric Convection*, Oxford University Press, New York.
11. Holtslag, A.A.M., and Moeng, C.H. (1991) Eddy diffusivity and countergradient transport in the convective atmospheric boundary layer. *J. Atmos. Sci.* **48**, 1690-1698.
12. Holtslag, A.A.M., and Boville, B.A. (1993) Local versus nonlocal boundary layer diffusion in

a global climate model., *J. Climate* **5**, 1013-1043.

13. Louis, J.F. (1979) A parametric model of vertical eddy fluxes in the atmosphere, *Boundary-Layer Meteorol.* **17**, 187-202.

14. Mellor, G.L., and Yamada, T. (1974) A heirarchy of turbulence closure models for planetary boundary layers, *J. Atmos. Sci.* **31**, 1791-1806.

15. McFarlane, N.A., Boer, G.J., Blanchet, J-P., and Lazare, M. (1992) The Canadian Climate Centre second-generation general cirulation model and its equilibrium climate, *J. Climate* **10**, 1013-1043.

16. Moeng, C.H., and Wyngaard, J.C. (1986) An analysis of closures for pressure-scalar covariances in the convective boundary layer, *J. Atmos. Sci.* **43**, 2499-2513.

17. Monin, A.S, and Obukhov, A.M. () Basic regularity in turbulent mixing in the surface layer of the atmosphere, *Akad. Nauk. S.S.S.R. Trud. Geofiz. Inst.* **24**, 163-187.

18. Miyakoda, K, Gordon, T., Caverly, R., Stern, W., Siruis, J., and Bourke, W. (1983) Simulation of a blocking event in January, 1977, *Mon. Wea. Rev.* **111**, 846-869.

19. Paulson, C.A. (1970) The mathematical representation of wind speed and temperature profiles in the unstable atmospheric surface layer, *J. Applied Meteorol.* **8**, 856-861.

20. Rotta, J.C. (1951) Statistische theorie nichthamogener turbulenz, *Z. Phys.* **129**, 547-572.

21. Sorbjan, Z. (1989) *Structure of the Atmospheric Boundary Layer*, Prentice Hall, Englewood Cliffs, N.J.

22. Troen, I., and Mahrt, L. (1986) A simple model of the atmopheric boundary layer: sensitivity to surface evaporation, *Boundary-Layer Meteorol.* **37**, 129-148.

MOIST CONVECTION

N. MCFARLANE
Canadian Centre for Climate Modelling and Analysis
Atmospheric Environment Service
University of Victoria
Victoria, B.C., V8W 2Y2, CANADA

1. Basic considerations

Most of the precipitation in the tropics and a portion of it in middle latitudes is convective in origin. The associated latent heat release is a major component of the diabatic heating of the troposphere in the tropics. Since individual convective cells, and even organized convective systems, have characteristic horizontal scales that are much smaller than those usually resolved by atmospheric general circulation models (GCMs), it has long been recognized that the effects of cumulus clouds must be parameterized. Because of their significant vertical extent, the transfers of heat, moisture, and perhaps also momentum, within cumulus clouds are an important part of their overall effects on the larger-scale circulation. These effects appear in the large-scale thermodynamic and moisture equations as terms directly attributable to condensation (C) or evaporation (E) and terms associated with sub-grid scale vertical transfers which are, in pressure coordinates, approximately as follows:

$$\frac{D\bar{\theta}}{Dt} = \left(\frac{\bar{\theta}}{\bar{T}}\right)\left[\frac{L}{C_p}(C-E) + Q_R\right] - \frac{\partial(\overline{\omega'\theta'})}{\partial p} = \left(\frac{\bar{\theta}}{\bar{T}}\right)Q_1 \tag{1}$$

$$\frac{D\bar{q_v}}{Dt} = (E-C) - \frac{\partial(\overline{\omega'q_v'})}{\partial p} = Q_2/L \tag{2}$$

where overbars denote averages over sub-grid scale quantities and $\bar{\theta}, \bar{T}, \bar{q_v}$ are respectively the large-scale means of potential temperature, temperature and specific humidity, and Q_R is the radiative heating rate. Primed quantities are (unresolved) deviations from the corresponding large-scale mean values and, in accord with standard notation, ω is the total rate of change of pressure, sometimes also referred to as the vertical pressure velocity because it appears mathematically in expressions for the vertical advection in a manner similar to the

P. Mote and A. O'Neill (eds.), Numerical Modeling of the Global Atmosphere in the Climate System, 239–262.
© 2000 *Kluwer Academic Publishers. Printed in the Netherlands.*

vertical velocity when geometrical height is used as a vertical coordinate.

In the cumulus parameterization literature it has become common to denote the net diabatic heating and the latent heating/cooling due to the moisture loss/gain associated with moist convection by the symbols Q_1 and Q_2. As noted above, the processes which give rise to these terms occur on spatial scales that are much smaller than those typically resolved in GCMs. Consequently, the parameterization problem is to account for the effects of these processes in terms of the prognostic variables that are typically used in GCMs. As is usual in parameterization problems, approaches to solving this problem involve using simplified representations of the basic physical features of convective processes that are considered important.

Research in this topic has been active for several decades and it is beyond the scope of this chapter to provide more than a cursory view of the historical context in which modern cumulus parameterization schemes have evolved. We will, however, mention some basic developments in order to provide a clear basis for the discussion of this topic.

2. Conditional Instability

A basic requirement for the occurrence of moist convection within a typical GCM grid volume is that conditional instability must exist somewhere within the grid volume. Since the prognostic variables in a GCM are representative of the horizontal mean within the volume, it has traditionally been assumed that the occurrence of convection requires the existence of conditional instability in this mean sense. In recent numerical weather prediction applications, particularly in models used for mesoscale prediction, attempts to account for the effects of sub-grid scale variability in terms of "convective trigger functions" are used to determine the potential for release of convective instability [9]. The optimal choice of such trigger functions is currently a subject of active research and it is beyond the scope of these notes to deal with this topic adequately, so we will confine our attention to the concepts that have been widely used in representing the effects of moist convection in GCMs.

The parcel method is commonly used as a means of assessing the existence of conditional instability (see [7], p.168). In a typical application of this method, a parcel of air, with properties typical of mean conditions at some reference level in the lower troposphere, is assumed to be conditionally unstable if adiabatic lifting could carry it above its level of free convection (LFC) if it exists. Once displaced beyond its LFC the parcel could, in principle, rise freely to a level where it is no longer buoyant with respect to the mean thermodynamic state (i.e., the density of the air in the parcel equals or exceeds large-scale mean air density.).

This is easily depicted on a thermodynamic diagram as illustrated in Figure

1. Assume (as is typical) that the relative humidity of the air in the parcel is initially less that 100 percent. As a first approximation it is assumed that there is no difference between the pressure within the parcel and that of the surrounding air. However, we assume that the parcel is otherwise isolated from the surrounding air as it ascends so that it does not become diluted by entrainment of environmental air as it ascends. Since the mean thermodynamic state is usually statically stable (lapse rate less than dry adiabatic) the parcel temperature is initially less than that of the environment because it rises dry-adiabatically until the water vapor within it condenses. This occurs at the lifting condensation level (LCL). The parcel rises moist adiabatically above the LCL. If the environmental lapse rate is larger than moist adiabatic for some distance above this point, the density of the ascending parcel may eventually become smaller than that of the surrounding air. The parcel becomes positively buoyant at this point (the LFC) and may begin to rise freely. This point is usually close to that at which the temperature of the parcel begins to exceed that of its environment and this criterion is commonly used as a means of identifying it on a thermodynamic diagram. In reality the evaluation of parcel density (and hence its buoyancy) is properly based on its virtual temperature and this quantity depends on the amount of condensed water present in the parcel. The ascent is said to be reversible if all of the condensed water remains within the parcel. In such a case the liquid water loading reduces buoyancy somewhat. For reversible ascent the LFC is, in reality, the level where the virtual temperature of the parcel becomes equal to that of the surrounding air.

If it is assumed that all of the water that condenses during ascent drops out of the parcel, its ascent is said to be pseudo-adiabatic. In this case the parcel buoyancy is determined entirely by its temperature, which is that of the moist adiabat that passes through the lifting condensation level (LCL). This assumption is often used in evaluating the presence of conditional instability and the possible maximum depth of cumulus clouds by reference to a thermodynamic diagram, as depicted in Figure 1. Because the test parcel is typically negatively buoyant with respect to the mean state during the initial part of its ascent, a certain mount of kinetic energy must be given to it to carry it upward through its LFC. Once rising freely, the parcel may accelerate as a result of having become buoyant. This may occur between the LFC and the level of neutral buoyancy (LNB) at which the virtual temperature of the parcel becomes and remains less than that of the environment.

A useful parameter in this regard is the convective available energy (CAPE). This is just the vertically integrated buoyancy of the parcel between the reference level where it first begins to ascend and the LNB. A reasonable and commonly used necessary condition for convection to occur within a GCM grid volume is

242

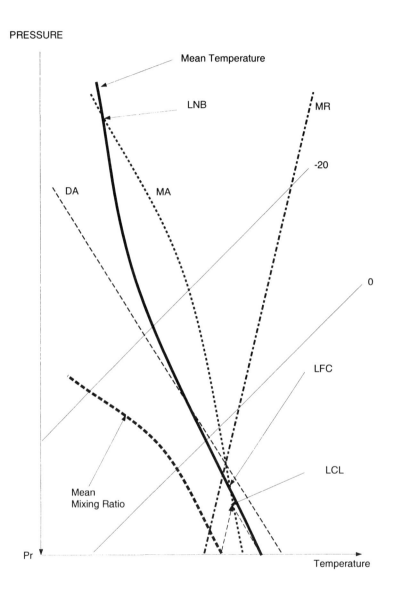

PRESSURE

Mean Temperature

LNB

MR

DA

MA

-20

0

LFC

LCL

Mean
Mixing Ratio

Pr

Temperature

Figure 1. A schematic view of a conditionally unstable atmospheric sounding showing the mean temperature and water vapor mixing ratio profiles , typical orientations of dry (DA) and moist (MA) adiabats, and isopleths of mixing ratio (MR). Also shown are the 0C and -20C isotherms, locations of the lifting condensation level (LCL) and the levels of free convection (LFC), and neutral buoyancy (LNB) for an undiluted parcel that does not contain liquid water.

that the parcel CAPE is positive. The parcel CAPE can be defined for either reversible or irreversible ascent (where it is assumed that the condensed water does not remain within the parcel). For irreversible ascent the parcel CAPE is proportional to the area between the curves denoting the temperature of an ascending parcel and the environment between the reference level where the parcel begins its ascent and the LNB. Here, unless otherwise stated, we assume reversible ascent for the definition of CAPE.

The situation that exists in reality within cumulus clouds is of course very complicated. Individual parcels typically do not rise in isolation. They may be diluted by mixing with adjacent air, both within and outside of the clouds, and of course the existence of convective precipitation is contingent upon the loss of condensed water from ascending air within cumulus clouds. The process of dilution with environmental air (i.e. air outside of the cumulus clouds) results in evaporation which can produce negative buoyancy. Production and descent of rainwater also reduces buoyancy and drives downdrafts within cumulus clouds. Thus, at a given height, an individual cumulus cloud may contain air parcels with a wide range of properties. Some of these may be positively buoyant and ascending, with much less than the amount of liquid water than they would have if ascending reversibly. Others may be negatively buoyant, and descending, because of combined effects of evaporative cooling associated with dilution and condensed water drag.

Later in this chapter methods for taking into account some of these complicating factors are discussed in the context of parameterization of moist convection using the mass flux approach. In this approach cumulus cloud properties are commonly formulated in terms of the dry and moist static energy variables

Dry Static Energy: $\qquad s = C_p T + gz$ $\qquad\qquad$ (3)

Moist Static Energy: $\qquad h = C_p T + gz + Lq$ $\qquad\qquad$ (4)

These variables are widely used in the literature on cumulus parameterization. An unsaturated parcel that ascends dry adiabatically retains its initial values of both dry and moist static energy. A saturated parcel of air that ascends moist adiabatically retains its initial moist static energy. Thus an initially unsaturated parcel ascending adiabatically from a reference level within the ABL to its level of neutral buoyancy in the upper troposphere, as discussed above, would retain its moist static energy during its entire ascent. Its dry static energy would remain unchanged during the unsaturated portion of its ascent.

This process is illustrated in Figure 2, which depicts the typical structures for dry and moist static energy variables in conditionally unstable atmospheres, and also the typical profile of h^*, the moist static energy for air that is saturated at the

244

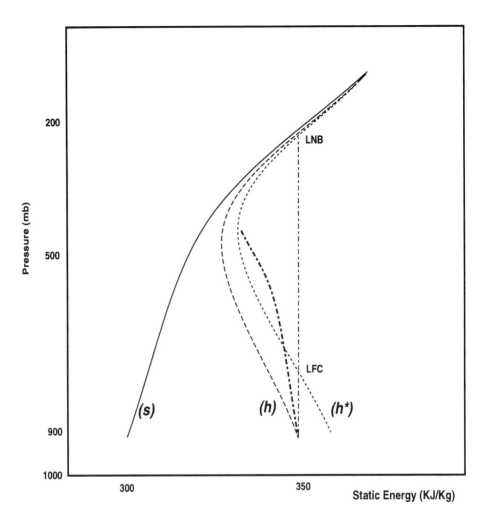

Figure 2. Schematic view of a conditionally unstable atmospheric sounding in terms of the mean dry (*s*) and moist (*h*) static energy profiles. Also shown are the moist static energy profiles for saturated air with the same temperature as the mean sounding (*h**), the moist static energy profiles for undiluted (vertical long-short dashed) and diluted (bold long-short dashed) cumulus cloud soundings, the levels of free convection (LFC) and neutral buoyancy (LNB) for an undiluted parcel which ascends pseudo-adiabatically (i.e. does not retain condensed water).

environmental temperature and pressure. In this figure an undiluted parcel ascending from the atmospheric boundary layer (ABL) follows the vertical straight (long-short dashed) line. Its temperature exceeds that of the environment between the points where its moist static energy exceeds the saturated value for the environment. In the absence of liquid water loading it will be positively buoyant in that region. If the parcel is diluted to some extent by entrainment of environmental air during its ascent, its moist static energy profile will be drawn toward that of the environment as depicted in the slightly curved (long-short dashed) profile in Figure 2. Accounting for the effects of dilution in cumulus parameterization is discussed further in section 3.4 below.

It is also noteworthy that the dry and moist static energy variables are closely related to those of potential temperature and equivalent potential temperature respectively. Potential and equivalent potential temperature and both dry and moist static energy values are conserved in unsaturated and undiluted ascent of an air parcel. A saturated parcel does not retain its potential temperature and dry static energy during adiabatic ascent but its moist static energy and its equivalent potential temperature are conserved.

3 Approaches to Parameterization

While cumulus parameterizations have been most commonly used in the context of modeling, they have found broader applications as well. They have been used very effectively as a diagnostic tool for analysing the interaction of moist convection with larger scale processes in the atmosphere (for example see [18]). This chapter focuses on the modeling aspects of cumulus parameterization.

There are two basic aspects of the cumulus parameterization problem. The first of these is that of choosing a conceptual model as a basis for developing a parameterization. Inevitably, there are free parameters of the conceptual model that must be specified in order to obtain a properly closed parameterization scheme. This second aspect, invoking of closure assumptions, is of fundamental importance in modeling applications.

3.1 BASIC CONVECTIVE ADJUSTMENT SCHEMES

Adjustment schemes are based on the principle that convection (moist or dry) acts to remove instability by adjusting the larger scale atmospheric state toward one that is neutrally stable. A simple and still commonly used example of this is dry convective adjustment, which is usually done iteratively, two layers at a time. The assumption of dry convective adjustment is that convective overturning occurs if a parcel, when lifted dry adiabatically from the mispoint of the lower layer to the midpoint of the upper layer, is found to be buoyant with respect to the upper layer. Convective overturning then occurs, exchanging air between the layers in such a way as to neutralize the lower layer with respect to the upper one while ensuring conservation of the sum of internal, potential, and latent energy. The neutral adjusted state is such that the layer mean potential temperatures and specific humidities of the two layers (assumed to coincide with values near the mid-points of the layers) are equal to each other. In the absence of condensation it is usually required that the vertical integrals of the temperature (more correctly of $C_p T$) and specific humidity must be conserved during the adjustment.

When the relative humidity of the lower layer is sufficiently large, an air parcel lifted from its midpoint may pass through its lifting condensation level (LCL) before reaching the midpoint of the upper layer. If so, latent heat release must be taken into account when the adjustment is carried out. In such circumstances a parcel rising moist adiabatically may become unstable. In this case the convective adjustment process is carried out much as for dry adjustment except that the relevant lapse rate is the moist adiabatic one, or possibly a weighted average of the dry and moist adiabatic lapse rates if the relative humidity of the lower layer is initially less than 100 percent. The relative humidity for the adjusted state is specified. The simplest such moist convective adjustment scheme assumes that the adjustment maintains the relative humidity of the lower layer at the ambient value or at 100 percent if the air in the lower layer is initially supersaturated. Condensed water is assumed to fall out as convective precipitation, and the mass weighted vertical integral of the moist enthalpy ($C_p T + Lq$) is conserved during the adjustment process.

The adjustment process is carried out iteratively, two layers at a time, as illustrated in Figure 3, which depicts schematically the adjustment between a saturated lower layer whose initial mean temperature and specific humidity are T_1 and q_1 respectively, and an upper layer with corresponding values of T_2 and q_2. The initial lapse rate implied by the difference between the layer mean temperatures exceeds the moist adiabatic value so saturated air with temperature and relative humidity values typical of the lower layer will become positively buoyant if lifted. After adjustment it is assumed that both layers are saturated and that the adjusted temperatures of the two layers lie on the same moist adiabat. Because the mass weighted sum of the moist enthalpy values in the two layers is held constant during adjustment, the lower layer cools and so holds less moisture while the upper layer warms. Some of the moisture lost from the lower layer may be transferred into the upper one to bring it to saturation and/or maintain that state. Excess condensed water drops out as precipitation.

Since the lower layer cools and the upper layer warms during an adjustment event, an adjusted pair of layers (such as layers 1 and 2 in Figure 3) may be destabilized by subsequent adjustment of an adjacent pair in which one of the layers is common (such as layer 2 and the layer immediately above it). Hence iteration may be required to achieve an over-all removal of instability within the column.

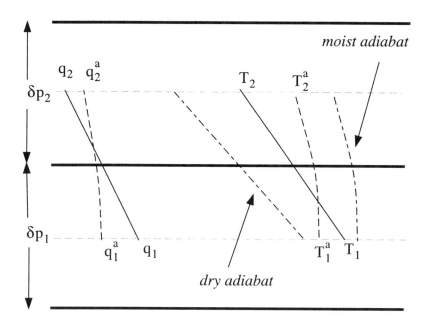

Figure 3. Schematic view of moist convective adjustment of an initially unstable state in an atmospheric model with lower and upper (layer mean) temperatures T_1 and T_2 and specific humidities q_1 and q_2. In this case the lower layer is initially saturated. The adjustment process restores the layer temperatures and humidity values to a saturated state in which adjusted values of the layer mean temperatures (T_1^a, T_2^a) are located on a moist adiabat.

Moist convective adjustment as discussed in the preceding paragraph, was first used by Manabe *et al.* [12] in a GCM. They chose to invoke moist adjustment of a pair of layers only if the lower one is saturated. Since it is applied only to saturated layers, considerable instability may build up before convective adjustment is invoked. Also, since adjustment is to a saturated moist adiabatic state, grid cells that are convectively active may become saturated and moist adiabatic over unrealistically large areas. Such a structure may be realistic on the scale of convective systems but these are usually much smaller in area than a typical GCM grid cell. Thus even in circumstances where moist convection is active it is likely that the humidity profile remains unsaturated and the temperature lapse rate is likely to be larger than moist adiabatic. A variety of variations on the above adjustment scheme have been proposed so as to make the results of applying it more realistic. For example [13], the adjusted lapse rate can be chosen to be between the moist and dry adiabatic lapse rates as a function of the adjusted relative humidity, which can also be chosen to be less than 100 percent.

However, even with such modifications, this type of convective adjustment scheme tends to produce relatively shallow convective heating layers because it does not adequately account for the penetrative nature of moist convection. It also largely fails to account for the role of shallow convection in which little precipitation reaches the surface and the net effect of convection is to produce warming and drying in the lower part of the convective layer and cooling and moistening in the upper part, opposite to what occurs during moist convective adjustment of saturated layers as outlined above.

3.2 BETTS-MILLER SCHEME

The convective adjustment scheme of Betts and Miller [4] addresses the above-noted deficiencies of the earlier schemes by specifying reference profiles of both temperature and humidity that are consistent with observed structures in the tropics and satisfy the same constraints as the traditional adjustment schemes discussed above. Deep and shallow convection are treated separately.

The reference temperature profile for deep convection is chosen on the basis of the empirical finding [3] that, in the tropics when deep convection is active, the mean temperature profile is close to that of a virtual moist adiabat up to the freezing level and then tends toward that of a pseudoadiabat above this level. Such a temperature profile is used as a first guess for a reference profile. It is assumed that the associated reference humidity profile has a humidity structure that is specified in terms of the saturation pressure. This is the ambient pressure at which a parcel of air, with the reference temperature and specific humidity, would become saturated if lifted adiabatically. In the Betts-Miller scheme, the difference between the saturation pressure and the ambient pressure at a given level is specified for the first-guess profile.

The reference profiles must also satisfy the constraint that

$$\int_{p_T}^{p_B} [C_p(T_R - T) + L(q_R - q)] dp = 0 \tag{5}$$

where p_T and p_B are, respectively, the pressure levels of the base and top of the convective layer. It may be necessary to adjust the reference temperature and/or the humidity profile to satisfy this constraint. There does not appear to be a unique way to do this.

The Betts-Miller scheme invokes a deep adjustment if the depth of the convective layer exceeds a specified value (about 300hPa in a pressure based coordinate). The top of the convective layer is the level above which the temperature on a moist pseudoadiabat (passing through the ambient state at the base of the layer) falls below that of the ambient mean state. If the depth of the convective layer is less than the specified value, shallow convection may occur. It is assumed that this type of convection does not produce precipitation. Hence the reference profiles for temperature and humidity must be chosen so that there is no net heating or drying in a vertically integrated sense.

In regimes where shallow convection predominates (such as the trade wind cumulus regime) it is commonly observed that the mean potential temperature and humidity profiles have certain characteristic features. In the atmospheric boundary layer (ABL), when not cloudy, they both vary slowly in the vertical, consistent with the well-mixed character of that region. The ABL is often capped by a weak potential temperature inversion. Shallow convection commonly occurs between this level (typically near the 950hPa level) and the top of the trade wind inversion (typically near the 850hPa level). The potential temperature and specific humidity profiles are close to a mixing line structure in which the equivalent potential temperature and specific humidity in the convective layer are proportional to each other. Often in fact they vary in a way that is close to what is expected for mixtures of air from near the base and top of the convective layer. Such a structure is consistent with the idea that the partially mixed feature of the trade-wind boundary layer is maintained by shallow convection.

The reference profiles for shallow convection in the Betts-Miller scheme are chosen to represent such a state. In the shallow convective part of the Betts-Miller scheme, first guess profiles of temperature and humidity are constructed on the basis of this assumption. Here also the saturation pressure enters into the definition of reference profiles. The reference potential temperature structure is specified by comparing potential temperature and humidities at the base of the convective layer and the top of the inversion capping it after lifting each to their respective saturation levels. It is assumed as a first approximation that, within the convective layer,

$$\frac{\partial \theta_R}{\partial p} = \beta \alpha \langle \frac{\theta_T - \theta_B}{p_B{}^* - p_T{}^*} \rangle \tag{6}$$

where $(\theta_B, p_B{}^*)$ and $(\theta_T, p_T{}^*)$ are respectively the potential temperature and saturation pressure at the base of the convective layer and the inversion top (which may have to be assigned arbitrarily in practice). The parameters α and β are both specified and are typically near unity in the convectively active region($p_T < p < p_B$); β specifies the vertical gradient of the saturation pressure, and hence the first guess humidity profile for the reference state. In particular, $(\partial p^* / \partial p)_R = \beta$ in the convective layer. The first guess profiles must be adjusted so as to ensure that

$$\int_{p_T}^{p_B} C_p (T_R - T) dp = \int_{p_T}^{p_B} L(q_R - q) dp = 0. \tag{7}$$

The foregoing is essentially the original version of the scheme proposed by

Betts [3]. Later modifications include adding effects of unsaturated downdrafts [5].

3.3 MASS FLUX SCHEMES

An increasingly popular approach to cumulus parameterization involves use of simple cloud models as a basis for representing the effects of moist convection on the larger scales. Since this approach attempts to account for the vertical mass flux that occurs in convective updrafts and downdrafts, they are commonly referred to as mass flux schemes.

We will consider here mainly the dynamical and thermodynamical effects of cumulus clouds on the heat and moisture budgets of the atmosphere. In general these effects, in the context of a mass flux scheme, are often written in the form:

$$Q_1 - Q_R = \left(\frac{L}{C_p}\right)(C - E)_c - \left(\frac{1}{\bar{\rho}}\right)\frac{\partial}{\partial z}[M_c(s_c - \bar{s})] \tag{8}$$

$$Q_2 = (E - C)_c - \left(\frac{1}{\bar{\rho}}\right)\frac{\partial}{\partial z}[M_c(q_c - \bar{q})] \tag{9}$$

where the diabatic heating contribution is written in terms of the dry static energy.

In these equations M_c is the net cumulus mass flux and the last terms in each of the above equations represent the vertical fluxes of heat and moisture by cumulus clouds. These terms have been written in the approximate form that results from neglecting terms that are small when the fractional area covered by convective circulations (updrafts and downdrafts) is small, an assumption that we make and discuss further below. Note also that terms labelled with the subscript ( )$_c$ in general represent the net effects of both updrafts and downdrafts.

3.3.1 Updrafts

Consistent with observations and common practice we assume that convective scale updrafts occupy a small fraction (a) of the horizontal area of a typical GCM grid volume. We assume also that all updrafts originate within the atmospheric boundary layer (ABL) but may penetrate to varying heights between the LCL and the LNB for undiluted parcels or plumes that ascend from a reference level near the top of the ABL. In what follows we will outline the basic features of parameterizations that are based on the entraining plume concept in which the depth of plume penetration depends on the degree of dilution that results from entrainment of environmental air into the sides of the plume. This concept was commonly employed in one-dimensional cloud models that were widely used in research in the physics and dynamics of convective clouds before the advent of modern computers permitted development of comprehensive three dimensional convective cloud models. It is still widely used in parameterizations for larger scale models. The paper of Arakawa and Schubert [1] (hereinafter AS) is the seminal work on this approach for parameterization. Most of the mass flux schemes in current use are based on variants of it. However, limitations to this concept have been noted in recent years, especially for shallow cumulus clouds, and alternative mass flux based formulations have been proposed recently and will be discussed briefly near the end of this chapter.

The basic conceptual features of entraining plumes that form convective updrafts and downdrafts are illustrated schematically in Figure 3. A commonly used assumption, central to the AS scheme, is that each updraft entrains air through its sides and detrains cloudy air only at its top. Detrainment over a broad range of depths is possible in convectively active regimes if an ensemble of plumes with a range of depths is present.

For the sake of simplicity in deriving mathematical representations of the most important properties of these entities we assume that the anelastic approximation can be used to derive equations governing convective scale quantities. In this approximation the horizontal and temporal variation of density is ignored except where it enters into the determination of buoyancy. Hence solutions to the analastic equations do not include acoustic modes. Consistent with this approximation the background density ($\bar{\rho}$) is used everywhere except in the definition of buoyancy. Equations governing the budgets of mass, heat, water vapor, condensed water, and mean updraft velocity for a single updraft are obtained by integrating over the fractional area occupied by the updraft, making use of Leibnitz's rule. The result is:

$$\bar{\rho}\left(\frac{\partial a}{\partial t}\right) + D_u - E_u + \frac{\partial M_u}{\partial z} = 0 \tag{10}$$

$$\bar{\rho}\frac{\partial(as_u)}{\partial t} + D_u(s_u)_D - E_u s_e + \frac{\partial(M_u s_u + (M's')_u)}{\partial z} = LC_u \tag{11}$$

$$\bar{\rho}\frac{\partial(aq_u)}{\partial t} + D_u(q_u)_D - E_u q_e + \frac{\partial(M_u q_u + (M'q')_u)}{\partial z} = -C_u \tag{12}$$

$$\bar{\rho}\frac{\partial(al_u)}{\partial t} + D_u(l_u)_D + \frac{\partial(M_u l_u + (M'l')_u)}{\partial z} = C_u - P_u \tag{13}$$

$$\frac{\partial M_u}{\partial t} + D_u w_u + \frac{\partial(M_u w_u + (M'w')_u)}{\partial z} = -\frac{\partial p_u}{\partial z} + \frac{\bar{\rho}g((T_v)_u - \bar{T}_v)}{\bar{T}} \tag{14}$$

252

where $(\ )_u$, $(\ )_e$ refer respectively to updraft and environmental quantities,

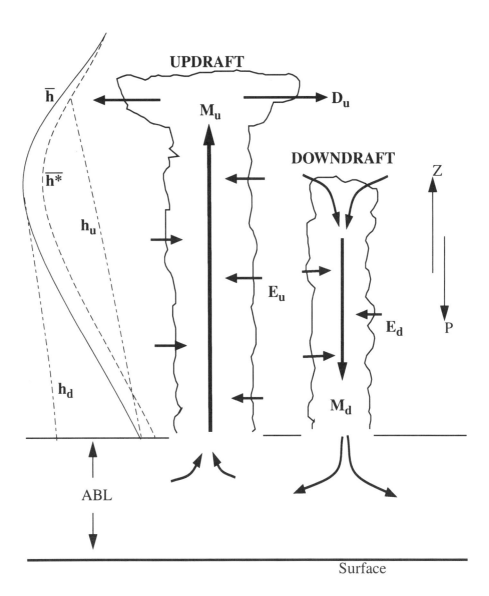

Figure 4. Schematic view of a typical individual entraining plume updraft and a corresponding saturated downdraft as used in mass flux parameterization schemes of the Arakawa-Schubert type. The updraft entrains environmental air into its bottom and sides and detrains at its top, at the point where the temperature of the updraft is equal to that of the environment. The downdraft entrains environmental air into its top and sides and detrains within the atmospheric boundary layer.

and $(\ )_D$ to quantities evaluated in the detrainment region of the updrafts. Following common practice in the cumulus parameterization literature The updraft quantities($Mu = \bar{\rho} a w_u,\ s_u,\ q_u,\ l_u$) are averages over the updraft area. The mass entrainment (E_u) and detrainment (D_u) rates for the updraft must be specified independently.

The above equations are usually simplified further for use in deriving parameterizations by invoking the following basic assumptions:

- "Top hat" profiles for updraft quantities ($(M'\chi')_u$ can be ignored, where χ represents an updraft quantity).
- Environmental quantities are nearly equal to those for the grid volume mean. In the absence of downdrafts (case to be considered below) the relationship between the grid volume mean, updrafts and mean environmental values of a scalar variable is $\bar{\chi} = \chi_e(1-a) + a\chi_u$. However since a is small this implies that $\chi_e \cong \bar{\chi}$
- Updraft quantities are quasi-steady, so derivatives with respect to time are ignored. In reality, this assumption is coupled with the assumption that there is an ensemble of updrafts within the grid volume which are at various stages of their life-cycles. It is assumed that the result of averaging over a time period long enough to encompass a few life cycles (but shorter than the typical time scales for the large-scale flow) is reasonably well approximated by using a steady state model for convective scale processes.

The entraining plume concept assumes that mass is entrained laterally into the sides of the plume at a rate that is proportional to the rate at which it flows upwards in the updraft. A similar assumption applies to detrainment, which however is typically thought to be confined mainly to regions in which the plume becomes neutrally buoyant with respect to the large-scale mean. Thus the entrainment and detrainment rates are expressed as

$$E_u = \lambda_u M_u \tag{15}$$

$$D_u = \delta_u M_u \tag{16}$$

In the original formulations the plume was taken to be axially symmetric and the fractional entrainment rate (λ_u) was taken to be inversely proportional to its radius, consistent with the assumption that the inflow velocity is proportional to the updraft velocity. Although this formulation does have some support from laboratory experiments, which also served to establish a constant of proportionality) most atmospheric applications assume that the fractional entrainment rate is independent of the local plume radius. Arakawa and Schubert assume that the fractional entrainment rate is a constant for given type of updraft and serves to characterize the updraft.

Let us assume that the convective activity within a given grid volume can be represented as an ensemble of entraining plumes, each of which has a different fractional entrainment rate. The plumes all have a common base level but differing top levels which are found near the levels where the plumes are neutrally buoyant but positively buoyant below. Detrainment is assumed to occur within a narrow region surrounding this level and brings about a complete depletion of the updraft mass flux. In these circumstances the updraft mass flux vanishes above the level $z = z_t$ and below that level it is given by

$$M_u = M_b exp[\lambda_u(z - z_b)]; z_b < z < z_t \qquad (17)$$

It is typically assumed that, below the updraft base, the mass flux would increase from zero to its base value (M_b) within a finite number of model layer depths and would decrease to zero within a single layer depth at the top of the entrainment region. In this case $D = (M_u(z_t))/(\Delta z_t)$ at the top of the plume and is zero elsewhere.

To understand how λ_u is determined it proves convenient to consider the equation governing the moist static energy of the updraft, $h_u = C_p T_u + Lq_u + gz$. In general, h_u behaves similarly to the equivalent potential temperature. In particular, it is typically uniform in height within the ABL and then decreases with height in the lower troposphere, reaching a minimum which, in the tropics, is usually in the middle of the troposphere. It increases with height again above this level. This characteristic vertical structure is a consequence of the combined effects of the latent energy (Lq) term, which dominates in the lower troposphere but decreases rapidly with height above the ABL, and the dry static energy term which increases with height in the same region.

Under the above assumptions the moist static energy of the updraft is determined from the following equation:

$$\frac{\partial h_u}{\partial z} = -\lambda_u(h_u - \bar{h}) \qquad (18)$$

It is reasonable to assume that the air in the updraft is saturated through most of its depth, certainly near the top of the updraft. In these circumstances the difference between the temperature in the updraft and the larger scale mean is given approximately by considering the first two terms in a Taylor expansion about the large scale mean, ignoring the difference between the pressure in the updraft and that of the large scale mean state, to give

$$T_u - T \cong (h_u - h^*)/[1 + (L/C_p)(\partial q^*/\partial T)_p] \qquad (19)$$

where h^* is the moist static energy for saturated air that has the same temperature and pressure as the large scale mean, and q^* is the corresponding specific humidity.

As stated above, we assume that the air in the plume is nearly neutrally buoyant near its top. As a first approximation the level at which this occurs is likely to be near that at which the temperature of the plume is nearly equal to that of the environment. In reality, of course, virtual temperatures should be compared and the effects of liquid water loading in the updraft taken into account in doing so. This can be done in practice. However, for simplicity here we assume that the top (detrainment level) of the plume is where the difference in moist static energies $(h_u - h^*)$ vanishes. At the base of the updraft, h_u must be specified. We will assume that the base of the updrafts (hereinafter referred to as the cloud base) is near the top of the ABL and that h_u is equal to the moist static energy of the boundary layer air. With this choice, both h_u and the associated fractional entrainment rate can be determined. The fractional entrainment rate is a function of the detrainment level and given by the following equation

$$\lambda_u(z_t) = \frac{h(z_b) - h^*(z_t)}{\int_{z_b}^{z_t} (h_u(z;\lambda_u) - \bar{h}(z))dz} , \qquad (20)$$

which can be solved iteratively for the value of λ_u that is required to give a plume top at a specified detrainment level.

The above equations form the basic ingredients of the AS parameterization scheme. It is assumed that most convective cloud systems are in a state of equilibrium with the larger-scale circulation in which they are embedded and that such a system could be represented in terms of an ensemble of entraining plume updrafts (and in later work associated downdrafts) which detrain at different levels in the atmosphere. All updrafts in a grid cell have a common base level near the top of the ABL. Thus each type of plume can be characterized either by a peculiar entrainment rate or by the level at which it detrains. In general, plumes become more diluted with environmental air as the fractional entrainment rate increases. In principle, a continuous spectrum of plumes, ordered in terms of the fractional entrainment rate, is possible. In practice a discrete number of plumes, each with its top in a model layer, is typically used in implementation of the AS scheme.

In the AS formulation each plume detrains mass only at its top, where it detrains all of its mass within a single model layer. However, since there is an ensemble of plumes, and the total mass flux is the sum of the contributions from each, the net detrainment can occur over several layers. Since

$$M_u = \sum_{\lambda_u(z_t)} M(z;\lambda_u(z_t)) \qquad (21)$$

the detrainment rate can be expressed as

$$D_u = \left(\frac{\partial M}{\partial \lambda_u} \frac{d\lambda_u}{dz} \right)_{z_t} \tag{22}$$

A simple case of the ensemble approach is one in which each member of the ensemble has the same mass flux at cloud base, as in the scheme proposed by Zhang and McFarlane [21] (hereinafter referred to as ZM). In this case the total updraft mass flux, for a continuous distribution of fractional entrainment rates can be expressed in the form:

$$M_u = \left(\frac{M_b}{\lambda_u^{max}} \right) \int_0^{\lambda_u^*} e^{\lambda_u(z-z_b)} d\lambda_u$$

$$= \left(\frac{M_b}{\lambda_u^{max}} \right) \left(e^{\lambda_u^*(z-z_b)} - 1 \right) \tag{23}$$

where $\lambda_u^*(z)$ is the fractional entrainment rate of plume whose top is at level z. For this formulation the net entrainment and detrainment rates are easily derived as:

$$E_u(z) = \left. \frac{\partial M_u}{\partial z} \right|_{\lambda_u^*} = \frac{M_b}{\lambda_u^{max}} \int_0^{\lambda_u^*} \left(\lambda e^{\lambda(z-z_b)} \right) d\lambda \tag{24}$$

$$D_u(z) = -\left(\frac{\partial M_u}{\partial \lambda_u^*} \right) \frac{d\lambda_u^*}{dz} = -\left(\frac{M_b}{\lambda_u^{max}} \right) \left(\frac{d\lambda_u^*}{dz} \right) e^{\lambda_u^*(z-z_b)} \tag{25}$$

In the ZM scheme the maximum fractional entrainment rate, λ_u^{max}, is chosen so that the shallowest plume detrains at the level, z_d, where $h^*(z)$ is at its minimum. This level then also marks the base of the detrainment layer. Since it typically occurs near middle tropospheric levels this choice ensures that all plumes have a certain minimum depth.

3.3.2 Downdrafts
In the original Arakawa-Schubert Parameterization, convective scale downdrafts were ignored. However, more recently [2] they have been included using an approach that attempts to account for a number of features of downdraft dynamics and microphysics that are typically left out of the more commonly used treatments. In these simpler formulations, convective downdrafts are typically treated in a manner that is conceptually similar to updrafts. It is usually assumed that convective scale downdrafts are driven by evaporation of precipitation into the

air that is drawn from the environment of the convective system into the top and sides of the downdraft. The relative humidity in these downdrafts is specified, usually 100%. A downdraft initiation level must be chosen. In principle this should be chosen to be at a level such that evaporative cooling induces negative buoyancy. If a single updraft-downdraft pair is chosen, one way of selecting this level is to assume that it is somewhere between the base and top of the updraft at a level such that a saturated mixture of updraft and environmental air would be negatively buoyant. This air is then allowed to descend with additional lateral entrainment taking place at a specified fractional rate (typically proportional to that of the associated updraft) until it loses buoyancy or reaches the boundary layer where it is assumed that all of its mass is detrained. If the air in the downdraft remains saturated the evaporation rate needed to maintain it is a function of its mass flux and properties (moist static energy) at the initiating level.

In the ZM scheme downdrafts are treated as a bulk ensemble using a mass flux formulation that is similar to that for updrafts. In particular, it is assumed that downdrafts are initiated at or below the base of the updraft detrainment layer. The downdrafts remain saturated and entrain air laterally at rates that are usually larger than the associated updraft entrainment rates. However, since downdrafts are initiated at or below the base of the detrainment layer, they remain negatively buoyant at all levels between initiation and the base of the updraft.

A further limitation that is imposed is that the magnitude of the downdraft mass flux at cloud base is no larger than that in the corresponding updrafts. This places a limit on the magnitude of the fractional entrainment rate that is used in the downdraft. The initiating downward mass flux is taken to be proportional to the ensemble mean updraft mass flux at cloud base with the proportionality factor that is a function of the precipitation rate.

The ensemble downdraft mass flux is given by:

$$M_d = \left(\frac{-\alpha M_b}{\lambda_d^{max}(z_d - z)} \right)[e^{\lambda_d(z_d - z)} - 1] \tag{26}$$

where M_b is the updraft mass flux of the ensemble at cloud base and

$$\alpha = \mu\left[\frac{PCP}{PCP + EVP} \right] \tag{27}$$

where PCP is the precipitation rate generated in the updraft ensemble and EVP is the evaporation rate that would be required to give the downdraft ensemble a mass flux at cloud base that is equal in magnitude to that of the updraft ensemble. the above choice for α ensures that the net mass flux at cloud base is positive. The quantity μ is a tuning parameter that can be chosen empirically to ensure a desired precipitation efficiency for the ensemble.

3.3.3 *Closure*

In the foregoing, we discussed the basic formulation needed to account for the effects of cumulus clouds in terms of ensembles of plumes representing convective scale updrafts and downdrafts. A common parameter in all of the above is the mass flux in updrafts at cloud base. The quantity must be specified independently in order to form a closed system for the mass flux components, updraft and downdraft properties.

Perhaps the most elaborate and ambitious closure scheme that has been proposed to date is that of the AS scheme wherein a cloud work function for each updraft plume is defined as the work per unit cloud base mass flux done by the thermal buoyancy of the given updraft. It is assumed that an equilibrium exists between the generation of convective instability in the large scale flow and the stabilizing effects of the cumulus ensemble. In the basic AS scheme it is assumed that the ensemble is comprised of a continuum of updrafts ordered according the their characteristic fractional entrainment rates or, equivalently, their detrainment levels. The cloud base mass flux associated with these updrafts can be taken to be a function of the fractional entrainment rate. The assumed equilibrium is characterized by quasi-steadiness of the work function for each updraft, an assumption that leads to an integral equation for the cloud base mass flux, as a function of the entrainment rate. This equation must be solved subject to the constraint that the cloud base mass flux is positive. In practice, since the cumulus ensemble is discretized based on the number of model layers, the integral equation can also be written in a discretized form. The problem of solving this equation is not always well posed so it has sometimes been found difficult to solve in practice and this has led to several simpler schemes that do not have this difficulty.

The ZM scheme represents one such simplification. Since all members of the updraft ensemble have the same cloud base mass flux, closure is achieved by specifying the value of this basic quantity. The closure assumption is simply that the effects of convection, acting alone, are to relax CAPE (including the effects of condensed water loading) toward zero (or a specified threshold value) with a specified time scale. Zhang and McFarlane [21] have shown, in the context of single column model experiments, that this simple closure scheme allows an equilibrium state to develop in circumstances where large-scale processes act persistently to produces conditional instability (i.e., to increase the local CAPE). The equilibrium state is one in which the temporal variability of the CAPE is small. This is the result of a balance between the destabilizing effect of the large-scale forcing and the convective stablization. Such a balance might be expected to occur in convectively active regions of the tropics for example. Recently

Zhang *et al.* [20] have confirmed this conclusion in simulations with CCM3 which have been found to be in quasi-equilibrium over the tropical Pacific warm pool.

In CCM3 the ZM and Hack [8] schemes are used together in a complementary manner. The ZM scheme is designed to account mainly for the effects of deep penetrative precipitating convection and does not realistically for the effects of shallow cumulus clouds that produced little precipitation. The Hack scheme is also based on a mass flux approach but is invoked as a type of adjustment scheme in which shallow convective updrafts are assumed to occur in succession from bases within several model layers within the lower troposphere (not necessarily just those within the ABL). The depth of each updraft is assumed not to exceed three model layers, and entrainment rates are chosen to ensure satisfaction of this constraint. This scheme, acting alone, does not produce the deep heating profiles that are expected for penetrative precipitating convection. It does however account for the effects of shallow convection much more realistically than does the ZM scheme.

3.4 SCHEMES BASED ON MOISTURE CONVERGENCE

In periods of deep convection in the tropics it is often happens that the rate of precipitation is closely balanced by the rate at which moisture is supplied to the convectively active region through net horizontal convergence and surface evaporation. This type of balance would of course be expected in a time averaged sense but it has been observed to hold approximately locally as well during periods of deep precipitating convection in the tropics. The idea of using this observed behaviour as a basis for parameterizing deep convective heating has been used most extensiveley in a class of cumulus parameterization schemes based on those proposed originally by Kuo ([10], [11]). In these schemes a basic assumption is that

$$P = \frac{F_s - \int_0^{p_s} (\nabla \bullet \mathbf{V} q) dp / g}{1 + b} \qquad (28)$$

where P is the rate of precipitation due to deep convection, F_s is the flux of moisture at the surface, q is the specific humidity, p_s is the pressure at the surface and $\mathbf{V}$ the horizontal velocity vector. All of these quantities are in principle known at any time for a given GCM grid cell. The parameter b must be specified also. Kuo [11] assumed that it is small and positive. In his formulation it represents the ratio of the rate of moistening of the atmosphere to the rate of precipitation in regions of deep convection. In later work several proposals, based on this concept, have been advanced for determining this quantity.

Specification of the precipitation rate and b in a GCM grid cell determines the net heating and moistening in the cell due to penetrative convection. However, some additional assumptions must be made to specify the vertical profiles of these quantities. Kuo [11] proposed that they be made proportional, respectively, to the magnitudes of the temperature and moisture excess for a saturated parcel of air lifted adiabatically from the top of the ABL.

The Kuo scheme has been used most extensively in numerical weather prediction models. A variety of variants of the original Kuo scheme have been proposed in attempts to improve its performance in this context, including coupling of a Kuo-type of closure assumption, similar in concept to that expressed in equation 28 with a mass flux type of representation for the convective heating and moistening/drying. A comprehensive review of the literature on the Kuo scheme and its descendants is beyond the scope of this chapter. Emanuel [7] includes a critical review of work over the past two decades on this topic.

3.5 BUOYANCY SORTING

The entraining plume model has been widely used as a basis for cumulus parameterization. In large part this is because it provides a relatively simple but self consistent framework for constructing practically useful parameterizations of the large scale effects of convective clouds and cloud systems. In this respect it has enjoyed considerable success. Many if not most of the existing GCMs include convective mass flux parameterization schemes that are based on it.

As mentioned above, one-dimensional cloud models based on entraining plume formulations were used initially in cloud modelling research. However limitations of these simple models in simulating actual convective clouds were discovered relatively soon after they came into use. A frequently cited example is the well known finding of Warner [19] that observed cumulus cloud top heights and average (within updrafts) liquid water content profiles could not be consistently predicted using an entraining plume cloud updraft model. More recently observational studies ([14], [16]) have revealed that samples of air in typical shallow cumulus clouds (which usually have low precipitation rates) often have properties that suggest that they are comprised of mixtures of air from cloud base and levels at and above that at which samples were taken. Also, even in circumstances, common in shallow convective regimes, where substantial quantities of environmental air are entrained at many levels within clouds it is still possible to find samples of undiluted air from near cloud base at all levels within the clouds.

Such findings are not consistent with the entraining plume model since it implies that air at any level within the plume is a mixture of air from cloud base and levels below the given level. This is a consequence of the organized nature of the entrainment process and the assumption of horizontal homogeneity within the entraining plume. The observations are more consistent with the idea, first suggested by Telford [17] and later further refined by various authors, that air is entrained in a less organized way into rising parcels or turrets within clouds. The initially positively buoyant parcels affected by such entrainment *episodes* will usually have their buoyancy reduced as a result of the mixing with environmental air. They may then either continue to ascend if still positively buoyant, or descend if negatively buoyant, possibly with further *episodic* mixing within the clouds until they reach levels at which they become neutrally buoyant. This process is now referred to as *buoyancy sorting*. The above cited paper of Taylor and Baker in particular provides convincing evidence that this process is likely ubiquitous in cumulus clouds.

It is beyond the scope of this chapter to provide a complete presentation of the buoyancy sorting hypothesis. It is discussed extensively in the textbook by Emanuel [7], which also includes a bibliography of the recent literature on the subject. The buoyancy sorting concept underlies implicitly the shallow cumulus adjustment component of the Betts-Miller scheme and has been used explicitly in the episodic mixing scheme proposed by Raymond and Blythe [15], refined and elaborated on by Emanuel [6] who also accounts for the generation of precipitation and evaporative driven downdrafts. Although quite comprehensive, Emanuel's scheme, or adaptations of it, has not yet been used extensively in GCM applications.

Acknowledgements

I am grateful to the students who read and commented on the first draft of the notes on this chapter while in Il Ciocco. Special thanks are due to Dominique Paquin, Sebastien Biner, Virginie Lorant, and Mijke Zachariasse for the reading and making detailed comments on a later draft of the chapter. Their contributions have been of much value in preparation of the final version.

References

1.Arakawa, A., and Schubert, W.H. (1974) Interaction of a cumulus cloud ensemble with the large-scale environment, Part I., *J. Atmos. Sci.*, **31**, 674-701.
2. Arakawa, A., Cheng, M-D. (1993) The Arakawa-Schubert Cumulus Parameterization, *The Representation of Cumulus Convection in Numerical Models*, (K.Emanuel, D.Raymond, ed.), Meteorological Monographs, American Meteorological Society, Vol. 24, Ch. 10.
3. Betts, A.K., (1986) A new convective adjustment scheme. Part I: Observational and theoretical basis, *Quart. J. Roy. Meteor. Soc.*, **112**, 693-709.

262

4. Betts, A.K., and Miller, M.J., (1986) A new convective adjustment scheme. Part II. Single column tests using GATE wave, BOMEX, ATEX and arctic air-mass data sets, *Quart J. Roy. Meteor. Soc.*, **112**, 693-709.

5. Betts, A.K., and Miller, M.J. (1993) The Betts-Miller Scheme, *The Representation of Cumulus Convection in Numerical Models*, (K.Emanuel, D.Raymond, ed.), Meteorological Monographs, American Meteorological Society, Vol. 24, Ch. 9.

6. Emanuel, K.A. (1991) A scheme for representing cumulus convection in large-scale models, *J. Atmos. Sci.*, **48**, 2313-2335

7. Emanuel, K.A. (1994) *Atmospheric Convection*, Oxford University Press, New York

8. Hack, J.J. (1994) Parameterization of moist convection in the National Center for Atmospheric Research community climate model (CCM2), *J. Geophys. Res.*, **99**, 5551-5568

9. Hong, S-Y, and Pan, H-L. (1998), Convective Trigger Function for a Mass-Flux Cumulus Parameterization Scheme, *Mon. Wea. Rev*, **126**, 2599-2620.

10. Kuo, H-L. (1965) On formation and intensification of tropical cyclones through latent heat release by cumulus convection, *J. Atmos. Sci.*, **22**, 40-63.

11. Kuo, H-L. (1974) Further studies of the parameterization of parameterization of the influence of cumulus convection on the large-scale flow., *J. Atmos. Sci.*, **31**, 1232-1240.

12. Manabe, S., Smagorinsky, J., and Strickler, R.F. (1965) Simulated climatology of a general circulation model with a hydrological cycle, *Mon. Wea. Rev.*, **93**, 769-798

13. McFarlane, N.A., Boer, G.J., Blanchet, J.P., and Lazare, M. (1992) The Canadian Climate Centre second-generation atmospheric general circulation model and its equilibrium climate, *J. Climate*, **5**, 1013-1044.

14. Paluch, I.R. (1979) The entrainment mechanism in Colorado cumuli, *J. Atmos. Sci.*, **36**, 2467-2478.

15. Raymond, D.J., and Blythe, A.M. (1986) A stochastic mixing model for nonprecipitating cumulus clouds, *J. Atmos. Sci.*, **43**, 2708-2718.

16. Taylor, G.R., and Baker, M.B. (1991) Entrainment and detrainment in cumulus clouds, *J. Atmos. Sci.*, **48**, 112-121

17. Telford, J.W. (1975) Turbulence, entrainment and mixing in cloud dynamics, *Pageoph.*, **113**, 1067-1084

18. Yanai, M., and Johnson, R. (1993) Impacts of cumulus convection on thermodynamic fields, *The Representation of Cumulus Convection in Numerical Models*, (K.Emanuel, D.Raymond, ed.), Meteorological Monographs, American Meteorological Society, Vol. 24, Ch. 4.

19. Warner, J. (1970) On steady-state one-dimensional models of cumulus convection, *J. Atmos. Sci.*, **27**, 1035-1040

20. Zhang, G.J., Kiehl, J.T., and Rasch, P.J. (1998) Response of climate simulation to a new convective parameterization in the National Center for Atmospheric Research community climate model (CCM3), *J. Climate*, **11**, 2097-2115

21. Zhang, G.J., and McFarlane, N.A. (1995) Sensitivity of climate simulations to the parameterization of cumulus convection in the Canadian Climate Centre general circulation model, *Atmos.-Ocean*, **33**, 407-446

CLOUDS AND CLOUD WATER PREDICTION

J.-J. MORCRETTE, CH. JAKOB, J. TEIXEIRA
European Centre for Medium-Range Weather Forecasts
Shinfield Park
Reading
Berkshire RG2 9AX
UK

1. Clouds: What is their role in the climate system?

Clouds are a common atmospheric feature and cover large parts of the globe. According to various climatologies, based on either surface [6] or satellite observations, between 50% and 65% of the globe is covered by clouds [23], with a most likely figure around 60-62%.

By their presence, they strongly affect the radiative fluxes throughout the atmosphere. Although there is agreement on their overall cooling effect in the shortwave part of the radiative spectrum ([18], [7], [24]), and on their heating effect in the longwave part, for the total Earth-atmosphere system, there are still some uncertainties. For example [9], does absorption of shortwave radiation in clouds actually have any sizeable heating effect in the vertical within the atmosphere ([3], [17], [19], ([14], [2], [12])? Whatever the origin of these uncertainties, clouds are without doubt the main modulators of the three-dimensional deposition of radiative energy within the Earth-atmosphere system.

At first, the distribution of clouds simulated by climate models was diagnosed from simple relationships with the model humidity fields. Today most general circulation models (GCMs) used in weather forecasting and climate studies are moving towards a prognostic approach to the representation of cloudiness. More effort is being spent in connecting cloudiness with latent heat release and consumption, either directly inside clouds or, as in the case of evaporation of precipitation, with the water/ice originating from clouds. An additional process now being accounted for in some models is the role of moist convection in transporting not only heat and moisture, but also momentum and chemical species over large distances in the vertical.

P. Mote and A. O'Neill (eds.), Numerical Modeling of the Global Atmosphere in the Climate System, 263–280.
© 2000 *Kluwer Academic Publishers. Printed in the Netherlands.*

However, the difficulties in cloud representation are many. First, there is a huge variety of cloud types, the result of complex interactions of a number of processes, e.g., moist convection, turbulence, large-scale vertical motion, and microphysics, of which many are only poorly understood. Second, many of the observed clouds are of sub-grid scale size (both horizontally and vertically) for the current generation of general circulation models. In this respect, the major cloud effect, i.e. their impact on radiation, is determined by many different cloud parameters, most of them determined on scales not resolved by the models (phase, refractive index, and particle size, which depend on the microphysical processes, but also the in-cloud distribution of the condensates within a cloud element, and/or the usually heterogeneous distribution of real clouds within the spatial domain considered in a GCM grid box).

For the more modern GCMs to-date, cloud parameterization tries to describe the area coverage and vertical extent of cloudiness, the amount of condensate that is formed, dissipated and converted to precipitation, and the phase, the size of particles, and in-cloud distribution of the condensate. In the following, a brief description of the history of clouds in the ECMWF model is given as an illustration of the efforts towards a better representation of the role of clouds in the general circulation of the atmosphere.

2. Cloud fraction

Many GCMs now allow for a partial coverage of a model grid box with clouds. This implies in principle the existence of a non-uniform distribution of temperature and/or specific humidity within that volume element, resulting in condensation occurring locally when q becomes larger than the saturation q at the local temperature and pressure. However, the distribution idea, although appealing, is difficult to use in practice, as it is very likely that different types of clouds will be connected to different shapes of distributions; that the distribution is a result of the presence of clouds rather than its precursor; and finally that those distributions are very poorly known.

Another implication of the partial cloud cover as a way of parameterizing cloudiness is that condensation has to occur before the grid mean relative humidity reaches 100%. How can we define a "critical" relative humidity above which we start to consider cloud formation? This is a difficult question as clouds can exist (at least for some time) in very low relative humidity environments [26]), and again, the relative humidity is determined by the clouds (and not the other way around).

Finally, what do we do about partial cover in the vertical? In the absence of other possibilities, the easiest solution is to assume that clouds completely fill a model layer. This might be reasonable provided that the model has a reasonably high vertical resolution (the ECMWF model has 31 levels on the vertical giving about 500 m thick layers). However, problems in vertical advection may arise for high vertical resolution even with small time-steps, as the numerical stability criterion on the vertical might be reached before the horizontal one.

3. A quick history of cloud representation at ECMWF

The very first GCMs externally prescribed the cloud fields following a zonal mean climatology (e.g., [8]). From its inception in 1978, the ECMWF forecast model had interactive cloudiness empirically diagnosed from large-scale model variables using the basic premise that condensation on smaller scales is part of the larger scale condensation regime. In such a method, the empirical functions are chosen to represent the probability of clouds occurring under certain atmospheric conditions.

However successful this approach might have been, the clouds are decoupled from the rest of the model physics. The only impact such clouds can have on the rest of the model is through their one-way effect on the radiation fields. Furthermore, the cloud radiative properties have to be somehow prescribed or diagnosed separately from a condensed water content, which is itself prescribed or diagnosed. This whole approach clearly lacks a sound physical basis, as the clouds are determined by the outcome of the physical processes rather than by the physical processes themselves.

In the ECMWF model, the layer cloud fractions were first diagnosed as a quadratic function of the departure of the model large-scale relative humidity from a height-dependent critical relative humidity following Geleyn [5]. Then in 1985, a diagnostic cloud scheme based on relative humidity and other large-scale parameters (static stability for low-level clouds, intensity of the convective precipitation within the last three hours for cover by convective towers and associated anvils, vertical velocity for other clouds) was introduced by Slingo [28].

With a prognostic cloud scheme, the cloud water content is explicitly calculated taking into account the formation and evaporation of cloud and rain drops. Although such an approach involves one (or more) additional model variable(s), it has the advantages of providing a proper representation of the thermodynamical effects of the sub-grid condensation and a better link between radiative, dynamical and hydrological processes, particularly with cloud radiative properties being now linked to the prognosed cloud water. Two of the main problems are related: first, to verification, as there is still little data on 3-D distribution of cloud water; and second, to the treatment of the transitions between mixed ice/liquid water and the rôle of the microphysical processes in them.

The last problem is related to whether the layer cloud fraction should be diagnosed or prognosed. The best solution is to consider an all-or-nothing cloud as in, for example, Fowler *et al.* [4], but one can wonder whether such an approach might provide, for example, the proper temporal evolution of forcing in the deep tropics when within a diurnal cycle a convective cloud might have a small fractional cloud cover while it rises, and then have a more extensive cover once it detrains.

The pioneering paper by Sundqvist in 1978 [32] proposed a diagnostic formulation of the cloud fraction, which has then been followed in a number of subsequent studies ([33], [34], [13], [29], [22], [20]). In these schemes, the cloud cover is related either to the relative humidity or to the distribution of some conserved variables. In the first case (Sundqvist-type), as in the previously mentioned diagnostic schemes, the cloud cover depends on the relative humidity of the grid box, and the cloud fraction approaches unity as the relative humidity approaches 100%. The second category of schemes ([29], [20]) follows a basic idea first proposed by Sommeria and Deardorff in 1977 [30] for cloud models. Within a grid box, temperature and moisture are not the same everywhere but are somehow distributed; thus for given q and T, some areas of the grid box might be saturated and hence cloudy. Therefore the cloud fraction depends on the joint distribution of q and T, and thus conserved variables (for example, θ_L and q_T) are required when dealing with the condensation processes.

In these prognostic schemes "of the first kind", the evolution of the condensate is linked to the evolution of the net outcome of the processes, and the scheme invokes a critical relative humidity for the existence of any cloud. On the other hand, Tiedtke [35] ("a prognostic scheme of the second kind" for the ECMWF model) insists that the evolution of the condensate be linked to each of the various processes themselves, and that critical relative humidity be used only for specific processes.

For the ECMWF model, Tiedtke [35] makes the following basic assumptions:

1. The clouds fill a whole model layer on the vertical.
2. As horizontal fractional cover is considered, some of the model variables have to be defined for cloudy and clear parts of the grid box.
3. Clouds have the same thermal state as the environmental air, so that any latent heat release is uniformly applied.
4. There is no distinction between the flow inside and outside the cloud, except for the convectively induced circulation.
5. The rain water is not suspended in the air but falls out immediately. It might be recycled further down.
6. The scheme distinguishes cloud ice and water only on the basis of temperature, so that only two prognostic equations (one for condensed water, one for cloud fraction) are required.

A full description of the scheme is given in Tiedtke [35], and a detailed documentation [10] is available from ECMWF.

4. Validation of cloud products

Within the context of a weather forecasting system, the validation of the model and of its related parameters might be somewhat easier than in the context of a climate model, since starting from analyzed conditions, during the first hours or days of simulation, the model should still be close to the observations. Therefore the verification of cloud products in such an environment, where the forcing terms, i.e. the dynamics, have not had time to stray away, should allow a clear assessment of the strengths and deficiencies of the various physical parameterizations involved in the production of cloud parameters. The following illustrates the cloud-related quantities currently produced and verified within the ECMWF forecasting system.

Figure 1 presents a verification of the total cloud cover over the European area after 72 hours of the operational $T_L 319$ forecast starting on 19980511 12UTC. A good forecast is seen in all areas where the model cloud cover is similar to the synoptic observations marked by a round symbol. Such a verification is performed every 6 hours for all reporting synoptic stations. In comparison, the validation of the cloud liquid water is much more prone to uncertainties. First, the observations from SSM/I are done from satellites whose tracks cover the full globe only every 11 days. Figures 2 and 3 respectively compare, over the first 11 days of May 1998, the total water vapor and total cloud water produced by the ECMWF model during its 6-hour first guess simulations with those derived from the SSM/I observations using either the algorithm of Alishouse et al. [1] for the water vapor and that of Weng and Grody [36] for the cloud water, or the operational ECMWF 1-D variational algorithm [16]. Whereas the vertically integrated water vapor that can be derived from these microwave observations appears reliable (Figure 2), there are large uncertainties on the retrieved cloud water (Figure 3). In particular, the rôle of the precipitation in saturating the radiance measurements is not fully understood. Moreover, this microwave measurement can give only the vertically integrated amount of liquid water, with no provision for cloud ice. As pointed out by Stephens et al. [31], the lack of current satellite measurement of cloud ice is particularly limiting, as this is an area where the current cloud parameterizations sometimes differ largely between themselves [4] and because the ice clouds, especially the tropical anvil clouds, are known to have a large radiative (greenhouse) effect. Moreover, the water vapor budget in the upper troposphere, part of it linked to the dissipation/evaporation of high-level cloudiness, is known to have a large influence on the emission of radiation to space and thus on the water vapor feedback [15].

Various other validation efforts are currently under way for verifying as much as possible the cloud parameters produced by the ECMWF model during its 10-day forecasts and their impact on the atmospheric profiles. One such effort is the comparison of the longwave window channel radiance observed by the geostationary satellites and produced by the model. Figure 4 illustrates such a verification

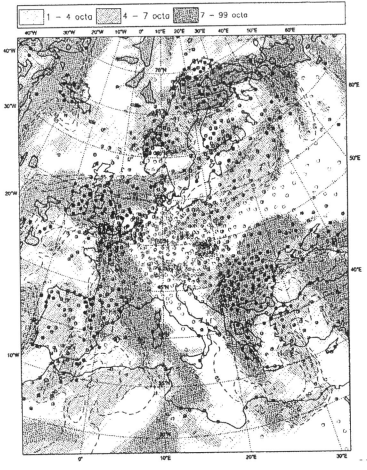

Total Cloud Cover [octa]   - - - RH 700;   —— LOW CL (7 octa)
FC 98050812 STEP 72 VT: 98051112   SYMBOLS: observations
N=1181 BIAS= 0.30 STDEV= 2.82 MAE= 2.11

Figure 1.   The cloud cover produced by the operational ECMWF $T_L319L31$ model after 72 hours of forecast compared to the synoptic observations over Europe.

for the ERICA storm (04-05 January 1989) off the U.S. East Coast, during a 48 hour simulation with the model run at the T_L639 resolution (grid box of $[0.28125°]^2$). Figure 4(a) shows the observed radiance as compiled in the ISCCP-DX dataset [25], whereas the radiance simulated from the model distributions of temperature, humidity, cloud fraction, cloud liquid and ice water appears in

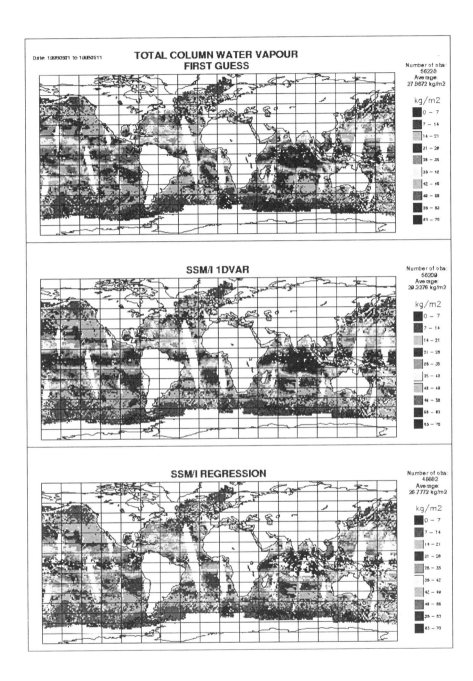

Figure 2. The total column water vapor in the first 11 days of May 1998, averaged over the corresponding ECMWF 6-hour first guess forecasts, analyzed from SSM/I observations with the ECMWF 1-D VAR algorithm and with the SSM/I regression.

270

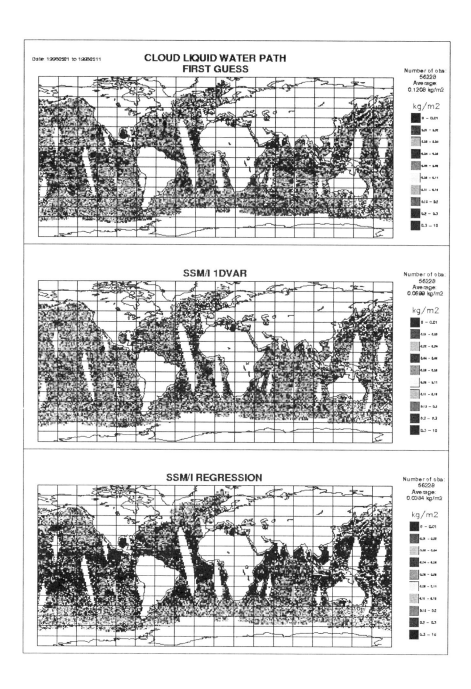

Figure 3. As in Figure 2, but for the vertically integrated cloud water.

DX IR BRIGHTNESS 18Z 04Jan1989

```
210  220  230  240  250  260  270  280  290  300  310
```

Figure 4a. The GOES longwave window channel brightness temperature over the ERICA storm. (a) the ISCCP-DX image for 19890104 18UTC.

Figure 4(b). Although the main structure of the cloud system is well captured by the model over its lifetime, the main deficiency is the lack of cloud amount behind the perturbation, making the frontal cloud usually appear too narrow. The brightness temperature over most clouds also appears too high, indicative of too small an ice content in the model, already seen in other comparisons ([21], [11]).

As part of the model development, the impact of the model horizontal and vertical resolution is assessed. The model cloud scheme does not normally display a strong sensitivity to the model horizontal resolution. Figure 5 presents comparisons of the total cloudiness after a 36-hour forecast starting 19911124 00 UTC

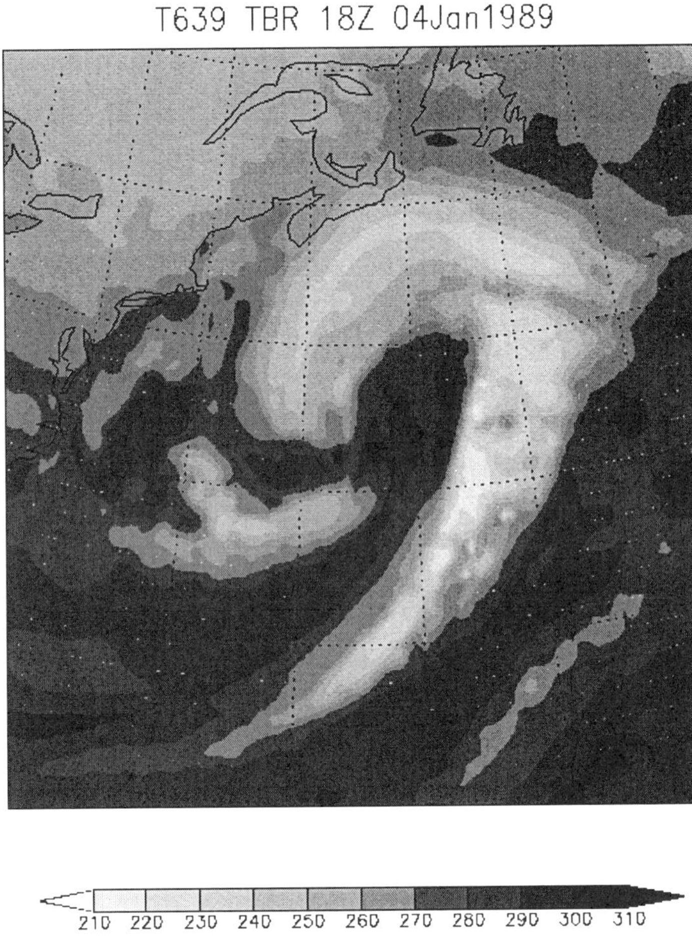

272

T639 TBR 18Z 04Jan1989

210 220 230 240 250 260 270 280 290 300 310

Figure 4b. The GOES longwave window channel brightness temperature over the ERICA storm. (b) the radiance simulated from the ECMWF T_L639 model after a 18 hour forecast starting 19890104 00UTC.

over the FIRE Ci-2 area. Results are presented for the four horizontal resolutions T63, T106, T213 and T_L639 (grid box of $[1.875°]^2$, $[1.125°]^2$, $[0.5625°]^2$ and $[0.28125°]^2$ respectively). Over the FIRE-Ci-2 area, the dominating process for creating cloudiness is mainly linked to the advection of humidity by the jet stream, so that the agreement between the various resolutions simply denotes a consistency in model dynamics when changing the horizontal resolution. Over the TOGA-COARE area (not shown), we could expect more discrepancies especially in areas not covered by an extensive anvil shield, as most of the clouds are becoming smaller when the resolution is increased. However the agreement

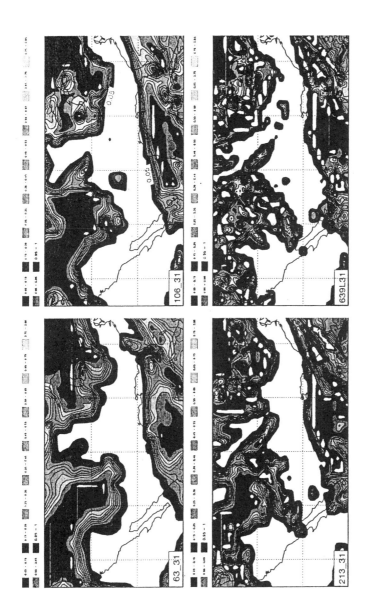

Figure 5. The total cloud cover after a 36-hour forecast from 19911124 00UTC with the ECMWF model at 4 different horizontal resolutions, over the FIRE-Cirrus 2 area.

between resolutions is still good. In this case, the dominating process is the deep convection which depends much more on the model horizontal grid size, with stronger ascents in cloudy columns and larger subsidence in between clouds.

Figure 6 presents the cloudiness off the coast of California during FIRE-Marine Stratus (19870705). In these areas, where low-lying stratocumulus over cold currents goes higher the further from the coast, the representation of the stratiform clouds depends much more on the vertical resolution, as these low-level clouds usually occur at the top of the planetary boundary layer, and would fill a small number of model layers (1 at L31, 2 at L40).

Another validation effort is directed at the clouds occurring over the various Atmospheric Radiation Measurement (ARM) sites [27]. Given the wealth of data now available for the South Great Plains (SGP) and Tropical West Pacific (TWP) sites (conventional meteorological observations, spectrometer, pyrgeometer and pyranometer radiative observations, microwave radiometer, micropulse lidar, ceilometer, energy-balance surface systems, satellite imagery), the synergy between all these observations should permit the bracketing of the deficiencies in the model representation of the clouds above these sites. Figure 7 compares the cloud fraction operationally produced every hour between forecast hours 12 and 36 over the TWP site during July 1997, together with 1-hour averaged ceilometer data. Similarly, the model total water vapor and total cloud liquid water, over the SGP site during December 1997, are compared with observations from the microwave radiometer. Figure 8 for the total column water vapor (top panel) shows a good agreement, except for some situations where the observations might have been contaminated by precipitation events. On the contrary, in the bottom panel of Figure 8, the agreement in liquid water path is much worse, showing in particular some bad timing by model clouds between days 7 and 12, and a sizeable uncertainty (~ -30 g/m^2) in the "zero value" taken as the measured clear-sky background.

What is the best strategy for validating a cloud scheme? One can think of various frameworks for performing such a task, from a full-blown 3-D GCM run in climate mode to a 1-dimensional set of physical parameterizations where horizontal forcings are specified. Here we have tried to show that the development of a cloud scheme (in fact of any physical parameterization) within the framework of a weather forecasting system certainly offers a number of advantages, particularly those using analyzed dynamical fields which should be very close to reality and thus provide a reasonable estimation of the vertical and horizontal temperature and humidity forcings. Whereas a 1-D framework with specified (constant or time-evolving) advection terms also appears attractive, it prevents the occurrence of any potential feedbacks.

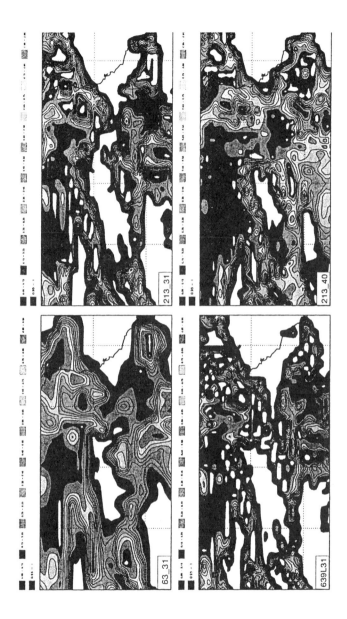

Figure 6. The low-level cloudiness (below $\sigma = 0.8$) after a 12-hour forecast from 19870705 00UTC at different horizontal and vertical resolutions (T63L31, T213L31, T213L40, and $T_L 639L31$) over the FIRE Marine Stratus area.

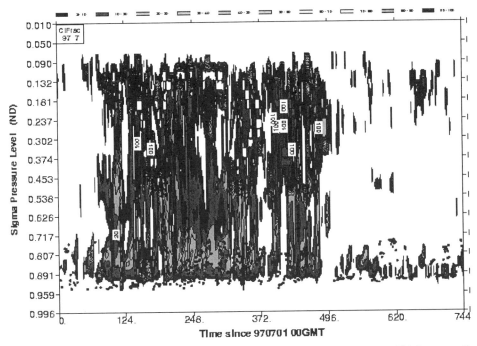

Figure 7. The cloud fraction over the ARM-TWP site in July 1997 from a series of 31 forecasts all starting at 12UTC. Fields are plotted for each day from hourly forecast values between hours 12 and 36. Points represent 1-hour averaged values of the cloud base from the ceilometer measurements.

5. Concluding remarks

Whatever the framework, comparisons of as many aspects as possible of the cloud scheme with observations is an absolute requirement. These comparisons have to be made in combination, as there is very little methodological sense in comparing (as was often done in the past), for example, cloud parameters with similar ISCCP-derived quantities, without tackling, within the same validation effort, the impact of clouds on the radiation fields, at the surface, and top of the atmosphere. The impact on the vertically integrated water vapor and cloud water, and on the precipitation falling from those clouds, should also be assessed.

In this respect, the development of improved cloud schemes (and other physical parameterizations) is in unison with the availability of observational datasets. Figure 9 shows a successful comparison of model cloud with profiles derived from a cloud lidar on the space shuttle (courtesy of S. Miller and G. Stephens, CSU). It would be comforting to imagine that the future can provide us with additional and more sophisticated ways of validating our parameterizations. However, in the near future, the main stumbling block with respect to cloud parameterizations will certainly be the lack of global information on the vertical distribution of cloud

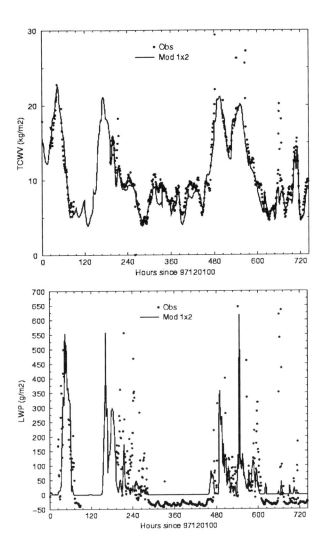

Figure 8. As in Figure 7, but for the total column water vapor (top) and total column cloud liquid water over the ARM-SGP site in December 1997. Points represent hourly averaged values of TCWV and TCCLW derived from Microwave Radiometer measurements.

liquid water, and of any global information on even the vertically integrated ice content of clouds.

278

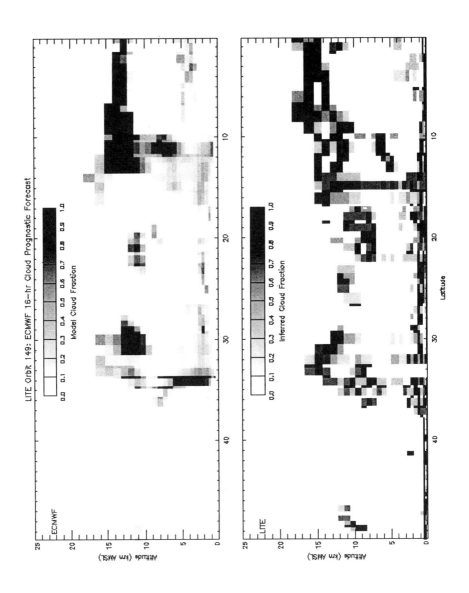

Figure 9. Comparison of the cloud vertical distribution forecast by the ECMWF operational model (top) and derived from LITE observations (from Steve Miller, CSU)

References

1. Alishouse, J.C., Snyder, S.A., Vongsathorn, J. and Ferraro, R.R. (1990) Determination of oceanic total precipitable water from the SSM/I, *IEEE Trans. Geosci. Remote Sensing* **28**, 811-816.
2. Arking, A. (1996) Absorption of solar energy in the atmosphere: Discrepancy between model and observations, *Science* **273**, 779-782.
3. Cess, R.D., Zhang, M.H., Minnis, P., Corsetti, L., Dutton, E.G., Forgan, B.W., Garber, D.P., Gates, W.L., Hack, J.J., Harrison, E.F., Jing, X., Kiehl, J.T., Long, C.N., Morcrette, J.-J., Potter, G.L., Ramanathan, V., Subasilar, B., Whitlock, C.H., Young, D.F. and Zhou Y. (1995) Absorption of solar radiation by clouds: Observations versus models, *Science* **267**, 496-499.
4. Fowler, L.D., Randall, D.A. and Rutledge, S.A. (1996) Liquid and ice cloud microphysics in the CSU general circulation model: Parts I & II, *J. Climate* **9**, 489-529, 530-560.
5. Geleyn, J.-F. (1981) Some diagnostics of the cloud/radiation interaction in ECMWF forecasting model, ECMWF Workshop on Radiation and Cloud-Radiation Interaction in Numerical Modelling, 15-17 Oct. 1980, ECMWF, Reading, U.K., 135-162.
6. Hahn, C.J., Warren, S.G., London, J., Chervin, R.M. and Jenne, R. (1982, 1984) Atlas of Simultaneous Occurrence of Different Cloud Types over the Ocean/Land, NCAR Technical Note TN-201+STR/TN-241+STR, Boulder, Co., 212 pp./216 pp.
7. Harrison, E.F., Minnis, P., Barkstrom, B.R., Ramanathan, V., Cess, R.D. and Gibson, G.G. (1990) Seasonal variation of cloud radiative forcing derived from the Earth Radiation Budget Experiment, *J. Geophys. Res.* **95D**, 18,687-18,704.
8. Holloway, J.L. and Manabe, S. (1971) Simulation of climate by a global general circulation model. 1. Hydrologic cycle and heat balance,*Mon. Wea. Rev.* **99**, 335-370.
9. Imre, D.G., Abramson, E.H. and Daum, P.H. (1996) Quantifying cloud-induced shortwave absorption: An examination of uncertainties and of recent arguments for large scale excess absorption, *J. Appl. Meteor.* **35**, 1991-2010.
10. Jakob, C. (1994) The impact of the new cloud scheme on the ECMWF's Integrated Forecasting System (IFS), Proceedings ECMWF/GEWEX Workshop on Modelling, Validation and Assimilation of Clouds, ECMWF, November 1994.
11. Jakob, C. and Rizzi, R. (1997) Evaluation of model OLR in cloudy regions using TOVS-1B data. Proc. of the Ninth International TOVS Study Conf., Igls, Austria, 197-206.
12. Kato, S. Ackerman, T.P., Clothiaux, E.E., Mather, H.H., Mace, G.G., Wesely, M.L., Murcray, F. and Michalsky, J. (1997) Uncertainties in modeled and measured clear-sky surface shortwave irradiances, *J. Geophys. Res.* **102**, 25881-25898.
13. Le Treut, H. (1991) Sensitivity and validation studies with a prognostic cloud generation model, ECMWF/WCRP Workshop on Clouds, Radiation and the Hydrological Cycle, ECMWF, Reading, U.K., 12-15 November 1990, 223-240.
14. Li, Zhanqing, Barker, H.W. and Moreau, L. (1995) The variable effect of clouds on atmospheric absorption of solar radiation, *Nature* **376**, 6540, 486-490.
15. Lindzen, R. (1990) Some coolness concerning global warming, *Bull. Amer. Meteor. Soc.* **71**, 288-299.
16. Phalippou, L. (1996) Variational retrieval of humidity profile, wind speed and cloud liquid water path with the SSM/I: Potential for numerical weather prediction, *Q. J. R. Meteorol. Soc.* **122**, 327-355.
17. Pilewskie, P. and Valero, F.P.J. (1995) Direct observations of excess solar absorption by clouds, *Science* **267**, 1626-1629.
18. Ramanathan, V., Cess, R.D., Harrison, E.F., Minnis, P., Barkstrom, B.R., Ahmad E. and Hartmann, D. (1989) Cloud-radiative forcing and climate: Results from the Earth Radiation Budget Experiment, *Science* **243**, 57-63.

19. Ramanathan, V., Subasilar, B., Zhang, G.J., Conant, W., Cess, R.D., Kiehl, J.T., Grassl, H.and Shi, L. (1995) Warm pool heat budget and shortwave cloud forcing: A missing physics? *Science* **267**, 499-503.

20. Ricard, J.L. and Royer, J.-F. (1993) A statistical cloud scheme for use in an AGCM, *Annales Geophysicae* **11**, 1095-1115.

21. Rizzi, R. (1994) Raw HIRS/2 radiances and model simulations in the presence of clouds, ECMWF Technical Report No. 73, 29 pp, September 1994.

22. Roeckner, E., Rieland, M. and Keup, E.(1991) Modeling of cloud and radiation in the ECHAM model, in ECMWF/WCRP Workshop on CLouds, Radiative Transfer and the Hydrological Cycle, ECMWF, Reading, U.K., 199-222.

23. Rossow, W.B., Walker, A.W. and Garder, L.C. (1993) Comparison of ISCCP and other cloud amounts, *J. Climate* **6**, 2394-2418.

24. Rossow, W.B. and Zhang, Y.-C. (1995) Calculation of surface and top of the atmosphere radiative fluxes from physical quantities based on ISCCP data sets, 2: Validation and first results, *J. Geophys. Res.* **100**, 1167-1198.

25. Rossow, W.B., Walker, A.W., Beuschel, D.E. and Roiter, M.D. (1996) International Satellite Cloud Climatology Project (ISCCP) Documentation of New Cloud Datasets. WMO/TD-No. 737, World Meteorological Organization, 115 pp (http://isccp.giss.nasa.gov/documents.html)

26. Saito, K. and and Baba, A. (1988) A statistical relation between relative humidity and the GMS observed cloud amount, *J. Meteor. Soc. Japan, Ser. II* **66**, 187-192.

27. Stokes, G.M. and Schwartz, S.E. (1994) The Atmospheric Radiation Measurement (ARM) Program: Programmatic background and design of the cloud and radiation testbed, *Bull. Amer. Meteor. Soc.* **75**, 1201-1221.

28. Slingo, J.M. (1987) The development and verification of a cloud prediction scheme for the ECMWF model, *Q. J. R. Meteorol. Soc.* **113**, 899-928.

29. Smith, R.N.B. (1990) A scheme for predicting layer clouds and their water content in a general circulation model, *Q. J. R. Meteorol. Soc.* **116**, 435-460.

30. Sommeria, G. and Deardorff, J.W. (1977) Sub-grid scale condensation in models of non-precipitating clouds, *J. Atmos. Sci.* **34**, 344-355.

31. Stephens, G.L., Jakob, C. and M. Miller (1998) Atmospheric ice: A major gap in understanding the effects of clouds on climate. GEWEX News, 8, 1-8.

32. Sundqvist, H. (1978) A parameterization scheme for non-convective condensation including prediction of cloud water content, *Q. J. R. Meteorol. Soc.* **104**, 667-690.

33. Sundqvist, H. (1988) Parameterization of condensation and associated clouds in models for weather prediction and general circulation simulation, in M.E.Schlesinger (ed.), *Physically Based Modelling and Simulation of Climate and Climate Change*, Kluwer, pp. 433-461.

34. Sundqvist, H., Berge, E. and Kristjansson, J.E. (1989) Condensation and cloud parameterization studies with a mesoscale numerical weather prediction model, *Mon. Wea. Rev.* **117**, 1641-1657.

35. Tiedtke, M. (1993) Representation of clouds in large-scale models, *Mon. Wea. Rev.* **121**, 3040-3061.

36. Weng, F. and Grody, N.C. (1994) Retrieval of cloud liquid water using the SSM/I, *J. Geophys. Res.* **91**C, 2289-2307.

RADIATION

J.-J. MORCRETTE[1], S.A. CLOUGH[2]
[1]*ECMWF*
Shinfield Park
Reading RG2 9AX
UK
[2]*Atmospheric and Environmental Research, Inc.*
Cambridge, MA
USA

1. Introduction

Among the various processes responsible for the diabatic heating of the atmosphere (convection, turbulence, large-scale condensation, heat, moisture and momentum transfer at the surface, and radiative transfer), which have to be described by physical parametrizations in a general circulation model (GCM), radiative transfer (RT) is prominent because its impact is felt at all times and over the whole 3-D domain. Compared with other processes, radiative transfer has a long history of theoretical developments. From the statistical and quantum mechanics at the end of the 19th and beginning of the 20th century, Boltzmann, Stefan, Wien, Planck and Einstein made pioneering advances in the spectral description of the radiation emitted by a black body. Afterwards, spectroscopic studies of the gases important for the radiative budget of the atmosphere, and the development of various approximations, made the calculation of the radiative transfer in the atmosphere a tractable problem.

The very first such approximation is the separation between shortwave (SW) and longwave (LW) schemes owing to the obvious wavelength difference between a black-body at the Sun's temperature (around 5800 K, with most of its energy below 4 microns, and the energy source outside the atmosphere) and that of the atmosphere (globally averaged equivalent temperature of about 255 K as seen from the top of the atmosphere, with most of its energy above 4 microns, and the sources being the surface, the radiatively active gases, and the clouds within the atmosphere). Chandrasekhar ([4], [5], [6]) provided the first approximations to the radiative transfer in a scattering atmosphere. Elsasser ([19], [20]), Goody ([25],

P. Mote and A. O'Neill (eds.), Numerical Modeling of the Global Atmosphere in the Climate System, 281–296.
© 2000 *Kluwer Academic Publishers. Printed in the Netherlands.*

282

[26]), Curtis ([13], [14]), Godson [24], and Malkmus [32] defined the first usable simplifications to the spectral representation of the radiative transfer including its dependence on two key atmospheric parameters, pressure and temperature, allowing atmospheric heating/cooling rates to be computed [45].

By the mid-1970s, it was said (WMO/Global Atmospheric Research Program, 1975) that radiative transfer was a solved problem, provided big enough computers were available. The only developments required were for fast parametrizations. The basic steps, required when designing a radiation scheme for use in a GCM, include (i) the development of a formal solution of the radiative transfer equation; (ii) the vertical integration, i.e. accounting for the variations in temperature, pressure, and density of the radiatively active absorbers and scatterers over the vertical, to give a directional quantity called monochromatic radiance; (iii) the integration over the zenith angle, to give a monochromatic irradiance; (iv) the spectral integration, i.e. an integration over the relevant part of the electromagnetic spectrum (LW or SW), to give a total flux; and finally (v) the differentiation over the vertical coordinate to get the radiative heating/cooling rate. A good description of the various approximations used at the different stages in the development of accurate and computer-efficient radiation schemes (and of the necessary trade-offs between these two contradictory arguments) is given in Stephens [49] and Fouquart [22].

Unfortunately, the InterComparison of Radiation Codes for Climate Models (ICRCCM) at the end of the 1980s ([18], [23], [1]) showed that if an overall good agreement in clear-sky longwave computations by line-by-line models existed, large discrepancies were found in the results of both the LW parametrized schemes and all types of SW schemes. Hereafter, at least for clear-sky atmospheres, these were tracked down to deficiencies, mainly in the implementation of the approximations used to deal with the spectral and vertical integrations.

In terms of time-scales, radiative transfer is likely to show its intrinsic impact over long-time scales in parts of the atmosphere where radiative equilibrium prevails (50 to 100 days in the winter polar stratosphere, [42]), so that systematic errors in the RT parametrization slowly accumulate. The same errors, however, will also affect GCM results over much shorter time-scales mostly through their interactions with cloudiness or the deposition of radiative energy at the surface, with almost immediate impact on surface skin temperature and precipitation through destabilization of the atmosphere.

Since ICRCCM, a number of parametrized RT schemes have been developed from line-by-line models (LbLs) so that their accuracy, at least in clear-sky atmospheres, should be very close to that of LbLs (e.g. [43], [17]). Such a scheme developed by Mlawer et al. [34] from a line-by-line model and validated against numerous spectrometric measurements as part of the Atmospheric Radiation Measurement (ARM) program has recently been tested within the ECMWF forecasting system, and some results of these tests are presented in the following.

As for the cloud parametrization (see Morcrette *et al.*, Chapter 10 of this book), the validation of a radiative transfer parametrization is more easily done over short time and space scales within a weather forecasting environment. In the rest of the chapter, we will present further comparisons of radiation parameters computed during first-guess and short term forecasts with observations.

2. Sensitivity of the ECMWF model to the representation of the longwave radiative transfer

The AER Rapid Radiative Transfer Model (RRTM) longwave radiation code is described in detail in Mlawer *et al.* [34]. Here we concentrate on the aspects that differ most from the ECMWF operational LW scheme. The operational LW scheme ([36], [37]; see also ECMWF Model Documentation for a complete detailed description of the scheme) accounts for absorption by water vapour, carbon dioxide, ozone, CH_4, N_2O, CFC-11, and CFC-12. It is based on a solution of the RT equation known as the emissivity method, i.e. the fluxes over relatively wide spectral intervals are obtained by computing the matrix of the transmission function between any two levels within the discretized atmosphere. The transmission functions for water vapour and carbon dioxide over the six spectral intervals of the scheme have been fitted using Pade approximants on narrow-band transmissions obtained with Malkmus and Goody statistical band models [35]. At the time of its development, it was validated against the Laboratoire de Météorologie Dynamique line-by-line model. However, some of its features make it somewhat outdated, particularly its parametrization of the water vapour continuum absorption, based on Roberts [44]. In this scheme, semi-transparent clouds are treated using an effective cloud cover, the product of the actual cloud cover by the cloud emissivity.

As stated in Mlawer *et al.* [34], the objective in the development of RRTM has been to obtain an accuracy in the calculation of fluxes and cooling rates consistent with the best line-by-line models. As RRTM deals with the RT over much narrower spectral intervals (a total of 256), each of these spectral intervals can be assumed monochromatic. Therefore, a vertical integration, where the radiance at the boundary of one layer is simply obtained as the radiance incoming at the other boundary plus the internal contribution of the layer itself, translates readily into a two-stream solution for the RT equation. RRTM utilizes the correlated-k method (16 spectral intervals, each including 16 values for the mean line intensity) and shows its affiliation to the Atmospheric & Environmental Research, Inc. line-by-line model (LBLRTM, [10], [11], [12]) through its use of absorption coefficients for the relevant k-distributions derived from LBLRTM. Therefore the k-coefficients in RRTM include the effect of the CKD 2.2 water vapour continuum [10]. The accuracy of these absorption coefficients has been established by

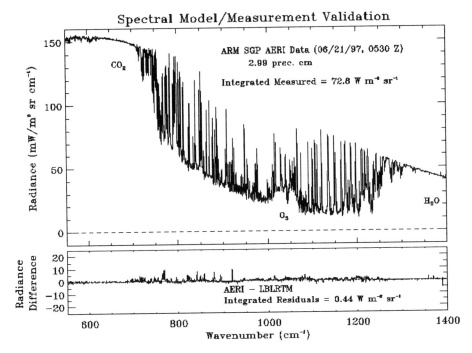

Figure 1. The upward looking radiance from the surface meaured by the AERI interferometer (top) and the difference in radiance between the AERI observation and the simulation with AER's LBLRTM.

numerous and continuing high-resolution validations of LBLRTM with spectroscopic measurements, in particular those performed during some ARM Intensive Observation Periods (see such a comparison in Figure 1, particularly the radiance difference at the bottom of the figure). Compared with the original RRTM [34], the version used in this study has been slightly modified to account for cloud optical properties and surface emissivity defined for each of the 16 bands over which spectral fluxes are computed. Other changes are the use of a diffusivity approximation (RRTM_1ang in Figure 2), instead of the 3-angle integration over the zenith angle (RRTM_3ang in Figure 2), to derive upward and downward fluxes from the radiances, and the modification of the original cloud random overlapping assumption to include (to the same degree of approximation as used in the operational SW scheme, Fouquart and Bonnel [21], Morcrette [36], OPE_SW in Figure 2) a maximum-random overlapping of cloud layers.

For the relevant spectral intervals of both schemes, ice cloud optical properties are derived from Ebert and Curry [16], and water cloud optical properties from Fouquart [22], but, whereas in the operational scheme the cloud emissivity used to compute the effective cloud cover is defined over the whole LW spectrum from spectrally averaged mass absorption coefficients and the relevant cloud water

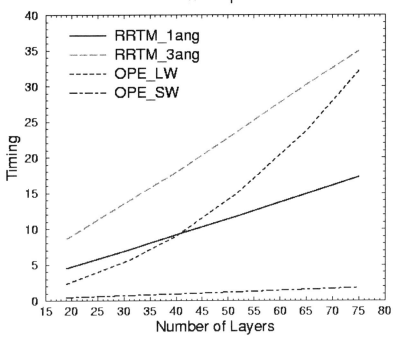

Figure 2. Efficiency of radiation schemes with respect to vertical number of layers.

and/or ice paths (following Smith and Shi [48]), in RRTM the cloud optical thickness is defined as a function of spectrally varying mass absorption coefficients and relevant cloud water and ice paths, and is used within the true cloudy fraction of the layer.

As RRTM is based on a two-stream solution of the LW radiative transfer, its efficiency depends linearly on the number of levels used to discretize the atmospheric profiles, whereas the operational LW scheme, based on an emissivity method, includes a matrix solution of the RT and thus displays a quadratic dependence on the number of levels. The impact of this difference on the efficiency of the LW calculations is illustrated in Figure 2, where RRTM and the operational scheme are compared for atmospheres with the number of levels varying between 19 and 75. Fluxes and cooling rates computed by RRTM and LBLRTM are compared in Mlawer *et al.* [34]. Here we compare fluxes and cooling rates from RRTM with those from the operational scheme. Figure 3 compares the cooling rates for the standard tropical, mid-latitude summer, mid-latitude winter, and sub-arctic winter atmospheres of McClatchey *et al.* [33]. The main difference between the two schemes is the increase in the intensity of the maximum cooling

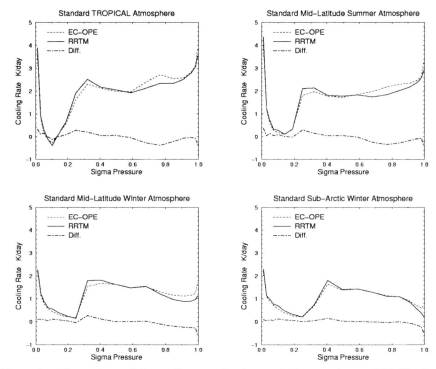

Figure 3. Comparison of the cooling rate for the standard atmospheres of McClatchey [33] computed by the ECMWF operational scheme and RRTM (K/day).

that RRTM brings to the clear-sky cooling rate for pressures between 200 and 400 hPa, and the decrease of the clear-sky cooling rate for pressures larger than 600 hPa, corresponding to an increase of the greenhouse effect of the longwave absorption, related to the improvement and strengthening of the water vapour p-type absorption, particularly important in the pure rotation part of the water vapour spectrum around 30 microns.

A detailed comparison of the impact of RRTM on both the climate of the ECMWF model as seen in ensembles of six 4-month simulations and on the quality of the forecast as seen in sets of 12 T213 10-day forecasts for the 15th of each month over a year is given in Morcrette *et al.* [40]. Here we simply compare the clear-sky initial forcing at the top (Figure 4) and the surface (Figure 5). The effect of the increased LW opacity of the atmosphere with RRTM is seen through a statistically significant decrease of the outgoing longwave radiation and of the surface net LW radiation (both negative quantities in the ECMWF model). Overall, these two changes somewhat combine but result in a decrease of the net clear-sky cooling of the atmosphere. This decreased cooling has a beneficial impact on some of the systematic errors of the model, as can be seen in Figure 6, which presents,

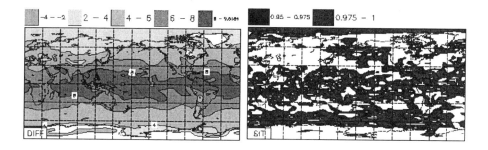

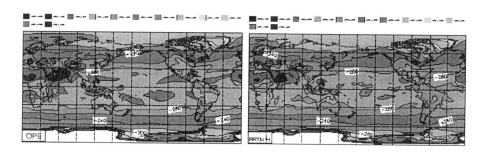

Figure 4. The clear-sky outgoing longwave radiation at the top of the atmosphere over the first 24 hours of integrations starting 19870501 00 UTC. Global mean values are: Control -270.7 W/m2, RRTM is -264.5 W/m2, RRTM-Control is 6.2 W/m2.

for the Northern hemisphere at 4 levels, the r.m.s errors in temperature of a set of twelve 10-day forecasts starting from the 15th of the month over one year. In Figure 6, RRTM is clearly an improvement over the operational LW scheme at 850, 200, and 50 hPa.

3. Validation

As for the representation of clouds, there is an imperative need to perform at least some of the validation on short-time and limited space scales, before the drift in the GCM climate makes comparison with observations very difficult to interpret. Relative to other physical processes, the existence of reference models (LbLs) that have been or can be validated against highly spectrally detailed measurements

288

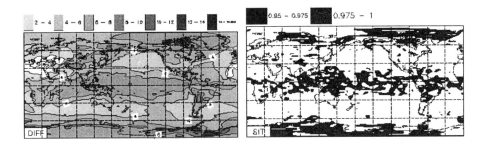

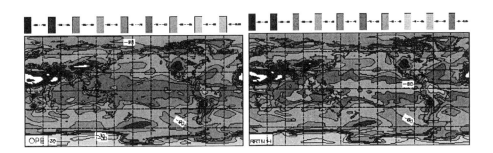

Figure 5. As in Figure 4, for the net clear-sky longwave radiation at the surface. Control is -88.6 W/m2, RRTM is -83.6 W/m2, RRTM-Control is 4.9 W/m2.

(such as those of spectrometer and interferometer, present on ARM sites, or embarked on Nimbus-3) offers a very useful tool for validation. However, although currently done for LW, this approach is not really used for SW, because of the lack of highly spectrally detailed measurements in the SW, and therefore of carefully validated LbL models in the SW.

The approach discussed above could also be used, albeit with more difficulties, for cloudy atmospheres. However, for years most of the validation effort for radiative transfer in cloudy atmospheres has dealt with comparison of radiation fluxes at the top of the atmosphere (TOA) with satellite-derived fluxes (Nimbus-7 Earth Radiation Budget, Earth Radiation Budget Experiment). More often than not, this was done without considering at the same time a validation of the cloud characteristics so that tweaking the cloud properties (particularly, the relationships giving the cloud fraction and cloud water loading in diagnostic cloud schemes) allowed an easy agreement with TOA fluxes, especially over monthly mean time-scales.

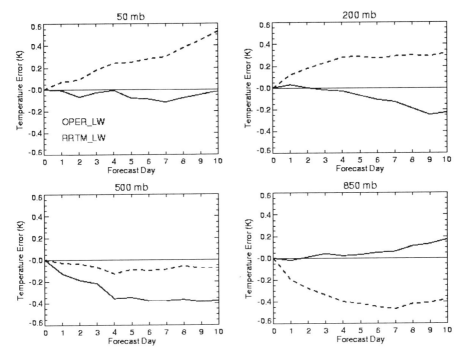

Figure 6. The r.m.s. error in temperature at 50, 200, 500 and 850 hPa in the Northern hemisphere from sets of 12 T213 L31 10-day forecasts starting on the 15th of each month between 15 April 1996 and 15 March 1997. The control set uses the operational LW scheme, while the other uses RRTM.

With the availability of carefully calibrated measurements of the radiation fluxes at the surface — Baseline Station Radiation Network (BSRN) [41], Global Energy Balance Archive, NOAA-ARL SURFace RADiation network (SURFRAD), or as is the case with the ARM program of radiation fluxes and relatively detailed information on the state of the overlying atmosphere and some information on the cloud structure — systematic verification of the radiation fields produced by the ECMWF model at various stages in the forecasts can now be carried out.

In the following, we present one example of such validation for a site where good quality surface radiation measurements are available with a high frequency (such sites are available as part of the BSRN, SURFRAD or ARM projects). For a weather forecasting system, it is often preferable to use measurements not too remote in time, as the analysis and forecast system undergoes regular improvement over the years. In this respect, the NOAA-ARL SURFace RADiation network offers real-time availability of surface radiation measurements over 6 sites in the U.S. Figure 7 compares the downward and upward LW radiation over Goodwin Creek, Mississippi, over the period 17 November–15 December 1997, when a new physical parametrization package was being tested. Figure 8 shows the cor-

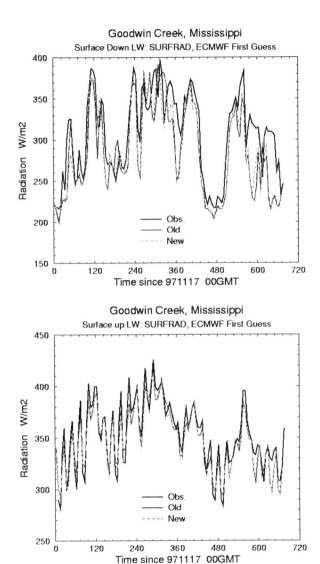

Figure 7. Downward (top) and upward LW radiation from the ECMWF 6-hour first-guess fore-casts compared with the SURFRAD measurements in Goodwin Creek, Mississippi over the period 971117 00 UTC to 971215 00 UTC.

responding downward SW radiation. As can be seen from the LW comparisons, the model is successful at producing the observed variability in temperature at the surface (as seen from the upward LW radiation in Figure 7, bottom panel) and in temperature, humidity, and cloud fraction in the lowest layers of the atmosphere (as seen from the downward LW radiation in Figure 7, top panel). On the other

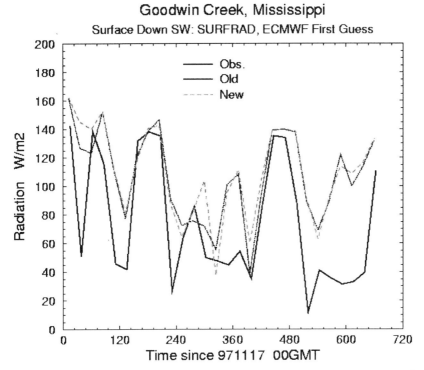

Figure 8. As in Figure 7, but for the downward SW radiation at the surface averaged over 24 hour-periods.

hand, it appears less successful at handling the amount of condensed water in clouds, with too small optical thickness translating in too large downward SW radiation at the surface (Figure 8).

4. Perspectives

Radiative transfer parametrization is generally the most expensive in computer time among the physical parametrizations used in a GCM. In the ECMWF operational forecast model, full radiation computations are performed only every 3 hours and on a reduced grid (1 point out of 4 along the longitude direction). Even so, radiation accounts for about 15% of the total cost of the model. Although such spatial and temporal sampling of the radiation forcing does not appear detrimental on the short time scales encompassed in the operational analyses of meteorological observations and the 10-day forecasts made with the high resolution (TL319) ECMWF weather forecast model, the impact becomes systematic and detrimental on seasonal time-scales and/or at lower resolution [38]. New algorithms are being

292

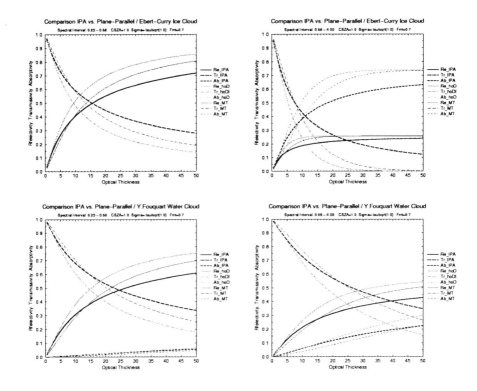

Figure 9. The transmissivity, reflectivity and absorptivity for ice (top) and water (bottom) clouds computed with Barker's IPA [2], the Delta-Eddington approximation and a Delta-Eddington approximation accounting for Tiedtke's inhomogeneity factor [54].

tested based on the neural network approach of Cheruy *et al.* [7] and Chevallier *et al.* [8], or on the linearized approach proposed by Chou and Neelin [9], both of which introduce major savings in computer time.

Other areas of development concern the role of the cloud inhomogeneity on the model horizontal sub-grid scale, and that of the vertical overlapping of cloud layers. On the first point (not equivalent to a full account of the three-dimensional cloud-induced radiative effects, but going away from the plane-parallel approach used over the last 25 years), Tiedtke [52] proposed a simple parametrization of this effect, while Barker [2], [3] modified a SW scheme similar to the ECMWF one and incorporated an approximate treatment of the impact of the liquid water inhomogeneities via a gamma-distribution approach. Figure 9 presents the differences in transmissivity, reflectivity and absorptivity for some standard liquid and ice water clouds in the two spectral intervals of the SW scheme. The impact is important over most of the range of optical thickness and for all quantities. Such a "horizontal" effect, to be accounted for also in the LW part of the spectrum, is likely to change the effect of the clouds on the radiative distribution within the whole atmospheric column. In view of the large effects of the sub-grid scale distribution of the condensed water on the radiative fluxes, it is likely that in the coming years GCMs will incorporate an equation at least to diagnose this sub-grid scale variability and account for its effect on the radiation fields.

Furthermore, studies by Wang and Rossow ([53], [54]) and Stubenrauch *et al.* [51] have recently focused on the potentially large impact on the atmospheric circulation linked to uncertainties in the vertical distribution of clouds and related radiative heating/cooling rates. Similar results were also obtained with the ECMWF model [39].

Until now, the validation of the large-scale radiative fields produced by a climate or a weather forecast GCM has mainly consisted of checks on the total cloud cover, and related top of the atmosphere and surface longwave and shortwave radiation fields. These fields (for example, [15], [27], [30], [31], [47], [56]) are provided by dedicated satellite observations such as ERB [50], ERBE [28], SCARAB [29], CERES [55] and/or operational satellite observations as part of ISCCP [46]. Although very useful, these essentially two-dimensional (3D if time history is considered) validation efforts are only a first step towards what would really be required to ascertain the adequacy of the representation of the cloud-radiation interactions: the 4-dimensional distribution of cloud volume and cloud water loading together with relevant 4-dimensional radiation parameters: a real challenge for the future.

References

1. Baer, F., Arsky, N., Charney, J.J. and Ellingson, R.G. (1996) Intercomparison of heating rates generated by global climate model longwave radiation codes, *J. Geophys. Res.* **101D**, 26,589-26,603.
2. Barker, H.W. (1996) A parameterization for computing grid-averaged solar fluxes for inhomogeneous marine boundary layer clouds. Part I: Methodology and homogeneous biases, *J. Atmos. Sci.* **53**, 2289-2303.
3. Barker, H.W., Morcrette, J.-J. and Alexander, D. (1998) Broadband solar fluxes and heating rates for atmospheres with 3D broken clouds, *Q. J. R. Meteorol. Soc.* **124**, 1254-1271.
4. Chandrasekhar, S. (1935) The radiation equilibrium of the outer layer of a star with special reference to the blanquening effect of the reverse layer, *Mon. Not. Roy. Astron. Soc.* **95**, 21.
5. Chandrasekhar, S. (1958) On the diffuse reflection of a pencil of radiation by a plane parallel atmosphere, *Proc. Nat. Acad. Sci. USA* **44**, 933-940.
6. Chandrasekhar, S. (1960) *Radiative transfer*, Dover Publications, New York.
7. Cheruy, F., Chevallier, F., Morcrette, J.-J., Scott, N.A. and Chedin, A. (1996) A fast method using neural networks for computing the vertical distribution of the thermal component of the Earth radiative budget, *C. R. Acad. Sci.* **322**, IIb, (in French).
8. Chevallier, F., Cheruy, F., Scott, N.A. and Chedin, A. (1997) A neural network approach for a fast and accurate computation of longwave radiative budget, *J. Climate* submitted.
9. Chou Chia and Neelin, J.D. (1996) Linearization of a longwave radiation scheme for intermediate tropical atmospheric models, *J. Geophys. Res.* **101D**, 15,129-15,146.

10. Clough, S.A., Kneizys, F.X. and Davies, R.W. (1989) Line shape and the water vapor continuum, *Atmos. Res.* **23**, 229-241.

11. Clough, S.A., Iacono, M.I. and Moncet, J.-L. (1992) Line-by-line calculations of atmospheric fluxes and cooling rates: Application to water vapor, *J. Geophys. Res.* **97D**, 15,761-15,786.

12. Clough, S.A. and Iacono, M.I. (1995) Line-by-line calculation of atmospheric fluxes and cooling rates, 2: Application to carbon dioxide, ozone, methane, nitrous oxide and the halocarbons. *J. Geophys. Res.* **100D**, 16,519-16,536.

13. Curtis, A.R. (1952) Discussion, *Q. J. R. Meteorol. Soc.* **78**, 638-640.

14. Curtis, A.R. (1956) The computation of radiative heating rates in the atmosphere, *Proc. Roy. Soc.* A **236**, 148-156.

15. Darnell, W.L., W.F. Staylor, S.K. Gupta, N.A. Ritchey and A.C. Wilber (1992) Seasonal variation of surface radiation budget derived from ISCCP-C1 data. *J. Geophys. Res.* **97D**, 15741-15760.

16. Ebert, E.E. and Curry, J.A. (1992) A parametrisation of ice cloud optical properties for climate models, *J. Geophys. Res.* **97D**, 3831-3836.

17. Edwards, J.M. and Slingo, A. (1996) Studies with a flexible new radiation code: I: Choosing a configuration for a large-scale model, *Q. J. R. Meteorol. Soc.* **122**, 689-720.

18. Ellingson, R.E., Ellis, J. and Fels, S.B. (1991) The Intercomparison of Radiation Codes used in Climate Models: Longwave results, *J. Geophys. Res.* **96D**, 8929-8954.

19. Elsasser, W.M. (1938) Mean absorption and equivalent absorption coefficient of a band spectrum, *Phys. Rev.* **54**, 126-129.

20. Elsasser, W.M. (1942) Heat transfer by infrared radiation in the atmosphere, *Harvard Meteorological Studies No.6*, 107 pp.

21. Fouquart, Y. and Bonnel, B. (1980) Computations of solar heating of the Earth's atmosphere: a new parameterization, *Beitr. Phys. Atmosph.* **53**, 35-62.

22. Fouquart, Y. (1987) Radiative transfer in climate models, NATO Advanced Study Institute on Physically-Based Modelling and Simulation of Climate and Climatic Changes. Erice, Sicily, 11-23 May 1986. M.E. Schlesinger, Ed., Kluwer Academic Publishers, 223-284.

23. Fouquart, Y., Bonnel, B. and Ramaswamy, V. (1991) Intercomparing shortwave radiation codes for climate studies, *J. Geophys. Res.* **96D**, 8955-8968.

24. Godson, W.L. (1953) The evaluation of infrared radiative fluxes due to atmospheric water vapour, *Q. J. R. Meteorol. Soc.* **79**, 367-379.

25. Goody, R.M. (1952) A statistical model for water vapour absorption, *Q. J. R. Meteorol. Soc.* **78**, 165-169.

26. Goody, R.M. (1964) *Atmospheric Radiation. 1: Theoretical Basis*, Clarendon Press, Oxford.

27. Gupta, S.K., Staylor, W.F., Darnell, W.L., Wilber, A.C. and Ritchey, N.A. (1993) Seasonal variation of surface and atmospheric cloud radiative forcing over the globe from satellite data, *J. Geophys. Res.* **98D**, 20,761-20,778.

28. Harrison, E.F., Brooks, D.R., Minnis, P., Wielicki, B.A., Staylor, W.F., Gibson, G.G., Young, D.F., Denn, F.M. and the ERBE Science Team (1988) First estimates of the diurnal variation of longwave radiation from the multiple-satellite Earth Radiation Budget Experiment (ERBE), *Bull. Amer. Meteor. Soc.* **69**, 1144-1151.

29. Kandel, R.S., Viollier, M., Raberanto, P., Duvel, J.-Ph., Pakhomov, L.A., Golovko, V.A., Trishchenko, A.P., Mueller, J., Raschke, E., Stuhlmann, R. and the International ScaRaB Scientific Working Group (1998) The ScaRaB Earth Radiation Budget Dataset, *Bull. Amer. Meteor. Soc.* **79**, 765-783.

30. Laszlo, I. and Pinker, R.T. (1993) Shortwave cloud-radiative forcing at the top of the atmosphere, at the surface and of the atmospheric column as determined from ISCCP C1 data, *J. Geophys.Res.* **98D**, 2703-2718.

31. Li, Zhanqing and Leighton, H.G. (1993) Global climatologies of solar radiation budgets at the surface and in the atmosphere from 5 years of ERBE data, *J. Geophys. Res.* **98D**, 4919-4930.

32. Malkmus, W. (1967) Random Lorentz band model with exponential tailed 1/S line intensity, *J. Optic. Soc. Amer.* **57**, 323-329.

33. McClatchey, R.A., Fenn, R.W., Selby, J.E.A., Volz, F.E. and Garing, J.S. (1972) Optical properties of the atmosphere. AFCRL 72-0497, Environ. Res. Paper No. 411, Bedford, Mass., 108 pp.

34. Mlawer, E.J., Taubman, S.J., Brown, P.D., Iacono, M.J. and Clough, (1997) Radiative transfer for inhomogeneous atmospheres: RRTM, a validated correlated-k model for the longwave, *J. Geophys. Res.* **102D**, 16,663-16,682.

35. Morcrette, J.-J., Smith, L. and Fouquart, Y. (1986) Pressure and temperature dependence of the absorption in longwave radiation parameterizations, *Beitr. Phys. Atmosph.* **59**, 455-469.

36. Morcrette, J.-J. (1990) Impact of changes to the radiation transfer parameterizations plus cloud optical properties in the ECMWF model, *Mon. Wea. Rev.* **118**, 847-873.

37. Morcrette, J.-J. (1991) Radiation and cloud radiative properties in the ECMWF operational weather forecast model, *J. Geophys. Res.* **96D**, 9121-9132.

38. Morcrette, J.-J. (1998a) On the effects of the temporal and spatial sampling of radiation fields on the ECMWF forecasts and analyses, ECMWF Technical Memorandum.

39. Morcrette, J.-J. (1998b) Response of the ECMWF model to changes in cloud overlap assumption, ECMWF Technical Memorandum.

40. Morcrette, J.-J, Clough, S.A., Mlawer, E.J. and Iacono, M.J. (1998) Impact of a validated radiative transfer scheme, RRTM, on the ECMWF model climate and 10-day forecasts, ECMWF Technical Memorandum No. 252, 47 pp.

41. Ohmura, A., Dutton, E.G., Forgan, B., Frohlich, C., Gilgen, H., Hegner, H., Heimo, A., Konig-Langlo, G., McArthur, B., Muller, G., Philipona, R., Pinker, R., Whitlock, C.H., Dehne, K. and Wild, M. (1998) Baseline Surface Radiation Netweork (BSRN/WCRP): New precision radiometry for climate research, *Bull. Amer. Meteor. Soc.* **79**, 2115-2136.

42. Ramanathan, V., Pitcher, E.J., Malone, R.C. and Blackmon, M.L. (1983) The response of a spectral general circulation model to refinements in radiative processes, *J. Atmos. Sci.* **40**, 605-630.

43. Ramaswamy, V. and Freidenreich, S.M. (1992) A study of broadband parameterizations of the solar radiative interactions with water vapor and water drops, *J. Geophys. Res.* **97D**, 11,487-11,512.

44. Roberts, R.E., Selby, J.E.A. and Biberman, L.M. (1976) Infrared continuum absorption by atmospheric water vapor in the 8-12 micron window, *Appl. Opt.* **15**, 2085-2090.

45. Rodgers, C.D. and Walshaw, C.D. (1966) The computation of the infrared cooling rate in planetary atmospheres, *Q. J. R. Meteorol. Soc.* **92**, 67-92.

46. Rossow, W.B., Kinsella, E., Wolf, A. and Garder, L. (1987) International Satellite Cloud Climatology Project (ISCCP): Description of Reduced Resolution Radiance Data. July 1985 (revised July 1987). WMO/TD-No.58, World Meteorological Organization, Geneva, 143 pp.

47. Rossow, W.B. (1993) Satellite observations of radiation and clouds to diagnose energy exchanges in the climate, in E. Raschke and D. Jacob (eds.), *Energy and Water Cycles in the Climate System*, NATO ASI Vol. 15, Springer-Verlag, Berlin, Heidelberg, pp. 123-141 (Part I); pp. 143-164 (Part II).

48. Smith, E. A. and Lei Shi (1992) Surface forcing of the infrared cooling profile over the Tibetan plateau. Part I: Influence of relative longwave radiative heating at high altitude, *J. Atmos. Sci.* **49**, 805-822.

49. Stephens, G.L. (1984) The parameterization of radiation for numerical weather prediction and climate models, *Mon. Wea. Rev.* **112**, 826-867.

50. Stowe, L.L. *et al.* (1989) Nimbus-7 Global Cloud Climatology. Part II: First year results, *J. Climate* **2**, 671-709.

51. Stubenrauch, C.J., Del Genio, A.D. and Rossow, W.B. (1997) Implementation of subgrid cloud vertical structure inside a GCM and its effect on the radiation budget, *J. Climate* **10**, 273-287.

52. Tiedtke, M. (1996) An extension of cloud-radiation parameterization in the ECMWF model: The representation of sub-grid scale variations of optical depth, *Mon. Wea. Rev.* **124**, 745-750.

53. Wang Junhong and Rossow, W.B. (1995) Determination of cloud vertical structure from upper-air observations, *J. Appl. Meteor.* **34**, 2243-2258.

54. Wang Junhong and Rossow, W.B. (1998) Effects of vertical structure on atmospheric circulation in the GISS GCM, *J. Climate* **11**, 3010-3029.

55. Wielicki, B.A., *et al.* (1996) Clouds and the Earth's Radiation Energy System (CERES): An Earth Observing System experiment, *Bull. Amer. Meteor. Soc.* **77**, 853-868.

56. Zhang, Y.-C., Rossow, W.B. and Lacis, A.A. (1995) Calculation of surface and top of the atmosphere radiative fluxes from physical quantities based on ISCCP data sets, 1: Method and sensitivity to input data uncertainties, *J. Geophys. Res.* **100**, 1149-1166.

GRAVITY-WAVE DRAG

N. MCFARLANE
Canadian Centre for Climate Modelling and Analysis
Atmospheric Environment Service
University of Victoria
Victoria, B.C., V8W 2Y2 CANADA

1. General aspects of the parameterization problem

General circulation models (GCMs) do not correctly represent all of the processes that may influence those that are explicitly resolved. Examples of such unresolved processes include turbulent transfer in the planetary boundary layer, penetrative (cumulus) convection, and the large-scale variations in the rates of transfer of momentum, heat, and atmospheric constituents (e.g., water vapor) associated with organized unresolved motions in the atmosphere. Gravity-wave drag (GWD) is an example of this latter type of process.

The problem of parameterization is to account for the effects of such unresolved processes in ways that are physically sound but simple enough to be practical in GCMs. Typically this is achieved by

- employing a simple model, or representation, of the relevant unresolved process and
- imposing (if necessary) constraints that may be empirical and/or based on observations in order to achieve closure.

GCMs solve numerically the time and space equations governing the large-scale flow in the atmosphere, for which the typical spatial scales are 100 km in the horizontal and 1 km in the vertical, and the typical temporal scale is hours to days. These equations govern the evolution of winds, temperatures, moisture and other trace constituents in the atmosphere and the associated interactions with the underlying surface. Gravity-wave drag parameterizations usually account mainly for the convergence of momentum flux vectors (Reynolds stress) associated with the dissipation of unresolved gravity waves. The momentum deposition associated with this process gives rise to an effective body force (GWD) for the larger scale flow.

P. Mote and A. O'Neill (eds.), Numerical Modeling of the Global Atmosphere in the Climate System, 297–320.
© 2000 Kluwer Academic Publishers. Printed in the Netherlands.

Since hydrostatic balance is usually assumed in GCMs, only the horizontal components of the GWD are important. Using pressure as a vertical coordinate and the usual meteorological notation, the vector equation for the resolved part of the horizontal momentum is

$$\frac{d\overline{\mathbf{V}}}{dt} + f\mathbf{k} \times \overline{\mathbf{V}} = -\nabla\overline{\Phi} - g\frac{\partial\tau}{\partial p} + \text{(other)} \tag{1}$$

where (other) denotes neglected terms, overbars indicate averages over unresolved scales, primes indicate deviations from those averages and the total time derivatives on the left sides include advection by the resolved flow. Here τ is often called the eddy stress vector. It is defined in terms of the vertical momentum flux associated with unresolved motions and can be expressed approximately as

$$\tau = -\overline{\rho}\overline{w'\mathbf{V}'} = \mathbf{i}\tau_{xz} + \mathbf{j}\tau_{yz} \tag{2}$$

where $(\mathbf{i}, \mathbf{j})$ are unit vectors in the eastward and northward directions. The contributions associated with the unresolved part of the density field are ignored. This assumption does not result in a significant loss of accuracy for spatial and temporal scales usually resolved by GCMs.

The effects of GWD are taken into account by parameterizing the component of τ that is associated with unresolved gravity waves.

2. Basic Gravity Wave Dynamics

Suppose that the large-scale flow $\overline{\mathbf{V}}$ is unidirectional in the eastward (positive x) direction and slowly varying in time (over a typical gravity-wave period) and in the horizontal (over a typical horizontal gravity-wave length). Although this large-scale flow is also assumed to be in hydrostatic balance, we need not assume at the outset that gravity-wave perturbations are also hydrostatic.

We do assume that horizontal wavelengths are sufficiently short that the Coriolis force can be ignored for the gravity-wave dynamics as a first approximation and that perturbations in air density can be ignored in the mass continuity equation and in the horizontal momentum equation. They must be taken into account because of their contribution to variations in buoyancy in the vertical momentum equation; this is the anelastic approximation. We also ignore the effects of horizontal and temporal variations in background quantities in the equations for the waves.

We will also assume that the scale heights for the background quantities—density, velocity, temperature—are longer than a vertical wavelength of the gravity waves that may be excited in response to (hitherto unspecified) forcing. This assumption enables simplification of the gravity wave dynamics by neglecting certain terms that involve higher derivatives of background variables through use of a systematic approximation procedure similar to an approximation method used in geometrical optics known as the WKBJ or Liouville-Green method [1]. This type of approximation is commonly used and is acceptable much of the time, but fails for vertically propagating mountain waves in circumstances where reflections may result from inhomogeneity in the troposphere [16].

2.1. SIMPLIFIED WAVE DYNAMICS FOR A MONOCHROMATIC SOURCE

We will consider a single wave excited at some reference level z_r with mean flow U in the eastward direction, an eastward phase speed c and horizontal wave number k. The approximate perturbation equations appropriate for this simple case are:

$$\frac{\partial}{\partial t}u' + U\frac{\partial}{\partial x}u' + w'\frac{\partial U}{\partial z} = -\frac{\partial}{\partial x}\Pi' \tag{3}$$

$$\frac{\partial}{\partial t}w' + U\frac{\partial}{\partial x}w' = -\frac{\partial}{\partial z}\Pi' + g\left(\frac{\theta'}{\Theta}\right) \tag{4}$$

$$\frac{\partial}{\partial x}u' + \frac{1}{\bar{\rho}}\frac{\partial}{\partial z}(\bar{\rho}w') = 0 \tag{5}$$

$$\frac{\partial}{\partial t}\theta' + U\frac{\partial}{\partial x}\theta' + w'\frac{\partial\Theta}{\partial z} = 0 \tag{6}$$

where Θ is the background potential temperature and $\Pi = p'/\bar{\rho}$, the ratio of the pressure perturbation to the mean density. Let

$$\chi = kx - ct + \int_{z_r}^{z} m(z')dz' \tag{7}$$

This quantity is the phase of the gravity wave that is excited. We may then define perturbation quantities in terms of this function and an amplitude function $\delta(z)$ as

$$w' = k(U - c)\delta(z)\cos\chi \tag{8}$$

$$u' = -\frac{1}{\bar{\rho}}\frac{\partial}{\partial z}[\bar{\rho}(U - c)\delta(z)\sin\chi] \tag{9}$$

$$\theta' = -\delta(z)\frac{\partial\Theta}{\partial z}\cos\chi \tag{10}$$

Here the quantity $\delta(z)$ is the amplitude of the displacement of streamlines from their undisturbed position due to the presence of the wave. At the reference level z_r we will assume as a boundary condition that the amplitude of this displacement is specified so that $\delta(z_r) = h$ where h is the specified amplitude of the wave. As will be seen later, for topographically excited waves, this specified quantity is just the amplitude of the topographic perturbation.

A radiation condition ensuring upward propagation of the wave must also be applied. Using the WKBJ assumption that background quantities do not vary

300

significantly in the vertical over length scales of the order of $1/m$ it can be shown that, with $N^2 = \frac{g}{\Theta} \frac{\partial \Theta}{\partial z}$,

$$k^2 + m^2 = N^2 / (U - c)^2. \tag{11}$$

In general, waves that transport momentum in the vertical must have a real-valued m, which requires $k^2 < N^2 / (U - c)^2$.

More generally, if the Coriolis force is taken into account, vertical momentum transport is possible when $f^2 / (U - c)^2 < k^2 < N^2 / (U - c)^2$, that is, when the intrinsic frequency of the waves is bounded below by the inertial frequency f and above by the buoyancy frequency N. Waves that initially propagate vertically away from their source regions through vertically varying background conditions may sometimes have their intrinsic frequencies Doppler-shifted to values that exceed the buoyancy frequency. Such waves are likely to be reflected back toward their source regions and do not in general contribute significant net values of vertical momentum flux. We ignore such complications here but such effects are taken into account in more general treatments [32].

Here we assume that k is roughly in the middle of the above range which, since the inertial frequency is typically much smaller than the buoyancy frequency above the atmospheric boundary layer, gives $m^2 \cong N^2 / (U - c)^2$ and, to the same order of approximation,

$$\delta = h \left[\frac{\bar{\rho}(z_r) N(z_r)(U(z_r) - c)}{\bar{\rho}(z) N(z)(U(z) - c)} \right]^{\frac{1}{2}} \tag{12}$$

The vertical momentum flux is then, also to the same order of approximation,

$$\overline{\rho u' w'} = -\bar{\rho} k N (U - c) \delta^2 = -\tau_{xz} \tag{13}$$

which, using equation (12) is easily seen to be independent of height, as expected in general from the Eliassen-Palm Theorem for non-dissipative wave motion ([5]; see also [1]).

2.2. EFFECT OF DISSIPATION THROUGH CONVECTIVE INSTABILITY

The Eliassen-Palm theorem applies in the absence of dissipation. If the wave is being dissipated, the vertical momentum flux will decay with height in the region where dissipative processes are active. This in turn gives rise to a local drag force that acts on the mean flow. The essence of the parameterization problem is to determine where and how strongly waves are excited and how strongly they are dissipated.

Although there are several processes that could lead to damping of gravity waves as they propagate away from their source regions, it is reasonable to suppose that the strongest dissipation will occur in regions where the wave becomes unstable and breaks down into turbulence. The process by which damping of the wave occurs as a result of the onset of turbulence is often referred to as wave breaking. A well known dynamical instability process is associated with the development of shear zones within the wave where the gradient Richardson number is less than the critical value (0.25) for the onset of instability of the Kelvin-Helmholtz type. Another process that may lead to wave breaking is the onset of convective instability when the wave amplitude becomes so large that it causes relatively cold, dense air to rise over less dense, warm air. Both of these instability processes depend on both the amplitude and phase of the wave and, when the amplitude is sufficiently large, may often occur in the vicinity of each other. This is so in part because the gradient Richardson number for the mean flow is typically substantially larger than unity in the free atmosphere above the atmospheric boundary layer. Convective instability may predominate in circumstances where it is possible, and it is thought to play a crucial role in the breakdown of gravity waves that propagate into the middle atmosphere from tropospheric source regions. This instability occurs when the vertical gradient in potential temperature becomes negative, i.e., when $\partial(\Theta + \theta')/\partial z < 0$.

Consistent with the WKBJ approximations used in the above analysis, the criterion for convective instability is most easily satisfied in the vicinity of the points where $N\delta/(U-c) > 1$. Lindzen [17] postulated that the onset of instability brings about turbulent dissipation of the wave, which reduces its amplitude sufficiently to neutralize the instability without significantly affecting its phase structure. This implies that the dissipation is just sufficient to ensure that

$$\delta^2 = \delta_s^2 = (U - c)^2/N^2. \tag{14}$$

This is known as the saturation hypothesis. Under this hypothesis momentum flux divergence (and hence GWD) is confined to those regions where $\delta = \delta_s$. In these regions equations (13) and (14) give

$$\tau_{xz} = \bar{\rho}k\frac{(U - c)^3}{N}. \tag{15}$$

Differentiating (15) with respect to z gives the well-known Lindzen-Holton [12] formula for the GWD due to a single vertically propagating gravity wave. In pressure coordinates it has the form

$$-g\frac{\partial \tau_{xz}}{\partial p} = -\frac{k(U - c)^3}{N}\left[\frac{1}{H} - \frac{3}{U - c}\frac{\partial U}{\partial z} + \frac{1}{N}\frac{\partial N}{\partial z}\right] \tag{16}$$

where H is the density scale height.

3. Parameterization of Orographic Gravity-wave drag

3.1. BASIC MONOCHROMATIC, COUNTER-FLOW PARAMETERIZATION

The simple concepts outlined above have been used extensively in mechanistic models of the middle atmosphere (see [1]). One of the first applications in a fully three-dimensional GCM was to represent the effects of waves excited by flow over mountains. In this case the waves are stationary ($c = 0$) and $\delta \cos \chi$ is the vertical displacement of streamlines from their undisturbed position. The reference level is taken to be near the large-scale mean surface where it is assumed that the perturbed streamlines conform to the local topographic perturbation. The formulations of Palmer *et al.* [28] and McFarlane [22] employed equation (13) and a saturation hypothesis. It is also assumed that the relevant component of the large-scale flow is parallel to that at the reference level.

Thus at the reference level the stress vector is given by

$$\tau = -\bar{\rho} N U \frac{h^2}{L} \mathbf{n} \tag{17}$$

where the amplitude h^2 is proportional to the variance of the unresolved topography (that is, the part with horizontal scales typical of those associated with vertically propagating gravity waves). The large-scale velocity is defined as $U = (\mathbf{V} \cdot \mathbf{V_R}) / \|\mathbf{V_R}\|$ where $\mathbf{V_R}$ is the velocity at the reference level and $\mathbf{n} = \mathbf{V_R} / \|\mathbf{V_R}\|$ is a unit vector in the direction of the velocity at the reference level. The length scale L is inversely proportional to the product of an effective horizontal wave number and an uncertain efficiency factor whose value is probably in the range of 10 to 20 percent. In practice it is an empirical parameter. L is typically near 100 km.

The saturation condition that is used is similar in form to equation (14). The saturation amplitude is taken to be of the form

$$\delta_s^2 = F_c^2 U^2 / N^2 \tag{18}$$

where F_c is a specified parameter of order unity in magnitude. McFarlane [22] chooses it to be slightly less than unity to account roughly for the effects of non-linearity of the lower boundary condition in enhancing the likelihood for the onset of convective instability. Palmer *et al.* [28] choose F_c to depend on the local large-scale gradient Richardson number to account approximately for the combined effects of shear and convective instability. A variant of this formulation is discussed more recently by Kim and Arakawa [13] who also summarize the main features of orographic gravity-wave schemes in current use.

The saturation condition is easily implemented in a discrete manner (from one model level to the next) by noting that, in the absence of dissipation between any two levels, equation (12) applies and the lower level is chosen as the reference level. The amplitude at the upper level is then limited by the saturation condition (equation (18)).

3.2. EFFECTS OF ANISOTROPY AND NON-LINEARITY

The orographic gravity-wave drag scheme outlined above is rather simple in many ways. In recent years a considerable amount of research has led to improved parameterizations for GCM applications. It is beyond the scope of these notes to review all of these studies. Instead we will draw attention to some basic short-comings and attempts that have been made to deal with them.

3.2.1. *Anisotropy and azimuthal structure*
Linear theory makes the questionable assumption that the effects of gravity wave emission are adequately represented in terms of a single wave oriented opposite to the large-scale flow at the reference level, regardless of the nature of the under-lying topography. In many locations the underlying terrain is highly anisotropic. Major mountain ranges are often comprised of series of elongated ridges aligned in a particular direction. Little excitation of gravity waves would be expected in circumstances where the lower tropospheric wind blows along the ridges. In these circumstances it is perhaps reasonable as a simplification to attempt to rep-resent the gravity wave emission in terms of a single representative wave; this is called the monochromatic assumption. However, it is the component of the flow that is normal to the long axis of the ridges that should be used to determine the magnitude and orientation of the surface stress associated with gravity wave emission.

The monochromatic assumption is less realistic in circumstances where the underlying terrain is more isotropic, as it is, for example, in rolling hilly terrain. In general for isotropic forcing, the terrain can excite all waves whose azimuthal orientation lies within 90° of the direction opposite to the flow (the counter-flow direction) at the reference level. Hitherto it has not been considered practical to attempt to account for such a broad spectrum of waves in the context of a param-eterization scheme designed for use in a GCM. An additional complication that arises for linear saturation theory when an azimuthal spectrum of waves is con-sidered is that the effects of superposing these waves may be difficult to take into account without introducing further simplifying assumptions. Hines ([9], [11]) proposes to account for these effects in isotropic cases by limiting the assumed spectrum to two azimuths that flank the counter-flow direction. These azimuths are chosen so that each accounts for the ensemble of waves that may be excited in the two sectors of angular width 90° flanking the counter-flow direction. The momentum flux deposition associated with each of these waves is determined by imposing a modified saturation condition on each that accounts crudely for the presence of two waves by reducing the value of F_c^2 by a factor of two.

Neither this formulation nor others that attempt to account explicitly for the presence of an azimuthal spectrum of orographically excited waves have been used extensively hitherto in AGCM applications. However, the parameterization of orographic drag recently proposed by Lott and Miller [18] does account for the integrated surface stress due to an azimuthal spectrum of waves and also for the effects of anisotropy by using a formulation for elliptically shaped mountains. Gregory *et al.* [7] accomplish this same goal by using a simplified spectral representation of the sub-grid scale topography.

3.2.2. *Non-linearity*

Most gravity-wave drag parameterizations that are currently used in GCMs employ linear wave dynamics and a linearized representation of the lower boundary condition, mainly because the problem is therefore analytically tractable, enabling development of parameterization schemes that are easily implemented. There are, however, important effects associated with non-linearity, particularly in respect of the lower boundary condition. The non-linearity arises because the correct boundary condition is that the gravity-wave streamline displacement must conform to the surface undulations at the actual (rather than the larger scale mean) level of the topography. The linear version of this condition, which applies the boundary condition at the large-scale mean height of the topography, is acceptably accurate when Nh/U (often referred to as the inverse Froude number or the nondimensional mountain height) is relatively small.

In practice, however, the nondimensional mountain height is rarely small, and a variety of effects may come into play to affect both the onset and vertical structure of the GWD when Nh/U is of order unity or larger. A well-known basic effect of non-linearity of the lower boundary condition for 2D flows is that it may enhance or inhibit the condition for the onset of convective instability depending on the nature and degree of upstream-downstream asymmetry [14]. However, a number of important additional effects have been found recently in numerical experiments. Among those most often considered potentially important for parameterization are the high drag states that have occurred in numerical simulations when the non-dimensional mountain height is sufficiently large ([29] and numerous later studies; see [27] and references cited therein). These states are characterized by the occurrence of strong downslope winds and surface stress (drag) values that are substantially larger than expected on the basis of linear theory. There is also some observational support for the occurrence of such flow regimes in nature [14].

Most simulations of these phenomena employed 2D models, relatively simple upstream flow conditions and the associated idealization of infinitely long ridges. These idealized simulations have revealed a connection between the occurrence of high-drag states and convective overturning first noted by Peltier and Clark [29]. A theory using an internal hydraulic formulation that seems to account reasonably well for many of the results obtained from these 2D studies was proposed by Smith [31]. More recent fully 3D studies have shown that high-drag states are not confined to predominantly 2D flows regimes [26]. These studies have motivated recent attempts to include this enhanced low-level drag in parameterizations ([2], [18], [7]). These parameterizations are, however, still of rather uncertain validity. While there is both modelling and observational evidence that points to the potential importance of accounting for the effects of such high-drag states in GCMs, there remains much that is not well understood about them. The recent work of Miranda and Valente [27] indicates that there are some significant differences among these states for predominantly 2D and 3D flow regimes.

It is also important to note that the subgrid-scale topography is never fully two-dimensional even though there may be a predominant orientation to its salient elements. When the inverse Froude number exceeds unity there will be a tendency for the air within a depth of order U/N to flow around rather than over the topographic undulations. This blocking effect reduces the effective mountain height for gravity wave emission. Baines [2] notes that it may also produce a region of weak flow downstream from the predominant topographic undulations that could be taken into account by introducing an additional drag force (similar in nature to form drag) within the blocked layers in an AGCM [18].

4. Non-orographic waves

4.1. SIMPLE PARAMETERIZATIONS

Holton [12] recognized that the zonal mean circulation in the mesosphere, particularly the zonal mean wind structure, could be simulated more realistically by supposing that waves with non-zero phase speeds in the zonal direction are ubiquitous in the middle atmosphere.

Garcia and Solomon [6] employed a simple (latitudinally dependent) spectrum of vertically propagating gravity waves comprised of a discrete set of phase speeds in the zonal direction. The initial amplitudes of these waves are set by specifying the magnitude of the vertical momentum flux for each wave at a reference level, which is often chosen to be in the lower stratosphere. Above that level the wave amplitudes are determined on the basis of simple wave propagation and saturation concepts as outlined above. Some middle-atmosphere GCMs employ a similarly simple specification for non-orographic waves.

4.2. CONTINUOUS SPECTRA AND NON-LINEARITY

Observations of gravity waves in the middle atmosphere in recent years have revealed that there is an abundance of gravity wave activity in the middle atmosphere that is excited in the troposphere from non-orographic sources. The (nearly continuous) vertical wave-number spectrum is for these waves appears to have a nearly universal structure. Theoretical attempts to explain these observations have led to recent proposals for parameterization of the momentum deposition due to such a continuous spectrum. Two of these parameterizations that have been published recently are currently being used in the Canadian Middle Atmosphere Model (CMAM), namely the Doppler spread parameterization of Hines ([10], [11]) and the non-linear diffusion parameterization of Medvedev and Klaassen [23]. The Doppler spread parameterization is also used in the middle-atmosphere version of the ECHAM4 general circulation model [19]. Both of these invoke non-linearity as well as instability in determining the conditions for gravity wave dissipation and momentum deposition.

Relaxing the monochromatic wave assumption gives rise to more mathematical complexity (see for example [28]) that is beyond the scope of these notes. The addition of non-linearity compounds this complexity, and simplifying assumptions must be made to permit analytical tractability. The distinguishing theoretical bases of the non-linear schemes alluded to above are in the assumptions made concerning the role of non-linearity and its mathematical representation.

5. Effects of Parameterized Gravity-wave Drag

5.1. OROGRAPHIC GRAVITY-WAVE DRAG

5.1.1. *The Location of Orographic Gravity-Wave Drag*
It is useful to consider the vertical structure of the quantity $F = N\delta/U$. If convective instability is used as the criterion for the onset of GWD then, as discussed above, it occurs approximately where F exceeds unity. In the absence of dissipation this quantity can be computed using the formula for δ given in equation (12). McFarlane [22] discusses the vertical structure of F as given by the resulting formula using typical values of zonal wind N and air density for middle latitudes in the Northern Hemisphere. The decrease of density with height causes F to increase with height. However, in the troposphere this effect is frequently offset by the effect of increasing winds with height so that F often decreases with height in the lower troposphere in winter, thus preventing wave breaking in that region. It increases rapidly in the lower stratosphere, particularly in the region just above the jet-stream level, where the wind speed and density decrease with height and N typically increases with height; U, $\bar{\rho}$, and N all act to cause F to increase with height. In middle to high latitudes in the winter, the establishment of the cold polar vortex causes zonal winds to increase with height again in the stratosphere and reach their largest values in the upper mesosphere. This feature of the wind profile can also cause F to decrease with height in the middle to upper stratosphere. Above this level the effects of the decrease of density with height begin to predominate and this typically causes F to increase again in the mesosphere.

Of course, wave breaking sets in when F first exceeds F_c and above that level δ is limited by the saturation condition and the saturated value δ_s is such as to ensure that F_s (which equals $N\delta_s/U$) does not increase with height. Thus GWD occurs in regions where (in the absence of dissipation) F is increasing with height because the wave damping required to maintain saturation in this region leads to a decrease of the magnitude of the vertical momentum flux with height. In contrast, GWD does not occur in regions where F decreases with height because a saturated wave at the base of that region becomes subsaturated within it. Thus it often happens that using a simple orographic gravity-wave drag parameterization in a GCM gives rise to localized regions of orographic GWD in the lower stratosphere and in the mesosphere. This is illustrated in Figure 1, which depicts the zonally averaged orographic gravity-wave drag from a simulation made with the CMAM (see section 4.2). Not shown in Figure 1 is the substantial longitudinal variation in the drag force that arises both because of spatial variation in the topography and because of spatial variability in filtering of gravity waves due to spatial variability of the wind field.

5.1.2. *Effect of OGWD on the Zonal Mean Circulation of the Atmosphere*
Although OGWD typically occurs in the lower stratosphere and the mesosphere, the zonal mean response to it extends all of the way to the surface. Typically, OGWD effects are most pronounced in the northern hemisphere at middle and high latitudes, where the most extensive ranges of mountains are found, and during winter (December-February, or DJF) when westerly winds extend from the surface to the mesopause in the zonal mean sense. Orographic gravity waves are excited at all times of the year but typically the zonally averaged zonal wind changes sign in the lower stratosphere in the summer months. Thus gravity waves that are excited in westerly winds near the surface usually have critical levels in the upper troposphere-lower stratosphere region in these months. Gravity-wave breaking in the troposphere does occur in the summer months but contributes effects that are typically smaller than those associated with other processes, such as turbulent transfer in the vertical.

Some typical basic features of the effects of orographic gravity-wave drag on the large-scale wind and temperature fields are illustrated in Figures 2 and 3, which depict zonal mean temperature and zonal wind fields for simulations made with the CCCMA third-generation atmospheric general circulation model (GCMIII). The GCMIII extends from the surface to 1 mb. It is a spectral model with a triangular truncation at 47 waves (T47) and 32 vertical levels. The vertical coordinate is a hybrid that is defined, as discussed by Laprise and Girard [15], to be terrain-following at the surface and to coincide with constant pressure surfaces in the upper part of the model. The CMAM mentioned in section 4.2 is an upwardly extended version of GCMIII with a top level close to the mesopause, a lower horizontal resolution (T32), and additional physical parameterizations

308

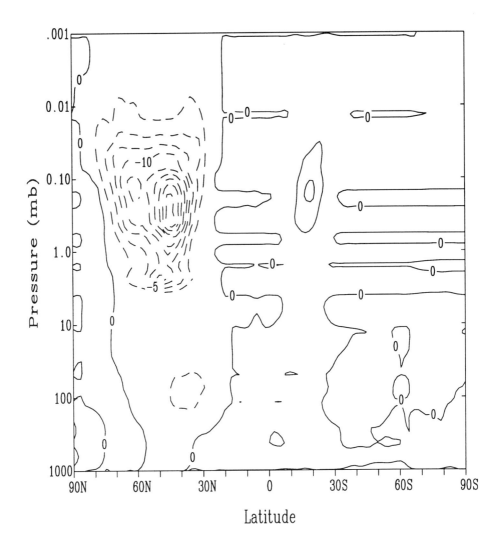

Figure 1.   Zonally averaged body force per unit mass due to orographic gravity wave drag for a typical January simulation with the Canadian Middle Atmosphere Model. The contour interval is 2.5 m/s/day.

appropriate for the upper stratosphere and mesosphere [3]. Both models employ the same orographic gravity-wave drag scheme.

The results shown in Figures 2 and 3 are from two different 5-year simulations using GCMIII; one simulation did not include the orographic gravity-wave drag parameterization that is normally part of GCMIII. They are averaged for the DJF period and over the ensemble of simulated years for each of the simulations. The upper panels in Figures 2 and 3 depict the mean zonal wind and temperature fields

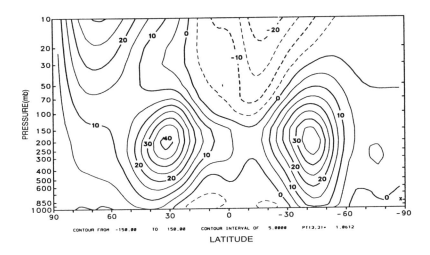

RUN PDHSPNOGWD5DJF. DAYS DJF. ZONAL VELOCITY (U)R. UNITS M/SEC.

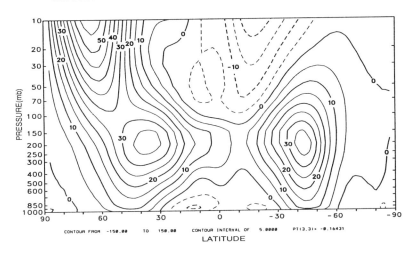

Figure 2. Zonally averaged zonal wind for simulations with (upper) and without (lower) gravity-wave drag.

for the simulation that was made with normal GCMIII configuration while the lower panels are for the simulation without the GWD parameterization.

Because of its lower top, GCMIII only resolves part of the upper stratospheric and mesospheric component of the typical gravity wave drag force illustrated in Figure 1. However, the lower stratospheric component is resolved and has a structure (not shown) that is very similar in location and magnitude to that il-

310

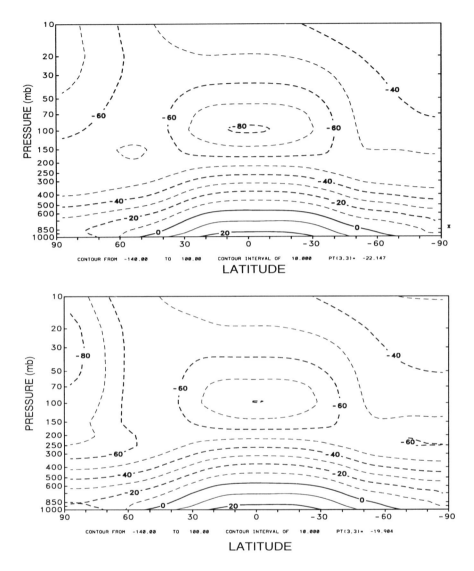

Figure 3. Zonally averaged temperature for simulations with (upper) and without (lower) gravity-wave drag.

lustrated in Figure 1. In particular, there is a component of the drag force in the lower stratosphere in middle latitudes that is typically centered near the 100mb pressure level in the latitude band between 30 and 50 degrees north and acts to decelerate the zonally mean westerlies predominantly in the region between the 50mb and 200mb levels. Although the implied zonal mean deceleration rate associated with this drag force is relatively small (2-3 m/s/day), it is responsible for most of the effects in the lower stratosphere and troposphere. The partially resolved component in the upper stratosphere and mesosphere is, for the GCMIII simulations, typically centered near the top of the model and acts predominantly in the region above the 10mb level in the latitude band between 45° and 60°N. Peak zonal mean decelerations for this component are in the range of 4 to 5 m/s/day near the highest model level. Of course, it must be remembered that the drag force is applied locally and is largest over mountainous regions where the implied deceleration rate may be as large as an order of magnitude larger than the zonal mean value.

As shown in Figure 4 for the typical zonal mean differences between simulations with and without GWD, in simulations with GWD the zonal winds are significantly weaker in the latitudinal bands where the drag is active. The weakening of the winds is accompanied by a general increase in the temperature in the region poleward and below the region where the drag is active. This poleward-and-below pattern of the large-scale response to an elevated drag force has been termed "downward control" by Haynes et al. [8]. However, the response is more complicated for a number of reasons including the fact that it is unsteady, the fact that it involves changes in planetary wave structure, and the fact that feedbacks from other processes are operating, both above and below the region where the imposed drag operates. A quadrupole feature is often found in the temperature response [24] which may be in part associated with countervailing effects from these other processes [29].

Despite these complications, the "downward control" part of the response can be understood qualitatively by considering the steady-state equations governing the zonally averaged difference between the simulations with and without GWD. In middle latitudes these are most conveniently written in terms of the transformed Eulerian mean (TEM) formulation [1]. In the application of this formulation used here the horizontal and vertical components (v^*, w^*) of the "residual mean" meridional circulation are closely related to the diabatic component of the meridional circulation that balances zonally averaged components of diabatic heating and drag forces.

We apply the TEM formulation to analyze the changes in the zonal mean circulation that result from introducing an orographic gravity-wave drag parameterization. In order to simplify this analysis we assume a quasi-steady equilibrium in which the mean zonal momentum equation is simplified to a balance between the Coriolis force and drag forces, the mean zonal flow is in thermal wind balance with the zonal mean temperature field, and diabatic heating is predominately balanced by vertical advection in the thermodynamic equation. With these assumptions the following equations are obtained:

$$-f \Delta v^* = \left(\frac{1}{\bar{\rho} \cos \phi} \right) (\nabla \cdot \Delta \mathbf{F}) - \Delta G \qquad (19)$$

312

$$f\frac{\partial}{\partial z}\Delta U = -\left(\frac{R}{aH}\right)\frac{\partial}{\partial \phi}\Delta T \tag{20}$$

$$\left(\frac{1}{a\cos\phi}\right)\frac{\partial}{\partial \phi}(\Delta v^* \cos\phi) + \left(\frac{1}{\bar\rho}\right)\frac{\partial}{\partial z}(\bar\rho\Delta w^*) = 0 \tag{21}$$

$$\left(\frac{HN^2}{R}\right)\Delta w^* = -\alpha\Delta T \tag{22}$$

where $\Delta \mathbf{F}$ is the change in the Eliassen-Palm flux associated with changes in the resolved planetary wave structure and ΔG is the change of parameterized forces. In this case ΔG is mainly due to changes in the parameterized vertical momentum flux (due to GWD) and the changes in the boundary layer momentum flux divergence in response to GWD. In all cases the symbol Δ denotes the difference in a quantity between the two simulations. Other symbols have the usual meaning. The background density field $\bar\rho$ is assumed, for simplicity, to be independent of latitude ϕ. Here the change in the diabatic heating is taken to be mainly due to a change in the radiative cooling, represented in terms of an effective Newtonian cooling coefficient α.

In the following, we will neglect for the present the effect of changes in the planetary wave Eliassen-Palm flux and assume that the ΔG term is entirely due to the change in the unresolved component of the vertical momentum flux divergence due to the contribution of the imposed gravity-wave drag force and the associated response to it. Consistently, the changes in horizontal heat and momentum transport are also ignored. With these assumptions we find that the change in the zonally averaged parameterized drag force is balanced by a corresponding change in the zonal component of the Coriolis force associated with a change in the residual mean meridional component of the velocity field. The corresponding change in the residual mean vertical motion, which is obtained by imposing mass continuity as expressed in equation (21), is given approximately by

$$\Delta w^* = -\left(\frac{1}{a\cos\phi}\right)\frac{\partial}{\partial \phi}\left(\frac{\overline{\Delta(u'w')}}{f}\cos\phi\right) \tag{23}$$

where the differentiated quantity on the right hand side is the change in the parameterized vertical momentum flux and the overbar now denotes a zonal mean. The thermal response to this change is such that the diabatic heating or cooling associated with the radiative response is balanced by the change in the vertical advection of the zonal mean potential temperature associated with this change in the vertical component of the diabatic circulation. This balance is expressed in equation (22) from which it is easily deduced that the zonal mean temperature change is given approximately as

$$\Delta T = \frac{1}{\alpha}\left(\frac{HN^2}{R}\right)\left(\frac{1}{a\cos\phi}\right)\frac{\partial}{\partial \phi}\left(\frac{\overline{\Delta(u'w')}}{f}\cos\phi\right). \tag{24}$$

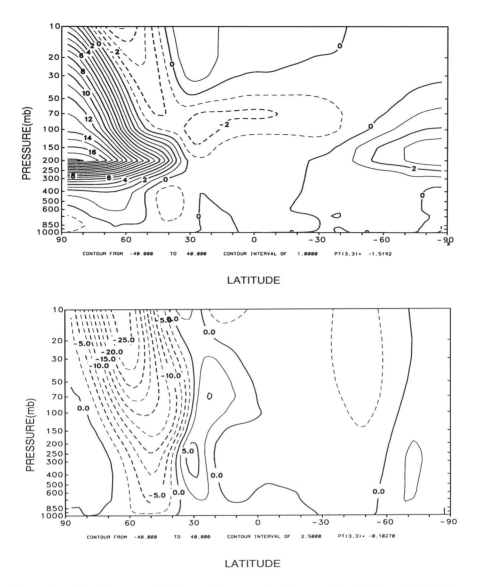

Figure 4. Zonally averaged difference between simulations with and without GWD. Upper: temperature difference (C). Lower: Zonal wind difference (m/s).

The associated change in the mean zonal wind is

$$\Delta U = -\left(\frac{1}{fa^2}\right)\int_0^z\left(\frac{N^2}{\alpha}\right)\frac{\partial}{\partial\phi}\left(\frac{1}{\cos\phi}\right)\frac{\partial}{\partial\phi}\left(\frac{\overline{\Delta(u'w')}}{f}\cos\phi\right)dz' + U(\phi)$$

(25)

where the function $U(\phi)$ can be introduced without violating the thermal wind

314

relationship. This function can in principle be determined by boundary conditions at the surface. However, since the assumption of steadiness is an idealization, initial conditions may play a role as well. Note that the form of Δw^* results from applying the upper boundary condition that the vertical mass flux associated with the residual mean circulation must vanish. This does not imply of course that the zonal wind response must also vanish.

For the sake of illustration, assume as a first approximation that there is no significant change in the structure of the planetary waves and/or the non-zonal component of the flow as a result of switching on GWD. Consistent with this simplification, it is reasonable to require that Δw^* must vanish at the lower boundary as well, which implies that the zonally averaged meridional gradient of the change in the parameterized vertical momentum flux must vanish at the lower boundary. This includes the sum of the gravity-wave momentum flux and the response to GWD of the boundary-layer component of the surface stress. Thus the added drag due to gravity-wave breaking must be at least partially compensated by a change in the boundary layer surface stress. In the context of the simplified analysis here, this compensation is required in the zonally averaged sense but does not in principle have to be localized to regions where GWD occurs. However, assuming that it is localized and using a simple drag formulation for the surface stress due to boundary layer turbulence, one can show that the local change in surface stress is approximately proportional to $2C_dU_s\delta u + Nh^2(U_s + \delta u)/L$ where δu is the local change in U_s (the zonal wind at the surface), C_d is the drag coefficient, and h and L are defined below equation (17). Setting this expression to zero and zonally averaging determines $U(\phi)$ and gives an approximate form for ΔU as

$$\Delta U = -\left(\frac{1}{fa^2}\right)\int_0^z\left(\frac{N^2}{\alpha}\right)\frac{\partial}{\partial\phi}\left(\frac{1}{\cos\phi}\right)\frac{\partial}{\partial\phi}\left(\frac{\overline{\Delta(u'w')}}{f}\cos\phi\right)dz' - \left(\frac{\overline{U_s}}{1+2\gamma}\right)$$
(26)

where $\gamma = C_dU_sL/(Nh^2)$. The assumption of complete compensation in this case is equivalent to requiring that the zonal wind response is mainly confined laterally to the region of gravity-wave emission. Over high mountain ranges γ is of the order of unity in circumstances where orographic gravity-wave emission is significant but becomes large elsewhere as the amplitude h of the orographic wave decreases.

Note that in these equations the change in the vertical momentum flux due to GWD is negative; that is, the vertical momentum flux is directed toward the surface for these waves. Moreover, since the imposed drag force is confined to a latitude band, the magnitude of this quantity has a local maximum in latitude. The associated thermal response is predominantly due to the meridional divergence of the increased downward momentum flux associated with the imposed GWD, i.e., $\frac{\partial}{\partial\phi}(\cos\phi\overline{\Delta(u'w')}/f)$. In the Northern Hemisphere in winter, this quantity is positive on the poleward side of the latitude band where the drag force is imposed and negative on the equatorward side, giving rise to warming on the poleward side of that region and cooling equatorward of it. The zonal wind response, assumed to be in thermal wind balance with this temperature response, extends through the depth of the atmosphere and is negative.

The zonal wind response at the surface is possibly big enough to offset the stress increase due to GWD with a reduction of the stress due to boundary layer turbulence. Of course, many simplifying assumptions have been made in the above analysis and some of them, for example the assumption used to derive the form of $U(\phi)$, are of doubtful validity in general. But the conclusion (that reduction of surface stress due to boundary-layer turbulence offsets the increase due to GWD) is confirmed by GCM simulations indicating that introducing an orographic GWD scheme does not necessarily result in a larger magnitude net surface stress [4].

In fact, the GCMIII simulations with the GWD parameterization have slightly smaller net stress at the surface in the northern hemisphere than the simulations without GWD. This is illustrated in Figure 5, which depicts the zonally averaged zonal component of the surface stress for the December-February season for the two simulations. Clearly, the zonally averaged surface stress is very similar in structure and magnitude for the two simulations in the tropics and most of the southern hemisphere where the effects of orographic GWD are weak. However, in the northern hemisphere there is a general weakening in middle latitudes and a slight southward shift in the simulation with GWD. In this case the contribution of the gravity-wave component is a substantial fraction of the total zonally averaged surface stress.

The above analysis is a simple application of the more general theory of downward control discussed by Haynes et al. [8]. In the above discussion it has been used to gain insight into the response of the large-scale circulation to the introduction of GWD. This more general formulation can be used as well to determine directly the contribution of resolved (Eliassen-Palm flux divergence) and unresolved (mainly GWD and turbulent transfer in the boundary layer) components of the eddy forcing terms in the zonal mean momentum equation to the residual mean circulation.

The above analysis is provided merely as a qualitative explanation of the response to orographic GWD. Among its limitations are the neglect of the effect of changes in the planetary wave forcing, which are not negligible, and the assumption of steadiness in the response. It can be shown that the assumption of steadiness is not valid if the scale of temporal variability of the gravity wave forcing τ_G is sufficiently small that $(\tau_G/\tau_R) < (N\Delta Z/(f\Delta Y))^2$ where τ_R is the radiative relaxation time scale and is inversely proportional to the Newtonian cooling coefficient used above to parameterize the radiative heating response. Here the length scales ΔZ and ΔY are, respectively, the effective depth and lateral extent of the GWD forcing. The steady state assumption is acceptable for the response to the mesospheric component of this forcing but perhaps less so for the lower

316

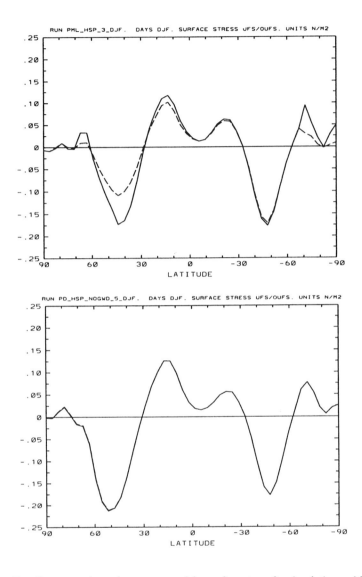

Figure 5. Zonally averaged zonal component of the surface stress for simulations with (upper) and without (lower) gravity-wave drag. Solid: total surface stress. Dashed, stress due to boundary layer turbulence.

stratospheric component.

The lower boundary condition used here is not appropriate in general since it ignores a number of factors, some of which are alluded to above. Orographic GWD in particular is highly non-zonal in its spatial structure as is the boundary layer contribution to the surface stress. Also ignored are changes in the surface pressure and the horizontal heat transport near the surface in response to the introduction of a gravity-wave drag parameterization. Haynes et al. [8] discuss this point, noting in particular that the surface pressure change is possibly a significant part of the total response. This is borne out in general by GCM simulations ([22], and Figure 5 here), which show that introducing orographic gravity-wave drag leads, on average, to higher surface pressures in high latitudes of the northern hemisphere and (as expected from considerations of conservation of mass in a hydrostatic model atmosphere) lower surface pressures at lower latitudes. It must be noted, however, that there is substantial inter-annual variability in the surface pressure field in the extra-tropics.

5.2. NON-OROGRAPHIC GWD

Although parameterizations of non-orographic GWD have been used in zonally averaged mechanistic models of the middle atmosphere for many years [1], they have only recently been used in comprehensive atmospheric general circulation models. In general, the most pronounced large-scale effects of the wave drag due to breaking of non-orographic gravity waves is thought to be in the upper stratosphere and mesosphere where it is potentially the dominant wave-driving process.

A number of recent studies of the effects of non-orographic GWD on the general circulation and climate of the middle atmosphere have been published recently. The response is in general quite sensitive to the (highly uncertain) specification of the source spectra ([20], [25]). Few direct comparisons of different schemes exist. However, given similar input spectra, the schemes in current use tend to produce similar responses ([19], [24]). All of these studies confirm that non-orographic GWD may be of great importance throughout the upper stratosphere and mesosphere.

6. Uncertainties

Although gravity-wave drag parameterizations are now widely used in AGCMs, they are still rather rudimentary in many respects and there are many uncertainties in their design and implementation. Of these uncertainties in GWD parameterizations the most important ones are probably associated with two broad categories:

1. inadequate knowledge and/or specification of wave sources;
2. uncertainties arising from over-simplification of the underlying theory.

The first of these is at present a significant stumbling block to further progress in representing the effects of non-orographic GWD in AGCMs. The second is in principle always a concern in designing parameterizations for larger-scale models of the atmosphere, not just for GWD but for other processes as well. Simplified representations of the basic physical processes of interest must be used for practical purposes. In principle, the validity and significance of these assumptions can be checked as improved observations and advances in theoretical understanding become available.

Acknowledgements

I am grateful to the students who read and commented on the first draft of these notes while in attendance at the NATO ASI on Numerical Modelling of the Global Atmosphere in Il Ciocco (John Knox, Mijke Zachariasse, Euain Drysdale, Rita Cardoso and Karine Sartelet). A special thanks to John, Mijke, Rita and Euain for subsequently reading later drafts and providing detailed comments that led to substantial improvement of the final version. Thanks are also due to my colleagues Richard Harvey and John Scinocca for making available to me results from recent simulations that they carried out with GCMIII.

References

1. Andrews, D.G., Holton, J.R., and Leovy, C.B. (1987) *Middle Atmosphere Dynamics,* Academic Press, Orlando.
2. Baines, P.G. (1995) *Topographic Effects in Stratified Flows,* Cambridge University Press, Cambridge.
3. Beagley, S.R., deGrandpré, J., Koshyk, J.N., McFarlane, N.A., and Shepherd, T.G. (1997) Radiative-dynamical climatology of the first-generation Canadian Middle Atmosphere Model, *Atmosphere-Ocean,* 35, 293–331.
4. Boer, G.J., and Lazare, M. (1988) Some results concerning the effect of horizontal resolution and gravity-wave drag on simulated climate, *J. Climate,* 1, 789–806.
5. Eliassen, A., and Palm, E. (1961) On the transfer of energy in stationary mountain waves, *Geofys. Publ.* 22(3), 1–23.
6. Garcia, R.R., and Solomon, S. (1985) The effect of breaking waves on the dynamics and chemical composition of the mesosphere and lower thermosphere, *J. Geophys. Res.,* 90, 3850–3868.
7. Gregory, D., Shutts, G.J., and Mitchell, J.R. (1998) A new gravity-wave drag scheme incorporating anisotropic topography and low-level wave breaking: Impact on the climate of the UK Meteorological Office Unified Model, *Q. J. R. Meteorol. Soc.,* 124, 463–493.
8. Haynes, P.J., Marks, C.J., McIntyre, M.E., Shepherd, T.G., Shine, K.P. (1991) On the "downward control" of extratropical diabatic circulations by eddy-induced mean zonal forces, *J. Atmos. Sci.* 48, 651–678.
9. Hines, C.O. (1988) A modeling of atmospheric gravity waves and wave drag generated by isotropic and anisotropic terrain, *J. Atmos. Sci.,* 45, 309–326.
10. Hines, C.O. (1997a) Doppler spread parameterization of gravity wave momentum deposition in the middle atmosphere, Part I: Basic formulation, *J. Atmos. Solar. Terr. Phys.,* 59, 371–386.

11. Hines, C.O. (1997b) Doppler spread parameterization of gravity wave momentum deposition in the middle atmosphere, Part II, Broad and quasi-monochromatic spectra and implementation, *J. Atmos. Solar Terr. Phys.*, 59, 387–400.

12. Holton, J.R. (1982) The role of gravity-wave induced drag and diffusion in the momentum budget of the mesosphere, *J. Atmos. Sci.*, 39, 791–799.

13. Kim, Y.-J., and Arakawa, A. (1995) Improvement of orographic gravity wave parameterization using a mesoscale gravity wave model, *J. Atmos. Sci.*, 52, 1875–1902.

14. Klemp, J.B., and Lilly, D.K. (1980) Mountain waves and momentum flux, Ch. 4 in *Orographic Effects in Planetary Flows*, GARP-WMO Pub. Ser. 23, 116–141.

15. Laprise, R., and Girard, C. (1990) A spectral general circulation model using a piecewise-constant finite element representation on a hybrid vertical coordinate system, *J. Climate*, 3, 32–52.

16. Laprise, R. (1993) An assessment of the WKBJ approximation to the vertical structure of linear mountain waves: implications for gravity-wave drag parameterization, *J. Atmos. Sci.*, 50, 1469–1487.

17. Lindzen, R.S. (1981) Turbulence and stress owing to gravity wave and tidal breakdown, *J. Geophys. Res.*, 86, 9707–9714.

18. Lott, F., and Miller, M. (1997) A new subgrid-scale orographic drag parameterization: Its formulation and testing, *Q. J. R. Meteorol. Soc.*, 123, 101–127

19. Manzini, E., McFarlane, N.A., and McLandress, C. (1997) Impact of the Doppler spread parameterization on the simulation of the middle atmosphere circulation using the MA/ECHAM4 general circulation model, *J. Geophys. Res.*, 102, 25751–25762.

20. Manzini, E., and McFarlane, N. (1998) The effect of varying the source spectrum of a gravity wave parameterization in a middle atmosphere general circulation model, Report No. 252, Max-Planck-Institute For Meteorology, Hamburg, Germany.

21. McFarlane, N., and Manzini, E. (1997) Parameterization of gravity-wave drag in comprehensive models of the middle atmosphere. *Adv. Space. Res.*, 20, 1241–1251.

22. McFarlane, N.A. (1987) The effect of orographically excited gravity wave drag on the general circulation of the lower stratosphere and troposphere., *J. Atmos. Sci.*, 44, 1775–1800.

23. Medvedev, A.S., and Klaassen, G.P. (1995) Vertical evolution of gravity-wave spectra and the parameterization of associated wave drag, *J. Geophys. Res.*, 100, 25,841–25,853.

24. Medvedev, A.S., Klaassen, G.P., and Boville, B.A. (1997) The Parameterization of Gravity Wave Drag Based on Nonlinear Diffusion of Wave Spectra, *NATO ARW on Gravity-Wave Processes and Their Role in Climate*, (K. Hamilton, ed.), NATO ASI Ser. Vol. I 50, 309–325.

25. Medvedev, A.S., Klaassen, G.P., and Beagley, S.R. (1998) On the role of an anisotropic wave spectrum in maintaining the circulation of the middle atmosphere, *Geophys. Res. Lett.*, 25, 509–512.

26. Miranda, P.M.A., and James, I.N. (1992) Non-linear three-dimensional effects on gravity-wave drag: Splitting flow and breaking waves, *Q. J. R. Meteorol. Soc.*, 118, 1057–1081.

27. Miranda, P.M.A., and Valente, M.A. (1997) Critical level resonance in three-dimensional flow past isolated mountains, *J. Atmos. Sci.*, 54, 1574–1588.

28. Palmer, T.N., Shutts, G.J., and Swinbank, R. (1985) Alleviation of a systematic westerly bias in general circulation and numerical weather prediction models through an orographic gravity wave drag parameterization, *Q. J. R. Meteorol. Soc.*, 112, 1001–1031.

29. Peltier, W.R., and Clark, T.L. (1979) The evolution and stability of finite-amplitude mountain waves. Part II: Surface wave drag and severe downslope windstorms. *J. Atmos. Sci.*, 36, 1498–1529.

30. Shepherd, T.G., Semeniuk, K., Koshyk, J.N. (1996) Sponge-layer feedbacks in middle atmosphere models, *J. Geophys. Res.*, 101, 23447–23469.

31. Smith, R.B. (1985) On severe downslope winds, *J. Atmos. Sci.*, 42, 2597–2603.

32. Warner, C.D., and McIntyre, M.E. (1996) On the propagation and dissipation of gravity-wave spectra through a realistic middle atmosphere, *J. Atmos. Sci.,* 53, 3213–3235.

LAND SURFACE PROCESSES AND HYDROLOGY

J.-F. ROYER
Météo-France, CNRM
42 Av. G. Coriolis
31057 Toulouse Cedex 1
France

1. Introduction

The surface of the earth receives from the overlying atmosphere energy, for the most part in radiant form (solar and thermal radiation), and water, mainly in the form of liquid (rainfall) and solid (snowfall) precipitation. Various physical processes convert this energy and water, and give back part of it to the atmosphere in the form of upward diffused shortwave radiation, thermal radiation and turbulent fluxes of sensible heat and water vapor, while the remaining part is retained and stored below the surface. The processes of energy and water exchanges between the atmosphere and the surface are inherently interdependent, since energy needs to be provided for the vaporization of the received liquid or solid water before the water can be transferred back to the atmosphere in gaseous form and carried upward away from the surface by diffusion and turbulent motions. The fundamental difference between oceans and continents is that the oceans are by definition an enormous water reservoir. As water is relatively transparent, solar radiation can penetrate and be absorbed gradually inside the oceanic waters, and the resulting heat spread out into a large mass of water can be further redistributed by the oceanic circulation. Compared with the atmosphere, the oceans have virtually unlimited heat and water storage capacity.

By contrast, over the continents, though some materials like foliage or snow cover may be semitransparent to the shortwave radiation, the solar radiation is absorbed within the first few millimeters inside the soil. As the thermal conductivity of the materials found on the continental surfaces is generally lower than the conductivity of water, large thermal gradients and temperature variations can easily arise. Also, the water storage capacity of the continents is limited, and the existence of streams and rivers that carry excess rainfall to the ocean demonstrates that the soil surface cannot absorb all the rainfall landing on it. The

P. Mote and A. O'Neill (eds.), Numerical Modeling of the Global Atmosphere in the Climate System, 321–351.
© 2000 *Kluwer Academic Publishers. Printed in the Netherlands.*

storage capacity is highly variable according to the composition of the soil and the topographic controls on the water at the surface or underground. This results in high spatial and temporal variability of the storage and exchanges of energy and water at the surface of the continents, and thus the interactions between the land-surface and the atmosphere are intrinsically much more complex and varied, than the interactions between the oceans and the atmosphere.

Land evaporation is important as a source of moisture for "precipitation recycling", defined as the fraction of water evaporated from the land surface within a specified region that falls back as precipitation within the same region. Brubaker *et al.* [7] estimated precipitation recycling in several regions on different continents and found that it ranges from 0.11 for the Eurasian region, 0.24 for the North and South American regions, and up to 0.31 for the African regions. These figures give an approximate measure of the importance of land-surface hydrology in maintaining the climate of these regions. The recycling of regional precipitation over large continental areas represents an important positive feedback mechanism, which could contribute to the intensification and persistence of large-scale droughts. A thorough review of precipitation recycling has been given by Eltahir and Bras [26], who point out that large-scale human activities that involve changes in land cover, such as tropical deforestation, are likely to modify climate through changes in the water cycle. As the land-surface supports the terrestrial biosphere, including plants, animals and the human population, there is considerable interest in studying the interactions between land-surface and climate.

The primary function of a land surface model (LSM) is to use information from the atmospheric general circulation model (GCM) to simulate surface processes and provide correct feedback to the atmospheric model in terms of fluxes of radiation, momentum, heat and moisture [80]. The essential tasks are the partitioning of incoming solar radiation into absorbed and reflected radiation; the partitioning of the net radiation balance into sensible heat flux, latent heat flux and soil heat flux; and the partitioning of precipitation into runoff and storage. This last task may be particularly important for applications of climate models since correct estimations of soil moisture and runoff may be essential for many human activities.

2. Water and Energy Budgets at the Surface

2.1. THE SURFACE RADIATION BUDGET

The basic energy input to the land-surface is in form of radiant energy of downward shortwave solar radiation R_S^+ (sum of direct and diffuse components) and longwave thermal radiation R_A^+ emitted by the overlying atmospheric column. The solar radiation has a marked diurnal cycle due to its variable incidence angle during the course of the day. Both these fluxes are computed by the radiation subroutine as a function of the vertical atmospheric profiles, as detailed in Chapter 10. Part of the incident solar radiation R_S^+ reaching the surface is sent back upwards by reflection or diffusion at the surface. The upward diffuse shortwave radiation R_S^- is usually computed by specifying the surface albedo $\alpha_s = R_S^- / R_S^+$. The surface albedo is strongly variable in space according to the composition and structure of the soil, the type and state of the land cover, the development of vegetation, and the presence and evolution of a snow cover.

There are two possible strategies for specifying the albedo: either to use an "observed" albedo map reconstructed from satellite data, or to compute the albedo from a classification of soil and land cover categories. The first strategy can be useful for the geographical variations of the surface albedo that are invariant in time, such as those related to the soil properties, or those that have a regular seasonal variation such as permanent vegetation covers. Satellite data have been processed and inverted to construct surface climatological albedo maps used to specify surface albedo in GCMs. The inherent difficulty with this approach is that the albedo is no longer an interactive variable, so that radiative feedbacks that involve the surface albedo are eliminated from the model. Such feedbacks are so important for the evolution of a snow cover, however, that all models include a parameterization to modify the surface albedo in the presence of snow, even when they make use of a climatological albedo map. As a consequence, one has to specify in the model a snow-free albedo, which requires processing the satellite data in order eliminate the influence of the snow cover, as was first attempted in the pioneering production of the minimal albedo map by Preuss and Geleyn (1980) [68]. However, besides the obviously important snow cover feedback, other surface conditions can potentially influence the albedo. Surface moisture exerts an important control on the state and development of the vegetation cover, and can also modify the albedo of bare soil. The specification of albedo maps independently from other surface boundary conditions can lead to inconsistencies. In the case of climate change scenarios or palaeoclimatic reconstruction the present climatological albedo is likely to become unrealistic as climate conditions and vegetation both change.

In order to overcome these difficulties the second strategy, which has been prevalent in recent land-surface parameterization schemes, has been to reconstruct albedo values on the GCM grid as a function of the distribution of soil and vegetation types distribution inside the grid. The soil and vegetation datasets of Matthews [51] and Wilson and Henderson-Sellers [90], who pioneered this approach, have since been used in many GCMs with a variable number of vegetation categories. The albedo is computed by assigning typical values for each vegetation and soil type in the classification. The use of a correspondence table between soil/vegetation types and albedo makes it easy to introduce in the land-surface albedo parameterization the influence of various factors such as soil moisture, the seasonal cycle and even the diurnal cycle, once typical albedo sensitivities to

324

these factors have been specified for the different types of land-cover. Spectral variations of the albedo may be taken into account according to the requirements of the radiation subroutine, as well as angular variation of the reflectance with the inclination angle of the incident insolation.

As the spectral reflectance of the surface in the infrared is generally very low, the downward thermal atmospheric radiation is absorbed very efficiently. The surface loses radiant energy in the form of longwave thermal radiation R_A^- according to the well known emission law: $R_A^- = \epsilon_s \sigma_{SB} T_r^4$, where ϵ_s is the emissivity of the surface, σ_{SB} is Stefan-Boltzmann constant, and T_r the radiative (blackbody) temperature of the land-surface. The surface emissivity, which according to Kirchoff's law is equal to the surface absorptivity, is very close to unity but can vary slightly with the nature of the land cover. This can be taken into account by the specification of a surface emissivity map in the boundary conditions of the model. Summing up all the radiative fluxes the net radiation balance R_N at the surface is

$$R_N = R_S^+(1 - \alpha_s) + \epsilon_s(R_A^+ - \sigma_{SB} T_r^4). \tag{1}$$

2.2. PRECIPITATION AND RUNOFF

Atmospheric precipitation can occur in liquid form P_L (rainfall) or solid form P_S (snowfall or, less commonly, hailfall) depending on the air temperature. When air temperature is below the freezing point of water, solid precipitation can accumulate on the ground forming a snow (or ice) cover. Whenever the energy balance of the snowcover becomes sufficiently positive to provide the energy needed for the phase change of water from the solid to the liquid phase, melting happens and provides an additional contribution in the form of snowmelt P_m to the underlying soil or vegetation layer. The snow cover provides a variable storage reservoir of solid water upon the surface of the ground that may be released as a kind of "delayed" precipitation at the time of melting. This storage is controlled in a highly nonlinear way by temperature, and thus energy balance, since melting occurs at a well defined constant temperature threshold. In favorable conditions this storage reservoir of solid water can become huge, as evidenced by mountain glaciers and the monumental ice caps covering Greenland and Antarctica.

Exchanges between the land surface and the atmosphere over a given area are usually studied locally in a one-dimensional vertical column framework representing a GCM grid cell. While the horizontal transport of heat inside the ground can be neglected, it cannot be neglected for surface waters or for underground waters. The lateral movements of water tend to follow the terrain slope, due to topographic features of various scales which collect and channel it into the hydrographic network formed by streams and rivers. The flux of water that escapes laterally from a soil column is called "runoff", while the vertical downward flux into deeper layers of the ground is called "infiltration" or "drainage."

Total liquid precipitation ($P = P_{\mathrm{L}} + P_{\mathrm{m}}$) at the surface can be partitioned into an infiltration part that is stored in the soil column (I_{s}) and a surface runoff part that is carried away by the hydrographic network (R_{s})

$$P = I_{\mathrm{s}} + R_{\mathrm{s}}. \tag{2}$$

The efficiency of the runoff process can be quantified by defining a dimensionless runoff ratio $\alpha_{\mathrm{w}} = R_{\mathrm{s}}/P$, which plays for rainfall a similar role as the albedo for solar radiation. Runoff is of major interest for hydrologists whose task is to monitor and forecast the flow of rivers, so as to allow efficient management of water resources and provide timely flood warning. Runoff represents the essential input source term to hydrographic models and is sometimes referred to as "effective rainfall." From the perspective of land-surface atmosphere modeling it represents the part of rainfall that is not stored locally and is thus no longer available for local evaporation. The flux of water absorbed by the soil is

$$I_{\mathrm{s}} = (1 - \alpha_{\mathrm{w}})P. \tag{3}$$

2.3. SURFACE STORAGE OF HEAT AND WATER

The surface is not in radiative balance, and on global average it gains energy by radiative transfer, while the atmosphere loses energy. This tends to give rise to unstable thermal stratifications of the atmospheric air close to the surface and promote the development of turbulent and convective motions in the atmospheric boundary layer, as discussed in Chapters 8 and 9. The resulting turbulent eddies contribute to renew the thin layer of air that can exchange heat directly with the surface by molecular conduction and diffusion, and thus sustain a turbulent flux of sensible heat (H). If water is available near the surface, and if energy is available to provide the latent heat of vaporization L required for the phase change of water from liquid to vapor, then the process of evaporation can take place and diffusion and turbulent eddies will transfer, together with the flux of sensible heat, a flux of water vapor E, which in terms of energy can be counted as a latent heat flux LE lost by the surface.

Representing by Q and W, respectively, the quantity of heat and water stored in a soil layer of thickness Δz starting from the surface, the heat and water balance equations for this layer can be written in the following form that underscores their coupling through the evaporation flux:

$$dQ/dt = R_{\mathrm{n}} - G_{\mathrm{b}} - H - LE \tag{4}$$

$$dW/dt = P - R_{\mathrm{s}} - I_{\mathrm{b}} - E = (1 - \alpha_{\mathrm{w}})P - I_{\mathrm{b}} - E, \tag{5}$$

where $G_{\mathrm{b}} = G(\Delta z)$ and $I_{\mathrm{b}} = I(\Delta z)$ are the vertical downward heat and water fluxes due to heat conduction and water infiltration in the ground through the

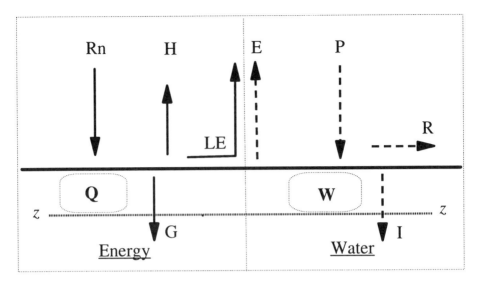

Figure 1. Simplified representation of the surface fluxes of energy and water Rn: net surface radiation, G: heat flux into the ground, H: sensible heat flux, LE: latent heat flux; P: precipitation, I: infiltration into the ground, R: surface runoff, E: evaporation

bottom boundary of the layer. The storage of heat in the ground is limited by the slow downward penetration of heat due to the poor conductivity of the soil material. Likewise the amount of water that can be stored is limited to a maximum storage capacity W_{max} by the available space inside the pores of the soil matrix. The moisture state of the soil layer can be characterized by a "wetness ratio" $\beta_w = W/W_{max}$. One can expect that both evaporation and runoff may be affected by soil water content, and most likely will be increasing functions of the wetness ratio, reaching their maximum when the soil is saturated with water ($\beta_w = 1$). The maximum evaporation that takes place when water is freely available at the surface is called potential evaporation E_p. When the soil can no longer absorb water all precipitation will have to go to runoff ($\alpha_w = 1$).

2.4. TURBULENT FLUXES

The vertical turbulent fluxes for a quantity X are usually parameterized by means of standard aerodynamic bulk formulas (Chapter 8) that can be written as

$$\overline{w'X'}_s = -K_X \frac{\partial X}{\partial z} = C_X \|V_a\| (X_s - X_a) \tag{6}$$

where C_X is the drag coefficient for the quantity X, V is the wind velocity, the index a refers to and atmospheric level (usually the lowest model level), and index s to a value at the roughness height z_X. The coefficient C_X is a function of the height of the atmospheric level z_a, of roughness height z_X and of stability and is usually expressed empirically as a function of the bulk Richardson number Ri [43] or the Obukhov length [88]. This formula can be rewritten in a form similar to Ohm's law by making the analogy between a flux and an electric current, between differences in the concentration of X and differences in electric potential, and by introducing an aerodynamic resistance to turbulent transfer $r_{Xa} = r_{Xa}(z_a/z_X, Ri)$, or its inverse the conductance g_{Xa}:

$$\overline{w'X'}_s = \frac{X_s - X_a}{r_{Xa}} = g_{Xa}(X_s - X_a). \tag{7}$$

This electrical analog has been found useful to represent various processes that interfere with turbulent transfer at the surface as supplementary resistances that can be connected in series or in parallel. The case of the water vapor flux will provide a good illustration.

Using the resistance formulation, the sensible heat flux and the evaporation can be written as

$$H = \rho(C_p T_s - C_p T_a - g z_a)/r_a \tag{8}$$

$$E = \rho(q_s - q_a)/r_a \tag{9}$$

where the same aerodynamic resistance ($r_a = r_{Ta} = r_{qa}$) has been used for both to simplify the notation, as is done in most parameterization schemes. The surface roughness can be specified as a function of small-scale topography, vegetation height and snow cover. Some sensitivity studies [86] have shown that the large-scale circulation can be extremely sensitive to the roughness height. Roughness can be computed from land cover distributions by assigning characteristic values to land cover types and computing weighted averages. However, the averaging method needs to be chosen carefully.

The computation of the turbulent fluxes requires the determination of the values of T_s and q_s (at roughness height). In the case of temperature this can be done by assimilating the value T_s with the radiative temperature T_r, considering both as a kind of "skin-temperature" and solving the energy balance equation for an interface layer of negligible heat capacity. As a lower boundary condition for this skin layer one needs to introduce the heat flux G into the ground, which can be computed by a similar resistance formulation assuming the temperature T_g is known at some small depth inside the ground

$$G = \rho_g C_g (T_s - T_g)/r_g = \Lambda(T_s - T_g). \tag{10}$$

Such a formulation has been applied in the ECMWF model [88] with T_g the temperature of a 0.07 m layer of ground and a constant value of $\Lambda = 7\ Wm^{-2}K^{-1}$.

2.5. POTENTIAL EVAPORATION

In order to solve the heat balance equation for the skin temperature T_s one needs to compute the latent heat flux which depends on the specific humidity q_s. When the surface is saturated with water (as is the case during and just after rainfall, or in artificially irrigated ground surface) the specific humidity is the saturation specific humidity q_{sat} which is a well defined function $q_{sat}(T_s, p_s)$ of temperature T_s and surface pressure p_s. In this case the heat balance equation becomes a closed implicit nonlinear equation for T_s, and can be solved by appropriate efficient numerical methods. An approximate method to obtain an analytic solution was proposed by Penman in 1948 [63] and generalized later by Monteith [58] by linearizing the nonlinear terms (mostly the saturation vapor pressure) around a reference temperature, usually chosen as the atmospheric temperature T_a. The main advantage of this solution, which is referred to as the Penman-Monteith equation, is that the potential evaporation can be written in a convenient analytical form as a function of net radiation and atmospheric measurements, which is convenient for physical analysis and for applications. The small inaccuracies due to the linearization can be problematic in some cases, and there has been some discussion of possible improvements [62]. For GCMs it seems worthwhile to avoid the inaccuracies of this approximation since it might hamper energy conservation. The evaporation computed in this way is a function of the surface temperature and is the potential evaporation E_p, as it represents the maximum evaporation possible in saturated moisture conditions (evaporation limited by atmospheric demand and energy supply). Potential evaporation representing the maximum evaporation available on a well watered ground patch is an important concept for agronomic applications, and can be measured in the field by lysimeters. Global maps of potential evaporation have been produced from remote sensing and large-scale reanalysis data [9].

2.6. ACTUAL EVAPORATION

Under more limited water supply the specific humidity near the surface is usually less than its saturation value, and the actual evaporation is restricted to smaller values than potential evaporation by the maximum sustainable water flux that the soil can transfer to the surface (moisture limited regime). One needs to find some method to relate the specific humidity at the surface to the soil water content and to the surface temperature. Two formulations have long been in use [59]. The first, referred to as the "α method" [45], is to assume that the relative humidity of the air is a constant α_q determined by the moisture condition of soil

$$q_s = \alpha_q \, q_{sat}(T_s). \tag{11}$$

The second method ("β method") can be formulated as a linear interpolation of specific humidity between the saturation value and the atmospheric value

$$q_s = \beta_q \, q_{sat}(T_s) + (1 - \beta_q) \, q_a. \qquad (12)$$

Inserting this expression into the evaporation equation one obtains

$$E = \beta_q \, \rho \, (q_{sat} - q_a)/r_a = \beta_q \, E_p. \qquad (13)$$

In this way the coefficient β_q appears as a reduction factor from potential evaporation, or "evaporation efficiency parameter."

Both methods need to relate their parameters to the quantity of water available for evaporation (the "wetness function"). Numerous empirical formulas have been used in different models to relate factors α_q and β_q to the moisture state of the surface [40]. Comparisons of the various currently used formulations against in situ evaporation data over a loamy bare ground have been made by Mahfouf and Noilhan [45]. Comparable results with the two methods have been found during daytime, but significant differences appeared at night with the α method tending to give excessive formation of dew over dry soils, while the β method has the tendency to overestimate the nocturnal evaporation [45]. This has led Ye and Pielke [94] to propose a combination of the two methods "α and β method" for non-vegetated surfaces, allowing an explicit dependence on soil porosity.

2.7. THE BUCKET MODEL

If evaporation cannot keep pace with precipitation, at some point the soil becomes saturated and excess water starts to run off. The simplest way of parameterizing runoff, used in the first GCM with a complete hydrologic cycle at GFDL (Manabe, 1969) [46] is the concept of the "bucket" type of reservoir, which stores all the rainwater until it is filled up to the brim and then overflows. The bucket hydrology can be expressed mathematically as

$$R/P = \alpha_w = H(\beta_w - 1) \qquad (14)$$

in which the Heaviside function ($H(x < 0) = 0$; $H(x \geq 0) = 1$) has been used so as to emphasize the functional (though discontinuous) dependence between runoff ratio and wetness ratio. The water in the bucket can evaporate at an evaporation rate E which is specified by a functional relationship ("wetness function") between the ratio of actual to potential evaporation $\beta = E/E_p$ ("evaporation efficiency") and the wetness ratio $\beta_w = W/W_{max}$. In the GFDL bucket model this relationship was specified, on the basis of the work of Budyko, as

$$\beta = \max(1, \beta_w/\beta_c). \qquad (15)$$

β_c represents a wetness threshold below which evaporation starts to decrease from its potential rate. In the GFDL parameterization $\beta_c = 0.75$. The storage capacity of the bucket is $W_{max} = 0.15$ m of water, uniformly over the globe. The bucket model needs as input the precipitation P and potential evaporation E_p, and provides as output the soil water content W and the actual evaporation E. The potential evaporation can be computed from the surface energy balance equation.

The bucket method has also been used in the NCAR model CCM1, even in recent publications concerning the sensitivity of the hydrologic cycle to CO_2 [72], and as a benchmark for comparison of other schemes in many studies [73] [84] [69]). Its main well-known defects are infrequent runoff (only when moisture is at full capacity and rainfall occurs), and in that case runoff that is too variable [91]. It is well suited to monthly time-scales and to models without a diurnal cycle of insolation. Milly and Dunne [54] have explored the sensitivity of the simulated global water cycle to the specification of the water holding capacity W_{max} by performing seven simulations varying W_{max} from 1 cm to 240 cm and additional simulations with zero storage and a storage of 15 cm maintained at saturation, similar to the Shukla and Mintz [82] dry and wet simulations. Their simulation revealed that even a small storage capacity of 1 cm was half as effective as an arbitrarily large capacity in producing evaporation from the continents. Evaporation increased linearly with the logarithm of storage capacity up to a storage capacity of 60 cm. The changes were larger in the tropics and had an effect on the Hadley circulation.

Several variants and possible improvements of the bucket hydrology have been proposed. A different functional relationship for the runoff ratio has been used in the GISS model $\alpha_{w(GISS)} = \beta_w/2$ when β_w in the top soil layer is unsaturated, and $\alpha_{w(GISS)} = 1$ when it is saturated. The factor $1/2$ was chosen so as to obtain agreement of the computed annual global mean runoff with observations [34] [38].

Wood et al. [91] have shown that the bucket model can be improved by the inclusion of a base flow (groundwater) term (represented by the term I in the water budget equation), making it a "leaky" bucket, and this conclusion was supported by sensitivity experiments [84]. A threshold was introduced in a bucket type model of subsurface runoff used in the simple water balance (SWB) of Schaake et al. [79] making subsurface runoff a linear function of the lower layer moisture content in excess of a minimum threshold W_R:

$$I = I_{max}\max(0, (W - W_R)/(W_{max} - W_R)),$$

where I_{max} is the potential subsurface runoff occurring when the lower layer is saturated, and W_R the soil water content below which subsurface runoff ceases.

Besides its use in GCMs, the bucket method has found applications in the estimation of climatologies of soil moisture and surface evapotranspiration (Mintz and Serafini [56], Mintz and Walker [57] with a choice of a different β ($\beta = 1 - exp[-6.8W/W_{max}]$). Dunne and Milly [23] considered the possibility of introducing geographical variations in the storage capacity of soil water by taking into account soil granulometry, water retention, and rooting depth of plants, so as to determine extractable soil water.

The limitations of the simple bucket reservoir come from the uncertainties in the appropriate choice of the wetness function and the lack of a physical basis to relate it to the properties of the soil and vegetation cover. The evaporation of summertime soil moisture is underestimated by single-layer models and requires the use of at least a two-layer soil model. This has motivated the development of more physically based soil-vegetation-atmosphere transfer schemes (SVATS). Deardorff (1978) [12] was the first to propose a parameterization including a layer of vegetation in the computation of soil temperature and moisture. More complex vegetation schemes like BATS [17] or SiB [81] have been developed in order to include in a comprehensive way the interactions among the soil, the vegetation and the snow cover in the treatment of the land-surface energy and water budgets. Other GCM groups have tried to develop simplified schemes that retain only the essential processes, like ISBA at Meteo-France [60], SECHIBA at LMD [21], BEST at BMRC [66], and CLASS at CCC [87].

3. Heat and Water in the Soil

A short summary of the main physical properties of soils is necessary to understand heat and water transport within the ground. Soils, as distinct from bare rock, are a porous material formed by the aggregation of mineral particles of various sizes and organic mater (humus, roots). Small spaces called "pores" are left between the solid particles. Their size distribution and volume is variable according to the granulometry and texture of the constituents (clay, sand, loam) and the degree of compaction of the soil. The "porosity" χ_p measures the fraction of volume occupied by the pores. The pores are filled with air, water or ice in variable proportions, and the soil physical properties (density, heat capacity, thermal and hydraulic conductivities) depend on the quantity of water stored in the pores, usually expressed by the volumetric water content η (volume of water per volume of soil in m^3/m^3). The maximum volumetric water content η_{max} is reached when all the pores are filled with water ($= \chi_p$), in which case the soil is completely saturated with water and can absorb it no more. The water holding capacity (in kg/m^2 of surface area) of a soil layer between depths z_1 and z_2 is

$$W = \int_{z_1}^{z_2} \rho_w \eta(z) \, dz \qquad (16)$$

where ρ_w is the density of liquid water (kg/m^3).

Water is held on to the solid particles by surface tension (capillary forces) and attracted downward by gravity (drainage). The field capacity W_{fc} is the maximum amount of water that can be retained against gravity (gravitational equilibrium). It can be determined by saturating the soil with water and letting it drain for several days. Precise thermodynamical expressions of water infiltration in the ground have been formulated. Water is removed by evaporation through the unfilled pores, by absorption by the root system of plants, and by gravitational drainage to lower layers. The wilting point W_{wilt} represents the water content below which plants can no longer extract water from the soil and wilt permanently. Bare soil evaporation is limited by the soil column's ability to conduct water to the land surface. Subsurface flow physics, based on Darcy's law, is controlled by hydraulic conductivity and the hydraulic head gradient. The depth of the water table is an important parameter for determining the subsurface flow.

3.1. THERMAL PROPERTIES

Heat conduction in the ground is expressed by the Fourier equation [35]

$$G(z) = -\lambda_g \, d(T_g)/dz, \tag{17}$$

giving the heat flux $G(z)$ (W/m^2) at depth z in the soil as proportional to the vertical gradient gradient of soil temperature T_g. The coefficient *lambda*$_g$ is called the "thermal conductivity" (Wm^{-1}K^{-1}).

The evolution of a profile of subsurface temperature $T_g(z)$ is given by

$$\rho_g c_g dT_g/dt = -dG/dz \tag{18}$$

with the appropriate boundary conditions, such as constant flux at large depth (geothermal flux), where ρ_g is soil bulk density (kgm^{-3}) and c_g is soil specific heat capacity (Jkg^{-1}K^{-1}). Their product is called the volumetric heat capacity $C_g = \rho_g c_g$ (Jm^{-3}K^{-1}).

When the above parameters are independent of depth it is convenient to introduce the soil thermal diffusivity $D_g = \lambda_g/C_g$ (m^{2}s^{-1}), in which case the heat diffusion equation becomes a linear diffusion along the thermal gradient:

$$d(T_g)/dt = D_g \, d^2T/dz^2. \tag{19}$$

Analytic solutions of the equation are easily found for a sinusoidal forcing, and are the base of the so-called force-restore method, which is used to select the appropriate top soil layer thickness for a correct representation of the diurnal cycle of surface temperature [16]. In order to capture the diurnal range, the thickness of the top soil layer needs to be smaller than the thermal damping depth of the diurnal oscillation. The force-restore equation for a thin soil layer overlaying a thicker layer at temperature T_B is

$$C_s dT_s/dt = G(0) - \gamma (T_s - T_B). \tag{20}$$

The soil heat capacity C_s (Jm^{-2}K^{-1}) of the top soil layer is related to the soil properties. The following relation has been proposed as giving accurate results [4]:

$C_s = 0.95C_g(D/2\Omega)^{1/2}$, and $\gamma = 1.18\Omega C_s$, where Ω is the diurnal frequency. The force-restore parameterization has been found to be rather accurate for resolving the diurnal cycle. Even in an extreme case in which the goal was to simulate the upper limit of attainable soil temperature (366 K, or 93°C) with a detailed multilayer model, the error by the force-restore method was only 5 K [31].

Volumetric heat capacity can be computed from the volumetric fraction of soil constituents (including liquid and gaseous phases). Thermal conductivity involves shape factors for soil particles and air inclusions, with contributions from phase changes of water (evaporation and condensation). The thermal conductivity can vary with the mineral composition of the soil, the size of its particles, the porosity, and the water content. Other mechanisms of heat exchange are heat carried into a layer by water infiltration, heat loss due to root extraction, and base flow runoff. More elaborate solutions for soil properties that vary with depth require the use of numerical solutions by discretizing the equations in several soil layers. Richardson in 1922 [70] had already recommended logarithmically spaced levels in the ground!

3.2. MOISTURE DIFFUSION IN THE GROUND

Infiltration can be defined as the entry into unsaturated soil of water available at the soil's surface [65]. Infiltration is affected by gravity and soil capillary forces. Unsaturated soil is a porous medium that contains both air and water. In porous media the water is submitted to capillarity, which is dominant in coarse-textured media, and adsorption, which is particularly important in dry media; water is retained in the smaller pores of the soil matrix. Taking z as vertical coordinate (defined so that it is positive downward relative to surface elevation), a total potential $\Phi = \psi - z$ is defined as the sum of a moisture potential ψ, arising from interactions at the interface between water and solid particles, and a gravitational potential. The custom is to divide the potentials by g so as to express them in length units equivalent to the height of a water column. By convention ψ is 0 at saturation and is a function of the volumetric moisture content η, decreasing to large negative values in unsaturated soils, and ψ is proportional to the work needed to extract water molecules from a unit mass of soil against the surface tension forces between soil and water. Because of the interactions with soil particles, the equilibrium water vapor pressure in soil pores is lower than the saturation pressure over a plane free-water surface. Its relative humidity h can be expressed as [64]:

$$h = \exp(g\psi(\eta)/R_v T) \tag{21}$$

where R_v is the gas constant of water vapor and T is temperature. Within the large soil pores molecular diffusion of water vapor from the water vapor source to a point just outside the surface of the soil can be described by a surface resistance to water diffusion r_g that is strongly dependent on soil moisture, texture, and structure [45].

The flow velocity V of water in unsaturated media is proportional to the gradient of the total potential and is expressed by Darcy's law:

$$V = -K(\eta)\nabla\Phi \tag{22}$$

where the proportionality constant $K(\eta)$, known as the hydraulic conductivity, is a rapidly decreasing function of the volumetric water content. The above two relationships can be combined with the continuity equation to formulate differential equations for the computation of infiltration in porous media. A frequently used form gives the evolution of the volumetric water content

$$\frac{d\eta}{dt} = \nabla \cdot (D\nabla\eta) - \frac{dK}{d\eta}\frac{d\eta}{dz} \tag{23}$$

where $D = K d\Psi/d\eta$ is the moisture diffusivity. The one-dimensional version of this equation is the basis for the computation of vertical infiltration in several land-surface schemes. An analogy between the solutions of steady quasi-linear flow in porous media and the scattering of plane harmonic waves has been established [65].

Functional equations are needed to provide relationships between soil moisture η, moisture potential ψ, and hydraulic conductivity K. The following relations [10] are frequently used:

$$\psi(\eta) = \psi_{sat}(\eta/\eta_{sat})^{-b} \tag{24}$$

$$K(\eta) = K_{sat}(\eta/\eta_{sat})^{2b+3}, \tag{25}$$

where the index *sat* represents the values at saturation, and b is a constant (depending on soil type) that represents the slope of the retention curve on a logarithmic graph. Values of these coefficients have been determined for the main textural classes based on the sand, silt, and clay content of soils [11]. Diffusion equations in the soil can be solved numerically by discretization in a number of soil layers or by methods similar to the force-restore method.

4. Vegetation

4.1. ROUGHNESS LENGTH

The presence of a vegetation cover brings a deep modification to the land-surface properties. As plants protrude from the ground they interfere with the atmospheric flow in the turbulent surface boundary layer. This effect is generally taken into account by modifying the roughness lengths for momentum (z_0) and heat (z_H) as a function of the vegetation height. A frequently used empirical rule is to specify z_0 as a constant fraction (about 10%) of vegetation height.

4.2. ALBEDO

Vegetation also interferes with the absorption of solar radiation. The chemical processes of photosynthesis, which are the basis for plant metabolism, are powered by the absorption of a fraction of incident solar radiation known as photosynthetically active radiation (PAR). The solar energy absorbed by photosynthesis is only a small fraction of insolation and is usually not considered explicitly in the radiation balance. Usually a vegetation cover has a lower albedo than individual leaves. This is due to the fact that the total surface area of the leaves over a given plot of land can be much larger than the area of the land itself, since the leaves are spread vertically in multiple layers and can have different orientations. Then solar radiation penetrating a leaf cover can undergo absorption, transmission and diffusion on several leaf layers. Such multiple diffusions increase the efficiency of the trapping of radiation inside a vegetation cover. In fact a vegetation cover acts like a cloud of leaves, and radiation decreases gradually by being absorbed by the upper layers of leaves as the distance from the top of the canopy increases. Insolation at different heights inside the vegetation cover is an important quantity for models that account explicitly for leaf photosynthesis (carbon cycle models), and can be computed by applying theories of radiation transfer in semi-transparent media. Some of the more sophisticated vegetation models like SiB [81] compute the variations of albedo due to changes in the leaf properties during the diurnal cycle.

4.3. INTERCEPTION

Besides its modification of the radiative and turbulent characteristics of the surface, vegetation has also a deep influence on its hydrologic properties. It is common sense to look for shelter below trees at the beginning of a short shower. The leaves of the tree effectively intercept the raindrops and slow the downward progression of rainwater. Some of the intercepted water remains on the leaves held by surface tension at the leaf surface, while the excess drips down and can either be intercepted by lower leaves or reach the ground. Part of the water can also flow along the branches or the stem and reach the ground as "stemflow." The part of rainfall that manages to cross the vegetation layer is referred to as "throughflow." In this way the vegetation acts like a screen that retains part of the rainfall at the surface of the leaves. Conversely the foliage can act as an efficient substratum for dew formation. This intercepted water (rainfall or dew) that remains at the surface of the leaves can evaporate very efficiently (after the end of the rainfall episode, or after sunrise) because it is spread out over a large well-ventilated surface. The part of rainfall intercepted by vegetation is thus returned directly into the atmosphere, usually rapidly, at the potential evaporation rate without entering the water balance of the ground nor producing runoff, and is referred to as "interception loss." This interception loss can be very important in the case of forests because of their high value of roughness length and the large surface area of their foliage [36].

It has been found suitable to model interception by vegetation in terms of an interception reservoir W_r representing the quantity of liquid water on the surface of the foliage [12]. The capacity of water storage of the vegetation layer is limited both by the surface area of the leaves and by the equilibrium between surface tension, which tends to retain water at the surface of the leaves, and gravity, which tends to attract water downwards. This limit is usually conceptualized as the maximum capacity W_{rmax} of the interception reservoir. As it depends on the total leaf surface area, it is natural to make the maximum canopy storage capacity a function of the leaf area index (LAI) and of the maximum thickness h_{rmax} of the water film that can be retained on each leaf. A typical parameterization proposed by Dickinson [15] is

$$W_{rmax} = h_{rmax} \, LAI$$

with $h_{rmax} = 0.2$ mm of water. The coefficient h_{rmax} could depend on vegetation type since different leaf structures (geometry, shape, texture) may have different capacities for retaining water on their surface. Manzi and Planton [50] on a basis of a calibration of the ISBA scheme from the ARME dataset for the Amazonian forest proposed $h_{rmax} = 0.1$ mm as more appropriate for tropical forests. The interception reservoir (or canopy storage) is classically modelled as a simple bucket with maximum storage capacity W_{rmax} assuming that evaporation is at potential rate on the wet fraction δ of the canopy. The parameterization of Deardorff [12]

$$\delta = (W / W_{rmax})^{2/3} \tag{26}$$

has been used in several subsequent models. Viterbo and Beljaars [88] report that a simple linear relation has been found to provide good results in the ECMWF model, and that due to its small capacity the interception reservoir can evaporate within a single time step of the model. Mahfouf et al. [44] prefer to use a simple linear relation to slow down the drying of the foliage, but they introduce a subgrid fractional distribution of convective precipitation that effectively reduces the quantity of intercepted rainfall [24]. The above bucket model does not represent the process of dripping of leaves which can extend for a while after a rainfall episode. This process, which represents a drainage term (a leak in the bucket), has been taken into account in the Rutter model [77], which has been used in several recent studies [83] [28]. Scott et al. [80] have performed GCM simulations using a complex SVAT parameterization with and without the interception reservoir and have compared them with simulations using the Manabe bucket scheme. Their conclusion is that most of the differences between the "SVAT" and "bucket" climates is due to the effect of the interception reservoir, which reduces the residence time of surface moisture by rapidly recycling the intercepted water, thus decreasing the excessive persistence of hydrological anomalies produced by the bucket scheme.

4.4. TRANSPIRATION

Though less visible than the canopy, which can capture water from the atmosphere, the root system of plants has the important function of extracting water from the soil to satisfy the plants' physiological needs. The water extracted from the soil is carried upward as sap flowing in the vascular system of the plants to bring water and nutrients to their cells, and particularly to the most active cells located in the leaves. The water inside the leaves is rather well insulated from the outside air by a cuticular membrane. However, plants need to take from the atmosphere the carbon dioxide that they use in photosynthetic production of new biomass, and thus plants need to exchange air with the atmosphere. This exchange takes place at the surface of mesophyll cells inside small vesicles, the opening of which is controlled by the stomatal cells. The stomata open and close in order to regulate the uptake of CO_2 as a function of plant metabolism depending on CO_2 concentration, light intensity, and leaf water stress. As the air inside the substomatal cavities is in contact with the cells of the plant it becomes enriched in water vapor relative to outside air and depleted in CO_2 fixed by photosynthesis. When the stomata open to replenish the CO_2 concentration this air is expelled and takes away water vapor, producing a flux of water vapor called "transpiration." This transpiration flux is a byproduct of plant metabolism, a kind of tax that the plant has to pay to the atmosphere in exchange for new CO_2. The transpiration flux may, however, have some beneficial effects as a thermal regulating mechanism. As the fluxes of water and CO_2 are in opposite directions and both regulated by stomatal opening, photosynthesis and transpiration are closely associated and controlled by physiological mechanisms.

Since the exchange of air is tightly controlled by the stomata, the transpiration rate of the plant is lower than the potential rate. The most prevalent method used to formulate the reduction of transpiration from the potential evaporation rate E_p is to represent it in the resistance formalism by means of an additional resistance to the transfer of water vapor, which is usually called a "stomatal resistance." Though it is not a standard denomination we prefer to call it loosely "vegetation resistance" r_v so as to have a comprehensive term including, if needed, vegetative controls besides purely stomatal ones. With this definition the transpiration E_{tr} by unit area of transpiring vegetation is

$$E_{tr} = \frac{r_{av}}{r_{av} + r_v} E_p(T_v), \qquad (27)$$

where $E_p(T_v)$ is potential evaporation computed at the skin temperature T_v of the vegetation canopy, and r_{av} the aerodynamic resistance appropriate for the vegetation layer. According to the parameterization schemes, the overall resistance

r_v can be composed of various subcomponent resistances representing different kinds of processes operating in series or in parallel, and which can be conveniently summarized by electric network diagrams. Though the main component is the stomatal resistance r_s, some schemes, such as SECHIBA [21], use an additional architectural resistance r_0 to represent the aerodynamic resistance between the leaves and the canopy top, as an attempt to take into account the variations of the environment of the leaves at different levels of the canopy. Values of the architectural resistance can vary with the vegetation type from 2 sm^{-1} for grassland, up to 50 sm^{-1} for evergreen forest [21].

The stomatal conductance $g_s = 1/r_s$ of the canopy results from the operation in parallel of all the stomata at the surface of the leaves, and is thus the sum of the individual conductance of each stoma. This makes it proportional to the LAI (total surface area of the leaves) and the number of stomata per unit surface area of foliage, which may depend on vegetation type. The conductance of an individual stoma depends on the rate of stomatal opening, which can vary with the plant metabolism as a function of environmental factors. An empirical fit of the stomatal conductance of illuminated leaves as a function of temperature, vapor pressure deficit, leaf water potential, CO_2 level, and photosynthetically active radiation has been calibrated by Jarvis [37] from field observations. The basis of his parameterization is to assume that these factors (normalized to vary between 0 and 1) act independently in a multiplicative fashion to reduce the optimal stomatal resistance when they deviate from the optimal condition for plant photosynthesis. The stomatal conductance can be written

$$g_s = LAI \, g_{smax} \, k_s(Q_p) \, k_s(T) \, k_s(\delta e) \, k_s(\psi_1) \, k_s(C_a), \qquad (28)$$

where $g_{smax} = 1/r_{smin}$ is the maximal stomatal conductance per unit surface area of leaves (its inverse r_{smin} being the minimal stomatal resistance) and k_s are different empirical dimensionless functions ($0 \leq k_s \leq 1$) giving the reduction of the conductance as a function of photon flux density Q_p, leaf temperature T, water vapor deficit of the air δe, leaf water potential ψ_1 and carbon dioxide concentration C_a. Variants of the Jarvis type parameterization have been adapted and used in several SVATs. Avissar [2] has reviewed the problem of the generalization to large scales of such an empirical relationship established from measurements at leaf-scale. His conclusion was that the large variability of conductances observed in a homogeneous plant canopy cast serious doubts on the usefulness of such a leaf-scale relationship at larger scales, and that a parameterization using external environmental forcing conditions would be more appropriate at the canopy scale.

5. Snow Cover

The land-surface properties are strongly modified in the presence of a snow cover. The most visible modification is the change in the surface albedo. The bright white color of fresh snow is a clear manifestation of its high reflectance, which may exceed 90% for clean snow. The presence of dust or various impurities can lower the albedo, and though the albedo of snow cover decreases as the snow gets older it remains much higher than the albedo of the underlying ground or vegetation surface. As the snowflakes are very light they are easily displaced by the fluctuations of the wind close to the ground and tend to accumulate preferentially in the hollow parts of the ground surface, thus reducing its roughness. The accumulation of snow can store large quantities of water in solid state above the ground surface. When the energy balance of the snow cover becomes positive, the snow melts and releases liquid water to the land surface. In this way snow cover acts like a thermally controlled storage reservoir for water and is thus a very important component for the hydrology of the land surface.

In the canonical GFDL model [46], the snow cover was simply represented as an additional storage reservoir W_n. When air temperature is below 0°C, precipitation is assumed to occur as snowfall and to accumulate in this reservoir. The snow storage can be depleted by sublimation E_n at its surface, or by melting P_m when the surface temperature of the snow cover T_n rises above the melting point of ice T_m. The heat content Q_n of the snowmass depends on the absorbed net radiation R_N, the heat loss by conduction at the ground surface G_n, the latent heat of fusion (L_m) and of sublimation ($L + L_m$). The water and energy budgets of the snow cover can be written

$$dW_n/dt = P_n - E_n - P_m \tag{29}$$

$$dQ_n/dt = R_N - G_n - H - (L + L_m)E_n - L_m P_m. \tag{30}$$

The sublimation of snow can be computed from the bulk aerodynamic formula by assuming that the air close to the snow surface is at the saturation humidity of ice $q_{ice}(T_n)$. In the parameterization schemes, melting is usually formulated in terms of a temperature adjustment performed at each time step: if the temperature of the snow cover becomes larger than T_m, an amount of snow is melted to bring down the temperature to T_m. This can be written in an equivalent differential form by the following reasoning: when snow is melting, its surface temperature is kept at the melting point, and thus the energy budget at the snow surface in the absence of melting is

$$B_m = (1 - \alpha_m) R_S - \epsilon_n (R_A - \sigma_{SB} T_m^4) - G_n$$

$$-\frac{\rho_a}{r_{an}} [C_p(T_m - T_a) - gz_a + (L + L_m)(q_{ice}(T_m) - q_a)].$$

In this expression, α_m represents the albedo of melting snow, which can be set at a lower value than the albedo of the snow cover; and r_{an} is the aerodynamic resistance over the snow cover computed with the roughness length z_n of the snow cover, which in general is smaller than the roughness length of the snow-free land-surface. This energy budget can thus melt snow at the rate B_m/L_m. When the energy budget B_m is negative, snow cannot melt. The general expression for snowmelt rate is thus

$$P_m = \max(0, B_m/L_m). \tag{31}$$

The above expressions assume that the meltwater percolates immediately downwards and is not available for evaporation. The most variable term in the heat budget B_m is the insolation term R_s. Snowmelt occurs only when insolation is above a threshold value given by the other terms, and thus is restricted to a limited number of hours in the daylight according to the season.

Snow on the ground undergoes a series of transformations that lead to a gradual increase of its density ρ_n and a decrease of its albedo α_n. These effects are usually taken care of by empirical relationships in the snow cover parameterizations used in GCMs. In the Canadian land surface scheme CLASS [87] the snow albedo decreases linearly from 0.75 to 0.57 as a linear function of an aging factor, and the density increases linearly as a function of snowmass from 189 kg/m^3 up to an upper limit of 450 kg/m^3. Slightly different expression have been used, after calibration with observed data, in the ISBA new snow parameterization scheme [20]: snow density increases in time with an e-folding time of about 4 days from a value $\rho_{min} = 100$kg/m^3 for freshly fallen snow to $\rho_{max} = 300$kg/m^3 for aged snow, and albedo decreases from 0.85 to 0.50, with an exponential decrease during melting or a weak linear decrease during colder days.

Because a thin layer of snow usually will not cover the ground completely due to various protruding features, the fractional area of the ground covered by snow p_{ng} is usually made dependent on snow mass W_n through some empirical relationship. In SiB [81] (and in NCAR CCM3) p_{ng} increases linearly with snow mass reaching full coverage with a snow depth of 5 cm. The fractional coverage of vegetation p_{nv} should depend on vegetation type. The following expression has been used in several models:

$$p_{nv} = h_n/(h_n + k_{nv}z_{0v}) \tag{32}$$

where h_n is the average snow depth, z_{0v} the momentum roughness length of vegetation, and k_{nv} a dimensionless coefficient (taken as 5 in ISBA [20] and 10 in BATS [93]).

6. Soil Hydrology

As runoff is only a small component in the water balance, it is of marginal interest for the computation of surface energy and water budgets, and it has formerly not received much attention in the land-surface parameterization. Most of the land-surface schemes do not distinguish between surface runoff and leakage to underlying aquifers. However, the importance of runoff as an input for ocean models or for water resources management has been recognized and has spurred interest on how surface runoff processes might be better represented in land-surface models. The practical importance of flood forecasting and warning has long motivated hydrologists to study how the flow of streams and rivers responds to storm rainfall over their catchment basins [92]. The rise and fall of a river following a storm is referred to as the storm hydrograph and is due to the part of the incident rainfall that is not absorbed into the soil (the effective rainfall). Runoff can occur through several mechanisms [3] and may depend on the antecedent moisture condition, the hydraulic conductivity of soils in the catchment, and the storm rainfall intensities. The concept that storm response is due primarily to overland flow when rainfall intensity exceeds infiltration capacity of the soil over all the basin was developed by Horton. This mechanism, referred to as "infiltration excess runoff" or Hortonian runoff, may be important in areas of low vegetation cover and high rainfall intensity, but is not predominant in vegetated humid temperate areas where the infiltration capacities are high compared to rainfall intensities [3]. In these areas runoff can happen from rainfall falling on areas of soil that are saturated by a rising water table; this kind of runoff is sometimes called saturation excess runoff.

Conceptual rainfall-runoff models of various complexities have long been developed and used by hydrologists to estimate runoff from meteorological data on catchments of different sizes, but for hydrologists, surface energy balance is of marginal interest, and parameters have to be calibrated with hydrologic and meteorological data for each individual catchment. Chiew *et al.* [8] explored the similarities and differences between the two modeling approaches, the energy balance approach used in GCM parameterizations and rainfall-runoff models used in hydrology. They compared in detail, for two Australian catchments, a rainfall-runoff model used in hydrology and a land-surface parameterization scheme (BEST). They concluded that the land surface schemes used in GCMs can estimate runoff reasonably well for wet catchments, but are too simplified to model runoff adequately in ephemeral catchments, where more sophisticated algorithms representing river recharge, leakage, and baseflow are required. Hydrologists conceptualize a catchment as a number of interconnected storages with functions describing the movement of water into, between, and out of them.

6.1. SPATIAL VARIABILITY OF PRECIPITATION

Convective precipitation falls only over a fraction of the grid cell, while stratiform precipitation is distributed over the whole area. Some studies have focused on subgrid scale spatial variability in precipitation, treating precipitation as a random variable in space with a specified statistical distribution over a fraction of a grid cell [18] [39] [25]. Runoff is then generated only from the fraction of the grid cell where precipitation exceeds infiltration. The combination of mutual spatial variability in precipitation and land surface characteristics was considered by Entekhabi and Eagleson [27], who used a two-parameter gamma probability density function to represent the soil moisture distribution within a grid cell, and by Famiglietti and Wood [30], who used independent statistical distributions. Liang *et al.* [41] generalized the previous approaches by considering also the effects of spatial correlations of precipitation in the framework of a statistical model with only one dimension in the horizontal to represent the subgrid spatial variability of rainfall and soil moisture. The results indicate that the precipitation coverage is more important than the statistical distribution of its intensity (whether the precipitation is brief and heavy or prolonged and light), at least when the spatial variability of soil properties is taken into account as is the case for infiltration in the VIC-2L model. Their conclusion is that a model with subgrid variability of soil infiltration is adequate provided that the fractional coverage of precipitation is specified, and they pointed out the importance of incorporating also the subgrid variability of canopy interception and storage capacity.

Rainfall rates may be large enough to exceed the infiltration capacity of the soil and produce "infiltration excess runoff." An approximate analytical solution of the nonlinear Richards equation has been used by Wood *et al.* [92]. They took into account the time needed for rainfall to bring the soil surface just to saturation, called the time to ponding t_p.

6.2. VARIATIONS IN THE INFILTRATION CAPACITY OF SOIL

The single-layer version of the variable infiltration capacity (VIC) water balance model of Wood *et al.* [91] incorporates a spatial distribution of soils of varying infiltration capacities to partition precipitation into infiltration and runoff. The distribution of storage elements of various capacities is represented by a simple statistical probability distribution of infiltration capacity (maximum depth of water that can be stored in a soil column) over a grid scale with a single shape parameter b_i:

$$i = i_m[1 - (1 - A)^{1/b_i}] \tag{33}$$

i is infiltration, A is the fraction of the area where infiltration is less than i, b_i is a shape parameter of the distribution. The maximum soil water storage is $W_c = i_m/(1 + b_i)$. Direct runoff is the amount of precipitation P that falls on the fraction of the grid where the soil reservoir is filled.

A similar scheme, with a slightly different formulation of the coefficients, has been implemented and tested in the Max Planck Institute GCM by Dümenil and Todini [22]. Another version of this scheme was used within a one-dimensional version of the UKMO GCM by Rowntree and Lean [75].

The VIC scheme of Wood *et al.* [91] has been implemented without modification in the GFDL R15 GCM by Stamm *et al.* [84], who compared it with the bucket hydrology model and performed sensitivity studies to various geographical distributions of the parameters W_c and b_i. In a first experiment with W_c fixed at the same value (15 cm) as in the bucket simulation and $b_i = 0.3$, increased runoff was obtained in the unglaciated continents except North America leading to lower soil moisture (0.33) than in the bucket model (0.53), and to lower evaporation and precipitation. Other simulations were performed with smaller (5 cm) and higher (25 cm) soil moisture capacity, and with a geographically variable W_c and b_i parameters determined from a $0.5° \times 0.5°$ dataset on the holding capacity of soil moisture (published later by Dunne and Willmott [23]). The large midlatitude continents (Eurasia and North America) were more sensitive to altered land hydrology, as recycling of summer precipitation is a more important source of moisture. In their conclusion they suggested that a two-layer soil model is needed to better represent the recycling of moisture during dry periods. In summer, deep soil moisture is disconnected from the top layer soil moisture, which is more readily available for evaporation following summer rain. The model has been generalized to include two soil layers, with a computation of gravitational drainage from the upper layer to the lower layer, and spatially varying vegetation with several land cover types [42]. A nonlinear drainage was introduced as a function of the water content of the lower layer to represent cases with subsurface storm flow. Evaporation can occur at the potential rate over the saturated area A_s. Over the remaining unsaturated area Liang *et al.* [42] consider that it occurs at a rate proportional to the local wetness fraction i_0/i and approximate the resulting integral by a power series. Further developments have been made recently to incorporate the spatial variability of precipitation [41].

The VIC-2 model has been validated over the Arkansas-Red River basins by Abdulla and Lettenmaier [1] who have used a geographical database on vegetation and soil data, and a spatial interpolation from 44 calibration catchments for estimation of the model parameters on the 61 $1° \times 1°$ grid cells covering the study area. The model evaporation was shown to be in good agreement with evaporation derived from an atmospheric moisture budget. Using a simple routing scheme to represent the drainage network, the simulated hydrographs were in good agreement with the observed hydrographs, and long-term mean streamflows show an agreement within 1–2% of observations, though the seasonal extremes were underestimated [1].

6.3. TOPOGRAPHIC CONTROL OF INFILTRATION

Topography plays a major role in partitioning discharge into surface (runoff) or underground flow of water (baseflow), and was not explicitly included in the previous approaches. In order to improve the simulation of the hydrologic cycle, an approach viewing the fundamental unit as the watershed rather than the single soil column has been proposed recently by Stieglitz *et al.* [85]. Saturated regions occur mainly in valleys and lowlands where the water table rises to the surface. The flow resulting from precipitation over a saturated region, known to hydrologists as Dunne runoff or saturation excess runoff, accounts for the majority of surface runoff, while runoff resulting from infiltration excess over impermeable soil (Hortonian runoff), is only of secondary importance. Thus topography is a major factor controlling the distribution of soil moisture, in that lowland regions are zones of convergent flow and may be replenished by downslope drainage and thus have high moisture content, while uplands tend to be drier. The expansion and contraction of lowland saturated zones are determined by downslope redistribution of recharged subsurface soil water. Beven and Kirby [3] developed a model (TOPMODEL) making use of a spatially variable topographic index and a theory for local hydrological similarity whereby all locations with the same value of the topographic index are assumed to respond in a similar manner to identical inputs. Calculation of the topographic index can be computed from digital elevation models. The TOPMODEL equations have been incorporated in a single column framework to parameterize downslope redistribution of soil water and to predict the saturated fraction and baseflow that supports the saturation [29]. The basic parameters are the mean watershed water table depth and a probability distribution function (pdf) of topographic and soil properties. For shallow soils the water table is nearly parallel to the soil surface. The local hydraulic gradient is close to the local slope angle $\tan \beta$. Assuming an exponential decrease with depth of saturated hydraulic conductivity $K_s = K_s(0)e^{-z/z_d}$ with a decay rate $1/z_d$ assumed constant in the catchment and assuming that the water table is recharged at a spatially uniform and constant rate, this leads to an analytic relation between the mean water table $\bar{z}$ and local water table depth z' approximated as

$$z' = \bar{z} - z_d[\chi - \Lambda] \qquad (34)$$

where $\chi = \ln(a/\tan \beta)$ is the topographic index derived from digital terrain analysis, a is the upslope area that can potentially flow through this location, β is the local slope angle used to represent the hydraulic gradient at the water table (measuring how fast water can be transported downslope), and Λ is the grid average of χ. All points with similar value of the topographic index χ are considered hydrologically similar. According to the theory of TOPMODEL, the soil moisture deficit or table depth is proportional to χ. All regions with χ greater than $\Lambda + \bar{z}/z_d$ are in saturated regions. The baseflow is

$$Q_b = K_s(0)z_d e^{-(\Lambda+\bar{z}/z_d)}. \qquad (35)$$

The depth of the water table can be used to estimate the water available in the root zone. A variable distribution of the root zone depth could even be considered [69]. A simplification for the calculation of fluxes is based on a division of the watershed into saturated and unsaturated zones for the computation of evapotranspiration, surface fluxes, and surface runoff.

6.4. RIVER ROUTING SCHEMES

The return flow of water from rivers to the oceans is important in coupled atmosphere-ocean models, since the freshwater flux has an impact on salinity and stratification of coastal ocean surface waters. River flow data offer the possibility of validating the overall surface hydrology of GCMs, and global runoff data have been produced. Drainage basins for the rivers were extracted from world maps and defined at resolution 2° × 2.5° in the studies of Russell and Miller [76] and Miller and Russell [53]. However for a detailed validation of the seasonal cycle of surface hydrology, as well as for realistic coupling with the ocean, the time lag introduced by the rivers cannot be neglected and river routing schemes are needed. River routing schemes have recently been developed and implemented in GCMs [78] [33]. A validation of river routing schemes at a regional scale has been made to validate the aggregation method at the larger scales of a GCM [32].

7. Sensitivity Experiments

Many sensitivity experiments have demonstrated the importance of land-surface processes ([89], [71], [74], [55]). Shukla and Mintz [82] illustrated the strong dependence of precipitation, surface temperature and pressure on land-surface evaporation by performing two simulations: one in which the land-surface evaporation was set equal to the potential evaporation (wet-soil case) and another without any evaporation on the continents (dry-soil case). In the dry-soil case the precipitation was severely reduced over the continents and surface temperatures were about 15 to 25°C warmer due to reduced cloudiness, which in turn reduces evaporative cooling and increases solar radiation. Strong thermal lows, with surface pressure 5 to 15 hPa lower than the wet-soil case, developed over the land. They concluded that the storage of precipitation during winter and spring was an important determinant of the subsequent summer rainfall, and that observations of the soil moisture are a necessity for the prediction of precipitation. Delworth and Manabe ([14], [13]) showed that an interactive soil hydrology can contribute to the simulated climate variability by increasing the timescales associated with surface humidity and temperature fluctuations. Oglesby and Erickson [61] used NCAR CCM1 to investigate the effect of imposed soil moisture anomalies on North American summertime climate, and their simulation supported the hypothesis that reduced soil moisture during late winter and spring over a midcontinental region could induce a dry and hot summer anomaly. Intercomparison of 20 land surface process models in PILPS [67] with the same atmospheric forcing has shown large differences in surface fluxes related to different hydrologic parameterizations of infiltration and runoff schemes [5] [6]. The question of the impact of large-scale deforestation on climate has been considered in several publications. The potential

346

climate influence of snow cover anomalies has also been the object of numerous investigations [19].

The role of land-surface processes in climate scenarios has been considered [52]. Midlatitude summertime droughts were found to occur in simulations with the GFDL model ([47], [49], [48]). Similar responses were found with the NCAR CCM1 with precipitable water, evaporation, and precipitation increasing with increasing CO_2 while snow decreases. Soil moisture changes display a regional and seasonal dependence with generally an increase during the winter in midlatitudes and decrease during the summer [72]. There are, however, some uncertainties due to the excessive soil moisture at the end of winter in the control simulation preventing storage of the increased winter precipitation in these simulations. River runoff has been suggested as a useful diagnostic. Miller and Russell [53] used GISS GCM simulations to compute model-generated runoff of the world's major rivers for the present and doubled CO_2 climate, in which global air surface temperature increased by 4.2°C; they found an increase of runoff for 25 of the 33 major rivers examined. The runoff increases for all rivers in high latitudes by about 25% due to increased precipitation, while the largest decreases in runoff occur at low latitudes due to combined effects of increased evapotranspiration and decreased precipitation. Interactive vegetation models (biome models) have been developed to study the potential response of vegetation to climate change and their potential feedback on climate.

8. Conclusion

The land surface is an important boundary condition that influences the fluxes of radiation, momentum, heat and water exchanged with the atmosphere. An essential task of land-surface parameterization schemes is to provide accurate precipitation, evaporation and soil moisture storage which are needed to properly represent the hydrologic cycle. Land-surface parameterization schemes have been developed to represent the mutually dependent water and energy budgets at the surface, in particular the processes of water storage in the ground and its transfer to the atmosphere by evaporation, or transpiration of a vegetation cover. The vegetation canopy can intercept a significant amount of rainfall and evaporate it at a potential rate, thus reducing surface runoff. Numerous sensitivity experiments have confirmed the mutual interactions between climate and land-surface processes. The issue of the level of complexity needed for the land-surface representation in GCMs is still unresolved. Most parameterizations in current use in GCMs deal only with the vertical movements of moisture into and out of the soil and ignore groundwater flow. New parameterizations are being developed to represent the effect of surface heterogeneities and topographic variability on the production of runoff. The fate of runoff water starts to be monitored more carefully in river routing schemes, which are needed to balance the global water cycle. International programs such as GCIP (Gewex Continental Scale International Project) are trying to define a modeling strategy for the development and testing of macroscale hydrological models for modeling the water and energy budget on large-scale areas (10^5 to 10^6 km^2), and comparison with water budget estimates derived from atmospheric budget analyses for determination of space-time variability of water and energy. Improving the reliability of land surface and hydrologic parameterizations and their interactions with the atmospheric hydrological cycle will be a crucial and vital step for understanding and forecasting the impacts of climate change, particularly for future climate change scenarios which will have to consider large-scale changes in land cover due to man-made or natural vegetation.

Acknowledgments: The author thanks Sebastien Biner, Stephen Graham, Bernard Nemry and Sylwia Trzaska for their comments that helped to improve the manuscript, and Krista A. Dunne, Jean-Paul Goutorbe and Stephane Goyette for providing useful material for preparing this lecture.

References

1. Abdulla, F.A., Lettenmaier, D.P., Wood, E.F., and Smith, J.A. (1996) Application of a macroscale hydrologic model to estimate the water balance of the Arkansas-Red River basin, *J. Geophys. Res.* **101**, 7449–7459.
2. Avissar, R. (1993) Observations of leaf stomatal conductance at the canopy scale: An atmospheric modeling perspective, *Bound.-Layer Meteor.* **64**, 127–148.
3. Beven, K.J. and Kirkby, M.J. (1979) A physically based, variable contributing area model of basin hydrology, *Hydrological Sci. Bull.* **24**, 43–69.
4. Bhumralkar, C.M. (1975) Numerical experiments on the computation of ground surface temperature in an atmospheric general circulation model, *J. Appl. Meteor.* **14**, 1246–1258.
5. Bonan, G.B. (1994) Comparison of two land surface process models using prescribed forcings, *J. Geophys. Res.* **99**, 25,803–25,818.
6. Bonan, G.B. (1996) Sensitivity of a GCM simulation to subgrid infiltration and surface runoff, *Climate Dyn.* **12**, 279–285.
7. Brubaker, K.L., Entekhabi, D., and Eagleson, P.S. (1993) Estimation of continental precipitation recycling, *J. Clim.* **6**, 1077–1089.
8. Chiew, F.H.S., Pitman, A.J., and McMahon, T.A. (1996) Conceptual catchment scale rainfall-runoff models and AGCM land- surface parameterisation schemes, *J. Hydrology* **179**, 137–157.
9. Choudhury, B.J. (1997) Global pattern of potential evaporation calculated from the Penman-Monteith equation using satellite and assimilated data, *Remote Sens. Environ.* **61**, 64–81.
10. Clapp, R.B. and Hornberger, G.M. (1978) Empirical equations for some hydraulic properties, *Water Resour. Res.* **14**, 601–604.
11. Cosby, B.J., Hornberger, G.M., Clapp, R.B., and Ginn, T.R. (1984) A statistical exploitation of the relationships of soil moisture characteristics to the physical properties of soils, *Water Resour. Res.* **20**, 682–690.

12. Deardorff, J.W. (1978) Efficient prediction of ground surface temperature and moisture, with inclusion of a layer of vegetation, *J. Geophy. Res.* **83**, 1889–1903.

13. Delworth, T. and Manabe, S. (1989) The influence of soil wetness on near-surface atmospheric variability, *J. Clim.* **2**, 1447–1462.

14. Delworth, T.L. and Manabe, S. (1988) The influence of potential evaporation on the variabilities of simulated soil wetness and climate, *J. Clim.* **1**, 523–547.

15. Dickinson, R.E. (1984) Modeling evapotranspiration for three-dimensional global climate models, In J.E. Hansen and T. Takahashi (eds.), *Climate Processes and Climate Sensitivity* American Geophysical Union, Washington, pp. 58–72.

16. Dickinson, R.E. (1988) The force-restore model for surface temperatures and its generalizations, *J. Clim.* **1**, 1086–1097.

17. Dickinson, R.E., Henderson-Sellers, A., Kennedy, P.J., and Wilson, M.F. (1986) *Surface Physics Parameterization Package for the NCAR Community Climate Model*, NCAR Technical Note, National Center for Atmospheric Research, Boulder.

18. Dolman, A.J. and Gregory, D. (1992) The parametrization of rainfall interception in GCMs, *Quart. J. Roy. Meteor. Soc.* **118**, 455–467.

19. Douville, H. and Royer, J.F. (1996) Influence of the temperate and boreal forests on the Northern Hemisphere climate in the Meteo-France climate model, *Climate Dyn.* **13**, 57–74.

20. Douville, H., Royer, J.F., and Mahfouf, J.F. (1995) A new snow parametrization for the Météo-France climate model. Part I: Validation in stand-alone experiments, *Climate Dyn.* **12**, 21–35.

21. Ducoudré, N.I., Laval, K., and Perrier, A. (1993) SECHIBA, a new set of parameterizations of the hydrologic exchanges at the land-atmosphere interface within the LMD atmospheric general circulation model, *J. Clim.* **6**, 248–273.

22. Dümenil, L. and Todini, E. (1992) A rainfall-runoff scheme for use in the Hamburg climate model, In J.P. O'Kane (ed.), *Advances in Theoretical Hydrology, a Tribute to James Dooge*, Elsevier, Amsterdam, pp. 129–157.

23. Dunne, K.A. and Willmott, C.J. (1996) Global distribution of plant-extractable water capacity of soil, *Int. J. Climatol.* **16**, 841–859.

24. Eltahir, E.A.B. and Bras, R.L. (1993) A description of rainfall interception over large areas, *J. Clim.* **6**, 1002–1008.

25. Eltahir, E.A.B. and Bras, R.L. (1993) Estimation of the fractional coverage of rainfall in climate models, *J. Clim.* **6**, 639–644.

26. Eltahir, E.A.B. and Bras, R.L. (1996) Precipitation recycling, *Rev. Geophys.* **34**, 367–378.

27. Entekhabi, D. and Eagleson, P.S. (1989) Land surface hydrology parameterization for atmospheric general circulation models including subgrid scale spatial variability, *J. Clim.* **2**, 816–831.

28. Entekhabi, D., Rodriguez-Iturbe, I., and Bras, R.L. (1992) Variability in large-scale water balance with land surface-atmosphere interaction, *J. Clim.* **5**, 798–813.

29. Famiglietti, J.S. and Wood, E.F. (1994) Multiscale modeling of spatially variable water and energy balance processes, *Water Resour. Res.* **30**, 3061–3078.

30. Famiglietti, J.S., Wood, E.F., Sivapalan, M., and Thongs, D.J. (1992) A catchment scale water balance model for FIFE, *J. Geophys. Res.* **97**, 18997–19008.

31. Garratt, J.R. (1992) Extreme maximum land surface temperatures, *J. Appl. Meteor.* **31**, 1096–1105.

32. Habets, F. (1998) *Modélisation du cycle continental de l'eau à l'échelle régionale. Application aux bassins versants de l'Adour et du Rhône*, Thèse de Doctorat de l'Université Paul Sabatier, Toulouse III, France.

33. Hagemann, S. and Dümenil, L. (1998) A parameterization of the lateral waterflow for the global scale, *Climate Dyn.* **14**, 17–31.

34. Hansen, J., Russel, G., Rind, D., Stone, P., Lacis, A., Lebedeff, S., Ruedy, R., and Travis, L. (1983) Efficient three-dimensional global models for climate studies: Models I and II, *Mon. Wea. Rev.* **111**, 609–662.

35. Hillel, D. (1982) *Introduction to Soil Physics*, Academic Press, New York.

36. Horton, R.E. (1919) Rainfall interception, *Mon. Wea. Rev.* **47**, 603–623.

37. Jarvis, P.G. (1976) The interpretation of the variations in leaf water potential and stomatal conductance found in canopies in the field, *Phil. Trans. Roy. Soc. London B* **273**, 593–610.

38. Johnson, K.D., Entekhabi, D., and Eagleson, P.S. (1993) The implementation and validation of improved land-surface hydrology in an atmospheric general circulation model, *J. Clim.* **6**, 1009–1026.

39. Jones, P.G. and Thornton, P.K. (1993) A rainfall generator for agricultural applications in the tropics, *Agric. Forest Meteor.* **63**, 1–19.

40. Lee, T.J. and Pielke, R.A. (1992) Estimating the soil surface specific humidity, *J. Appl. Meteor.* **31**, 480–484.

41. Liang, X., Lettenmaier, D.P., and Wood, E.F. (1996) One-dimensional statistical dynamic representation of subgrid spatial variability of precipitation in the two-layer variable infiltration capacity model, *J. Geophys. Res.* **101**, 21,403–21,422.

42. Liang, X., Lettenmaier, D.P., Wood, E.F., and Burges, S.J. (1994) A simple hydrologically based model of land surface water and energy fluxes for general circulation models, *J. Geophys. Res.* **99**, 14,415–14,428.

43. Louis, J.F. (1979) A parametric model of vertical eddy fluxes in the atmosphere, *Bound.-Layer Meteor.* **17**, 187–202.

44. Mahfouf, J.F., Manzi, A.O., Noilhan, J., Giordani, H., and Déqué, M., (1995) The land surface scheme ISBA within the Météo-France climate model ARPEGE. Part I: Implementation and preliminary results, *J. Clim.* **8**, 2039–2057.

45. Mahfouf, J.F. and Noilhan, J. (1991) Comparative study of various formulations of evaporation from bare soil using in situ data, *J. Appl. Meteor.* **30**, 1354–1365.

46. Manabe, S. (1969) Climate and the ocean circulation. I: The atmospheric circulation and the hydrology of the earth's surface, *Mon. Wea. Rev.* **97**, 739–774.

47. Manabe, S. and Stouffer, R.J. (1980) Sensivity of a global climate model to an increase of CO2 concentration in the atmosphere, *J. Geophy. Res.* **85**, 5529–5554.

48. Manabe, S. and Wetherald, R.T. (1987) Large-scale changes of soil wetness induced by an increase in atmospheric carbon dioxide, *J. Atmos. Sci.* **44**, 1211–1235.

49. Manabe, S., Wetherald, R.T., and Stouffer, R.J. (1981) Summer dryness due to an increase of atmospheric CO2 concentration, *Climatic Change* **3**, 347–385.

50. Manzi, A.O. and Planton, S. (1994) Implementation of the ISBA parametrization scheme for land surface processes in a GCM - An annual cycle experiment, *J. Hydrology* **155**, 353–387.

51. Matthews, E. (1983) Global vegetation and land use: New high-resolution data bases for climate studies, *J. Climate Appl. Meteor.* **22**, 474–487.

52. Meehl, G.A. and Washington, W.M. (1988) A comparison of soil-moisture sensitivity in two global climate models, *J. Atmos. Sci.* **45**, 1476–1492.

53. Miller, J.R. and Russell, G.L. (1992) The impact of global warming on river runoff, *J. Geophys. Res.* **97**, 2757–2764.

54. Milly, P.C.D. and Dunne, K.A. (1994) Sensitivity of the global water cycle to the water holding capacity of land, *J. Clim.* **7**, 506–526.

55. Mintz, Y. (1984) The sensitivity of numerically simulated climates to land-surface boundary conditions, In J. Houghton (ed.) *Global Climate*, Cambridge University Press, Cambridge, pp. 79–105.

56. Mintz, Y. and Serafini, Y.V. (1992) A global monthly climatology of soil moisture and water balance, *Climate Dyn.* **8**, 13–27.

57. Mintz, Y. and Walker, G.K. (1993) Global fields of soil moisture and land surface evapotranspiration derived from observed precipitation and surface air temperature, *J. Appl. Meteor.* **32**, 1305–1334.

58. Monteith, J.L. (1981) Evaporation and surface temperature, *Quart. J. Roy. Meteor. Soc.* **107**, 1–27.

59. Nappo, C.J. (1975) Parameterization of surface moisture and evaporation rate in a planetary boundary layer model, *J. Appl. Meteor.* **14**, 289–296.

60. Noilhan, J. and Planton, S. (1989) A simple parameterization of land surface processes for meteorological models, *Mon. Wea. Rev.* **117**, 536–549.

61. Oglesby, R.J. and Erickson, III, D.J. (1989) Soil moisture and the persistence of North American drought, *J. Clim.* **2**, 1362–1380.

62. Paw U, K.T. (1992) A discussion of the Penman form equations and comparisons of some equations to estimate latent energy flux density, *Agric. Forest Meteor.* **57**, 297–304.

63. Penman, H.L. (1948) Natural evaporation from open water, bare soil and grass, *Proc. Roy. Soc. London Ser. A - Math. Phys. Sci.* **193**, 120–146.

64. Philip, J.R. (1957) Evaporation, and moisture and heat fields in the soil, *J. Meteor.* **14**, 354–366.

65. Philip, J.R. (1989) The scattering analog for infiltration in porous media, *Rev. Geophys.* **27**, 431–448.

66. Pitman, A.J. (1991) A simple parameterization of sub-grid scale open water for climate models, *Climate Dyn.* **6**, 99–112.

67. Pitman, A.J., Yang, Z.L., and Henderson-Sellers, A. (1993) Sub-grid scale precipitation in AGCMs, Re-assessing the land surface sensitivity using a single column model, *Climate Dyn.* **9**, 33–42.

68. Preuss, H.J. and Geleyn, J.F. (1980) Surface albedos derived from satellite data and their impact on forecast models, *Arch. Meteor. Geophys. Bioclimatol.* **29**, 345–356.

69. Quinn, P., Beven, K., and Culf, A. (1995) The introduction of macroscale hydrological complexity into land surface-atmosphere transfer models and the effect on planetary boundary layer development, *J. Hydrology* **166**, 421–444.

70. Richardson, L.F. (1922) *Weather Prediction by Numerical Process*, University Press, Cambridge. (Reprint, 1965, with a new introduction by Sydney Chapman, Dover Publications, New York.)

71. Rind, D. (1982) The influence of ground moisture conditions in North America on summer climate as modeled in the GISS GCM, *Mon. Wea. Rev.* **110**, 1487–1494.

72. Roads, J.O., Marshall, S., Oglesby, R., and Chen, S.C. (1996) Sensitivity of the CCM1 hydrologic cycle to CO2, *J. Geophys. Res.* **101**, 7321–7339.

73. Robock, A., Vinnikov, K.Y., Schlosser, C.A., Speranskaya, N.A., and Xue, Y., (1995) Use of midlatitude soil moisture and meteorological observations to validate soil moisture simulations with biosphere and bucket models, *J. Clim.* **8**, 15–35.

74. Rowntree, P.R. and Bolton, J.A. (1983) Simulation of the atmospheric response to soil moisture anomalies over Europe, *Quart. J. Roy. Meteor. Soc.* **109**, 501–526.

75. Rowntree, P.R. and Lean, J. (1994) Validation of hydrological schemes for climate models against catchment data, *J. Hydrology* **155**, 301–324.

76. Russell, G.L. and Miller, J.R. (1990) Global river runoff calculated from a global atmospheric general circulation model, *J. Hydrology* **117**, 241–254.

77. Rutter, A.J., Kershaw, K.A., Robins, P.C., and Morton, A.J. (1971) A predictive model of rainfall interception in forests, I. Derivation of the model from observations in a plantation of Corsican pine, *Agric. Meteor.* **9**, 367–384.

78. Sausen, R., Schubert, S., and Dümenil, L. (1994) A model of river runoff for use in coupled atmosphere ocean models, *J. Hydrology* **155**, 337–352.

79. Schaake, J.C., Koren, V.I., Duan, Q.Y., Mitchell, K.E., and Chen, F. (1996) Simple water balance model for estimating runoff at different spatial and temporal scales, *J. Geophys. Res.* **101**, 7461–7475.

80. Scott, R., Koster, R.D., Entekhabi, D., and Suarez, M.J. (1995) Effect of a canopy interception reservoir on hydrological persistence in a general circulation model, *J. Clim.* **8**, 1917–1922.

81. Sellers, P.J., Mintz, Y., Sud, Y.C., and Dalcher, A. (1986) A Simple Biosphere model (SiB) for use within general circulation models, *J. Atmos. Sci.* **43**, 505–531.

82. Shukla, J. and Mintz, Y. (1982) Influence of land-surface evapotranspiration on the Earth's climate, *Science* **215**, 1498–1501.

83. Shuttleworth, W.J. (1988) Evaporation from Amazonian rainforest, *Proc. Roy. Soc. London Ser. B* **233**, 321–346.

84. Stamm, J.F., Wood, E.F., and Lettenmaier, D.P. (1994) Sensitivity of a GCM simulation of global climate to the representation of land surface hydrology, *J. Clim.* **7**, 1218–1239.

85. Stieglitz, M., Rind, D., Famiglietti, J., and Rosenzweig, C. (1997) An efficient approach to modeling the topographic control of surface hydrology for regional and global climate modeling, *J. Clim.* **10**, 118–137.

86. Sud, Y.C., Shukla, J., and Mintz, Y. (1988) Influence of land surface roughness on atmospheric circulation and precipitation: A sensitivity study with a general circulation model, *J. Appl. Meteor.* **27**, 1036–1054.

87. Verseghy, D.L. (1991) CLASS - a Canadian Land Surface Scheme for GCMs. I: Soil model, *Int. J. Climatol.* **11**, 111–133.

88. Viterbo, P. and Beljaars, A.C.M. (1995) An improved land surface parameterization scheme in the ECMWF model and its validation, *J. Clim.* **8**, 2716–2748.

89. Walker, J. and Rowntree, P.R. (1977) The effect of soil moisture on circulation and rainfall in a tropical model, *Quart. J. Roy. Meteor. Soc.* **103**, 29–46.

90. Wilson, M.F. and Henderson-Sellers, A. (1985) A global archive of land cover and soils data for use in general circulation climate models, *J. Climatology* **5**, 119–143.

91. Wood, E.F., Lettenmaier, D.P., and Zartarian, V.G. (1992) A land-surface hydrology parameterization with subgrid variability for general circulation models, *J. Geophys. Res.* **97**, 2717–2728.

92. Wood, E.F., Sivapalan, M., and Beven, K. (1990) Similarity and scale in catchment storm response, *Rev. Geophys.* **28**, 1–18.

93. Yang, Z.L., Dickinson, R.E., Robock, A., and Vinnikov, K.Y. (1997) Validation of the snow submodel of the biosphere-atmosphere transfer scheme with Russian snow cover and meteorological observational data, *J. Clim.* **10**, 353–373.

94. Ye, Z. and Pielke, R.A. (1993) Atmospheric parameterization of evaporation from non-plant-covered surfaces, *J. Appl. Meteor.* **32**, 1248–1258.

ATMOSPHERIC CHEMISTRY AND AEROSOL DYNAMICS

J. FEICHTER
Max Planck Institute for Meteorology
Bundesstrasse 55
D-20146 Hamburg
Germany

1. Introduction

Many attempts have been made in recent years to understand the physical, chemical and biological processes that constitute the Earth-atmosphere system using global numerical models. Global climate changes are the product of both natural processes and a growing suite of relatively recent anthropogenic perturbations. Chemical processes in the troposphere and stratosphere are important for evaluating the extent of possible climate changes over the coming decades. The uptake and the release of various trace gases by the marine and the terrestrial biosphere determine to a large degree the chemical composition of the atmosphere. Trace gases such as CO_2, CH_4, N_2O and tropospheric O_3 (greenhouse gases "GHG") trap terrestrial infrared radiation and warm the troposphere and cool the stratosphere. Relatively small amounts of trace gases such as CO, NO_X and volatile organic compounds play a key role in atmospheric chemistry by affecting the tropospheric and stratospheric concentrations of ozone. It is now evident that agricultural, and more recently, industrial activities as well as land use practices have changed the composition of the atmosphere. As a consequence, the atmospheric volume mixing ratio of CO_2 has increased from 280 to 350 ppm, that of CH_4 from 0.7 to 1.7 ppm, and that of N_2O from 0.28 to 0.31, between the years 1860 and 1990. Ozone, also a GHG, has nearly doubled in the troposphere, whereas in the lower stratosphere it has decreased owing to anthropogenic halocarbon emissions. Since ozone absorbs solar ultraviolet radiation, which heats the stratosphere, stratospheric ozone depletion exerts a cooling effect in the stratosphere and the upper troposphere. Loadings of tropospheric aerosols (e.g. sulfate, nitrate and carbonaceous aerosols) have increased substantially over the past 150 years as a consequence of industrial activities. These aerosols enhance reflection of solar radiation both directly, by scattering light in clear air and, indirectly, by increas-

P. Mote and A. O'Neill (eds.), Numerical Modeling of the Global Atmosphere in the Climate System, 353–374.
© *2000 Kluwer Academic Publishers. Printed in the Netherlands.*

354

ing the reflectivity and life-time of clouds. Aerosols act as cloud condensation nuclei and control the formation and the optical and physical properties of clouds. Explosive volcanic eruptions inject volcanic ash and sulfur dioxide gas into the stratosphere. The ash particles quickly fall out of the stratosphere because of their large size. The sulfur dioxide gas is photochemically converted into sulfuric acid which condenses to form submicron-sized particles. These sulfuric acid particles remain in the stratosphere for a few years where the particles scatter and absorb sunlight, and absorb and emit infrared radiation. Additionally, a reduction in stratospheric ozone has been observed following volcanic eruptions, due to the loss of ozone through heterogeneous chemistry on the aerosol particles.

In turn, atmospheric dynamics and cloud processes control the concentration and distribution of atmospheric constituents. Winds transport gaseous and particulate matter and loft dust and sea-salt aerosols into the atmosphere. The intensity of the solar radiation and the temperature determine the chemical reaction rates. Cloud droplets are chemical reactors and contribute to the formation of aerosol particles, and precipitation removes gases and particles from the atmosphere. Vegetation and biogenic emissions are governed by meteorological parameters. In turn, vegetation plays a major role in regulating the hydrological cycle.

Global general circulation models (GCMs) of the atmosphere used for climate studies have been recently explored to evaluate the atmospheric transport and the interactions of gaseous and particulate constituents. First attempts have been made to use GCMs to study the carbon cycle of the marine system. Recently, a dynamic model of the terrestrial ecosystems was linked to the NCAR GCM CCM3 [5].

This chapter describes the processes coupling the atmospheric dynamics and cloud physics to models of atmospheric chemistry. First the concept of mass budgets and fluxes is explained. In the next section, transport processes in different domains of the atmosphere and for typical temporal scales are discussed. Thereafter, modeling of sources, sinks and chemical transformations of atmospheric constituents is addressed and some recommendations about the evaluation of transport and sink processes are given.

2. Budgets and Cycles

A useful way to describe the characteristic behavior of gaseous and particulate constituents of the atmosphere is in terms of the material balance. A cycle consists of connected reservoirs that exchange mass. If all the material cycles within the system, the system is closed. Budgets and cycles can be considered for different reservoirs (e.g. planetary boundary layer (PBL), troposphere, stratosphere, hemisphere, regional scale).

If M is the total amount of a species in the reservoir, F the rate of its release or formation, S the rate of its removal or chemical destruction, equilibrium of a reservoir in steady state (sources and sinks balance the budget) is defined by $F = S(M)$ [27]. S is in general a function of the content of the reservoir M. Often this flux is proportional to M which is referred to as a first-order process.

The different exchange processes can be characterized by typical time scales ([46], [22]). The first is the turn-over time τ_0 which is defined as

$$\tau_0 = M/S. \tag{1}$$

The turn-over time is the time it would take to empty the reservoir if the sink S remained constant while the sources were zero. If material is removed from the reservoir by different processes (e.g. different transport processes, chemical destruction, deposition), one can define turn-over times with respect to each such process i. The total turn-over time of a species in the reservoir is then given by: $\tau_0^{-1} = \sum (\tau_0^i)^{-1}$.

The residence time τ_r is the average time spent in a reservoir by an individual molecule or particle. If the only sink is transport out of the reservoir, the term transit time can be used. Even for the same species, different atoms will have different residence times in a given reservoir. The probability density function of residence times is denoted by $\Psi(t)$ where $\Psi(t)dt$ indicates the fraction of the molecules having a residence time between t and $\tau + d\tau$. The average residence time is then defined as

$$\tau_r = \int_0^\infty \tau \Psi(\tau) d\tau. \tag{2}$$

The average age of all molecules of a specific kind since they entered the reservoir is given by

$$\tau_a = \int_0^\infty \tau \phi(\tau) d\tau \tag{3}$$

where $\phi(t)$ is the age probability function.

For a reservoir in steady state, the turn-over time and the average residence time are equal. The turn-over time equals the average age if the sink process is of first order. But τ_0 can be very different from τ_a depending on the age distribution. Rodhe [46] gives a very illustrative example for this. The average age of all Swedes is between 30 and 40 years, whereas the average residence time, i.e. the average length of life, is more than 70 years.

A further useful concept is that of the response time (e-folding time) which is the time needed to reduce the effect of a sudden change in a reservoir to 1/e of the initial perturbation.

356

3. Transport of trace species

3.1. GLOBAL FEATURES

Transport processes are of particular importance from the standpoint of the chemical composition of the atmosphere and the life-time of specific constituents in different reservoirs. The atmosphere possesses a large spectrum of motions from planetary waves, synoptic scale disturbances, and mesoscale processes to turbulent exchange. The scales of motion that are important for the transport of a specific constituent depend on the atmospheric residence time of the species in question. Generally, the distribution of highly reactive species is dominated by chemical and microscale interactions at surfaces, while that of less faster reacting species is dominated by fast mixing processes, and that of slowly reacting species by large-scale transport. On larger spatial scales, the winds transport species with long lifetimes far away from the source region. Pollutants predominantly released in the northern hemisphere continents are moved across entire continents and also contribute by interhemispheric transport to the load of the southern hemisphere. Subgrid-scale processes, such as turbulent exchange and vertical transport in clouds, quite efficiently dilute polluted boundary layer air by mixing with free tropospheric air masses. The degree of vertical mixing controls the dry deposition at the ground, the transit time until a parcel enters a cloud or the rate of photochemical decomposition.

3.1.1. *Global circulation*

All the atmospheric motion is driven by the radiation from the sun. Solar radiation is either returned to space by reflection and scattering in the atmosphere and by reflection at the ground or absorption and radiation back towards space as terrestrial (long-wave) radiation. 36% of the solar radiation is reflected (planetary albedo) and 64% is used to heat the Earth-atmosphere system (17% the atmosphere and 47% the Earth's surface). Spatial and temporal variations in this energy balance drive the atmospheric motions. For example, the annual amount of energy received at the equator is larger by a factor of 2.4 than that received at the pole. Additionally, such variations are controlled by the land-sea distribution, by the soil type, by the vegetation cover, by clouds, and by the chemical composition of the atmosphere (water vapor, greenhouse gases, and aerosols).

The atmosphere can be classified, on the basis of the vertical profile of the global temperature, into the troposphere, from the surface up to 10 - 18 km (depending on latitude), the stratosphere (from 10 - 18 km to 50 km), and the mesosphere (from 50 - 80 km). The latter two are nowadays commonly referred to as the middle atmosphere. The troposphere accounts for about 85% of the mass of the atmosphere and virtually all of the atmospheric water. Processes occurring in the troposphere are primarily responsible for weather disturbances and climate variability. The troposphere and the middle atmosphere are linked through radiative and dynamical processes and through the exchange of gases that are important in the photochemistry of the ozone layer [25].

Chemically non-reactive atmospheric constituents, called tracers, are useful to demonstrate the transport characteristics. They are most valuable when they are conservative, so that their values remain constant following the motion. But no tracer is truly conservative, since even in the absence of diabatic effects and chemical sources and sinks there will always be some mixing on scales smaller than the model's resolution. In order to study transport processes, we use tracers whose chemical lifetimes are much longer than the transport timescale.

3.1.2. *Troposphere*

Three main circulation regimes can be distinguished in the troposphere:

The Hadley Cell. The Hadley Cell is a meridional circulation which is driven by the heating of air in the equatorial region. Equatorial air moves upwards and air from higher latitudes moves laterally towards the equator to produce the trade winds. These lower level winds, the most persistent wind system of the atmosphere, move over the sea and carry water vapor towards the equator. The trade winds from both hemispheres converge near the equator (Inter-Tropical Convergence Zone, ITCZ), and water vapor condenses within the ITCZ forming large cumulonimbus clouds. This flow is balanced by an outflow at higher altitudes. The Hadley Cell is closed by subsidence at about 30° in both hemispheres (horse latitudes). The Coriolis force, associated with the Earth's rotation, deflects moving air parcels to the right north of the equator and to the left south of the equator. Owing to this Coriolis force, air parcels in the upper branch of the Hadley Cell are deflected to the east, and air in the lower branch to the west. Convection in the ITCZ is very effective in transporting atmospheric constituents into higher altitudes. Convection provides also for downward transport within the "downdrafts" and by slow sinking processes in between clouds (see Chapter 9).

The Westerlies in the temperate latitudes. The differential heating between the equator and the poles creates a pole-to-equator temperature gradient which results in westerly thermal wind flow in the mid-latitudes. These zonal winds (jet streams) become baroclinically unstable, and in the free troposphere troughs and ridges of low and high pressures are formed. This cyclogenesis results in poleward-moving warm air which is lifted above the equatorward-moving cold air. This exchange of mass across the latitudes is far less regularly ordered than the regimes 1 and 3.

Polar region. The polar region is in particular in winter covered by a high pressure system, the Polar High. Within this Polar High, air subsidence occurs and at higher altitudes air moves poleward to take its place. This forms a meridional circulation cell.

GCMs calculate the large-scale transport of atmospheric constituents by wind (three-dimensional advection) (see Chapter 7, Section 4) and subgrid-scale vertical transport by turbulent exchange and within convective clouds. Horizontal diffusion of trace constituents is mostly neglected. Generally, these transport processes are calculated in the same way as the transport of water vapor. Table 1 shows typical transit times between different atmospheric reservoirs and of different transport processes.

TABLE 1.   Atmospheric transit times.

Processes	
Vertical transport within convective clouds	1 hr
Mixing between the PBL and the free troposphere (depending on latitude and season)	2 - 10 days
Large-scale vertical mixing in the troposphere	1 - 4 weeks
Mixing within latitude belts	2 - 4 weeks
Hemispheric mixing	2 - 6 months
Interhemispheric exchange	1 yr
Stratospheric-tropospheric exchange	1 - 3 yrs
Transport from the surface up to the mesosphere	5 - 8 yrs

To give an overview of the behavior of the different transport regimes on a global scale we present in Figure 1 a snapshot of the distribution of two artificial tracers. One tracer is released at the surface with a constant and globally uniform emission rate and decays with an e-folding time of 5.5 days. A second tracer is fixed in altitudes above 100 hPa and decays in the troposphere with an e-folding time of 4 weeks. This experiment has been performed with the T106 version of the Hamburg climate model ECHAM4 [43]. The dynamics of the model has been relaxed to observed wind, temperature and surface pressure fields (the method is called Newtonian relaxation or nudging). Figure 1 shows horizontal distributions of the mean sea-level pressure and of the stratospheric tracer at 700 hPa and the tropospheric one at 300 hPa, which correspond to a height of 3 and 9 km, respectively. The distribution of both tracers is largely controlled by vertical transport processes, and the concentrations vary regionally by two orders of magnitude. The tracer released at the surface (lower graph) exhibits low mixing ratios over the poles and in the subtropical high pressure systems where subsidence dominates. High mixing ratios indicating rising motion are found within the ITCZ and within the west wind belt of the mid-latitudes. Generally, the maximum concentrations

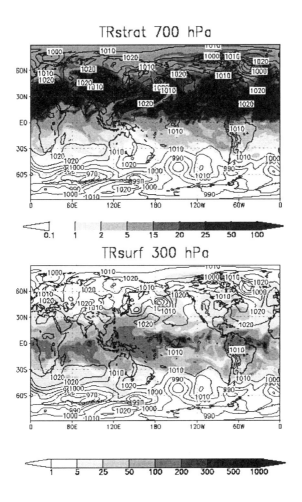

Figure 1. The horizontal distribution of the mean sea-level pressure (contours), together with the distribution of tracer at (top) 700 hPa and (bottom) 300 hPa.

exhibit band-like structures in the tropics which are connected to tropical cloud clusters.

3.1.3. *Stratosphere*

Before discussing the distribution of the stratospheric tracer, we will briefly review the transport regimes in the stratosphere and the exchange of mass between the troposphere and the stratosphere (Stratosphere-Troposphere Exchange, STE). In the middle atmosphere, radiative heating due to the absorption of solar radiation by ozone is balanced by infrared radiative cooling. As a result, the mean temperature increases up to 50 km (stratopause) and decreases higher up. A circulation only driven by this diabatic heating would exhibit upward motion in the summer hemisphere and downward motion in the winter hemisphere. But this diabatic circulation is not the circulation actually observed. Departures from this circulation are produced by eddy transports. Brewer-Dobson presented a model which describes the mean transport in the middle atmosphere on a meridional plane and the STE as well. The middle atmosphere meridional transport consists of rising motion across the tropical tropopause, a poleward drift in higher altitudes and subsidence over high latitudes. Mass conservation is maintained by a return flow into the troposphere in the extratropics.

Figure 1 (upper graph) shows the calculated distribution at 700 hPa of a tracer kept fixed in the stratosphere. The distribution indicates that downward transport of stratospheric air is most efficient within the subsidence regions of the subtropics in the winter hemisphere. Cross tropopause exchange in mid-latitudes is dominated by tropopause folding. The term "folding" is used to describe a process in which the dynamical tropopause intrudes deeply into the troposphere. The typical horizontal scale of such foldings is about 100 km.

4. Sources and sinks

Emission inventories are the basis for numerical model studies of the relationship between human activities and the environment. Terrestrial and marine biological processes, volcanic eruptions, biomass burning, and weathering generate emissions of numerous trace constituents. Anthropogenic activities, such as energy production, industrial production, traffic and waste disposal, have changed the chemical composition of the atmosphere. Chemical constituents released into the atmosphere are chemically transformed, deposited at the Earth's surface and removed by precipitation. Removal by rain and dry deposition at the ground controls the atmospheric residence time of gases and aerosol particles and contributes to the acidification and eutrophication of ecosystems (eutrophication is the enrichment of water courses by inorganic plant nutrients). In that way, gases and particles are not only released from, but also adsorbed and consumed by, vegetation, ocean and land surfaces.

4.1. EMISSION INVENTORIES

An atmospheric emission inventory is a data base of information on: emission factors (how much of a species is produced from a specific source per unit time), number and geographical distribution of emission sources and activity statistics [37]. Such emission data bases have been derived for the present, past and future. For application in global models, the Global Emissions Inventory Activity Project (GEIA), for example, provides emission data bases valid for the years 1985 and 1990 of some key species from anthropogenic and biogenic sources.

Production of volatile species and subsequent emission, through plant phys-iological and soil microbial processes, are influenced by environmental factors such as temperature, insolation, humidity, and precipitation. Global scale models describing the release of gases from and consumption by the biosphere have so far been developed only for CO_2. The rate of gaseous diffusion in soils and in forest canopies which affects the gas exchange with the atmosphere has been estimated based on ^{222}Radon flux measurements ([56], [6]).

The oceans are important sources of dimethyl sulfide, ammonia, hydrocar-bons, and halogenated organics of biological origin. Understanding the exchange between the atmosphere and the ocean requires knowledge about the species con-centration in the surface water and a parameterization of the exchange between the sea and the atmosphere. The basic principles governing transfer of gases across the sea surface have been summarized by Liss and Merlivat [33]. The flux F across the sea surface is driven by the concentration difference in air and water, ΔC, and by the transfer velocity k_w : $F = k_w \Delta C$. The concentration difference can be expressed as $\Delta C = C_a H^{-1} - C_w$, where C_a and C_w are the gas concentrations in air and water, respectively, and H is the dimensionless Henry's Law Constant. Transfer velocities of individual gases have been measured and modeled as well. It has been shown that k_w is proportional to the friction velocity in air and to the Schmidt number ($Sc=$ kinematic viscosity/molecular diffusivity) to the power -2/3. On the basis of wind tunnel and field experiments, Liss and Merlivat [33] proposed three relationships for the variation of k_w [cm s^{-1}] with the wind speed (Table 2):

TABLE 2.

$k_w = 0.17\,u$	smooth wind regime	$u < 3.6$ [m s^{-1}]
$k_w = 2.85\,u$ - 9.65	rough regime	$3.6 < u < 13$
$k_w = 5.9\,u$ - 49.3	breaking waves	$u > 13$

More recently, Wanninkhof [57] presented a parameterization based on measure-ments of an artificial tracer (SF6). Globally averaged DMS fluxes derived from this scheme are by a factor of two higher than fluxes calculated applying the Liss and Merlivat approach, indicating the uncertainty range of such calculations [29].

Agitation of sea and land surfaces by winds ejects large amounts of particulate matter into the atmosphere. Sea-salt aerosols originate from the bursting of air bubbles in wind-induced whitecaps. The formation of dust is caused by weath-ering of crustal material of the Earth and subsquent lofting into the atmosphere by winds. Both aerosol types together account for most of the aerosol mass intro-duced into the atmosphere. Parameterizations of the sources have been derived for application in global models from Gong *et al.* [21] for sea-salt aerosols, and from Tegen *et al.* [52] and Schulz *et al.* [47] for mineral dust.

4.2. DRY DEPOSITION

The dry deposition of gases and aerosol particles from the atmosphere to a receptor surface is governed by the concentration in air and by the turbulent transport processes in the boundary layer, as well as by the chemical and physical nature of the depositing species and the capability of the surface to capture or absorb gases and particles. Two layers are distinguished in the boundary layer (PBL): the turbulent layer and the surface layer (quasi-laminar boundary layer) which has a typical depth of 1/10 the PBL height. In the laminar layer, the transport of gases to the receptor surface is dominated by molecular diffusion. The surface uptake is frequently controlled by the ability of the surface to absorb the species. Reactive gases are collected immediately upon contact with all surfaces. Less reactive soluble gases tend to be taken up through stomata or through the leaf cuticle. These gases can also be absorbed in water layers at the leaf surface, or at the soil, as a result of dew fall or rain interception [9]. Less soluble species, like O_3 and NO_2, are still removed by stomatal uptake, so there must be some additional mechanisms to consume these species within the substomatal cavities. However, the most important factor is the gas water solubility. If the gas reacts in the aqueous phase, the resistance to additional mass transfer is low. However, if the gas dissolves without irreversible chemical reaction in the water, the mass transfer decreases as the amount of gas deposited increases.

For particles, the transport from the free atmosphere to the receptor surface is similar to gas transport, but transport processes through the laminar layer differ considerably. These processes are largely controlled by the microstructure of the receptor surface and by the size distribution and the density of the particles. Brownian diffusion dominates for particles with radii less than 0.1 μm, impaction for particles in the size range 0.1 - 1.0 μm, and gravitational settling (sedimentation) for supermicron particles.

A concept to describe the dry deposition has been developed by analogy with a current in an electrical circuit governed by Ohm's law. Assuming one-dimensional transfer over a homogeneous surface, the difference between concentration at a reference height z within the constant flux layer and the surface, ΔC, equals the product of the surface resistance R_d and the flux F.

$$\Delta C = R_d \times F \qquad (4)$$

Assuming that the surface is a perfect sink, implying that the surface concentration is zero at the surface and that the reciprocal of R_d is the dry deposition velocity V_d, we obtain the flux directed towards the surface

$$F = V_d \times C \qquad (5)$$

This approach has been widely used in chemical transport models assuming that V_d depends only on the deposited species and on the underlying surface and with C being the concentration of the lowest model level. Typical values of V_d over grass are given in Table 3. Dry deposition velocities are mainly derived from field and laboratory studies. The reference height for the deposition velocity is about 1 m over smooth land surfaces and 10 m for ocean surfaces. For airborne particles, the variation of deposition velocity with diameter can be as great as two orders of magnitude and is least efficient for particles in the accumulation mode (mass median diameter 0.1-1.0 μm).

TABLE 3.   Measured deposition velocities over grass ([13], [22])

Species	HNO$_3$	SO$_2$	O$_3$	Particulate sulfur
V_d [cm s^{-1}]	1.0-4.7	0.1-4.5	0.5-1.8	0.02-0.42

The analogy with Ohm's law has been extended by assuming that the total surface resistance consists of separate serial resistances: the aerodynamic resistance R_a, the laminar resistance R_b and the canopy resistance R_c ([24], [60]). The dry deposition velocity of gases is the inverse of the three resistances.

$$V_d(z) = \frac{1}{R_a(z) + R_b + R_c} \tag{6}$$

R_a depends mainly on the local atmospheric turbulence intensities and can be calculated as a function of the friction velocity, u^*, and the roughness height, z_o. Both are used to describe the wind speed profile above different surfaces. R_b quantifies the diffusive transport through the quasi-laminar layer and is for trace gases generally an order of magnitude smaller than the aerodynamic resistance. R_c depends on the solubility and reactivity of the deposited species and on surface cover, e.g. water, bare soil, snow, ice, and its characteristics, e.g. soil type, vegetation type, the plant activity and the canopy structure expressed by the canopy height and the total amount of standing biomass (Leaf area index).

For particles, the distinction between laminar and canopy resistance is not very useful. The deposition of particles smaller than 0.1 μm is controlled by the Brownian diffusion, whose efficiency is inversely related to the particle diameter. This Brownian diffusion is, in particular, important in the immediate vicinity of the receptor surface. Particles larger than 1μm undergo only turbulent exchange and sedimentation ([49], [50]).

Recently a dry deposition scheme for gases and particles designed for use in GCMs has been developed by Ganzeveld and Lelieveld [16] and Ganzeveld et al.

[17]. Similar approaches as those presented in this section are used to describe land surface processes in meteorological models (see Chapter 13).

4.3. WET DEPOSITION

4.3.1. *Process*
Wet deposition or wet scavenging is the process by which gases or particles are attached and dissolved in cloud droplets and precipitation, and as a result are deposited at the Earth's surface. Wet scavenging is a complex phenomenon that depends on a wide spectrum of spatial scales from the microphysical to the storm scale. Because the variables available from global GCMs are typically averaged over hundreds of kilometers, global modelers have generally not been able to parameterize these subgrid-scale processes in detail.

Only recently, GCMs have employed prognostic variables for cloud water and cloud ice (see Chapter 10). However, the assumptions about fractional cloud cover are still crude. Most GCMs diagnose cloud cover as a function of relative humidity. This approach is referred to as an explicit approach. It is assumed that clouds occupy the whole grid box in the vertical, an assumption which is not justified by observations. Another approach is the statistical approach, in which a distribution function of total water is used to obtain cloud amount and cloud water at the same time. This approach is more consistent. However, little is known about how to obtain a distribution function, especially in the absence of turbulence as in the free troposphere. Tiedtke [54] introduced a prognostic variable for cloud cover, which may solve the problem. Very recently, Ghan *et al.* [18] and Lohmann *et al.* [34] introduced a prognostic equation for the cloud droplet number concentration in order to account for the activation of cloud droplets. The parameterization of the activation of cloud droplets couples the aerosol cycle with the cloud cycle, as the activation of cloud droplets is a sink of aerosol particles.

We distinguish between two processes removing gases or particles from the atmosphere: (1) In-cloud scavenging is the result of the incorporation of species into cloud droplets. These droplets and also the aerosol particles are finally removed from the atmosphere by precipitation (in-cloud scavenging + removal = rainout). (2) Below-cloud scavenging is the collection of particles or the uptake of gases below clouds by raindrops and snowflakes (washout).

4.3.2. *Gases*
Gaseous species can be absorbed by cloud droplets or rain drops to the extent that these are soluble in water. If sufficient time is available for equilibrium to be reached, the concentration in the water droplets is given by Henry's Law: $C_{liquid} = C_{air}/H$, where C_{liquid} is the concentration in the water droplet [M L^{-3}], C_{air} is the concentration in the interstitial air ("interstitial" means within the cloud but not within a cloud droplet) and H is the dimensionless Henry's Law constant.

The dissolution of reactive gases proceeds in two steps. The first involves the hydration of the substance, and the second the formation of ions from the hydrate. Simple Henry's Law considerations ignore interactions between solutes and do not take into account the fact that the pH of individual cloud elements varies over a wide range [58]. Acid-forming gases are poorly absorbed by cloud water when the pH values are low; ammonia shows the opposite behavior.

The flux of material is determined by molecular diffusion, inside and outside the liquid water drop, by the transport resistance at the gas-liquid interface, and by the rate of hydrolysis and ion formation. Warneck [58] points out that small droplets adjust to Henry's Law equilibrium within a fraction of a second, which suggests that liquid water clouds are essentially always near equilibrium. The situation is less favorable for larger rain drops. In particular, strongly soluble gases have a relatively long equilibrium time and rain drops scavenge only a fraction of the material. Gas scavenging by ice-clouds occurs predominantly if the temperature is close to the freezing point and the ice particle is covered by a pseudo-liquid layer.

4.3.3. *Particles*

Aerosol particles serve as cloud condensation nuclei (CCN) and are scavenged from the atmosphere when the cloud precipitates. This nucleation scavenging is the most efficient process for removing particles from the atmosphere. When clouds evaporate, the aerosol particles released back to the atmosphere may have changed the chemical composition and physical properties due to in-cloud processes. Since drops form predominantly on large particles which contain most of the aerosol mass, about 80-99% of the aerosol mass is scavenged, but only 10-90% of the particle number, since small particles can be left unactivated and remain present as an interstitial aerosol [14]. Essentially, all particles with radii greater than 0.2 μm are expected to undergo nucleation scavenging [58]. However, the scavenging efficiency may be smaller in polluted than in clean air.

During their entire life-time, drops will capture unactivated aerosol particles in the air (impaction scavenging). This process is negligible within the cloud as the small particles contribute little mass, but it is the dominant scavenging process below cloud base in the rainfall area. According to a model study by Flossmann [14], about 5% of the aerosol mass is scavenged below clouds, but 30% of the aerosol mass in precipitation on the ground originates from below-cloud scavenging. Nucleation scavenging by ice-clouds is generally small. The impaction scavenging by ice crystals depends on temperature being most efficient around the freezing point and is enhanced for dendritic aggregates. For the ice phase, a third scavenging process exists through freezing of drops, which contain aerosol mass, and through collision of drops with ice particles and subsequent freezing (riming).

4.3.4. *Modeling*

Most of the approaches to modeling wet scavenging have descended from the work of Junge and Gustafson [28], who described the scavenging of the species whose mixing ratio in cloud is C_{cl}.

$$\frac{\partial C_{cl}}{\partial t} = -\frac{\varepsilon \dot{Q}}{L} C_{cl} = -\lambda C_{cl}, \tag{7}$$

where ε is the scavenging efficiency, $\dot{Q}$ is the rate of production of precipitation [kg m^{-3} s^{-1}] and L is the cloud liquid water content [kg m^{-3}]. Thus the scavenging frequency $\lambda = \varepsilon \dot{Q}/L$.

Scavenging parameterizations that have been used in global-scale models (mostly chemical transport models) are not much more sophisticated than the approach above. For example, Mahlman and Moxim [35] have specified the vertical variation of λ (a decrease with height), but allowed the model-predicted surface precipitation to determine the temporal and spatial variation of λ. Penner *et al.* [38] used a similar approach with λ also dependent on the solubility and reactivity of the species being scavenged. Giorgi and Chameides [19] developed a parameterization for the scavenging of very soluble species in which the spatial and temporal variation of λ was chosen by the $\dot{Q}$ predicted in each grid cell of the meteorological model. This scheme has also been applied in a GCM ([4], [11]) and performed quite well, at least for the scavenging in stratiform clouds. Scavenging in convective clouds occurs when air is pumped upwards within the convective updrafts. These updrafts cover only a small fraction of the grid-box and most of the soluble gases and aerosols are removed before the air disperses at the top of the cloud. For example, Folkins *et al.* [15] report a removal efficiency of greater than 90%, based on measurements in the biomass burning plume over South-East Asia, for sulfate aerosol particles.

When clouds or precipitation evaporates, the gases and particles dissolved in the cloud droplets or rain drops are released again. Removal by precipitation and evaporation below the cloud can be a quite efficient and fast downward transport mechanism.

The scavenging parameterization in advanced GCMs depends on the degree of complexity of the model's cloud physics. At least the following processes have to be considered: in-cloud scavenging, removal by precipitation, release of species mass back to the atmosphere via evaporation of clouds and precipitation, and below-cloud scavenging.

When particle number concentrations are calculated rather than the aerosol mass, processes like the nucleation scavenging become much less and impaction scavenging much more effective.

A general problem of parameterizing scavenging processes with GCMs is that all model variables represent a grid-box average. Thus even if the scavenging is only calculated within the cloud-covered fraction of the grid-box, the concentration changes within the whole volume implying an unrealistically strong mixing. Additionally, in the presence of broken clouds there is an asymmetry in respect of the scavenging between air undergoing subsidence and air transported upwards within the cloud which cannot be resolved by the relatively coarse-resolution global models [45].

5. Chemistry and aerosol micro-physics

5.1. PHOTOCHEMISTRY

Chemical reactions in the atmosphere are initiated mainly by photochemical processes. When a photon is absorbed by a molecule, the photon energy increases the internal energy exciting the rotational, vibrational and electronic motion within the molecule. If the energy of the absorbed photon exceeds the bond strength of the molecule, the interaction will result in photodissociation [22]. To estimate the solar flux available for photochemistry, one needs to know the flux outside the atmosphere (solar constant: 1370 Wm^{-2}), but also the extent of light absorption and scattering within the atmosphere. With regard to absorption processes, the spectrum of the sun may be subdivided into three wavelength regions. Radiation within the extreme ultraviolet portion of the spectrum (< 120 nm) is absorbed at altitudes above 100 km. Radiation within 120 nm - 300 nm is absorbed mainly in the middle atmosphere, and radiation within 300 nm - 1000 nm penetrates into the troposphere [58]. The photolysis rate coefficient J_x for a species X can be calculated from the spectral actinic flux (actinic means capable of causing photochemical reactions) $F(l)$ via the integral

$$J_x = \int \sigma_x(\lambda)\phi_x(\lambda)F(\lambda)d\lambda, \tag{8}$$

where λ is the wavelength, $\sigma_x(\lambda)$ the absorption coefficient and $\phi_x(\lambda)$ the quantum yield. The quantum yield gives the probability for the occurrence of a specific photolytic process and is, as well as the absorption coefficient, measured in laboratory experiments. In practice, this equation is solved by calculating the sum over discrete spectral intervals. To obtain accurate results the spectral range of interest has to be subdivided into over 100 intervals. Such calculations are very time consuming since they have to be performed at each time-step and gridpoint of the model. Recently, Landgraf and Crutzen [30] proposed a computationally efficient method for on-line calculations of photolysis rates. They subdivided the spectral range into eight wavelength bands and introduced a correction factor accounting for scattering by molecules, aerosols and clouds.

5.2. GAS PHASE CHEMISTRY

The atmosphere is an oxidative medium, and chemical constituents of the atmosphere tend to reach a more oxidized state. Chemical species do not react with molecular oxygen but react with free radicals (atoms or molecules which possess unpaired electrons and which react vigorously to reach a more stable state), with the hydroxyl radical OH being the most important one. Global models deal only with a limited number of species, with the focus on the background chemistry rather than on the chemistry of polluted air. Key species in the chemistry of the background troposphere are ozone, methane, carbon monoxide and nitrogen oxides.

The changes in concentration are described in terms of species-specific reaction rate coefficients and the number concentration of the various reactants. The reaction rate coefficients vary with temperature and some additionally with pressure and are derived from experimental data [2]. A set of rate equations forms a so-called stiff system of ordinary differential equations which can be solved numerically. Because chemical equations have to be solved for a large number of gridpoints, fast methods are required. Two techniques, which are widely used by global modelers, are the "Quasi-steady State Approximation" and the "Euler Backward Iterative Method" [23]. Chemical species are grouped together in families and the group is transported rather than the individuals. The chemical life time of such a group is commonly much longer than that of the individual compounds. This technique saves computer time, stabilizes the numerical schemes and makes the scheme mass conserving. Hydrocarbon chemistry schemes are often simplified by calculating explicitly only some key species, assuming that other species behave similarly (lumping).

5.3. HETEROGENOUS CHEMISTRY

Atmospheric chemical processes are not limited to the gas phase, but also occur on the surface of solid particles and within liquid particles, such as aerosols and cloud droplets. The important role of heterogenous reactions on aerosol surfaces has been shown in studies of the stratospheric ozone hole. Reactions on sea salt may also play an important role in the marine PBL [1]. Mineral dust particles react with sulfur and nitrogen particles to form sulfates and nitrates, respectively ([36], [7]). Clouds control the formation of aerosols and their removal by scavenging. For example, the oxidation of SO_2 to sulfate in cloud droplets is much more efficient than in the gas phase [12]. Generally, reaction pathways and rates differ considerably from those in cloud-free air [31]. Moreover, clouds also affect the photochemistry by enhancing the actinic fluxes above the cloud and by reducing it below the cloud compared to clear-sky conditions. Aqueous-phase reaction rates depend on the gas-phase concentrations, solubility and rate of mass transfer of oxidizing agents. The cloud receives trace gases from its inflow region, its vertical winds redistribute the gases, and the cloud transforms the gases through gas and aqueous-phase chemistry. Vertical mixing transports trace gases from the more polluted lower levels to less polluted upper levels of the cloud, and precipitation evaporated below the cloud base releases the dissolved constituents performing a fast downward transport.

5.4. AEROSOL PARTICLE INTERACTIONS

Liquid or solid aerosol particles are injected into the atmosphere from natural and anthropogenic sources (primary particles) or are formed in the atmosphere from condensable gases (secondary particles). The size of particles which undergo long-range transport ranges from a cluster with few molecules up to 10 μm. Processes controlling the temporal evolution of atmospheric aerosols are homogeneous and heterogenous nucleation, condensation, coagulation and removal by dry deposition and by precipitation. Homogeneous nucleation is the condensation to small droplets in supersaturated vapor. In polluted air, condensable gases are more likely to condense on preexisting particles rather than to form new particles (heterogenous nucleation). Coagulation is the main sink for small particles ($< 0.1 \mu$m) which undergo diffusion and collide and stick to one another, shifting the size distribution of a particle population to larger radii. Activation is the process of cloud droplet formation by condensation of water vapor on aerosol particles. Such particles which can initiate the formation of cloud droplets are called cloud condensation nuclei (CCN). Particles which consist of water-soluble, hygroscopic substances are most likely to act as CCN.

Parameterizations of these processes for use in global models have been developed by Tegen and Fung [52] and Schulz *et al.* [47] for mineral aerosols, by Gong *et al.* [21] for sea-salt aerosols, and by Wilson and Raes [61] and Timmreck and Graf [55] for sulfuric acid. Numerous models of the photochemistry and the heterogenous chemistry were developed during the last two decades. Most of them have been applied in transport models and only a few have considered feedback processes between chemistry and meteorology using a GCM (e.g. [41], [44]). Implementation of schemes which describe the evolution of aerosols and their coupling to cloud microphysics and chemistry in GCMs is under development and will be a major goal in atmospheric research during the next few years. (For more information regarding atmospheric chemistry and aerosol microphysics, we refer the reader to Seinfeld [48]; Warneck [58]; Graedel and Crutzen [22]; Pruppacher and Klett [40]).

6. Evaluation

To examine the role of atmospheric motions and deposition processes in the dispersal of atmospheric constituents, we perform model experiments using test tracers. Species used as test tracers should meet the following conditions: they should be chemically inert, sources and sinks should be well-known, and sufficient observational data should be available for comparison with model results. Table 4 (page 374) presents tracers widely in use for evaluating specific aspects of the model's transport characteristics. Not only tracer distributions, but also the ratios between different tracers should be compared with observations. For example, the ratio between the radionuclides Be-7 and Be-10 (both have the same source distribution in the stratosphere but very different radioactive life-times, and both are removed by wet deposition) is an ideal indicator for stratospheric-tropospheric exchange. Because the life-time of Be-10 in the stratosphere is much longer than for Be-7, stratospheric air is characterized by high values of Be-10/Be-7. The ratio between tracers released near the surface and those released in the upper troposphere and stratosphere reflects the vertical exchange processes (e.g. Pb-210/Be-7).

A principal problem of such comparisons between observations and GCM-calculated concentrations and deposition fluxes arises from the fact that observations are taken at a specific time and place, and may not be representative for a larger region, whereas models calculate volume averages. Furthermore, long-term measurements of chemical constituents are sparse, particularly outside the PBL.

Acknowledgements

The author would like to thank E. Manzini, L. Ganzeveld and H. Rodhe for useful discussions.

TABLE 4. Illustrating tracers widely used to evaluate specific aspects model transport characteristics.

Species	Sources	Sinks	Process	Ref.
Radon-222 radioactive noble gas	Product of Uranium-238 decay; exhaled mainly from land surfaces	Radioactive decay, half-life time 3.8d	Subgrid-scale vertical exchange	[8], [10], [26]
Lead-210 radionuclide attached to ambient aerosol	Radioactive decay of Radon-222	Wet and dry deposition	Continental scale transport and deposition	[3], [11]
Beryllium-7 radionuclide attacghed to ambient aerosol	Produced by cosmic ray neutrons in the upper troposphere and stratosphere	Radioactive decay, half-life time 53d; wet and dry deposition	STE and deposition	[4]
Strontium-90 radionuclide attached to ambient aerosol	Nuclear bomb tests; released into the stratosphere	Wet and dry deposition	STE and deposition	[42]
Fluorocarbons	Antropogenic	Photochemical decomposition in the stratosphere	Continental scale transport and interhemispheric exchange	[20], [39]
Krypton-85 radioactive noble gas	Nuclear fuel reprocessing	Radioactive decay, half-life time 10.76 y	Continental scale transport and interhemispheric exchange	[26], [62]
$^{14}CO_2$	Nuclear bomb tests; released into the stratosphere	Uptake by the marine and terrestrial biosphere	STE	[41]
SF_6 gas	Anthropogenic	Photochemical decomposition in the mesosphere; life-time ~3200 y	STE and transport in the middle atmosphere	[32], [59]
Methyl Chloroform gas	Anthropogenic	Chemical reaction with OH and photochemical decomposition in the stratosphere	Evaluation of model calculated OH radical distribution	[51], [53]

372

References

1. Andreae, M.O. and Crutzen, P.J. (1997) Atmospheric aerosols: Biogeochemical sources and role in chemistry, *Science* **276**, 1052-1058.
2. Atkinson, R., Baulch, D.L., Cox, R.A., Hampson Jr., R.F., Kerr, J.A. and Troe, J. (1992) Evaluated kinetic and photochemical data for atmospheric chemistry, *Atmos. Environ.* **26A**, 1187-1230.
3. Balkanski, Y.J., Jacobs, J.D., Gardner, G.M., Graustein, W.M. and Turekian, K.K. (1993) Transport and residence times of tropospheric aerosols inferred from a global 3-dimensional simulation of 210Pb, *J. Geophys. Res.* **98**, 20,573-20,586.
4. Brost, R.A., Feichter, J. and Heimann, M. (1991) Three-dimensional modeling of 7Be in a global climate simulation model, *J. Geophys. Res.* **96**, 22,423-22,445.
5. Craig, S. (in press) The response of terrestrial carbon exchange and atmospheric CO_2 concentrations to El Niño SST forcing. Report Chemical Meteorology CM-94, Meteorological Institute, University Stockholm, *J. Geophys. Res.*.
6. Davidson, E.A. and Trumbore, S.E. (1995) Gas diffusivity and production of CO_2 in deep soils of the eastern Amazon, *Tellus* **47B**, 550-565.
7. Dentener, F.J., Carmichael, G.R., Zhang, Y., Lelieveld, J. and Crutzen, P.J. (1996) Role of mineral aerosol as a reactive surface in the global troposphere, *J. Geophys. Res.* **101**, 22,869-22,890.
8. Dentener, F., Feichter, J. and Jeuken, A. (in press) Simulation of Radon222 using on-line and off-line global models, *Tellus*.
9. Erisman, J.W. and Draaijers, G.P.J. (1995) *Atmospheric Deposition in Relation to Acidification and Eutrophication, Studies in Environmental Science 63*, Elsevier, Amsterdam.
10. Feichter J. and Crutzen, P.J. (1990) Parameterization of the vertical transport due to deep cumulus convection in a global tracer transport model and its evaluation by 222Radon, *Tellus* **42B**, 100-117.
11. Feichter, J., Brost, R.A. and Heimann, M. (1991) Three-dimensional modeling of the concentration and deposition of 210Pb aerosols, *J. Geophys. Res.* **96**, 22,447-22,469.
12. Feichter, J., Lohmann, U. and Schult, I. (1997) The atmospheric sulfur cycle and its impact on the shortwave radiation, *Climate Dyn.* **13**, 235-246.
13. Finnlayson-Pitts, B.J. and Pitts, J.N. (1986) *Atmospheric Chemistry: Fundamentals and experimental techniques*, John Wiley & Sons, New York.
14. Flossmann, A. (in press) Air Pollution Modeling and its Application XII, in S.E.Gryning and N. Chaumerliac (eds.), *NATO Challenges of Modern Society 22*, Plenum Press.
15. Folkins, I., Chatfield, R., Baumgardner, D. and Proffitt, M. (1997) Biomass burning and deep convection in southeastern Asia: Results from ASHOE/MAESA, *J. Geophys. Res.* **102**, 13,291-13,300.
16. Ganzeveld, L. and Lelieveld, J. (1995) Dry deposition parameterization in a chemistry general circulation model and its influence on the distribution of reactive trace gases, *J. Geophys. Res.* **100**, 20,999-21,012.
17. Ganzeveld, L., Lelieveld, J., and Roelofs, G-J. (1998) A dry deposition parameterization for sulfur oxides in a chemistry and general circulation model, *J. Geophys. Res.* **103**, 5679-5694.
18. Ghan, S. J., Leung, L. R., Easter, R. C. and Abdul-Razzak, H. (1997) Prediction of cloud droplet number in a general circulation model, *J. Geophys. Res.* **102**, 21,777-21,794.
19. Giorgi, F. and Chameides, W.L. (1986) Rainout lifetimes of highly soluble aerosols and gases inferred from simulations with a general circulation model, *J. Geophys. Res.* **91**, 14,367-14,376.
20. Golombek, A. and Prinn, R. G. (1986) Global three-dimensional model of the circulation and chemistry of CFCl3, CF2 Cl2, CH3 CCl3, CCl4, and N2O, *J. Geophys. Res.* **91**, 3985-4001.

21. Gong, S. L., Barrie, L. A., and Blanchet, J. -P. (1997) Modeling sea-salt aerosols in the atmosphere, 1, Model development, *J. Geophys. Res.* **102**, 3805-3818.
22. Graedel, T. E. and Crutzen, P. J. (1992) *Atmospheric Change*, W. H. Freeman and Company, New York.
23. Hertel, O., Berkowicz, R., Christensen, J. and Hov O. (1993) Test of two numerical schemes for use in atmospheric transport-chemistry models, *Atmos. Environ.* **27A**, 2591-2611.
24. Hicks, B.B., Baldocchi, D.D., Meyers, T.P., Hosker Jr., R.P. and Matt, D.R. (1987) A preliminary multiple resistance routine for deriving dry deposition velocities from measured quantities, *Water Air Soil Pollut.* **36**, 311-330.
25. Holton, J. R. (1992) *An Introduction to Dynamic Meteorology*, Academic Press, San Diego.
26. Jacob, D.J., Prather, M.J., Rasch, P.J., Shia, R.-L., Balkanski, Y.J., Beagley, S.R., Bergmann, D.J., Blackshear, W.T., Brown, M., Chiba, M., Chipperfield, M.P., de Grandpre, J.G.E., Dignon, J.E., Feichter, J., Genthon, C., Grose, W.L., Kasibhatla, P.S., Kohler, I., Kritz, M.A., Law, K., Penner, J.E., Ramonet, M., Reeves, C.E., Rotman, D.A., Stockwell, D.Z., Van Velthoven, P.F.J., Verver, G., Wild, O., Yang, H. and Zimmermann, P. (1997) Evaluation and intercomparison of global atmospheric transport models using 222Rn and other short-lived tracers, *J. Geophys. Res.* **102**, 5953-5970.
27. Junge, C. E. (1963) *Air Chemistry and Radioactivity*, Academic Press, New York and London.
28. Junge, C.E. and Gustafson, P.E. (1957) On the distribution of sea salt over the United States and its removal by precipitation, *Tellus* **9**, 164-173.
29. Kettle, A. J., *et al.* (1996) A preliminary global data base of sea surface dimethyl sulfide measurements and a simple model to predict sea surface dimethyl sulfide measurements as a function of latitude, longitude and month, *EOS* **77**, 417.
30. Landgraf, J. and Crutzen, P.J. (1998) An efficient method for online calculations of photolysis and heating rates, *J. Atmos. Sci.* **55**, 863-878.
31. Lelieveld, J. (1989) Role of clouds in the photochemistry of the background troposphere, Cologne. Univ., Institut für Geophysik und Meteorologie, Mitteilungen 61, 117-120.
32. Levin, I. and Hesshaimer, V. (1996) Refining of atmospheric transport model entries by the globally observed passive tracer distributions of ^{85}krypton and sulfur hexafluoride (SF_6), *J. Geophys. Res.* **101**, 16,745-16,755.
33. Liss, P.S. and Merlivat, L. (1986) Air-sea gas exchange rates: Introduction and synthesis, in P. Buat Menard (ed.), *The Role of Air-Sea Exchange in Geochemical Cycling*, NATO-ASI Series C, Mathematical and Physical Sciences 185, D. Reidel, Dordrecht, pp. 113-125.
34. Lohmann, U., Feichter, J., Chuang, C.C. and Penner, J.E. (1999) Prediction of the number of cloud droplets activation in the ECHAM GCM, *J. Geophys. Res.* **104**: (D8), 9169-9198.
35. Mahlman, J.D. and Moxim, W.J. (1978) Tracer simulation using a global circulation model: Results from a mid-latitude instantaneous source experiment, *J. Atmos. Sci.* **35**, 1340-1378.
36. Mahmane, Y. and Gottlieb, J. (1989) Heterogenous reactions of minerals with sulfur and nitrogen oxides, *J. Aerosol. Sci.* **20**, 303-311.
37. Pacyna, J. M. and Graedel, T. E. (1995) Atmospheric emissions inventories: Status and prospects, *Ann. Rev. Energy Environ.* **20**, 265-300.
38. Penner, J. E., Atherton, C. S., Dignon, J., Ghan, S. J., Walton, J. J., and Hameed, S. (1991) Tropospheric Nitrogen: A Three-Dimensional Study of Sources, Distributions and Deposition, *J. Geophys. Res.* **96**, 959-990.
39. Prather, M. (1987) Chemistry of the global troposphere: fluorocarbons as tracers of air motion, *J. Geophys. Res.* **92**, 6579-6613.
40. Pruppacher, H. R. and Klett, J. D. (1997) *Microphysics of clouds and precipitation*, Kluwer Academic Publishers, Dordrecht, Netherlands.
41. Rasch, P. J., Tie, X., Boville, B. A., and Williamson, D. L. (1994) A three-dimensional transport model for the middle atmosphere, *J. Geophys. Res.* **99**, 999-1018.

374

42. Rehfeld, S. and Heimann, M. (1995) Three dimensional atmospheric transport simulation of the radioactive tracers 210Pb, 7Be, 10Be, and 90Sr, *J. Geophys. Res.* **100**, 26,141-26,162.

43. Roeckner, E., Arpe, K., Bengtsson, L., Christoph, M., Claussen, M., Dümenil, L., Esch, M., Giorgetta, M., Schlese, U. and Schulzweida, U. (1996) The atmospheric general circulation model ECHAM-4: Model description and simulation ᴖf present-day climate, Report Max-Planck-Institute for Meteorology, Hamburg.

44. Roeckner, E. Bengtsson, L., Feichter, J., Lelieveld, J, and Rodhe, H. (1998) Transient climate change simulations with a coupled atmosphere-ocean GCM including the tropospheric sulfur cycle, Rep. No. 266, Max Planck Institute for Meteorology, Hamburg, Germany.

45. Rodhe, H. (1984) Precipitation scavenging and tropospheric mixing, in Pruppacher *et al.* (eds.) *Precipitation Scavenging, Dry Deposition and Resuspension 1*, Elsevier Science Publishing Co. Inc., New York, pp. 719-729.

46. Rodhe, H. (1992) Modeling biogeochemical cycles, in S.S. Butcher, R.J. Charlson, G. Orians and G. V. Wolfe (eds.), *Global Biogeochemical Cycles*, Academic Press, pp. 55-72.

47. Schulz, M., Balkanski, Y. J., Guelle, W. and Dulac, F. (1998) Role of aerosol size distribution and source location in a three-dimensional simulation of a Saharan dust episode tested against satellite-derived optical thickness, *J. Geophys. Res.* **103**, 10,579-10,592.

48. Seinfeld, J. H. and Pandis, S.N. (1998) *Atmospheric chemistry and physics: From air pollution to climate change*, John Wiley & Sons, Inc., New York, U.S.A.

49. Slinn, S.A. and Slinn, W.G.N. (1980) Predictions for particle deposition on natural waters, *Atmos. Environ.* **14**, 1013-1016.

50. Slinn, W.G.N. (1982) Predictions for particle deposition to vegetative surfaces, *Atmos. Environ.* **16**, 1785-1794.

51. Taylor, J. A., Brasseur, G. P., Zimmerman, P. R., and Cicerone, R. J. (1991) A study of the sources and sinks of methane and methyl chloroform using a global three-dimensional Lagrangian tropospheric tracer transport model, *J. Geophys. Res.* **96**, 3013.

52. Tegen, I. and Fung, I. (1994) Modeling of mineral dust in the atmosphere: Sources, transport, and optical thickness, *J. Geophys. Res.* **99**, 22,897-22,914.

53. Tie, X., Alyea, F. N., Cunnold, D. M. and Kao, C.-Y.J. (1991) Atmospheric Methane: A global three-dimensional model study, *J. Geophys. Res.* **96**, 17,339.

54. Tiedtke, M. (1993) Representation of clouds in large-scale models, *Mon. Wea. Rev.* **121**, 3040-3061.

55. Timmreck, C. and Graf, H.-F. (in press) A microphysical model to simulate the development of stratospheric aerosol in a GCM, *Contrib. Atmos. Phys.*

56. Ussler, W., III, Chanton, J. P., Kelley, C. A. and Martens, C. S. (1994) Radon 222 tracing of soil and forest canopy trace gas exchange in an open canopy boreal forest, *J. Geophys. Res.* **99**, 1953-1963.

57. Wanninkhof, R. (1992) Relationship between wind speed and gas exchange over the ocean, *J. Geophys. Res.* **97**, 7373.

58. Warneck, P. (1988) *Chemistry of the Natural Atmosphere*, Academic Press Inc., San Diego.

59. Waugh, D.W., Hall, T.M., Randel, W.J., Rasch, P.J., Boville, B.A., Boering, K.A., Wofsy, S.C, Daube, B.C., Elkins, J.W., Fahey, D.W., Dutton, G.S., Volk, C.M. and Vohralik, P.F. (1997) Three-dimensional simulations of long-lived tracers using winds from MACCM2, *J. Geophys. Res.* **102**, 21,493-21,514.

60. Wesely, M.L. (1989) Parameterization of surface resistances to gaseous dry deposition in regional-scale numerical models, *Atmos. Environ.* **23**, 1293-1304.

61. Wilson, J. and Raes, F. (1996) M^3 a multi modal model for aerosol dynamics, in M. Kulmala and P.E. Wagner (eds.), *Nucleation and Atmospheric Aerosols*, Elsevier Science Limited.

62. Zimmermann, P.H., Feichter, J., Rath, H.K., Crutzen, P.J. and Weiss, H. (1989)A global three-dimensional source receptor model investigation using [85]Krypton, *Atmos. Environ.* **23**, 25-35.

ATMOSPHERIC DATA ASSIMILATION

ALAN O'NEILL
NERC Centre for Global Atmospheric Modelling
University of Reading
Department of Meteorology
Earley Gate, PO Box 243
Reading RG6 6BB
UK

1. Introduction

Observational data for the earth's atmosphere are derived from a variety of sources, including satellites. The data are of different kinds — e.g., temperature measurements by radiosonde or radiance measurements by satellites — and they suffer from different errors. To get the maximum benefit from the data, for weather forecasting or research, we need to find a way to combine the heterogeneous mixture of data we receive. Moreover, we should like to do this in a manner that leads to a sequence of "analyses" (three-dimensional states) that is consistent with our knowledge of how the atmosphere evolves (the equations of motion with a representation of physical and chemical processes). The modern method of achieving this goal is data assimilation, a technique at the heart of present-day weather forecasting. Data assimilation is increasingly being used to optimize the scientific return from expensively acquired satellite data. Although data assimilation is performed primarily for use in numerical weather prediction, it is also used for climate analysis and in validating general circulation models, and GCMs themselves are used in data assimilation.

This chapter presents a brief overview of the techniques used in data assimilation, focusing on its application to the atmosphere. Details can be found in the book by Daley [3]. Much of the mathematical basis of data assimilation is given in the book by Tarantola [14]. Some papers giving details of the application of data assimilation to weather forecasting are: Talagrand and Courtier [13], Lorenc *et al.* [9], Courtier [2].

P. Mote and A. O'Neill (eds.), Numerical Modeling of the Global Atmosphere in the Climate System, 375–386.

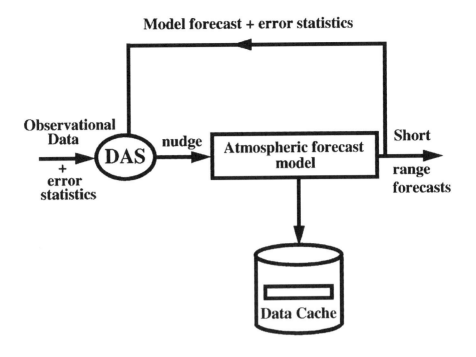

Figure 1. Schematic of the data assimilation process. DAS is the data assimilation system, or algorithm, used to combine observational and model data. See Section 2 for details.

2. What is data assimilation?

Data assimilation is the technique whereby observational data are combined with output (forecast fields) from a dynamical (i.e., evolving) numerical model to produce an optimal representation of the evolving state of the atmosphere. The word "optimal" means that information from instruments and model is combined in a manner that allows for the (supposedly known) errors in the observations and in the short-range forecasts made by the model. The model brings dynamical consistency to the observational data, and interpolates data, in a dynamically consistent manner, into data void regions. The observational data correct the trajectory through state space of the imperfect model, keeping it "on the rails".

This synergism is illustrated by Figure 1, a schematic of the data assimilation process. Atmospheric data assimilation has many features in common with the broader field of control theory, used for example to design automatic landing systems for aircraft.

Main elements:

- A dynamical model makes a short-range forecast of the future state of the atmosphere, x_B (B for "background field")
- The state x_B is passed to the data assimilation algorithm, where two things occur:

  - the observational data y are quality controlled by comparison with each other and with x_B. Statistical methods are used to reject data "outliers", i.e., bad data.
  - the observational data y are combined with the forecast data x_B to produce, by allowing for errors in y and x_B, an optimal estimate of the current state of the atmosphere x_A (A for "analysis").

- The state x_A is fed into the numerical model as the initial field to produce the next forecast, x_B, and so on.
- Periodically (e.g., every 12 hours) long-range weather forecasts are run from initial fields x_A.
- Typically four times per day, the states x_A are archived to produce global three-dimensional datasets for researchers. Also archived are two additional, important datasets:

  - statistics showing how well observational data y compared with the "truth" x_A. This information can be used to correct problems with instruments.
  - statistics showing how well model forecasts x_B compared with the "truth" x_A. This information can be used to correct problems with the model.

3. What are the benefits of data assimilation?

- Quality control.
  The model background state x_B provides a method of data quality control. By comparing y with x_B, one can determine whether a datum is likely to be erroneous.
- Combination of data.
  Data assimilation combines and reconciles observational data from different sources (e.g., radiosondes and satellites) with *a priori* data from a forecast model to get a best estimate of the atmospheric state x_A.
- Errors in data and in model.
  By comparing observational data y and forecast data x_B with our best estimate x_A of the true state, we build up valuable error statistics which can be used to improve instruments or retrievals, and to improve models.
- Filling in data-poor regions.
  The model provides a way to propagate information from data-rich regions of space and time to data-poor regions, e.g., over the oceans.

- Designing observational systems and networks.
  The data assimilation system can be used to design "observing system simulation experiments" to test the value of particular measurements and observing systems. Typically, the model is sampled by the observing system as if it were the real atmosphere. These synthetic measurements are then fed through the assimilation system to see if $\mathbf{x_A}$ comes closer to the actual state of the model.
- Maintaining physical and chemical consistency.
  The model ensures that physical and chemical relationships among variables are maintained, to good approximation, despite data errors that would lead to inconsistencies.
- Estimating unobserved quantities.
  Through the relationships expressed in the governing equations of the model, the model provides information about quantities that are unobserved or inadequately observed: e.g., temperature observations are used to deduce winds; concentrations of unmeasured chemicals can be inferred from the observed evolution of others.

4. How is data assimilation done?

How does one combine different estimates of atmospheric variables provided by a set of instruments and a model background field? An optimal estimation method is used to combine an *a priori* estimate $\mathbf{x_B}$ of the atmospheric state with an estimate based on observations $\mathbf{y}$ of that state. One method is to take a linear combination of these estimates weighted by the inverses of error covariance matrices that measure our confidence in the estimates. For this method to work as stated, any biases in $\mathbf{x_B}$ and $\mathbf{y}$ should be removed beforehand. Such a statistical method is widely used in data assimilation for the objective analysis of atmospheric data. When the statistics are accurate and when no approximations are used in applying them, the data assimilation method is called optimal interpolation (OI). In practice, methods are seldom optimal.

4.1. A SIMPLE STATISTICAL ALGORITHM

We consider for simplicity the case of combining estimates of a scalar quantity $f(\mathbf{r})$ (the univariate case) where $\mathbf{r}$ indicates the three-dimensional spatial coordinates (x, y, z). Let $f_A(\mathbf{r}_i)$ be the desired analyzed value of f at the analysis (model) gridpoint $\mathbf{r}_i$, f_B the background (first-guess or *a priori*) value of f at $\mathbf{r}_i$, and $f_o(\mathbf{r}_k)$ and $f_B(\mathbf{r}_k)$ the observed and background values at the observation points, the latter obtained by interpolation in space from gridpoint values ($i =$

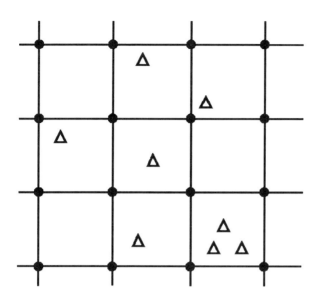

Figure 2. Schematic showing the horizontal positions, with respect to a model grid, of increments Δ. The increment is the difference between the observed value at the indicated position and the estimated model value, obtained by interpolating from the grid to the position of the observation.

gridpoint; k = observation point). Then our estimate for $f_A(\mathbf{r}_i)$ can be written:

$$f_A(\mathbf{r}_i) = f_B(\mathbf{r}_i) + \sum_{k=1}^{K} W_{ik}[f_o(\mathbf{r}_k) - f_B(\mathbf{r}_k)] \tag{1}$$

where K is the number of observation points and W_{ik} is the weight, to be determined, indicating the weight to be given to mismatches between the background field and the observations. The weights will depend on our confidence in $f_o(\mathbf{r}_k)$ and $f_B(\mathbf{r}_k)$ and on the correlation between quantities at $\mathbf{r}_i$ and $\mathbf{r}_k$. Weights are chosen to minimize the expected analysis error variance (see Daley [3], Chapter 4). Thus, if the model background value is believed to less accurate than the observed value, the model value is given the lower weight.

If we call the quantity in square brackets an increment, the procedure expressed by Equation (1) amounts to adjusting the model background field by a weighted field of increments, calculated at the observation points and then interpolated onto the model's gridpoints. Figure 2 shows schematically a possible horizontal distribution of increments with respect to the model's gridpoints. Figure 3 shows schematically the effect on the model's trajectory through state space

380

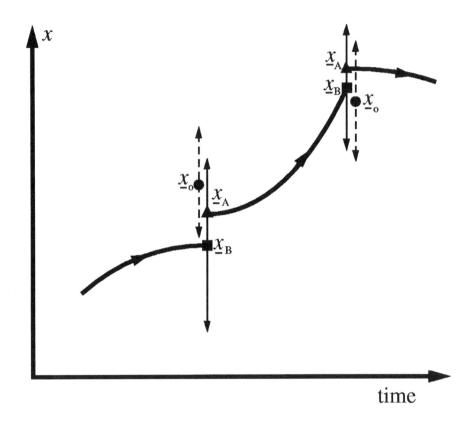

Figure 3. Schematic showing how a model's trajectory through its state space is adjusted to accommodate new observational data. The solid two-ended arrow indicates the expected error in the model forecast; the dashed two-ended arrow indicates the expected error in the observation. The adjusted model state falls within the ranges of these errors.

when this adjustment procedure is applied periodically. The adjusted model state, from which the next forecast is made, falls within the error bounds of the observed state and the model background state, lying closer to the state with the smallest errors. In practice, the adjustment may be spread out over a few hours to avoid a shock to the model, which would generate noisy fields by exciting inertia-gravity modes. Alternatively, the model may be "initialized" before a forecast is made. to prevent the excitation of inertia-gravity modes [10] [1].

4.2. A GENERALIZED ALGORITHM: MULTIVARIATE CASE

Because atmospheric variables are related to each other (e.g., temperature and winds; ozone and nitrous oxide) we need a multivariate algorithm that can treat all these variables simultaneously. We now write down a general method for combining observations and model background fields. Let $\mathbf{y}$ be the vector of observations available at (or attributed to) a certain time, and $\mathbf{x_B}$ the vector of forecast values of all atmospheric variables at all gridpoints ($\mathbf{x_B}$ has dimension 10^6 or 10^7!). Let $\mathbf{x_A}$ be the desired analysis that combines the available information. We construct a so-called penalty function that penalizes mismatches between our estimated state vector $\mathbf{x_A}$ and the observations $\mathbf{y}$, and between $\mathbf{x_A}$ and $\mathbf{x_B}$. We penalize ourselves according to our error estimates (error covariances) of $\mathbf{y}$ and $\mathbf{x_B}$: the greater the confidence we have in say $\mathbf{y}$, the more we penalize mismatches between our estimate $\mathbf{x_A}$ and $\mathbf{y}$. Thus:

$$J(\mathbf{x_A}) = [\mathbf{y} - \mathbf{K}(\mathbf{x_A})]^T \mathbf{O}^{-1}[\mathbf{y} - \mathbf{K}(\mathbf{x_A})] + [\mathbf{x_B} - \mathbf{x_A}]^T \mathbf{B}^{-1}[\mathbf{x_A} - \mathbf{x_B}] \quad (2)$$

$\uparrow$
scalar

$\mathbf{O}$ is the error covariance matrix of the observations telling us, for example, how observation errors are correlated spatially. See Section 6, which explains what an error covariance matrix is.

$\mathbf{B}$ is the error covariance matrix of the model forecast (background) field, providing us with analogous information.

$\mathbf{K}$ is the operator that takes us "from model space to observation space". For example, the elements of $\mathbf{y}$ refer to observation points; the elements of $\mathbf{x_A}$ to model gridpoints. To calculate the mismatch, we interpolate from model gridpoints to the observation points. Then $\mathbf{K}$ would be the interpolation operator.
Procedure:

To get our optimal estimate of $\mathbf{x_A}$, minimize J with respect to $\mathbf{x_A}$ by moving down the (local) gradient of J until the minimum in J is reached to a prescribed tolerance (Figure 4). A Newtonian descent algorithm can be used, or a variant thereof, to get J_{min}. See, for example, Gill et al. [8] for an account of descent algorithms. A complication is that if $\mathbf{K}$ is a nonlinear operator there may be more than one J_{min}.
Notes on the method:

(1) The procedure depends crucially on our knowledge of the error covariance matrices $\mathbf{O}$ and $\mathbf{B}$. Often these are not well known or may change as the atmosphere evolves. The error covariances are often modelled in terms of algebraic functions, e.g., of the distance between points: the greater the distance between points, the less errors are assumed to be correlated.

(2) This generalized algorithm allows us to assimilate the measured quantity directly as long as that quantity can be estimated from the model state. Thus if $\mathbf{y}$

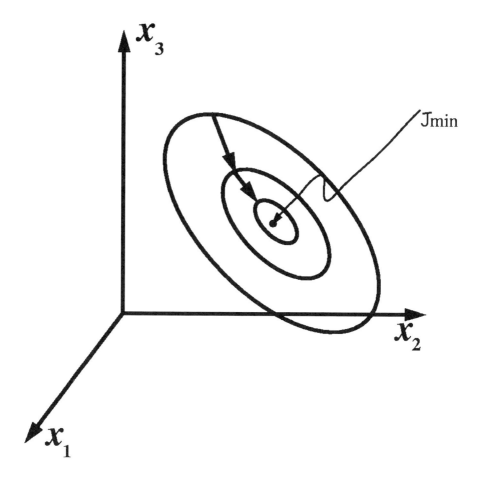

Figure 4. Surfaces of constant J in the model's state space at a given time. Arrows show the direction of movement as the state vector is adjusted to minimize J.

were a vector of satellite radiances, $\mathbf{K}(\mathbf{x_A})$ would be the corresponding estimate of radiances derived, using a radiative transfer algorithm, from the model state $\mathbf{x_A}$. Thus we avoid introducing errors by an off-line retrieval process.

(3) Procedures can be designed to accept data in finite windows of time (analysis times) or continuously (better for asynoptic satellite data). Some procedures lead to states $\mathbf{x_A}$ that are not in "balance" (e.g., geostrophic balance), in which case an initialisation step is included in the assimilation cycle. If suitable constraints are built into the penalty function J, an initialisation step can be avoided.

(4) The method just described takes in data only within a certain (short) time window to produce a three-dimensional analysis $\mathbf{x_A}$. Past data affects this anal-

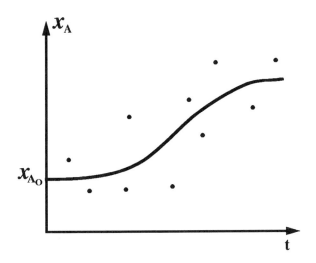

Figure 5. Schematic showing the model's trajectory through state space resulting from a given choice of initial conditions (and possibly of other control variables). The dots denote observed values. The initial state is determined to minimize the sum of the error-weighted mismatches between the trajectory and the observational data.

ysis implicitly, and information about the future state of the atmosphere is not included. Because x_A is varied at a given time to find the minimum of J, and hence the optimal analysis, the method is referred to as 3DVAR.

4.3. 4DVAR

The formulation of the 3DVAR method just described serves as the basis of a more general, and potentially more powerful, formulation of the atmospheric data assimilation problem. This is four-dimensional variational data assimilation (4DVAR). It demands a close fit to the data (within the error bounds of the data), plus consistency with a dynamical model over an extended period of time. Since the model is deterministic (given the initial state, future states are preordained), the method involves adjusting this initial state, x_{Ao}, to give a best fit to the data, weighted by error covariances (Figure 5). The state x_{Ao} is then a vector of "control variables" of the system, which controls the fit to the data. These control variables could include model parameters that we should like to estimate from the demand of consistency with the observational data (e.g., an unknown chemical reaction rate in a set of photochemical equations).

Procedure:

Construct a penalty function as before, but this time (a) summing over time to measure the misfit over an extended period of time, (b) treating the model as perfect, (c) minimising $J(\mathbf{x_{Ao}})$ with respect to the initial state $\mathbf{x_{Ao}}$.

The penalty function is

$$J(\mathbf{x_{Ao}}) = \sum_{n=0}^{N} [\mathbf{y_n} - \mathbf{K}(\mathbf{x^F})]^{\mathrm{T}} \mathbf{O}^{-1} [\mathbf{y_n} - \mathbf{K}(\mathbf{x^F})] \qquad (3)$$

where $\mathbf{O}$ is the error covariance matrix of the observations, $\mathbf{x^F}$ is the model forecast starting from state $\mathbf{x_{Ao}}$ (which therefore depends implicitly on $\mathbf{x_{Ao}}$), and other notation is as in Section 4.2.

How do we achieve this minimisation? Details will not be given here; a readable account can be found in Daley [3], Section 13.2. One method - not at all an obvious approach - is to find the minimum of J by constructing the linearized version of the model (the so-called tangent linear model), taking the adjoint of this tangent linear model and integrating it backwards in time. In order to ensure that the initial state (and consequently all later states in the time window) is consistent with the end state of a previous assimilation cycle, a background term may be added to Equation (3) to penalize mismatches between initial and end states [15].

Rabier *et al.* [12] find that 4DVAR has some advantages over other data assimilation methods. It produces more accurate analyses at the end of the data assimilation period; there is much smaller error growth in subsequent analyses; and it is more robust in controlling the spurious excitation of inertia-gravity modes. A particular benefit of the method is that it can be used to derive wind fields from the evolution of trace chemical fields, since these fields are implicitly related in time in Equation (3). Fisher and Lary [5] have used a variant of the method, using precomputed air parcel trajectories, to assimilate photochemically active trace chemical data.

4.4. KALMAN FILTER

The 4DVAR algorithm is very powerful but it has the drawback that the forecast model is introduced as a strong constraint on the minimisation of J - which is appropriate only if the model is perfect. This limitation is avoided by use of the Kalman filter technique. The model is assumed to be imperfect, and its forecast error covariance matrix is updated as the integration proceeds by using the governing equations of the model and error fields at previous timestep. Descriptions of the application of this method to the atmospheric data assimilation problem can be found in Ghil *et al.* [6], Dee *et al.* [4] and Parrish and Cohn [11]. Though theoretically very attractive, the method is very expensive to use because it requires many matrix operations with matrices of dimension $10^6 \times 10^6$!

5. Concluding remarks

The stored archives of meteorological analyses (three-dimensional state vectors) produced by operational data assimilation systems have provided a very valuable resource for climate modellers. The analyses enable detailed studies to be made of climate processes, including calculations of momentum and heat budgets. For instance, the European Centre for Medium Range Weather Forecasts (ECMWF) has produced a fifteen-year series of global analyses from 1979 to 1993 by assimilating historical meteorological data into a data assimilation scheme that was not changed during the period [7]. Keeping a fixed assimilation system reduced the risk of introducing changes in the resulting time series that might have been interpreted as a change in climate. ECMWF plans to produce a 40-year reanalyis of meteorological data, dating from 1958. There are similar activities at other operational weather centres.

New satellite programmes should provide a wealth of information on the climate system. The European Space Agency's Envisat satellite, scheduled for launch in late 2000, will carry instruments making measurements of the atmosphere, ocean and land. The U.S. National Aeronautic and Space Administration's Earth Observer System, deploying several satellites around the turn of the century, will also provide a comprehensive suite of climate measurements. Data assimilation will be an essential tool to maximize the scientific return from the expensively acquired data.

6. Supplement: the (error) covariance matrix S_x

The concepts of variance (and standard deviation) of a scalar quantity x will be familiar to the reader. When we work with a vector quantity $\mathbf{x}$, the analogue of variance is covariance. The covariance matrix S_x of vector quantity $\mathbf{x}$ is defined as:

$$S_x = E\{\mathbf{x}\mathbf{x}^T\}$$

where S_x is a square matrix, $E\{\ \}$ is the expectation of what is inside $\{\ \}$, $\mathbf{x}^T$ is the transpose of $\mathbf{x}$ (so $\mathbf{x}\mathbf{x}^T$ is a square matrix). For a sample with N elements,

$$S_x = N^{-1} \sum \{(\mathbf{x_i} - <\mathbf{x}>)(\mathbf{x_i} - <\mathbf{x}>)^T\}$$

where $<\mathbf{x}>$ is the mean of the sample. If not stated explicitly, assume the sample mean has been subtracted off. The diagonal elements of S_x are the variances of the elements of $\mathbf{x}$, i.e.,

$$\sigma_i^2 = (S_x)_{ii}$$

where σ_i^2 is the variance of x_i. The off-diagonal elements give a measure of the correlation between vector elements or of their errors (with respect to the sample mean). Error covariance matrices are used in Equations (2) and (3).

386

References

1. Baer, F. (1977) Adjustments of initial conditions required to suppress gravity oscillations in non-linear flows, *Beitr. Phys. Atmosph.* **50**, 350-366.
2. Courtier, P., Thépaut, J.-N., and Hollingsworth, A. (1994) A strategy for operational implementation of 4D-Var, using a incremental approach, *Q. J. R. Meteorol. Soc.* **120**, 1367-89.
3. Daley, R. (1991) *Atmospheric Data Analysis*, Cambridge University Press, Cambridge.
4. Dee, D., Cohn, S., Dalcher, A. and Ghil, M. (1985) An efficient algorithm for estimating noise covariances in distributed systems, *IEEE Trans. Automatic Control* **30** (11), 1057-65.
5. Fisher, M. and Lary, D.J. (1995) Lagrangian four-dimensional variational data assimilation of chemical species, *Q. J. R. Meteorol. Soc.* **121**, 1681-1704.
6. Ghil, M., Cohn, S., Tavantzis, J., Bube, K. and Isaacson, E. (1981) Applications of estimation theory to numerical weather prediction, in L. Bengtsson, M. Ghil, and E. Kallen (eds.), *Dynamic meteorology: data assimilation methods*, Springer Verlag, New York, pp. 139-224.
7. Gibson, J.K., Kallberg, P., Uppala, S., Hernandez, A., Nomura, A. and Serrano, E. (1997) ECMWF re-analysis project report series, 1. ERA description, European Centre for Medium Range Weather Forecasts, Shinfield, U.K.
8. Gill, P., Murray, W. and Wright, M. (1981) *Practical optimization*, Academic Press, London.
9. Lorenc, A.C., Bell, R.S., Macpherson, B. (1991) The Meteorological Office analysis convection data assimilation scheme, *Q. J. R. Meteorol. Soc.* **117**, 59-89.
10. Machenauer, B. (1977) On the dynamics of gravity oscillations in a shallow water model with application to normal mode initialization, *Contrib. Atmos. Phys.* **50**, 253-271.
11. Parrish, D. and Cohn, S. (1985) A Kalman filter for a two-dimensional shallow-water model: Formulation and preliminary experiments, National Meteorological Center, Office Note 34, Washington D.C., U.S. Dept. of Commerce, NOAA, National Weather Service.
12. Rabier, F., Courtier, P., Pailleux, J. Talagrand, O. and Vasiljevic, D. (1993) A comparison between four-dimensional variational assimilation relying on three-dimensional variational analyses, *Q. J. R. Meteorol. Soc.* **119**, 845-880.
13. Talagrand, O. and Courtier, P. (1987) Variational assimilation of meteorological observations with the adjoint vorticity equations. Part I. Theory, *Q. J. R. Meteorol. Soc.* **113**, 1311-1328.
14. Tarantola, A. (1987) *Inverse Problem Theory*, Elsevier, Amsterdam.
15. Zupanski, D. (1997) A general weak constraint applicable to operational 4DVAR data assimilation systems, *Mon. Wea. Rev.* **125**, 2274-2292.

SEASONAL PREDICTIONS

P.W. MOTE AND E.S. SARACHIK
Joint Institute for the Study of the Atmosphere and Oceans
University of Washington
Box 354235
Seattle, WA 98195 USA

M. DEQUE
Météo-France
Centre National de Recherche Météorologique
42 Avenue G. Coriolis
F-31057 Toulouse
FRANCE

1. Introduction

General circulation models (GCMs) were first developed for the purpose of augmenting existing weather prediction techniques, which were primarily empirical and heuristic. Miyakoda (Chapter 1) provides an insightful overview of the early years of GCM development with an emphasis on numerical weather prediction (NWP) and its extension to longer range. The first climate studies with a GCM were performed primarily to evaluate biases in the GCM's mean state, since biases could affect weather forecasts. Climate modeling[1], the primary subject of this book, matured into a self-sustaining discipline almost entirely separate from NWP.

Recently a third discipline using GCMs has developed: seasonal predictions. Seasonal predictions, which focus fairly qualitatively on a few key climate variables like surface temperature and precipitation, are distinct in purpose, in approach, and in timescale from both NWP and climate modeling. This chapter compares the goals and techniques used in these three disciplines and discusses

---

[1] In this book we use the term "climate modeling" to refer to studies of the long-term mean climate. We use the term "seasonal predictions" to mean predictions of seasonal anomalies, thereby avoiding the confusing nomenclature "climate predictions," which is often used to mean seasonal predictions.

P. Mote and A. O'Neill (eds.), Numerical Modeling of the Global Atmosphere in the Climate System, 387–402.
© 2000 *Kluwer Academic Publishers. Printed in the Netherlands.*

some aspects of the recent development in seasonal predictions, of which GCMs are but one tool in a growing toolbox of techniques. Palmer and Anderson [18] provided a comprehensive review of seasonal predictions, to which the interested reader is referred; our focus here is on the role of GCMs in seasonal predictions. We discuss the prediction of El Niño–Southern Oscillation (ENSO) as a basis for global seasonal predictions. We also present an example of a coordinated effort to evaluate and improve seasonal forecasts through "hindcasting," that is, by carrying out a forecasting exercise for periods in the past.

1.1. QUANTIFYING PREDICTIVE SKILL

Evaluating the skill of a prediction requires the following:

- adequately complete and accurate verifying observations;
- a quantitative measure of predictive skill;
- a threshold value of that measure, below which forecasts are considered to be useless.

For time-dependent measures of skill, the time at which the measure of predictive skill falls below the threshold value is often called the "limit of predictability." The predictive skill is a function of (1) the field considered, (2) the model used for the forecast, and (3) the initial state of the system. Limitations on predictive skill arise from both the imperfections of the forecast model (see Section 4 for examples of "perfect" boundary conditions) and the nonlinearity of the climate system. From the perspective of nonlinear dynamical systems (Chapter 2), predictive skill is low when the trajectory through phase space passes through a bifurcation point. Therefore, the predictive skill should be calculated for a large number of situations in order to make the most general statement possible. Also, any statement about predictive skill should include the caveat "for this model."

For NWP, the data used for verifying observations are usually the gridded analyses produced by assimilating data from surface, *in situ*, and satellite observations (Chapter 15). At most forecast centers, both the assimilation and the forecast are carried out using closely related versions of the same model. A common measure of forecast skill in NWP is the anomaly correlation (AC), i.e., the time-dependent correlation between the observed anomaly of a field (for example 500 hPa height) and the forecast of the anomaly (where the anomaly can be defined with respect to the model's mean field, removing model bias) over some region of the earth. The conventional threshold for midlatitude skill is AC=0.6 [16]. For many measures of predictive skill, the threshold can be taken as the level at which the skill of the forecast equals the skill of a "forecast" using persistence (i.e., the initial state).

Evaluations of the predictive skill at seasonal timescales are far less developed. Because seasonal forecasts are usually time means, the standard time-dependent AC does not apply. Furthermore, the AC is an inappropriate skill score for evaluating forecasts of probabilities (as is done, e.g., at the U.S. Climate Prediction

Center). A single AC does, however, have some value, especially when evaluated over a large number of forecasts (see Section 4), because it quantifies the pattern correlation while at the same time removing the model's systematic error which (for a seasonal forecast) may be comparable to the observed anomaly.

It is not yet clear what threshold constitutes a useful forecast for seasonal predictions, though some results suggest a value of 0.3 (Section 4). A threshold based on the skill of a persistence forecast is inappropriate, since the seasonal cycle changes both the means and the variance of the background state so significantly over the timescales of the prediction that a persistence forecast (even as a departure from climatology) soon has no skill at all.

The foregoing discussion concerns seasonal predictions of the surface fields (generally temperature and precipitation) themselves, but at the Climate Prediction Center the predictions are of shifts in the probability distributions of temperature and precipitation. That is, they describe the likelihood that the temperature or precipitation will fall in the upper, middle, or lower third of the distribution. This approach requires a different skill score, e.g., the Heidke skill score.

1.2. THE BASIS FOR SEASONAL PREDICTIONS

Why would we expect a seasonal prediction to have any skill at all? An illustration of the basis for seasonal predictions is given in Figure 1, which shows the AC (see section 1.1) of 500 hPa height. The solid curve in Figure 1 is the AC of daily "forecasts" (using observed sea surface temperature, or SST; see section 4), as a function of lead time. After 1 day, the AC is 0.98, but it falls with increasing rapidity to less than 0.50 and then approaches zero at about 20 days. Beyond 20 days, however, the average value of the daily AC is slightly positive, possibly indicating that a very small degree of skill remains. Further evidence for some skill at long lead times is given by the thin dotted curve, which shows the AC of running 30-day means. If the daily data had no skill at all, neither would the monthly data. Time means evidently reduce the unpredictable "noise" and emphasize the predictable features, which are doubtless associated with anomalies in the lower boundary conditions rather than with some deterministic predictability.

To further emphasize predictable features, one can perform ensemble forecasts of 30-day means, where the ensemble members are generated by initializing with observations on succeeding days [11]. The thick dotted curve in Figure 1, which corresponds to the scores of an average of nine forecasts starting at situations lagged by 24 h, is indeed above the dashed curve. When the size of the ensemble tends to infinity, the AC does not tend to unity, even if the model is perfect, but is bounded by the square root of the correlation with a single forecast [9]. This limit on the AC occurs partly because we cannot produce an ensemble of observed data, and the observed atmospheric flow can be considered as one possible realization

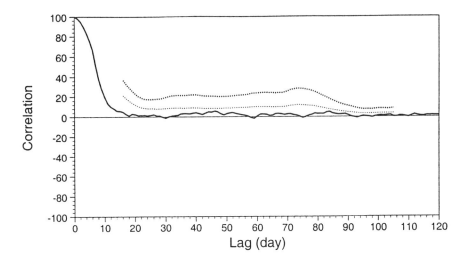

Figure 1. Anomaly correlation over the northern hemisphere for winter 500 hPa height as a function of forecast range (day): daily data (solid line), 30-day means (thin dotted line), 30-day ensemble means (thick dotted line).

among many equiprobable ones that could have been produced by identical SST forcing.

There is, however, a simple way to produce an ensemble of observed *months*: by considering a seasonal average. Instead of predicting the month-to-month evolution, as in the time-dependent extended-range forecast ([5], [7]), we simply seek to duplicate the mean response of the atmosphere to the mean SST forcing ([24], [4]). Indeed, the large-scale SST anomalies have a lifetime of several months, and there is little reason to expect that the pure *monthly mean* response to SST could change dramatically from one month to the next.

2. Three Types of Predictions

In the introduction, we noted that GCMs have been applied to three fundamentally different problems: weather prediction, climate studies (which include predictions of long-term climate changes), and seasonal predictions. While all use GCMs, the goals and techniques differ.

2.1. WEATHER PREDICTION

The goal in NWP is to forecast the exact state of the atmosphere from initial conditions, at high time resolution over several days. A large and demanding clientele, the general public, wants forecasts at high spatial and temporal resolution of temperature, precipitation (timing, quantity, duration, and phase), cloudiness (amount of cloud, altitude of cloud base), wind speed and direction, frost, and fog. For some applications (e.g., aviation), forecasts are needed not just at the surface but also above the boundary layer in the free atmosphere, where verifying observations are far fewer. To accommodate these demands, NWP must accurately forecast both the exact large-scale state of the free atmosphere and the interaction of the large scales with smaller-scale topography, which can have a profound influence on local weather, especially as perceived at the surface.

The need to model accurately the interaction of large-scale atmospheric flow with smaller-scale topography has led to the development of a variety of regional modeling approaches, which are outlined in Chapter 17. Operational NWP centers use such approaches to produce more detailed dynamical information, but in the final stage of most forecasts, human forecasters apply the dynamical information to the local environment using a combination of statistical techniques, experience, and intuition.

While it may come as a surprise to the skeptical public, there are aspects of the state of the atmosphere that are predictable (see Section 1.1) as much as 2 weeks ahead. These fairly predictable aspects are the aforementioned large-scale features in the free atmosphere. The limit of predictability at midlatitudes is usually considered to be 10–14 days at these large scales [14], up from just 2–3 days a few decades ago.

In NWP, the slowly evolving lower boundary conditions (SST being the most important) are often assumed to be constant, since short-term predictability arises solely from internal atmospheric dynamics. Successful predictions require detailed and accurate knowledge of the initial state of the atmosphere, and a variety of observing systems are in place to provide this knowledge. Owing to the nonlinearity of the atmosphere, however, small errors in the initial state can grow rapidly and lead to a poor forecast even with a perfect model. The range of possible outcomes can be estimated using ensemble forecasts, in which slightly different initial conditions are used for each of several forecasts. In order for the correct forecast to be somewhere within the envelope of the ensemble, the ensemble initial conditions must capture the most rapidly growing disturbances; various techniques ([17], [26]) have been developed to generate such initial conditions.

2.2. PREDICTION OF LONG-TERM CLIMATE CHANGE

For predicting climate changes that might result from changing greenhouse gas concentrations (see Chapter 21), or for reconstructing past climates (Chapter 20), a different set of requirements applies. The goal is to characterize changes in the long-term mean atmospheric and oceanic circulation and especially to characterize mean changes at the earth's surface. The appropriate tool is a coupled atmosphere-ocean-land-ice, or climate system, model (Chapter 18).

The climate system has been forced, since the industrial revolution, by the addition of radiatively active constituents, both gaseous and particulate, to the atmosphere. Although the atmosphere alone responds fairly rapidly to changes, many components of the climate system have a long response time. For this reason, a rigorous prediction of the effects of the changes in radiatively active constituents requires us to know both the history of the forcing and the initial state of the entire climate system, or at least those components with a long response time. In the absence of such knowledge of the initial state, a less rigorous prediction can be made by calculating the forced response. Only the statistics of the boundary conditions are predicted, so that only the gross statistics of the response can be known. For example, while we cannot predict the characteristics of a specific summer, we can say that summers 50 years hence will be warmer, on average, than they are now.

2.3. SEASONAL-TO-INTERANNUAL PREDICTION

Whereas NWP must predict the time evolution of the exact state of the atmosphere, and long-term climate predictions are concerned with gross features of a changed climate averaged over many years, seasonal predictions seek to describe statistical aspects of atmospheric anomalies averaged over 1–3 months. For example, at the time of writing, the U.S. Climate Prediction Center[2] and the International Research Institute for climate prediction[3] issue forecasts of the probability that 3-month mean temperatures and precipitation will be significantly above or below normal. These forecasts are issued for 3-month periods that are from a few weeks to a few months away.

The basis of these and all predictions at timescales longer than a month is the hypothesis that, on these timescales, the atmospheric statistics are in equilibrium with the surface boundary conditions and that a prediction of the boundary conditions will lead to some statistical knowledge of the atmosphere (see Figure 1).The boundary conditions of interest are the SST, the soil moisture, the sea ice extent, and the surface albedo. In general, these boundary conditions vary slowly and therefore impose a slow variation of atmospheric statistics. The boundary conditions themselves involve a strong interaction with the atmosphere: soil moisture can be calculated only by knowing the history of rainfall and evaporation, albedo by knowing the distribution of snow and ice, and sea surface temperature by knowing the fluxes of heat and momentum from the atmosphere. Along with these interactions with the atmosphere, each boundary condition evolves under its own dynamics. The proper tool for calculating their mutually consistent evolution is therefore (as with long-term climate predictions) a climate system model. Recent

---

[2] URL: http://nic.fb4.noaa.gov
[3] URL: http://iri.ucsd.edu/forecast/net_asmt/

work by Saravanan [23] suggests that the atmospheric statistics in middle latitudes cannot be predicted as well by using fixed (or separately forecast) boundary conditions as by using a fully coupled model.

The forecasting procedure, then, ideally would proceed by first initializing the state of the coupled system with observations of the coupled system at the initial time and, second, allowing the coupled model to run freely for the forecast time of a season or longer. In practice, predictions are compromised by the fact that the full climate system can neither be measured nor be modeled completely. Only some of the boundary conditions can be initialized, and only some are important globally. The key to global seasonal predictions lies in the tropical Pacific Ocean.

3. Predicting Global Anomalies by Predicting ENSO

Apart from the seasonal cycle, El Niño–Southern Oscillation is the dominant mode of variability of the ocean-atmosphere system (e.g., p. 415 of [19]). It is for this reason that efforts at seasonal prediction, outlined above, have focused on (1) predicting the SST variations characteristic of ENSO and (2) linking these SST variations to variations in seasonal anomalies around the world. GCMs have played an important role in the latter and are rapidly becoming important tools for the former as well, a task for which other tools have heretofore dominated.

SST anomalies in the tropical Pacific characteristic of ENSO have strong associations with atmospheric anomalies in the rest of the tropical Pacific basin, moderate to strong associations with atmospheric anomalies in other parts of the tropics, and at best moderate associations with anomalies in midlatitudes ([2], [10], [13]). Skill for seasonal predictions appears to have similar spatial variations.

3.1. PREDICTING ENSO

As a result of the international Tropical Ocean–Global Atmosphere (TOGA) Program, an observing system for the upper tropical Pacific Ocean has been put into place [15]. The real-time observations made by this system[4] form a crucial component of efforts to predict ENSO.

In contrast to the theoretical type of short-term climate forecasting described in Section 2.3, in which the entire climate system is initialized and modeled, most practical forecasting proceeds as follows. For step 1, the first task is to initialize just the tropical Pacific upper ocean by means of the TOGA observing system and a data assimilation procedure for the ocean, and the second task is to run a coupled atmosphere-ocean model from this ocean initial state with only the tropical Pacific active in the global ocean model (with the anomalies persisting in midlatitudes) to predict only the SST in the tropical Pacific. Step 2 is discussed in section 3.3.

---

[4] URL: http://www.pmel.noaa.gov/toga-tao/realtime.html

394

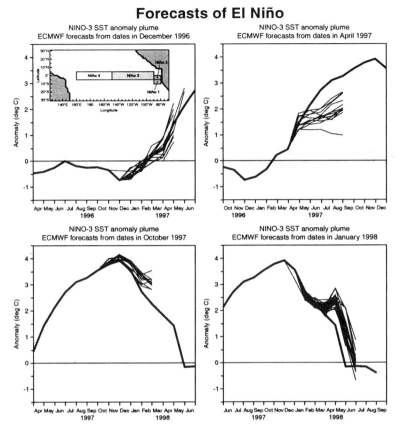

Forecasts of El Niño

Figure 2. Forecasts of the NINO3 index (average SST in the box shown) by the ECMWF coupled ocean-atmosphere model. In each panel, several forecasts are made from different atmospheric initial conditions. Figure courtesy of David Anderson, ECMWF.

An example of a dynamical ENSO SST prediction is shown in Figure 2, which shows forecasts of the NINO3 index by the European Centre for Medium-Range Weather Forecasts (ECMWF) coupled ocean-atmosphere model [25]. Model runs are initialized with new atmospheric initial conditions three times weekly, providing an ensemble of ENSO forecasts.

The earliest prediction of the tropical Pacific SSTs was accomplished by a simple diagnostic atmospheric model coupled to a simple ocean model and initialized by a history of the winds [6] without any internal ocean data. Because the tropical Pacific upper ocean thermal structure depends primarily on a history of wind forcing, there is a certain redundancy between the history of the winds and internal upper ocean temperature. Present improvements to the procedure include more sophisticated ocean data assimilation systems that exploit this redundancy by using the following: estimates of the internal state of the ocean from combinations of data, model forecasts, and wind histories; more sophisticated parameterizations of physical processes that more correctly describe the details of cloud processes and oceanic mixing; and innovative assimilation techniques that construct the state of the ocean's interior just from satellite measurements of the ocean's surface. A recent review of the state of the art of ENSO prediction is given by Latif *et al.* [12].

3.2. ENSO PREDICTABILITY

ENSO-related SST predictions are scored by correlating the observed time series of SST with the predicted time series in various portions of the Pacific and by calculating the rms differences between them. The skill of the SST forecast is a property of a long series of predictions. The predictions of various quantities either local or remote from the tropical Pacific, e.g., monthly averaged temperature or precipitation anomalies, are then scored similarly.

There are three basic sources of error in forecasts of tropical Pacific SST: (1) initial errors in both the atmosphere and ocean, (2) errors added to the system by unpredictable high-frequency noise in both the atmosphere and the ocean as the coupled system evolves, and (3) model errors. The first source can be minimized by more accurate and dense measurements in the ocean. Like the atmosphere, however, the ocean is a chaotic dynamical system, and even in the absence of the second source of errors, arbitrarily small initial errors will eventually grow large enough to render the prediction useless, at which point the system has passed the limit of predictability (see Section 1.1).

The potential limit of predictability of ENSO is not known and depends on unresolved questions about the dynamics of ENSO itself. If ENSO is due to the so-called "delayed oscillator" of the atmosphere-ocean system [1], the potential predictability will depend on how long an initial signal in the thermocline can retain its identity in the face of random noise from the atmosphere, most likely several years. If the dynamics of ENSO instead involve random forcing of transient growing signals in a stable coupled system [20], then the potential predictability is likely to be not much longer than a year. Since there is no observational method yet devised to distinguish whether or not the coupled system is unstable, and therefore no observational method of choosing between these mechanisms, the potential predictability of SST variations in the tropical Pacific has yet to be determined. As a practical matter, there appears to be predictability of SST at least a year in advance, but without knowing more about the potential predictability, it is impossible to say how close this practical limit approaches the theoretical limit and therefore how much improvement in skill can be expected in the future.

3.3. PREDICTING THE EFFECTS OF ENSO

ENSO-related changes in temperature and wind at the surface of the tropical Pacific have robust effects in the region directly adjoining the tropical Pacific and less robust effects throughout the rest of the world. Once the state of ENSO has been forecast (see section 3.1), its effects on the rest of the world are usually predicted either by running a statistical correlation model or by running a high-resolution atmospheric GCM a number of times to form an ensemble. The initial conditions of such an ensemble are chosen from different days within a month, during which the predicted SST can be assumed not to have changed.

During the warm phase of ENSO, deep cumulonimbus convection lying over the warmest water (SST > 28°C, usually confined to the far western part of the tropical Pacific) moves eastward into the central Pacific as the area of SST > 28°C expands eastward. For very large warm events (e.g., 1982–1983 and 1997–1998) the area of warmest water extends across the entire equatorial Pacific, and heavy rain falls in the normally arid regions of Ecuador and northwestern Peru and in some of the central Pacific islands. Droughts in the far western Pacific are common during the warm phase of ENSO, especially in the islands of the maritime continent like Borneo and Java, and in the Malaysian peninsula and parts of northern Australia. In all of these cases, the effects of SST changes are direct: the deep convection moves with the warmest water, so that a prediction of SST is akin to a prediction of time-averaged precipitation.

When the area of warmest SST and deep convection moves, it forces a train of planetary-scale waves that propagate into middle latitudes, where the results are muddied by other sources of variability. Remote effects ("teleconnections") of tropical Pacific SST changes have been noted by Ropelewski and Halpert ([21],[22]). Predicting climate anomalies in middle latitudes from SST anomalies in the tropical Pacific is difficult for two reasons. First, the background variability is greater in middle latitudes. Second, the mechanisms by which atmospheric modes are forced in the tropics and propagate to higher latitudes are not completely understood; the role of midlatitude eddies in this process is particularly obscure (see [27] for a recent review). Nevertheless, trusting both that models reproduce even those parts of the atmosphere that are not yet understood and that the sample of observed midlatitude patterns is adequate to characterize the "true" teleconnections, predictions are made of the remote effects of the tropical SST variations.

The ideal way of making these predictions is to initialize and run a high-resolution global coupled atmosphere-ocean model, and in fact this is how ECMWF makes seasonal forecasts (Figure 2). However, elsewhere a "two-tiered system" [3] is widely used. The tropical Pacific SST is predicted with a coarse-resolution coupled model in which only the tropical Pacific is initialized (or, for that matter, active), and the remote effects are predicted by ensemble forecasting with a higher-resolution atmospheric GCM using the predicted tropical Pacific SST as boundary conditions. This approach is taken for several reasons:

- Computer limitations require coupled models to be of relatively low resolution.

— There are not enough data to initialize the global ocean.

— Prediction of the tropical ocean is nearly deterministic on these timescales, and thus only a small number of coupled model runs may be needed, whereas prediction of anomalies of the midlatitude circulation is probabilistic, so that large ensembles are needed.

— Many coupled models experience climate drift (Chapter 18). By predicting SST *anomalies*, and then adding the anomalies to the climatological SST field for an ensemble of atmospheric GCM forecasts, the effects of climate drift can be removed.

While there are definite modeling indications that the tropical SST drives atmospheric circulation patterns in the extratropics ([27], [13]), the role of midlatitude SSTs on these patterns is less well understood. Recent work by Saravanan [23] indicates that the two-tiered system with midlatitude SST specified may produce less overall variability than a fully coupled system. The overall skill of predicting remote effects of ENSO variations has not yet been quantified.

4. A Multiyear Multimodel Experiment: PROVOST

A first step in quantifying predictive skill of midlatitude seasonal forecasts is the evaluation of hindcasts, that is, forecasting exercises carried out for periods in the past. Three such projects have appeared: "Dynamical Seasonal Prediction" in the USA, "Reforecasting" in Canada, and "PROVOST" in Europe. PROVOST is partly funded by the European Commission and coordinated by the ECMWF. The core of the project is a 15-year (1979–1993) hindcast experiment in which ECMWF reanalyses are used for the initial conditions, the boundary conditions, and the verification data for four modeling groups. While the use of specified SST may appear to be equivalent to a perfect prediction of SST, recent work [23] mentioned in the preceding section implies that specified SST may not be so perfect. Additional experiments have used persistent SSTs or a coupled ocean-atmosphere model, which are more similar to real-time forecasts.

For each of the 15 years, four 4-month forecasts are produced, starting in late February, late May, late August, and late November. Four models have been used: the ECMWF model at a resolution of T63, the Météo-France model at T63 and T42 (the T63 forecast is limited to the winter cases and was run by the electricity company EDF, which participates in PROVOST), and the United Kingdom Meteorological Office (UKMO) model (a gridpoint model with a resolution equivalent to T42). Each forecast consists of nine model integrations started on successive days. It is important that all forecasts for each model be performed with the same configuration of the model and with the same kind of initial and boundary conditions to ensure the homogeneity of the database.

The total simulated time for each model in PROVOST—180 years—strains the resources of computer time and disk space and limits the resolution and the number of ensemble members. An ensemble size of nine is insufficient to estimate the spread of a given year, but the score of a nine-member average is rather close to the asymptotic value with an infinite ensemble [9]. While operational short- and medium-range forecasts use much higher resolution, and higher resolution for seasonal forecasts would be desirable, the costs are not presently justified by the gains, which are small. Studies of extended-range forecasts have shown that increasing the resolution beyond T42 produced only modest gain. In PROVOST, the mean scores of the four models are similar; there are no statistically significant differences among them, although scores of the two models at T63 often, but not always, exceed the scores of the two models at T42 (see also Figure 4).

4.1. EVALUATING PREDICTIVE SKILL

Evaluating predictive skill requires a large number of runs for two reasons: first, to establish the model climatology, and second, to average over large year-to-year variability in skill (Section 4.2). The model climatology must be known with accuracy because a seasonal forecast is an anomaly forecast: one calculates the difference between the model-predicted value and the model climatology. The seasonal mean anomalies are often about 1 K for temperature, or a few tens of millimeters per day for precipitation, which in some places is less than the model systematic error.

The first result of PROVOST was that temperature and precipitation can be predicted with statistically significant skill in some parts of the world. Skill is higher in the tropics than in midlatitudes (as was suggested in the preceding section), higher in North America than in Europe, and higher in winter than in summer for the middle latitudes. Even with 15 years, it is difficult to get a robust estimate of the local predictability (see Section 1.1).

Experience in PROVOST suggests that an appropriate threshold value of the anomaly correlation for a successful seasonal forecast could be defined as 0.3. That value is much lower than the conventional NWP threshold (0.6) for a useful forecast and typical of the AC of a NWP forecast at 10 days (see Figure 1). The skill of a single forecast is low, but averaging several forecasts produces higher skill. Averaging the 36 ensemble forecasts produces higher skill than averaging the 9 forecasts by a single model. It is not yet clear, however, whether the improvement results from ensemble size alone or whether additional skill is gained by the use of different models. Figure 3 shows the correlation, averaged over 15 winters, of the precipitation field in winter (day 31–120 average) for the multi-model ensemble forecast (i.e., the correlations between 15 observations and 15 forecasts, each forecast the average of the 36-member ensemble forecast).

If we consider probabilistic forecasts, that is, forecasts of the probability that

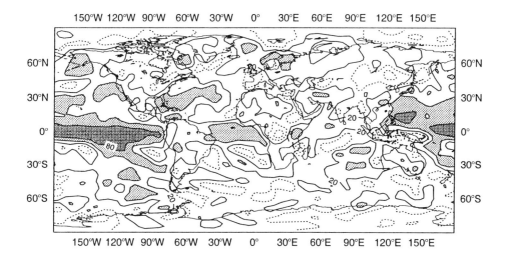

Figure 3. Anomaly correlation for winter (January–March) precipitation forecasts based on the four models of PROVOST.

the temperature or precipitation anomalies will fall in the upper, middle, or lower third of the probability distribution, then the impact of using different models seems to be larger. In the tropics, none of the models alone beats the climatological probabilistic forecast (which, for a given year, is for equal thirds based on the other 14 years), but the multimodel ensemble does. This is because an ensemble forecast with a single model leads to an underestimation of the uncertainty, since the model is assumed to be perfect. Using several models introduces an additional uncertainty due to the different modeling approaches, thereby widening the spread to more closely resemble the uncertainty.

The PROVOST runs discussed so far are those with prescribed SSTs. For the coupled ocean-atmosphere PROVOST runs, it is more difficult to evaluate the skill since the starting situations for the ocean are available for only the last 4 years of the PROVOST experiment from the TOGA observing system (Section 3.1). Using an autoregressive scheme for predicting the SST leads to poor skill scores in middle latitudes but better scores in the tropics. It appears better, in the tropics, to use observed SSTs and arbitrary initial conditions (as in AMIP; see Chapter 19) than to use forecast SSTs and observed initial conditions. In midlatitudes, both perfect initial and boundary conditions are necessary to get some skill.

4.2. INTERANNUAL VARIABILITY OF SKILL

For a number of quantities, interannual variability in skill clearly shows the influence of ENSO, apparently because the strength of ENSO guides the strength of the

400

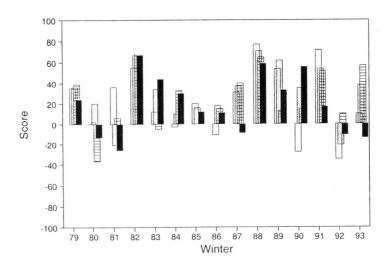

Figure 4. Anomaly correlation over the northern hemisphere for 15 winter 500-hPa-height seasonal forecasts: CNRM (blank), ECMWF (dot), EDF (dash) and UKMO (black).

response. For example, skill in hindcasting midlatitude winter precipitation is high for 1982–1983, 1988–1989, and 1991–1992, corresponding to both warm (1982–1983, 1991–1992) and cold (1988–1989) ENSO events. Similar results obtain for geopotential height, shown in Figure 4. Most of the models achieve the 0.3 threshold less than half the time, and scores in excess of 0.5 occur about as frequently as scores less than zero. On the basis of only a few years' experience, it would be easy to form—depending on the strength of ENSO events—too optimistic or too pessimistic a view of seasonal predictability. The interannual variability of skill underscores the point made in section 1.1 that evaluating predictability requires a large number of forecasts.

On the positive side, however, Figure 4 suggests that more accurate forecasts are possible, provided that one can predict *a priori* the "good years." Indeed, the time correlation between two series of scores is about 0.5, which shows that the time distribution of good and bad scores is not purely model dependent. The use of the ensemble spread in a single model's forecasts as a predictor of the skill is a natural choice: if the different forecasts agree, then there is an adequately strong response to the SST forcing. Unfortunately, this method can be deceptive, since an imperfect model may react to the forcing in the wrong way. On the other hand, strong forcing can generate some outliers, which increase the spread.

A more promising way of evaluating the skill is to consider several models. As can be seen in Figure 4, when at least one model has a score greater than 0.5, the other models also tend to do similarly well, whereas when at least one model has a negative score, the spread in skill is greater. Using several models in the ensemble is a way to take into account our uncertainty about the numerical representation of the laws of physics, since different models make different assumptions. By contrast, performing an ensemble forecast with a single model takes into account only the uncertainty about the initial state (the butterfly effect) and leads to biased probability forecasts [8].

Atmospheric reanalyses provide the consistent validation needed for multi-year hindcasting efforts like PROVOST. Coordination among all such hindcasting efforts offers a way to obtain multimodel ensembles and to address the problem of skill prediction. Experience gained in hindcasting with GCMs, and especially coupled ocean-atmosphere GCMs, will lead to better seasonal forecasting. GCMs are in some sense complementary to the empirical methods of forecasting in current use: empirical methods use the time evolution of a few fields, while GCMs use the accurate state of all the fields at a given date. Hybrid approaches to seasonal forecasting could lead to better results.

Acknowledgements

We wish to thank Franco Molteni, Nathan Mantua, Jeffrey Yin, and especially Ming Ji for comments on this chapter. We gratefully acknowledge Dr. David Anderson for providing figure 2. This is JISAO contribution number 640.

References

1. Battisti, D.S., and Hirst, A.C. (1989) Interannual variability in the tropical atmosphere/ocean system: Influences of the basic state, ocean geometry and nonlinearity, *J. Atmos. Sci.* **46**, 1687–1712.
2. Battisti, D.S., and Sarachik, E.S. (1995) Understanding and predicting ENSO, *Rev. Geophys.* **33 Supplement**, 1367–1376.
3. Bengtsson, L., Schlese, U., Roeckner, E., Latif, M., Barnett, T.P., and Graham, N. (1993) A two-tiered approach to climate forecasting, *Science* **261**, 1026–1029.
4. Brankovic, C., and Palmer, T.N. (1997) Estimates of seasonal predictability in ensemble integrations, *Mon. Weather Rev.* **125**, 859–874.
5. Brankovic, C., Palmer, T.N., Molteni, F., Tibaldi, S., and Cubasch, U. (1990) Extended-range predictions with ECMWF models: Time lagged ensemble forecasting, *Quart. J. Roy. Meteor. Soc.* **116**, 867–912.
6. Cane, M.A., Zebiak, S.E., and Dolan, S.C. (1986) Experimental forecasts of El Niño, *Nature* **321**, 827–832.
7. Déqué, M., and Royer, J.F. (1992) The skill of extended-range extratropical winter dynamical forecasts. *J. Clim.* **5**, 1346–1356.
8. Déqué, M., Royer, J.F., and Stroe, R. (1994) Formulation of Gaussian probability forecast based on model extended-range integrations, *Tellus* **46A**, 52–65.
9. Déqué, M. (1997) Ensemble size for numerical seasonal forecasts, *Tellus* **49A**, 74–86.
10. Graham, N. (1995) Simulation of recent global temperature trends, *Science* **267**, 666–671.

11. Hoffman, N.R., and Kalnay, E. (1983) Lagged average forecasting, an alternative to Monte Carlo forecasting, *Tellus* **35A**, 100–118.
12. Latif, M., Anderson, D., Barnett, T., Cane, M., Kleeman, R., Leetmaa, A., O'Brien, J.J., Rosati, A., and Schneider, E. (1998) A review of the prediction and predictability of ENSO, *J. Geophys. Res.* **103**, 14,375–14,393.
13. Lau, N.-C., and Nath, M.J. (1994) A modeling study of the relative roles of tropical and extratropical SST anomalies in the variability of the global atmosphere-ocean system, *J. Clim.* **7**, 1184–1207.
14. Lorenz, E.N. (1982) Atmospheric predictability experiments with a large numerical model, *Tellus* **34**, 505–513.
15. McPhaden, M.J., Busalacchi, A.J., Cheney, R., Donguy, J.-R., Gage, K.S., Halpern, D., Ji, M., Julian, P., Meyers, G., Mitchum, G.T., Niiler, P.P., Picaut, J., Reynolds, R.W., Smith, N., and Takeuchi, K. (1998) The Tropical Ocean–Global Atmosphere observing system: A decade of progress, *J. Geophys. Res.* **103**, 14,169–14,240.
16. Miyakoda, K., Hembree, G.D., Strickler, R.F., and Shulman, I. (1972) Cumulative results of extended forecast experiments, I, Model performance for winter cases, *Mon. Weather Rev.* **100**, 836–855.
17. Molteni, F., Buizza, R., Palmer, T.N., and Petroliagis, T. (1996) The ECMWF ensemble prediction system: Methodology and validation, *Quart. J. Roy. Meteor. Soc.* **119**, 269–298.
18. Palmer, T.N., and Anderson, D.L.T. (1994) The prospect for seasonal forecasting — A review paper, *Quart. J. Roy. Meteor. Soc.*, **120**, 755–793.
19. Peixoto, J.P., and Oort, A.H. (1992) *Physics of Climate,* American Institute of Physics, New York.
20. Penland, C., and Sardeshmukh, P.D. (1995) The optimal growth of tropical sea surface temperature anomalies, *J. Clim.* **8**, 1999–2024.
21. Ropelewski, C.F., and Halpert, M.S. (1987) Global and regional scale precipitation patterns associated with the El Niño/Southern Oscillation, *Mon. Weather Rev.* **115**, 1606–1626.
22. Ropelewski, C.F., and Halpert, M.S. (1989) Precipitation patterns associated with the high index phase of the Southern Oscillation, *J. Clim.* **2**, 268–284.
23. Saravanan, R. (1998) Atmospheric low frequency variability and its relationship to midlatitude SST variability: Studies using the NCAR Climate System Model, *J. Clim.* **11**, 1386–1404.
24. Stern, W., and Miyakoda, K. (1995) Feasibility of seasonal forecasts inferred from multiple GCM simulations, *J. Clim.* **8**, 1071–1085.
25. Stockdale, T.N., Anderson, D.L.T., Alves, J.O.S., and Balmaseda, M.A. (1998) Global seasonal rainfall forecasts using a coupled ocean-atmosphere model, *Nature* **392**, 370–373.
26. Toth, Z., and Kalnay, E. (1997) Ensemble forecasting at NMC and the breeding method, *Mon. Wea. Rev.* **125**, 3297–3319.
27. Trenberth, K.E., Branstator, G.W., Karoly, D., Kumar, A., Lau, N.-C., and Ropelewski, C. (1998) Progress during TOGA in understanding and modeling global teleconnections associated with tropical sea surface temperatures, *J. Geophys. Res.* **103**, 14,291–14,324.

REGIONAL MODELS

M. DEQUE
Météo-France CNRM
42 Avenue Coriolis
F-31057 Toulouse
France

1. Introduction

Simulating the global atmospheric general circulation with a few equations is a scientific challenge which is very satisfying for modelers, but has little impact on other citizens. But reproducing regional features, i.e., features at the scale of human activities, like the geographical distribution of rainfall over a country, provides a powerful tool to decision makers. It would be possible to evaluate quantitatively the impact of a decision like clearing a forest, or creating an artificial lake, before the decision is taken. It would also be possible to anticipate the regional effects of an expected change, like the greenhouse gas increase.

Regional climate simulation with a stand-alone local numerical model is impossible since the atmosphere has no borders and, for example, an atmospheric event in southern California affects a few days later the air temperature distribution in Italy. Even though the first numerical weather prediction models in the 1950s were regional, the first climate models were global [22], or at least hemispheric (e.g. [25]). The very low resolution of these first models did not allow the investigation of regional features. With time, increasing computer power has allowed models to simulate finer and finer structures. But the huge cost of high-resolution climate simulation is not always rewarded by a matching progress in realistic reproduction. In fact, high resolution covers two different cases:

1. High resolution allows a more accurate discretization of the Navier-Stokes equation, which cannot be solved analytically.
2. High resolution allows a better representation of the surface forcing, which is exerted at many scales, from 1 mm to 1000 km.

In case 1), the minimum resolution appears to be about 300 km, but further increases have not appeared to modify by much the large scale planetary circulation

P. Mote and A. O'Neill (eds.), Numerical Modeling of the Global Atmosphere in the Climate System, 403–418.

(e.g. [30]). The regional simulation is in fact linked to case 2. In particular, a better resolution of the orography may locally change the distribution of precipitation. In case 2, there is in principle no limit to the increase in resolution.

The above considerations have led modelers to uncouple the two aspects of the resolution: global resolution for solving the advection problems, local resolution for solving the local forcing problems. This separation is arbitrary, and the compromise imposed on the modelers by computer limitations is not completely satisfactory. The best answer (from a purely scientific point of view) to the regional climate problem would be a global climate model with an overall high resolution. In Section 2, we will examine some aspects of the resolution increase. In Sections 3, 4 and 5, we will examine solutions proposed to allow a local increase at a reasonable cost. Conclusions will be given in Section 6.

2. Resolution impact

As discussed in Chapter 19, the role of the horizontal resolution has been widely studied in model intercomparisons. The struggle for resolution is as old as the first GCMs (500 km to 250 km in Manabe *et al.* [26]). With an Eulerian model, multiplying the resolution by two implies multiplying the CPU cost by 8: indeed, the time step must be changed by the same factor as the horizontal grid interval, due to CFL considerations (see Chapter 7, Section 2). With a semi-Lagrangian model, the time step may be 30 minutes for a broad spectrum of resolutions, so the cost is multiplied only by 4. The most visible impact of increasing the horizontal resolution is the refinement of the orography. Figure 1 shows the orography of a spectral GCM at T42 (280 km grid) and at T106 (120 km grid). However, even T106 is not sufficient for regional climates: over Europe some important mountain ranges (Pyrenees, Apennines) are missing. One has also to take into account the fact that the grid size is smaller than the size of the smallest structures the model is able to reproduce with a physical accuracy.

The impact of increasing the horizontal resolution has been discussed by several authors. Boville [3] studied the NCAR model at resolution T21, T31, T42 and T63. He observed a stronger impact in the southern hemisphere than in the northern hemisphere. The improvement is monotonic with resolution, for example for the eddy kinetic energy. Boyle [4] studied the ECMWF model at resolution T21, T42, T63 and T106. He did not find any significant improvement beyond T42. His simulations were only one year long. A one-year simulation provides a poor statistical estimate (see Chapter 4); only huge differences can be found as significant. Déqué *et al.* [12] studied the Météo-France model at T21, T42 and T79. Since the simulations are 10 years long, it is easier to compare subtler differences. The pressure field improves generally with resolution. The tropospheric temperature increases with resolution, reducing cold biases, but also increasing

405

warm biases. The troposphere gets drier, with fewer clouds, and the precipitation increases with resolution.

Increasing the resolution not only increases the accuracy with which the Navier-Stokes equations are solved, but may also change the behavior of physical parameterizations. For example, horizontal diffusion depends on resolution. Boyle [4] and Déqué *et al.* [12] used the same diffusion coefficient for the smallest wavenumber of all resolutions. This does not mean that the diffusion is the same at all resolutions: if one considers wavenumber 21, the damping rate is 64 times smaller at T42 than at T21 (since the coefficient varies as n^6) and again almost 64 times smaller at T79 than at T42. Indeed, the horizontal diffusion takes into account the unresolved scales and must decrease as resolution increases. In Boville [3], the damping rate of the shortest scale by the horizontal diffusion varies from 1.5 days for T21 to 0.3 days for T63. Keeping the empirical parameters constant when increasing the resolution facilitates the comparison, but favors the resolution at which they have been tuned. In the case of the Météo-France model, the warming and drying as resolution increases are due to the fact that the convection scheme [2] is resolution sensitive. This model has been mostly tuned at T42 resolution. In practice, any resolution increase should be associated with an appropriate tuning of the physics.

The resolution impact with a more recent version of the Météo-France model was evaluated with 10 year simulations at T42 and T106 [11]. Here an additional tuning has been done in the cloud scheme of the T106 version, so that the global annual means of shortwave and longwave radiation at the top of the atmosphere are identical to the T42 values.

The atmospheric circulation over the northern Atlantic is improved by the resolution, whereas it is degraded over the northern Pacific in winter. In the southern hemisphere, in agreement with Boville [3], the circum-Antarctic low deepens with a higher resolution, which enhances the error at T106. The eddy kinetic energy increases with resolution, as found by Boville [3], but only in the low temporal frequencies.

Figure 2 presents the global winter precipitation pattern in December-January-February at T42 resolution (280 km) version, at T106 (120 km) version, and the observations [24]. The storm track over the Atlantic is moved northwards at T106, which is an improvement. The Inter-Tropical Convergence Zone (ITCZ) is tighter and more intense at the higher resolution. Other details are hardly visible, but improve the realism, for example the maxima over southern Norway or over the Dinaric Alps. The mean sea level pressure pattern (not shown) exhibits a deepening of the lows and an intensification of the anticyclones as resolution increases.

In zonal averages of the troposphere or stratosphere, the impact of the higher resolution is very modest. The temperature field (not shown) is slightly warmer. The wind field (Figure 3) is slightly improved in the lower stratosphere with a

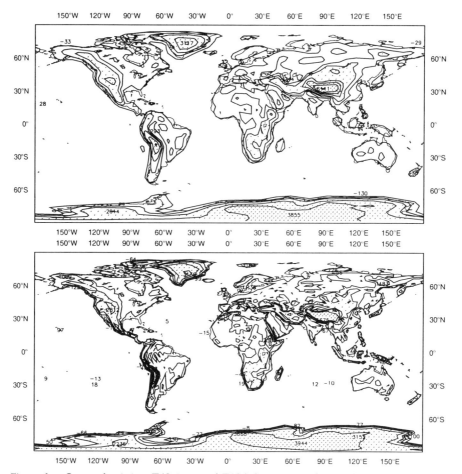

Figure 1. Orography (m) at T42 (top) and T106 (bottom) resolution. Contours 100, 500, 1000, 2000, 3000 and 4000 m, shading above 1000 m.

stronger westerly jet in the northern hemisphere and a weaker easterly jet in the southern hemisphere.

An increase in vertical resolution is not really a step toward regionalization. The cost increase is less huge than for the horizontal resolution, since the limiting factor for the time step is generally the horizontal resolution. The impact on the model climate is often modest [1].

To summarize the above discussion, increasing the resolution does not by itself make the model converge toward reality (see also Chapter 19 on model intercomparison). A global simulation at 50 km resolution over a sufficiently large period is not yet available. However, this corresponds to a T213 resolution which has been in use at ECMWF since 1991. So we can hope that, within a few years, expensive but feasible GCM simulations will be performed at a resolution compatible with

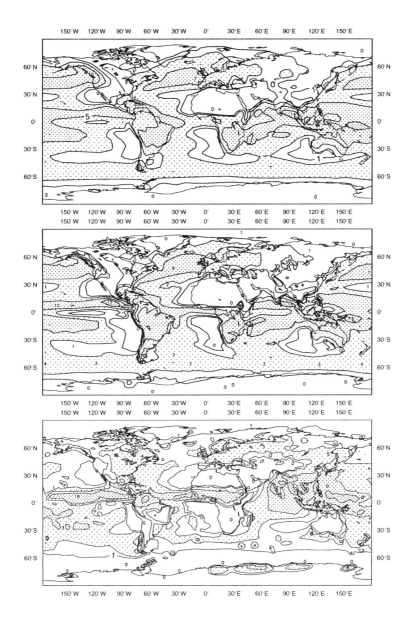

Figure 2. Winter precipitation with the Météo-France model at T42 (top) T106 (middle) resolution, and as observed (bottom). Contours 1, 2, 5 and 10 mm/day, shading above 2 mm/day.

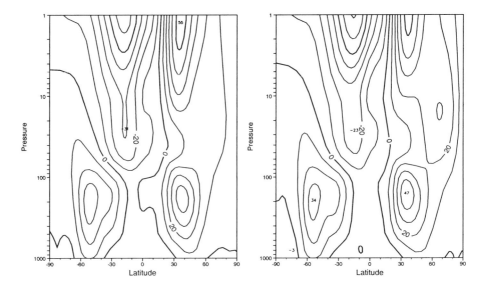

Figure 3. Winter zonal mean of zonal wind with the Météo-France model at T42 (left) and T106 (right) resolution. Contour interval 10 m/s.

regional features. The physical empirical parameters will probably be different from those for a T42 model, and the improvements will mainly come from improved physics. Rather than waiting for high computer resources, modelers have found other ways to simulate regional climates, and we will examine them in the following sections.

3. Limited Area Model

The most widely used solution is to restrict the model to a given domain with simple boundaries (rectangular with respect to the system of horizontal coordinates used). The Limited Area Model (LAM) needs the time evolution of the prognostic variables at the boundaries of the domain. They are provided by a global model (nesting) or a wider LAM including its domain (multiple nesting). In the case of climate validation (not prediction or scenario), the boundary data may come from analyzed observational data. These methods are called one-way nesting since there is no feedback from the LAM to the GCM. The two-way nesting [27] is less common in limited area modeling, since it is much more complicated to design.

The one-way nesting technique has been primarily used in numerical weather prediction ([5], [15], [20], to quote but a few). The lateral boundary condition used in this kind of model is neither a Dirichlet one (imposing the values of the fields at the border) nor a Neumann one (imposing the fluxes at the border), but a linear relaxation in a strip of a few gridpoints breadth of the LAM prognostic variables toward the interpolated values calculated by the large-scale model [10]. The relaxation coefficient is generally stronger at the outer boundary than at the inner one.

The LAMs are usually gridpoint models, but, as with the global atmospheric models, there exists a spectral alternative [19]. The method consists in using a Fourier expansion in both directions. Beyond the relaxation zone where the LAM is constrained by the large-scale data, a zone of periodicity is used to allow a continuous connection of the Fourier modes (which are periodic in both directions) and avoid the spurious effects of the Gibbs phenomenon. The domain of interest can be considered as a fraction of a torus. The advantage of this method is to provide an easy way of calculating horizontal derivatives and of solving the Helmholtz equation. With fast Fourier transforms in both directions, the spectral to grid transformations are cheaper than in spherical geometry.

More recently, the one-way nesting technique has been applied to climate simulation. Giorgi [16] used the NCAR Community Climate Model (CCM1) for the large-scale simulation, and the Pennsylvania State University/NCAR mesoscale model (MM4) for the regional simulation. The LAM domain covers the western United States with a resolution of 60 km, whereas the driving GCM has a T42 resolution. The large-scale average circulations over the western United States by the LAM are not substantially different from those of the GCM, but the nested model reproduces better regional details of temperature and precipitation, owing to the finer resolution of the mountains. A second version of the coupled system, named RegCM2, was proposed by Giorgi et al. [17]. This version uses an improved physical parameterization package. It has been used over the European domain with a 70 km resolution.

A regional model has been developed over eastern Asia by Kida et al. [23]. The grid size is 127 km at 60°N. The precipitation is improved by the use of a spectral boundary coupling [28].

In Europe, different LAMs have been developed, and the comparison of seven of them against observations when forced by atmospheric analyses has been made in the framework of the European Commission funded Regionalization project [6]. These models have a horizontal resolution of 50-60 km and a domain covering the whole of Europe, or 20-30 km and a domain covering only a part of Europe. The comparison with July 1990 and January 1991 shows an unrealistic summer drying of the soil over southeast Europe, causing a warm bias. In winter, the models exaggerate the moisture advection from the ocean, and a bias linked to this feature can be seen in the precipitation and in the mean sea level pressure fields. However, these LAMs capture most of the regional patterns in the scale 200-1000 km.

4. Variable resolution model

The problem of forcing a model at its lateral boundaries may be avoided by using a global model. Since a model is discretized on a computational grid, there is no reason why the resolution of this grid should be uniform over the sphere. The traditional gridpoint models in latitude-longitude coordinates undergo a dramatic variation of the mesh size in the east-west direction in moving from the equator to the pole (the critical problem at the pole is solved by a Fourier filtering). The grid size variation is also obtained when working in polar stereographic projection, but there the resolution is isotropic. The counterpart of the absence of a lateral boundary condition is that such models cannot be tuned or checked by the perfect boundary condition approach [6].

A simple way to implement a variable resolution in a gridpoint model is to vary the size of the latitude increment and of the longitude increment, so that Δx and Δy have a small value in the area of interest, and a large value far from this region, in particular at the antipodes ([31], [29], [7]). This method is very flexible, since the choice of the domain and the resolution is arbitrary. It is, however, more expensive than a nested LAM, since the transition of the resolution outside the domain must be smooth enough to avoid numerical problems, so the number of gridpoints for which numerical computations are performed but not exploited may be large. Another problem is that in some regions outside the domain of interest Δx may be much larger than Δy, or conversely. This is harmless in short-range prediction, but may degrade climate simulations by artificially enhancing some mountain ranges and damping some others, depending on their orientation.

The method developed at Météo-France, for short-range forecasting as well as for climate study, is fully compatible with a spectral representation over the sphere. Courtier and Geleyn [8] have shown that the family of conformal transforms between two spheres (that is for which Δx and Δy are multiplied by the same factor m) can be constructed with one point on the sphere (the pole of stretching) and one coefficient c (the stretching coefficient). A transform consists of three simple geometrical operations:

1. a stereographic projection from the sphere to the tangent plane
2. a scaling with the factor c
3. a stereographic projection back to the sphere

If we use a regular mapping on the target sphere to discretize the atmosphere equations, a higher resolution (if $c > 1$) is obtained in the vicinity of the pole of stretching, and the isotropy is preserved in any location on the sphere. If (λ, μ) are the longitude and sine of the latitude, the model is written with a new set of coordinates (λ', μ') calculated as a function of c and (λ, μ). Figure 4 shows the distribution of the gridpoints as a function of (λ, μ) in the case of a T63 resolution with the pole in the Tyrrhenian Sea and with $c = 3.5$.

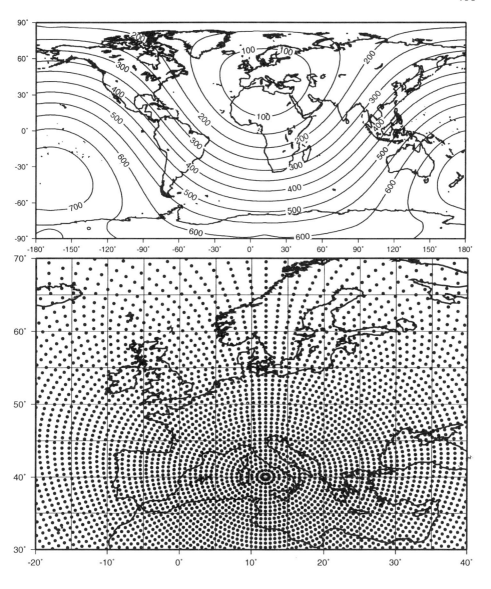

Figure 4. Horizontal resolution of the T63,$c = 3.5$ stretched model used at Météo-France. Top panel: distance between neighboring gridpoints, contour interval 50 km. Bottom panel: individual gridpoints in the area of maximum resolution.

The only modification to the model equation is the multiplication of the horizontal derivatives by the mapping factor:

$$m = \frac{1}{2c}\left((c^2 + 1) + (c^2 - 1)\mu'\right) \tag{1}$$

The fact that m varies as a linear function of the sine of the pseudo-latitude μ' implies that the equations are compatible with a spectral representation. The Helmholz equation can be solved by a tridiagonal algorithm (instead of a diagonal one when $c = 1$), and the aliasing due to the spectral resolution is negligible. Courtier *et al.* [9] have shown that the numerical simulation of a planetary rotation of a barotropic wave by a stretched adiabatic model yields the same results as a low (high) uniform resolution model in the low (high) resolution area. The case of the diabatic terms (physical parameterizations) is different, since we have seen in Section 2 that some of them are resolution dependent. We have reduced the sensitivity of the convection scheme by reducing the moisture convergence term which is an input of this scheme by an empirical factor depending on m.

A climate version of this model at T63 with $c = 3.5$ and the pole of stretching in the Tyrrhenian Sea has been described in [11]. The resolution varies from 60 km in the Mediterranean Sea to 700 km in the southern Pacific Ocean. A 10 year simulation has been performed and compared with a T106 simulation. Objective scores show that the variable resolution model is closer to the observed climatology over Europe. This is mostly explained by a better representation of the orography (Figure 5). Over the Atlantic Ocean, however, the pressure field and the variability are better at the T106 simulation.

A second version with the same horizontal resolution but an improved physics has been used by Déqué *et al.* [13] to study the impact of doubling the CO_2 concentration. Figure 6 shows the mean winter pressure field in the 10 year control simulation. Although we have not produced a T42 simulation with exactly the same version of the model, we can say that the features are very similar to what is produced by a T42 resolution. The local pattern of precipitation is improved, as can be seen when comparing the 10 year mean of the stretched model simulation (Figure 7) with the T42 results (Figure 2). One can see, however, that the position of the storm track too far to the south, and the lack of precipitation over the eastern Mediterranean basin, are not corrected by the high resolution.

The stretched method suffers, however, from one shortcoming. It does not allow the choice of a domain size and a resolution independently. Once the pole is determined, the only two degrees of freedom are the truncation and the stretching factor. If μ^* is the sine of the angular distance between a point and the pole of stretching, we have:

$$\mu' = \frac{(1 - c^2) + (1 + c^2)\mu^*}{(1 + c^2) + (1 - c^2)\mu^*} \qquad (2)$$

As a result, the area covered by the pseudo-northern hemisphere, i.e. corresponding to 50% of the gridpoints (and thus 50% of the computational expense), decreases as c increases. This implies that with c greater than 4, the area with low resolution is geographically close to the pole. For example a T42,$c = 6$ version with the same pole as before has a resolution equivalent to T21 in the northern

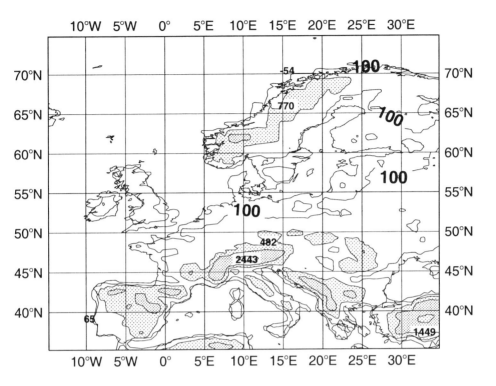

Figure 5. Orography of the T63,$c = 3.5$ stretched model used at Météo-France. Contours 100, 500, 1000, 2000, 3000 and 4000 m, shading above 1000 m.

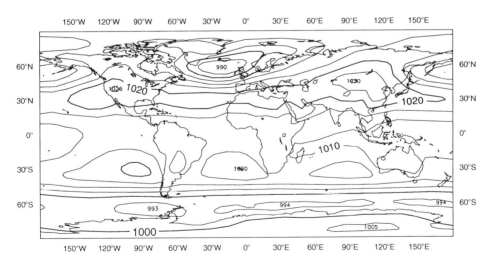

Figure 6. Winter mean sea level pressure of the T63,$c = 3.5$ stretched model used at Météo-France. Contour interval 5 hPa.

414

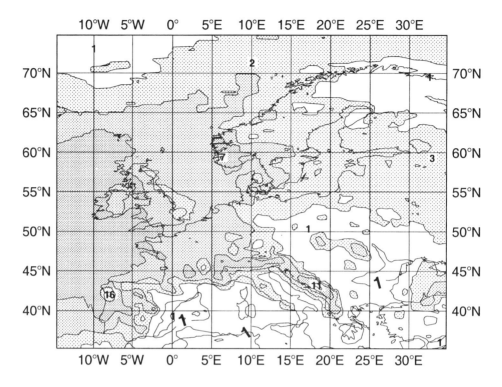

Figure 7. Winter precipitation simulated by the T63,$c = 3.5$ stretched model used at Météo-France. Contours 1, 2, 3, 5, 10, 15 mm/day, shading above 2 mm/day.

Atlantic, with all the inconvenience in terms of general circulation over Europe, even though the resolution over Europe is higher than with a T63,$c = 3.5$. Therefore the only way to get a higher resolution is to increase the spectral resolution, which makes the model very costly. In order to get a 20 km resolution over a part of Europe, the reasonable solution would be to nest a LAM with 20 km resolution inside a T63,$c = 3.5$ version which has a grid size of about 60 km over Europe. This method is applied by Météo-France in short-range forecasting.

5. Other solutions

The most frequently used methods are the LAM and the variable resolution. But other solutions to the regionalization problem have been proposed. One can quote here, although it is not a modeling activity, the technique of statistical downscaling ([21], [33]). This technique consists in using long series of observed values to establish links between large-scale circulation patterns and local meteorological values. Then these relations (linear regression or analogs) are used with GCM outputs as predictors. The problem, in the case of climate change, is the validity of these relations in a climate different from the one used to "train" the model.

Another method using statistical connections is proposed by Frey-Buness *et al.* [14]. It consists in performing short runs with a local high resolution model. Each run starts from a typical initial situation, determined by clustering a sample of observed large-scale data on a long period. Then the regionalization of a GCM experiment consists in determining the frequency, in the GCM run, of the different large-scale clusters, and then calculating a weighted average of the results from the high resolution short runs. No additional integration is needed.

The Physical-Regional (FIZR) approach [18] is an off-line numerical down-scaling. The GCM run provides the large-scale transport terms which are introduced in the right-hand terms of the evolution equations of the regional model. The latter model recalculates all the physical parameterizations, and an additional dynamical term (the so-called mesoscale forcing due to the GCM-unresolved orography) on a fine grid.

The delocalized physics method [32] is an on-line downscaling. Here, the physical parameterizations are calculated on a grid which is different from the one for which the prognostic variables are saved and the dynamical terms are calculated. This grid is finer and may have an irregular distribution. At each time step, the prognostic variables are interpolated from the GCM grid to the delocalized grid, and the diabatic fluxes are integrated from the delocalized grid to the GCM grid. This method does not allow treatment of subgrid scale orographic effects, but it is particularly well suited for coupling with an ocean model. Indeed, the delocalized physics may use the same grid as the ocean model (which is generally finer than the atmosphere one owing to the smaller scale of the leading structures in the ocean), and then no interpolation is done at the interface. The interpolation is done in the free atmosphere, where the small-scale structures are less dominant.

6. Conclusion

Several techniques have been presented here which serve as alternatives to the use of very high resolution global models to study the climate of a particular region. There is little chance that this kind of approach will become obsolete as computer power increases:

1. The need for small size grids grows faster than computer progress would allow.
2. This increase in power is absorbed by costly improvements of the GCMs. Ten years ago, the standard resolution was T42. Today, it is still T42, but with more computationally expensive models.
3. The realization of scenario experiments with ocean-atmosphere coupled models requires long integrations. A coupled fine resolution model is very expen-

sive, since the ocean time scales involve a long integration before the coupled system is stabilized.

However, the problem of numerical downscaling needs considerable further research, as shown by the results of two recent European projects. The HIRE-TYCS (High Resolution Ten Year Climate Simulations) project has concluded that the reduction of the model systematic errors by an increase in the horizontal resolution was marginal. The RACCS (Regionalization of Anthropogenic Climate Changes) project has concluded that the systematic errors of the driving GCMs are to a large extent reproduced by the LAMs. This last result leads the modelers to the following alternatives:

- if the size of the LAM domain is small, the errors of the GCM which are induced by its poor resolution are not corrected by the LAM since the constraint at the lateral boundary is too strong;
- if the size of the LAM domain is large, discrepancies may occur at the LAM boundaries, since the relative freedom of the LAM in simulating the weather may place a cyclone where the GCM places an anticyclone.

Fortunately several solutions exist, besides the use of a global variable resolution model. One can use multiple nesting, so that the conflicts are reduced. One can also use a large integration domain, a large fraction of which is covered by the relaxation area (with low constraints in the inner part of the relaxation area). This is simpler to implement, but more demanding in computer time.

The stronger challenge remains, for modelers, to reduce the systematic errors that are not due to resolution (first step) and the systematic errors that are not solved by resolution (second step). Indeed, high resolution is perhaps a necessary, but certainly not a sufficient, condition for an accurate climate simulation.

Acknowledgements

Thanks are due to P. Marquet for his help in preparing the manuscript. This work was partly supported by the Commission of the European Union (Contracts EV5V-CT94-0505 and ENV4-CT95-0184), and by the French Department for the Environment (ECLAT program).

References

1. Bossuet, C., Déqué, M. and Cariolle, D. (1998) Impact of a simple parameterization of convective gravity-wave drag in a stratosphere-troposphere general circulation model and its sensitivity to vertical resolution, *Ann. Geophysicae* **16**, 238-249.
2. Bougeault, Ph. (1985) A simple parameterization of the large-scale effects of deep cumulus convection, *Mon. Wea. Rev.* **113**, 2108-2121.

3. Boville, B.A. (1991) Sensitivity of simulated climate to model resolution, *J. Climate* **4**, 469-485

4. Boyle, J.S. (1993) Sensitivity of dynamical quantities to horizontal resolution for a climate simulation using the ECMWF (cycle 33) model, *J. Climate* **6**, 796-815.

5. Chen, J.M. and Miyakoda, K. (1974) A nested grid computation for the barotropic free surface atmosphere, *Mon. Wea. Rev.* **102**, 181-190.

6. Christensen, J.H., Machenhauer, B., Jones, R.G., Schä:r, C., Ruti, P.M., Castro, M. and Visconti, G. (1997) Validation of present-day regional climate simulations over Europe: LAM simulations with observed boundary conditions, *Climate Dyn.* **13**, 489-506.

7. Côté, J., Roch, M., Staniforth, A. and Fillon, L. (1993) A variable-resolution semi-lagrangian finite-element global model of the shallow-water equations, *Mon. Wea. Rev.* **121**, 231-243.

8. Courtier, Ph. and Geleyn, J.F. (1988) A global numerical weather prediction model with variable resolution: Application to the shallow water equations, *Q. J. R. Meteorol. Soc.* **114**, 1321-1346.

9. Courtier, Ph., Freydier, C., Geleyn, J.F., Rabier, F. and Rochas M. (1991) The ARPEGE project at Météo-France, *Proceedings of ECMWF workshop on Numerical methods in atmospheric modelling*, 9-13 Sept. 1991, Vol. **2**, 193-231.

10. Davies, H.C. (1976) A lateral boundary formulation for multilevel prediction models, *Q. J. R. Meteorol. Soc.* **102**, 405-418.

11. Déqué, M. and Piedelievre, J.Ph. (1995) High resolution climate simulation over Europe, *Climate Dyn.* **11**, 323-331.

12. Déqué, M., Dreveton, C., Braun, A. and Cariolle, D. (1994) The ARPEGE/IFS atmosphere model: a contribution to the French community climate modelling, *Climate Dyn.* **10**, 249-266.

13. Déqué, M., Marquet, P. and Jones, R.G. (1998) Simulation of climate change over Europe using a global variable resolution general circulation model, *Climate Dyn.* **14**, 173-189.

14. Frey-Buness, A., Heimann, D. and Sausen, R. (1995) A statistical-dynamical downscaling procedure for global climate simulations, *Theor. Appl. Climatol.* **50**, 117-131.

15. Geleyn, J.F., Bougeault, Ph., Rochas, M., Cariolle, D., Lafore, J.Ph., Royer, J.F. and André, J.C. (1988) The evolution of numerical weather prediction and atmospheric modelling at the French weather service, *J. Theoretical Applied Mechanics* **7**, 87-110.

16. Giorgi, F. (1990) Simulation of regional climate using a limited area model nested in a general circulation model, *J. Climate* **3**, 941-963.

17. Giorgi, F., Marinucci, M.R. and Bates, G.T. (1993) Development of a second-generation regional climate model (RegCM2). Part I: boundary-layer and radiative transfer processes, *Mon. Wea. Rev.* **121**, 2794-2813.

18. Goyette, S. and Laprise, J.P.R. (1996) Numerical investigation with a physically based regional interpolator for off-line downscaling of GCMs: FIZR, *J. Climate* **9**, 3464-3495.

19. Haugen, J.E. and Machenhauer, B. (1993) A spectral limited-area model formulation with time-dependent boundary conditions applied to the shallow-water equations, *Mon. Wea. Rev.* **121**, 2618-2630.

20. Juang, H.-M.H. and Kanamitsu, M. (1994) The NMC nested regional spectral model, *Mon. Wea. Rev.* **122**, 3-26.

21. Karl, T.R., Wang, W.C., Schlesinger, M.E., Knight, R.W. and Portman, D. (1990) A method of relating general circulation model simulated climate to the observed local climate. Part I: seasonal statistics, *J. Climate* **3**, 1053-1079.

22. Kasahara, A. and Washington, W.M. (1967) NCAR global general circulation model of the atmosphere, *Mon. Wea. Rev.* **95**, 389-402.

23. Kida, H., Koide, T., Sasaki, H. and Chiba, M. (1991) A new approach for coupling a Limited Area Model to a GCM for regional climate simulations, *J. Meteor. Soc. Japan* **69**, 723-728.

24. Legates, D.R. and Willmott, C.J. 1990. Mean seasonal and spatial variability in gauge-corrected global precipitation, *Int. J. Climatol.* **10**, 111-127.

25. Manabe, S., Smagorinsky, J. and Strickler R.F. (1965) Simulated climatology of a general circulation model with a hydrological cycle, *Mon. Wea. Rev.* **93**, 769-798.

26. Manabe, S., Smagorinsky, J., Holloway, J.L. and Stone M.M. (1970) Simulated climatology of a general circulation model with a hydrological cycle. III: Effect of increase horizontal computational resolution, *Mon. Wea. Rev.* **98**, 175-212.

27. Phillips, N.A. and Shukla, J. (1973) On the strategy of combining coarse and fine meshes in numerical weather prediction, *J. Appl. Meteorol.* **13**, 763-770.

28. Sasaki, H., Kida, H., Koide, T. and Chiba, M. (1995) The performance of long-term integrations of a Limited Area Model with the spectral boundary coupling method, *J. Meteorol. Soc. Japan* **73**, 165-181.

29. Sharma, O.P., Upadhyaya, H., Braine-Bonnaire, T. and Sadourny, R. (1988) Experiments on regional forecasting using a stretched coordinate general circulation model, *J. Meteorol. Soc. Japan, Spec. NWP. Symp. Vol.*, 263-271.

30. Sperber, K.R., Hameed, S., Potter, G.L., and Boyle, J.S. (1994) Simulation of the northern summer monsoon in the ECMWF model: sensitivity to horizontal resolution, *Mon. Wea. Rev.* **122**, 2461-2481.

31. Staniforth, A.N. and Mitchell, H.L. (1978) A variable-resolution finite-element technique for regional forecasting with the primitive equations, *Mon. Wea. Rev.* **106**. 439-447.

32. Vintzileos, A. and Sadourny, R. (1997) A general interface between an atmospheric GCM and underlying ocean and land surface models: delocalized physics scheme, *Mon. Wea. Rev.* **125**, 926-941.

33. von Storch, H., Zorita, E. and Cubasch, U. (1993) Downscaling of global climate change estimates to regional scales: an application to Iberian rainfall in wintertime, *J. Climate* **6**, 1161-1171.

TOWARD A COMPLETE MODEL OF THE CLIMATE SYSTEM

B.A. BOVILLE
National Center for Atmospheric Research[1]
PO Box 3000
Boulder, Colorado 80307 USA

1. Introduction

Most of this book has been concerned with atmospheric general circulation models (AGCMs). However, changes in climate, whether anthropogenic or natural, involve a complex interplay of physical, chemical, and biological processes not just of the atmosphere but also of the ocean and the land surface—the "climate system". As climate system research seeks to explain the behavior of climate over time scales of years to millennia, focus necessarily turns to behavior introduced by physical, chemical, and biogeochemical interactions among climate subsystems. The challenges of modeling the role of anthropogenic emissions of carbon dioxide, of reactive trace gases, and of changing land use in the earth system require a coupled climate system approach.

Several groups around the world are working towards comprehensive climate system models that allow for the interactions among physical, chemical, and biogeochemical processes. Since it will be many years before mature models of this type exist, this chapter will focus on the physical climate system. Even this simplified system requires coupling an AGCM to an ocean general circulation model (OGCM), a sea ice model, and a land surface model. These component models all communicate with each other through interfacial fluxes of momentum, energy, and water, which generally depend on the state of more than one model, so the structure of a climate system model is exceedingly complex. Figure 1 illustrates this point by showing the flow of variables in the NCAR Climate System Model (CSM-1). At the center of this figure is a "flux coupler" which is responsible for coordinating the flow of variables between components, regridding as necessary since the components typically do not use the same horizontal grid structure. Other

---

* The National Center for Atmospheric Research is sponsored by the National Science Foundation.

P. Mote and A. O'Neill (eds.), Numerical Modeling of the Global Atmosphere in the Climate System, 419–442.
© 2000 *Kluwer Academic Publishers. Printed in the Netherlands.*

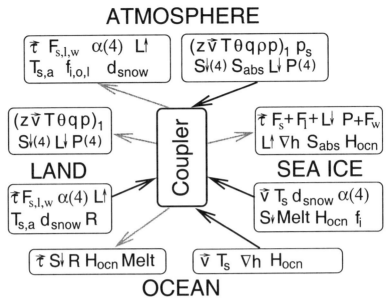

Figure 1. The flow of variables in CSM-1. Subscripts s, 1, and a refer to the surface, 1st AGCM level, and 2 m level, respectively. τ is the surface stress; $F_{s,l,w}$ are the turbulent fluxes of sensible and latent heat and water vapor; $f_{i,o,l}$ are the fractional areas of sea ice, ocean, and land, respectively; d_{snow} is the snow depth; R is drainage basin integrated runoff; p is pressure; T is temperature; v is velocity; z is geopotential height (above the surface); θ is potential temperature (with respect to p_s); q is specific humidity; ρ is density; α and S are the albedo and downward solar radiation at the surface with 4 components for direct and diffuse flux in the visible and near infrared bands, while S_{abs} is the total absorbed solar radiation at the surface; L is the longwave radiative flux (upward or downward); P is the precipitation with 4 components for convective and nonconvective fluxes in liquid and solid phases; ∇h is the gradient of the sea surface height; H_{ocn} is the ocean heat used to form or melt ice; MELT is the ice meltwater flux.

climate models use different coupling methods, but the basic principles are the same and CSM-1 will be used in the remainder of this chapter as an example of the present capabilities of physical climate models.

Coupled model experiments are now common enough that a Coupled Model Intercomparison Project (CMIP) has been established to compare standardized simulations (see Chapter 19). At present, results of two simulations have been collected from modeling groups around the world: one set of control simulations (CMIP-I) and one set of simulations (CMIP-II) with CO_2 increasing 1% per year. Control simulations have constant (usually present-day) forcing and durations of from 25 to 1100 years. The CMIP-II simulations are run at least until the time of CO_2 doubling ($\sim$ 70 years). In this chapter, the results of the CSM-1 simulations that NCAR contributed to CMIP are summarized, as examples of both equilibrium climate and of climate change experiments.

The remainder of this chapter discusses: some constraints imposed on atmospheric models when they are coupled to ocean and sea ice models; some relevant aspects of ocean and sea ice modeling; the coupled solution of CSM-1 with constant (present day) forcing; the response of CSM-1 to increasing CO_2; and the computational requirements for coupled modeling. Land surface models are discussed in chapter 13 and most AGCMs also contain some form of land model.

2. Constraints Imposed by Coupling

Coupling to ocean and sea ice models imposes a set of constraints on an AGCM. While these constraints are not actually new, they become greatly increased in importance when the coupling problem is considered. The first major constraint is energy conservation. When the sea surface temperatures are specified, as is typical in AGCMs, the ocean becomes an infinite reservoir for both heat and fresh water. The conservation of energy may not be a primary concern of model developers in that case. However, climate change problems typically involve perturbations of 1–10 W m^{-2}, on a global and annually averaged basis. For instance, doubling the carbon dioxide concentration in the atmosphere perturbs the longwave radiative fluxes by about 4 W m^{-2} (e.g., [28]).

2.1. GLOBAL ENERGY CONSERVATION

To be used in a CSM designed for studying such problems, an AGCM should conserve energy to better than 1 W m^{-2}, and preferably to a least 0.1 W m^{-2}. This level of conservation is actually difficult to achieve and the NCAR Community Climate Model, version 3 (CCM3) conserves energy only to within ~ 0.4 W m^{-2}. Even this level of conservation is somewhat fortuitous, involving cancellation of errors between individual processes that conserve only to within ~ 1 W m^{-2} (D. Williamson, personal communication). For comparison, the NCAR CCM0, completed in 1981, conserved energy only to ~ 15 W m^{-2}. This level of conservation was normal for AGCMs of that time and, while not a desirable feature, was not considered to be an urgent problem since the model was intended to be used with specified SSTs. Advances in numerical methods, particularly in formulating energetically consistent vertical approximations for various terms, have been the biggest factor in improving the energy conservation properties of the CCM.

2.2. GLOBAL ENERGY BALANCE

The Earth system is believed to be near equilibrium, with the global and annual average of the absorbed solar radiation being closely balanced by the longwave radiation escaping from the "top of the atmosphere" (TOA). Although this balance can only be verified by observations to within 5 W m^{-2} (e.g., [22]), it imposes a second global energy constraint when coupling an AGCM to ocean and ice models. The AGCM should balance the TOA radiative budget to within its internal conservation error when forced with observed SSTs. Simply conserving energy within the AGCM is not sufficient, since an imbalance at the TOA that is matched by an imbalance at the surface cannot be maintained when the SSTs are allowed to change in coupled simulations. In this case, the earth system will have to warm up or cool down until the imbalance at the TOA is equal to the total of the nonconservative processes in the coupled model.

2.3. LOCAL ENERGY BALANCE

In addition to the global energy constraints discussed above, there are local energy constraints that an AGCM should satisfy to be appropriate for coupling. The geographic distribution and mean annual cycle of absorbed solar and emitted longwave radiation can be determined from satellite measurements. The AGCM should reproduce these fluxes when forced with observed annual cycle of SSTs. The AGCM should also produce the same mean annual cycle of surface fluxes of energy, momentum and water as in the real world. Unfortunately, not only are the surface fluxes not observed, but also the turbulent fluxes depend very strongly on poorly observed quantities (e.g., low level wind speed, temperature, static stability, and humidity). In consequence, atmospheric analyses are not reliable sources for these quantities.

Similar constraints apply to the OGCM. When forced with observed atmospheric states and radiative fluxes, the ocean model should reproduce the observed fluxes and SSTs. The OGCM should, at least, reproduce the observed SSTs when forced with observed fluxes. We note again that surface fluxes are not actually observed, so the extent to which these balances are violated is difficult to determine. Using the turbulent flux formulae employed by CSM-1 together with climatological SSTs and analyses of low level atmospheric states, surface insolation, and precipitation, Large et al. [23] found a global and annual average energy imbalance of over 40 W m^{-2} at the ocean surface. This imbalance presumably has contributions from inaccuracies in the flux formulae, and from errors in the analyses of low level winds, temperatures and humidity, and surface radiation. Local errors are likely to be much larger than the global imbalance.

2.4. MERIDIONAL HEAT TRANSPORT

The annually averaged net surface fluxes of heat and fresh water over oceans must be balanced by divergence of the heat and water fluxes in the ocean if equilibrium is to be maintained. Over short (e.g., seasonal) time scales, the net surface heat fluxes can have a substantial contribution from the change in heat storage in the ocean. On multi-year time scales the change in heat storage is expected to be small. This implies that the poleward transport of heat in the ocean can be computed by integrating the surface heat flux

$$H(\phi) = \int_{-\frac{\pi}{2}}^{\phi} \int_0^{2\pi} (\text{net surface flux}) a \, d\lambda \cos \phi \, d\phi$$

where a is the radius of the earth, and it is convenient to begin the integration at the south pole since the meridional heat flux in the ocean is zero.

The implied ocean heat transport can be computed from the surface fluxes produced by an AGCM forced with specified SSTs and compared with analyses of implied ocean heat transport (e.g., [30], [31]). Ocean observations are generally insufficient to compute meridional heat transports directly, although estimates have been made for a few latitudes. Many AGCMs do not reproduce the analyzed heat transports very well. For example, Gleckler *et al.* [15] found that most AGCM simulations submitted to the Atmospheric Model Intercomparison Project implied heat transports of the wrong sign in the Southern Hemisphere. Deficiencies in the actual heat transports by OGCMs are addressed below.

2.5. CLIMATE DRIFT

Generally, AGCMs driven by observed SSTs do not produce surface fluxes that match those required by OGCMs in order reproduce the observed SSTs. Therefore, when the AGCM and OGCM are coupled together, an adjustment in the states of both models must take place. The SST will adjust to the surface fluxes produced by the atmosphere and the atmospheric circulation will adjust to the new SST on a fairly rapid time scale, of order one month to one year. The ocean circulation will change in response to the surface stresses, heat and water fluxes with adjustment time scales of less than one year (for the ocean boundary layer) to over 1000 years (for the deep ocean basins). The new state may be significantly different from present climate. Of course, it is not certain that the present climate is in equilibrium, or was before CO_2 started increasing due to human activities. It is not even certain that the earth system has an "equilibrium" state. Although orbital variations are often invoked to explain the ice ages, it is possible that the climate variations found in the paleoclimate record are largely of internal origin. This hypothesis cannot be disproved using available observations or models.

Regardless of the existence of an equilibrium state for the earth system, the adjustment of the atmospheric and ocean circulation in coupled models is often referred to as "climate drift" and attempts are made to reduce or eliminate it in (at least) three ways. The most obvious method is by improving the GCMs being coupled. However, even perfect GCMs would produce a long term drift if their initial states were not sufficiently close to reality. The ocean is not observed well enough to analyze its instantaneous state throughout its volume in order to generate an initial condition for an OGCM, as is common for AGCMs in numerical weather prediction. Errors in the initialization would produce initial errors in fluxes and circulation, and the ocean-atmosphere system would have to adjust back to equilibrium on the long ocean adjustment time scales. The second method of reducing climate drift in coupled models is to employ some form of spinup procedure to generate compatible initial states for the ocean, atmosphere, land, and sea ice. This method is used for nearly all coupled models to reduce initial drifts in the coupled solution, although the nature of the spinup varies widely. Stouffer and Dixon [29] summarize the general topic of spinup and climate drift and the methods used to reduce it in coupled models, while Boville and Gent [1], and references therein, discuss the procedure used to spin up CSM-1 in more detail.

2.6. FLUX ADJUSTMENT

The third method of attempting to eliminate climate drift in coupled simulations is to introduce a "flux correction" or "flux adjustment" (e.g., [25], and references therein) in an attempt to compensate for the systematic differences between (1) the fluxes produced by the AGCM when forced with observed SSTs ($\overline{F}_A$) and (2) the fluxes required by the OGCM in order to reproduce the observed SSTs ($\overline{F}_O$).

The required flux adjustment is then $\overline{\Delta F} = \overline{F}_O - \overline{F}_A$. It should be noted that computing $\overline{F}_A$ is straightforward, but that $\overline{F}_O$ is defined implicitly and computing it generally requires a combination of OGCM and coupled simulations with relaxation to observed states. The $\overline{\Delta F}$ are determined for each ocean point and may be determined for heat and water fluxes only, or may also include surface stresses. Usually, ensemble average fluxes over several years are computed and the annual cycle is retained in $\overline{\Delta F}$. During a coupled simulation the instantaneous flux F is determined from the atmospheric and oceanic states and is applied in the atmosphere, while $F + \overline{\Delta F}$ is applied in the ocean. The adjustment values $\overline{\Delta F}$ are often larger than than the original flux values F. For example, heat flux adjustments may locally be much greater than 100 W m^{-2}.

Flux adjustment is a controversial method that was employed by the majority of coupled models until recently and is still employed in many models. The NCAR CSM-1, in common with several newer models, does not employ flux adjustments. The following discussion will use some examples from Johns et al. [20] (J97 hereafter) discussion of the Hadley Centre HADCM2 simulation as illustrative of typical flux corrections. We note that recent HADCM3 simulations have been successfully performed without flux adjustments ([16]).

Even in the zonal mean, the $\overline{\Delta F}$ may exceed 100 W m^{-2} (e.g., J97, figure 21). Commonly the heat flux adjustments act to heat the North Atlantic in winter, since ocean models tend to underestimate the poleward heat transport there (see below), and act to cool the southern ocean around Antarctica since the ocean models tend to overestimate the heat transport across the Antarctic circumpolar current. Water flux adjustments in the high-latitude oceans are often comparable in magnitude to actual precipitation rates in the intertropical convergence zone (e.g., J97, figure

22), probably because the transports of fresh water in sea ice are not correct.

The zonally averaged heat flux adjustments constitute an implied poleward transport that does not take place in either the atmosphere or the ocean. For example, J97, figure 9, shows the zonal and annually averaged northward heat transports for the HADCM2 model. In that model the actual poleward ocean heat transports in low and middle latitudes are only about half of the implied transport determined from the net surface heat fluxes. The flux adjustment term makes up the difference, transporting almost 1 PW of heat poleward in both hemispheres.

3. Ocean General Circulation Models

In principle, OGCMs are very similar to AGCMs. In both cases, the hydrostatic primitive equations are solved in spherical coordinates and subgrid scale processes are parameterized. Most OGCMs in use today are derived from the original model of Bryan [4]. At the Geophysical Fluid Dynamics Laboratory, the Bryan model has evolved into the Modular Ocean Model (MOM, see [27]) and many coupled climate models use variations of MOM. For example, the OGCM in CSM-1 [12] is based on MOM 1.1. Most OGCMs use depth (z) coordinates, which is the equivalent of using height coordinates in an AGCM. However, isopycnal models (equivalent to potential temperature coordinates in AGCMs) are also used [26].

The principal differences between AGCMs and OGCMs arise because the ocean is nearly incompressible and its density depends strongly on salinity, in addition to temperature. The ocean domain also has holes, caused by continents and islands, which introduce additional numerical problems. Like AGCMs, OGCMs typically solve the primitive equations on a global latitude-longitude grid. This introduces a stability problem at the poles, as discussed in Chapter 7. In fact, the problem only occurs at the North Pole in OGCMs because the South Pole is located within Antarctica. Recently, the introduction of orthogonal curvilinear dipole coordinate systems have allowed the displacement of the North Pole into a land mass (usually North America, but Asia is also a possibility) (e.g., [24]). These mappings have some similarity to the coordinate stretching sometimes used for concentrating the resolution of an AGCM in a particular region (see Chapter 17).

3.1. LENGTH SCALE OF ENERGETIC EDDIES IN OCEAN

One of the major challenges in ocean modeling is representing the effects of the energetic eddies either directly, through extremely high resolution, or indirectly, through parameterizations. The length scale of the most energetic eddies in the ocean is only ~50–100 km. These eddies are the equivalent of the atmospheric 1000–5000 km synoptic scale waves, although their length scale corresponds approximately with the atmospheric mesoscale.

The small length scale of ocean eddies arises because the Rossby radius of deformation, L_R, is much shorter in the ocean than in the atmosphere.

$$L_R = \frac{N}{mf},$$

where N is the Brunt-Vaisala frequency, m is the vertical wavenumber, and f is the coriolis parameter (e.g., chapter 7 of [14]). N has a similar magnitude in both atmosphere and ocean ($\sim 10^{-2}$ s^{-1}) and f is the same in both. However, the appropriate vertical wavenumber is $\sim 10^{-3}$ m^{-1} for ocean eddies compared to $\sim 10^{-4}$ m^{-1} for atmospheric synoptic waves, so that L_R is an order of magnitude smaller for the ocean.

The atmosphere tends to be very stable at high latitudes and deep convection occurs primarily in the tropics. In contrast, the ocean convects and forms deep water at high latitudes so $N \rightarrow 0$ there. Consequently, the appropriate value of L_R in the high latitude ocean is extremely small (50 km or less). This increases the computational challenge of resolving ocean eddies, requiring resolutions of about $1/10°$. While OGCM simulations have been performed at such extremely high resolutions, they are not yet feasible for the long time scales required in climate modeling, where OGCM resolutions coarser than 1° are common. Some discussion of the differing resolution requirements for different problems appears in Chapter 1.

3.2. DEPENDENCE OF HEAT TRANSPORT ON RESOLUTION

The implied ocean meridional heat transport was discussed above, as a crucial quantity that AGCMs must get right in order to be successfully coupled to OGCMs. Of course, the OGCM must also simulate the correct ocean heat transport in order to avoid large climate drifts. Unfortunately, the poleward heat transport in OGCMs is strongly dependent on horizontal resolution (e.g., [7]). The heat transport across the Antarctic circumpolar current is overestimated in coarse resolution models because the horizontal diffusion that they require for stability causes large spurious transports due to the large temperature and density gradients in that region. In the North Atlantic, coarse resolution models tend to underestimate poleward heat transports and resolutions of $1/4°$, or better, are required to properly represent the heat transports there.

3.3. PARAMETERIZATION OF EDDY TRANSPORTS

The computational difficulty of resolving ocean eddies has prompted the development of eddy transport parameterizations. These parameterizations are intended for use in simulations that use relatively coarse (non-eddy resolving) meshes. The GM90 parameterization [13] used in the ocean component of CSM-1 [12] is probably the most popular of these parameterizations at present. In GM90, the horizontal diffusion on the model's height surfaces is replaced by horizontal diffusion on isopycnal surfaces. This had been done before, with varying degrees of success. The new component of GM90 was the introduction of eddy transport velocities into the tracer (e.g. potential temperature and salinity) equations. The tracer equations are then of the form:

$$\frac{\partial \theta}{\partial t} + \frac{\partial}{\partial x}[(u + u_E)\theta] + \frac{\partial}{\partial y}[(v + v_E)\theta] + \frac{\partial}{\partial z}[(w + w_E)\theta]$$

$$= K_I \nabla_I^2 \theta + \frac{\partial}{\partial z}\left[K_v\left(\frac{\partial \theta}{\partial z}\right)\right] + H,$$

where θ is the advected tracer; u, v, and w are the normal zonal and meridional and vertical velocities that appear in the momentum equations; K_I is the horizontal diffusivity on isopycnal surfaces; K_v is the vertical diffusivity; and H is a source/sink term. In the case that θ is potential temperature, H would be the diabatic heating rate. The eddy transport velocity components, u_E, v_E, and w_E, do not appear in the momentum equations and are defined by

$$u_E = \frac{\partial}{\partial z}\left[K_I \frac{\rho_x}{\rho_z}\right], \qquad v_E = \frac{\partial}{\partial z}\left[K_I \frac{\rho_y}{\rho_z}\right],$$

$$\frac{\partial u_E}{\partial x} + \frac{\partial v_E}{\partial y} + \frac{\partial w_E}{\partial z} = 0.$$

The introduction of the eddy transport velocities allows the use of much smaller horizontal and vertical diffusivities at the same resolution, while including the principal effects of the unresolved eddies. Incorporation of the GM90 parameterization has resulted in satisfactory simulations of the ocean heat transports, both across the Antarctic circumpolar current and in the North Atlantic, at horizontal resolutions as coarse as 3° (e.g., [8]). Boning et al. [3] showed that North Atlantic simulations at 1° horizontal resolution represented the North Atlantic heat transports very well with GM90, but significantly underestimated the transports with conventional horizontal diffusion.

3.4. REQUIREMENT FOR EQUATORIAL RESOLUTION

The equatorial zone in the ocean is of great importance for climate. Particularly in the Pacific Ocean, the SSTs vary substantially on a time scale of several years, with global consequences for temperature and precipitation. In fact, the relatively long time scale of ENSO is the principal basis for skill in seasonal forecasting (see Chapter 16). Possible changes in ENSO frequency and intensity are of great concern in climate change.

Simulating the equatorial oceans places different resolution requirements on ocean models than does simulating higher latitudes. The equatorial zone is characterized by long zonal scales and short meridional scales. There are very strong zonal currents that extend across most of the Pacific basin but are narrowly confined meridionally (within $\pm 5°$) and vertically. The equatorial undercurrent flows eastward, confined within $\pm 2°$ and in the upper 300 m, reaching speeds ~ 1 m s^{-1} between 100 and 200 m depth. The surface counter-currents flow westward, centered near 2° N and 2° S. These current systems support equatorial waves that are also long zonally and narrow meridionally.

Ocean grids commonly have fixed zonal resolution but variable latitudinal resolution. The latitudinal grid spacing is then reduced near the equator. The latitudinal grid spacing may also be reduced in high latitudes, where the convergence of the meridians reduces the effective zonal spacing, in order to keep approximately square grid boxes. For example, CSM-1 uses an ocean grid with uniform 2.4° spacing in longitude and an average of 2° in latitude. However, the latitudinal spacing reduces to 1.2° at the equator, increases to 2.3° in midlatitudes, and decreases again to 1.2° in high latitudes. It is not sufficient to simply decrease the meridional resolution near the equator while holding the horizontal diffusion coefficient constant. Since horizontal diffusion must be large enough to control noise in the coarsest resolution part of the domain, a uniform coefficient will not allow fine enough structure to form in the higher resolution part of the domain. Recently, the introduction of an asymmetric diffusion tensor with spatially varying coefficients in the CSM ocean model has resulted in vastly improved the equatorial ocean simulations (W. G. Large, personal communication). The improvements have been obtained with a modest increase in meridional resolution (0.6–0.9°), while retaining coarse zonal resolution (2.4–3.6°).

4. Sea Ice Models

Sea ice models are significantly different from oceanic or atmospheric general circulation models because ice does not behave as a Newtonian fluid. One of the principal assumptions on which the Navier-Stokes equations, governing fluid motion, are based is that stress is proportional to strain. In sea ice, the relationship between stress and strain is highly nonlinear. In fact, sea ice rheology cannot be expressed exactly, since there is not a simple constitutive law as in the case of ideal gases. Compact sea ice will diverge quite readily, but will resist convergence up to a certain point, at which the ice begins to ridge and buckle. Similarly, it will resist shearing stresses until they become large enough that the internal ice structure fails and relatively free (plastic) slippage may occur.

Sea ice models must also deal with the accumulation and melting of snow on the ice and the thermodynamic energy transfers between ocean and atmosphere through the ice and snow. These problems are actually similar to land modeling (see Chapter 13) and will not be addressed here. A more complete discussion of sea ice modeling can be found in [19] and references therein.

4.1. ICE RHEOLOGY

The motion of sea ice, in response to the forces imposed by sea surface tilts and the surface stresses on the atmosphere and ocean, is largely determined by the mechanical properties of sea ice. These properties are expressed in approximate constitutive laws, which give stresses that may be relatively independent of the strain rate, for realistic nonlinear viscous-plastic rheologies.

The momentum equations for sea ice can be expressed as:

$$m\frac{\partial \mathbf{V}}{\partial t} = mf\mathbf{k} \times \mathbf{V} = \tau_A + \tau_O - mg\nabla z_O + \nabla \cdot \sigma,$$

where m is the ice mass per unit area, $\mathbf{V}$ is the horizontal velocity vector, f is the coriolis parameter, $\mathbf{k}$ is the vertical unit vector, $\tau_{A,O}$ are the atmospheric and oceanic surface stresses, g is acceleration of gravity, z_O is the height of the ocean surface, and σ is the two dimensional ice stress tensor.

The relationship between the stress and strain tensors is specified by the constitutive law, which has the general form:

$$\sigma_{ij} = 2\eta\dot{\epsilon}_{ij} + [(\zeta - \eta)\dot{\epsilon}_{kk} - P]\delta_{ij},$$

where P is the internal ice pressure, η and ζ are the shear and bulk viscosities, $\dot{\epsilon}$ is the strain rate tensor, i, j and k are indexes in standard tensor notation and $\delta_{ij} = 1$ for $i = j$, or 0 for $i \neq j$.

The definitions of the P, η and ζ depend on the sea ice rheology. A particularly simple example is the cavitating fluid rheology of Flato and Hibler [10], in which ice has no shear strength ($\eta = \zeta = 0$), resulting in $\sigma_{ij} = -\frac{P}{2}\delta_{ij}$. This approximation is used in several climate models, including CSM-1 (see [32]). The more complete plastic-viscous rheology of Hibler [18] is also used in climate models, but is considerably more complex and will not be described here. However, in either the case, the ice pressure is determined by $P = P^*/2$, where $-P^*$ is the maximum compressive stress allowed before ice failure (e.g., ridging). We note that P^* has not yet been specified, since this requires knowledge of the distributions of ice thickness and compactness (fractional ice area).

4.2. TWO-LEVEL ICE MODEL

The sea ice should be described by a thickness distribution within a grid cell. However, the ice thickness distribution is often approximated with just two categories, thick and thin ice. The thick ice is assumed to be of uniform thickness and the thin ice is actually open water (zero thickness). In this case, there are only two variables to predict, the mean ice thickness h and the compactness a of the thick ice. The thickness of thick ice is h/a and the fractional area of open water is just $(1 - a)$. Since the mean ice thickness is actually computed (and the ice density is uniform), the equations for h and a are just

$$\frac{dh}{dt} = S_h$$
$$\frac{da}{dt} = S_a \quad (a \leq 1),$$

where S_h represents the total ice growth and melting due to thermodynamic processes and S_a represents lateral ice growth. Requiring $a < 1$ will result in ice thickening where compact ice is converging. However, only h/a, the thickness of the "thick ice", is affected. The mean ice thickness h is a conserved variable that is unaffected by a.

In the two-level ice model, the maximum compressive stress P^* is determined by a simple formulation for ice strength,

$$P^* = P_0 h e^{-c(1-a)},$$

where P_0 and c are constants. This formulation allows for ice strength that increases linearly with mean thickness and exponentially with compactness as $a \rightarrow 1$.

While the two-level model has proven useful in early climate model simulations, it does not really capture the nature of sea ice very well, particularly in the Arctic. In a few years, it is likely that most sea ice models used in climate models will use multilevel ice distributions and variations of plastic-viscous rheology.

5. NCAR CSM-1 Coupled Solution, Equilibrium Forcing

CSM-1 is a physical climate model, similar in nature to several other coupled models that have been used for climate studies. For example, Gates *et al.* [11] and Kattenberg *et al.* [21] discuss climate change experiments performed by climate models from several countries. The main new features in the CSM-1 compared to previous coupled climate models are the coupling strategy and the physical parameterizations in the ocean model.

5.1. MODEL FORMULATION

The ocean model is based on the GFDL MOM model, with several improvements in the numerics. What sets it apart from previous ocean models used in coupled applications is inclusion of the GM90 eddy parameterization, described above, and a true ocean boundary layer parameterization [23].

The AGCM is the NCAR Community Climate Model, Version 3 (CCM3, [22] at T42 horizontal resolution, with 18 levels in the vertical. One of the greatest improvements in CCM3, compared to the previous version (CCM2) and other AGCMs of the same period, is in the representation of surface fluxes and implied ocean heat transports. Until recently, AGCMs have not provided accurate depictions of the implied ocean heat transport. Gleckler *et al.* [15] showed that most AGCMs participating in the Atmospheric Model Intercomparison Project had implied ocean heat transports of the wrong sign in the Southern Hemisphere. Hack [17] showed that changes in both convection and cloud parameterizations between CCM2 and CCM3 were required to produce relatively accurate implied transports. The land surface model is LSM1 [5], which is run on the same grid as CCM3.

These models—ocean, atmosphere, and land surface—of physical and biophysical processes differ from several other current models in the nature of individual parameterizations, but not in the processes that are considered. The sea ice model [32] is also similar to other ice models, considering ice rheology (cavitating fluid in this case) and simple ice thermodynamics. This sea ice model, which runs on the ocean model grid, is comparable to that used in other climate models, but probably should not be considered state-of-the-art.

The philosophy has been adopted in the CSM-1 that the most appropriate boundary conditions for the component models are the fluxes at the Earth's surface. Where possible, those interfacial fluxes that depend directly on the state of more than one component model, e.g., turbulent fluxes of latent and sensible heat, are computed within the flux coupler (see [6] and Figure 1). No flux corrections in momentum, heat or fresh water are applied. The flux coupler is also responsible for interpolating and averaging between the different grids of the component models while conserving local and integral properties. The flux coupler currently allows two separate coupling intervals between itself and the component models. The atmosphere, land, and sea-ice models communicate at the faster interval, usually one hour, and the ocean model communicates at the slower interval, usually one day. Instantaneous values of state variables and interfacial fluxes are time averaged over the coupling interval and are passed to the flux coupler. Therefore, fluxes are computed from instantaneous state variables, and the time integrals of the fluxes applied in the different component models are the same.

5.2. SPINUP PROCEDURE

In order to perform relatively drift-free coupled simulations, compatible initial states for the component models are required. Stouffer and Dixon [29] discuss the variety of spinup procedures that have been used to obtain nearly balanced atmosphere, ocean, and ice initial conditions. CSM-1 simulations have used a relatively complex, multistage spinup ([1], [5]). Very recently, Gordon *et al.* [16]

432

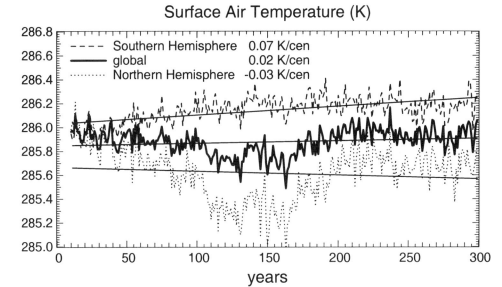

Figure 2. Time series of 12 month mean surface air temperatures from the CSM-1 control simulation averaged over the whole globe (solid), Southern Hemisphere (dashed), and Northern Hemisphere (dotted). Thin solid lines show the respective least squares trend lines and the slopes are listed in the legend.

have found that their latest coupled model is accurate enough that the spinup procedures can be largely omitted. The same is believed to be true of more recent versions of CSM.

5.3. RESULTS OF A 300 YEAR CONTROL SIMULATION

As mentioned above, it has been notoriously difficult to obtain stable coupled model solutions without resorting to flux corrections. The CSM-1 simulation ([1] and references therein) was the first multi-century experiment without significant trends in the surface temperatures. Figure 2 shows time series of annual average surface air temperature for the last 290 years of that simulation. Early in the simulation, there is a small tendency for the Southern Hemisphere to get warmer and for the Northern Hemisphere to get colder, resulting in somewhat larger trend magnitudes for the individual hemispheres than for the globe. However, the dominant signal in the time series is a centennial time scale variation in the Northern Hemisphere, which is also seen for the globe, associated with a variation in North Atlantic sea ice extent and ocean overturning (see [1]). The linear trends in these time series are insignificant.

A map of the difference of the simulated SST compared to present climatology (Figure 3) also shows that the the Northern Hemisphere is generally colder than observed, while the Southern Hemisphere is generally warmer. The cold bias is

annual mean SST error (K)

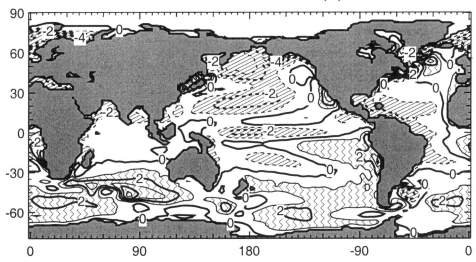

Figure 3. Difference between the simulated SST and climatological temperatures for 1950–1979. The contour interval is 1 K and magnitudes less than 1 K are unshaded.

most notable in the North Pacific and in the Barents Sea, while the warm bias is largest in the Southern Atlantic and Indian Oceans. Warm biases are also found in the regions of persistent subtropical marine stratocumulus off the western coasts of the Americas and Africa. These subtropical biases are common to most coupled models and result from an inability of present AGCMs to adequately simulate the stratocumulus. Another bias that is common to most coupled models is the excessively cold water in the central equatorial Pacific. This error results from an extension of the observed equatorial cold tongue from the eastern Pacific much too far into the central and western Pacific. The SST error gives rise to significant changes in the atmospheric circulation, such as the formation of a strong double ITCZ in southern summer [2], again a common problem in coupled models.

Very similar global time series and SST error maps can be found in [16] for the most recent version of the Hadley Centre for Climate Prediction and Research coupled model (HADCM3). HADCM3 also does not use flux corrections and has now been integrated even longer than CSM-1 (more than 800 years) without significant surface temperature trends.

Although the CSM-1 simulation has no significant trends in the surface temperatures, there are significant trends in other quantities. The largest trends are in deep ocean temperatures and salinities. Figure 4 shows the horizontally averaged temperature and salinity time series at 4 km depth in the ocean from the control simulation and from an additional simulation. Negative years refer to the spinup phase of the simulations, during which the ocean and ice models are driven with output from a CCM3 simulation with fixed SSTs. During the spinup phase for the

Ocean Horizontal Means on Level 38 (~4 km)

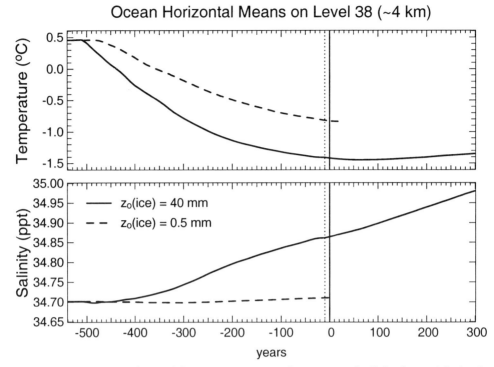

Figure 4. Area averaged potential temperature (top), and area averaged salinity (bottom) for level 38 in the ocean model, which is at about 4km depth. Negative years refer to the spinup phase. Solid lines are from the CSM-1 control run, dashed lines from a second simulation with reduced aerodynamic roughness for sea ice.

300-year control simulation, the deep ocean gets colder and saltier. The temperature has nearly equilibrated by the end of the spinup (in fact, this was one of the factors determining the length of the spinup), and the trend is very small through the coupled simulation. However, the salinity trend continues through the entire control simulation, during which the deep ocean continues to get saltier while the near surface ocean gets fresher.

The deep ocean trends are largely produced by excessive export of ice from Antarctica, forming too much very cold salty water that sinks and spreads in the deep ocean. The temperature trend only stops because the deep ocean is approaching the freezing point of sea water ($\sim$-1.8° C) and cannot get any colder. The second simulation in Figure 4 demonstrates the sensitivity of the coupled system to sea ice. In that simulation, the aerodynamic roughness length (see chapter 8) of sea ice was reduced from 40 mm, appropriate for very rough multi-year ice, to 0.5 mm, appropriate for very smooth first year ice. The latter is more representative of Southern Hemisphere sea ice, and changes the drag coefficient by a factor of $\sim$ 4, since it depends on the logarithm of the roughness length. A more realistic ice distribution was also used in the fixed SST CCM3 run used for driving the spinup. The reduced ice drag results in less ice export from the coast of Antarctica, less deep water formation around Antarctica and a slowing of the Antarctic circumpolar current. In consequence, the deep ocean temperature trend is reduced by ~25% and the salinity trend by at least an order of magnitude. This sensitivity to the roughness of sea ice is probably excessive, because of other problems in the sea ice model, but illustrates the very strong, long time-scale sensitivity of the climate system to sea ice processes.

6. 1%/year CO_2 Increase Experiment

The atmospheric CO_2 concentration has been increasing approximately exponentially for more than a century and is expected to continue to increase rapidly for at least another century. The present rate of increase is somewhat less than 1%/year, compounded. Concentrations of other greenhouse gases (e.g., methane, nitrous oxide and a variety of halocarbons, including CFCs) are also increasing with time. Since many models do not explicitly account for these other greenhouse gases in their radiation parameterizations, it has been common practice to include their effects through an effective CO_2 concentration. The present rate of increase of the effective CO_2 is close enough to 1%/year that many groups have performed idealized climate change experiments by increasing CO_2 at that rate, while holding all other gases constant. The CO_2 concentration doubles after 70 years and triples after 110 years at this rate of increase. Results of several such experiments have been collected in CMIP-II and are currently being compared. More realistic versions of such experiments are at the heart of the IPCC process as discussed in Chapter 21.

Figure 5 shows the time series of annually and globally averaged surface air temperature from a CSM-1 simulation with CO_2 increasing at 1%/year and no change in aerosol effects. For convenience, the time at which CO_2 begins to increase is labeled year 0 on this and subsequent plots. The simulation actually uses an initial condition from year 15 of the control simulation, but CO_2 remains constant for the first 10 years. The control time series is included for reference in Figure 5, where year 0 corresponds to year 25 in Figure 2. The globally averaged temperature increases nearly linearly in time, at a rate of 2.2 K/century, with variability about the trend line that is comparable to the variability about the time mean in the control simulation. At fairly high absorber concentrations the absorbed radiative flux varies with the logarithm of the concentration (which is increasing exponentially) so a linear temperature increase with time should be expected in this case. The rate of temperature increase in CSM-1 is actually somewhat lower than found in most other models, in fact, it is the second lowest among the simulations submitted to CMIP-II. It is also noteworthy that the temperatures over land are increasing faster than over ocean, so that the land-sea temperature contrast is increasing with time.

436

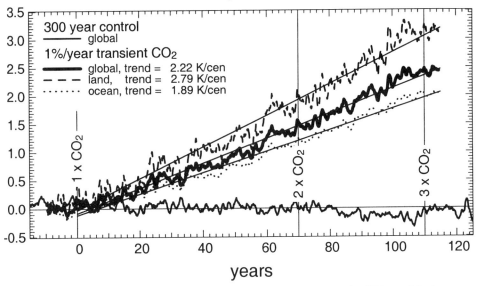

Figure 5. Time series of 12 month mean surface air temperature from the 1%/year CO_2 increase experiment averaged over all surfaces (thick solid), land only (dashed), and ocean and sea ice (dotted). Deviations are shown from the long term means of the control run (285.85, 280.21, and 289.06 K, respectively). Thin solid lines show the least squares trend lines and the slopes are listed in the legend. The global mean from the comparable period of the control run is also shown (medium solid).

Again this result is expected, since land points have much lower heat capacity than ocean points.

Feedbacks in the climate system help determine the surface temperature response to greenhouse gases changes; one of the most important is the ice-albedo feedback. Sea ice has much higher albedo than sea water (>0.6 compared to 0.07), so areas of decreased sea ice result in increased absorption of solar radiation, tending to warm the surface further. Therefore, any decreases in sea ice area as the global temperature increases will constitute a positive feedback. Models that give large temperature increases due to greenhouse gas increases tend to have large sea ice decreases. The effects of this feedback are limited by the relatively small area of sea ice, and the fact that it occurs primarily in high-latitude winter, where the insolation is low. Figure 6 shows the areas covered by sea ice in the Northern and Southern Hemispheres, separately. The annual range of ice area in the control simulation is shown by shading, while the monthly values in the CO_2 increase experiment are given by the thin line. One can see that the sea ice areas in both hemispheres decrease as the CO_2 concentration increases, with the effect being largest in the maximum (winter) areas. However, the effect is quite small until around the time of CO_2 doubling. One can also see that CSM-1 tends to produce considerably too much ice in the Northern Hemisphere, so that the simulated areas

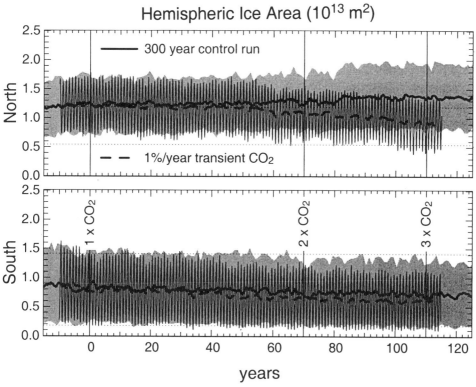

Figure 6. Time series of 12 month mean total ice areas in the Northern (top) and Southern (bottom) Hemispheres from 1%/year CO_2 increase experiment (dash) and the comparable period of the control run (thick solid). The thin solid line and shading show the annual range of monthly mean values for the 1%/year and control simulations, respectively.

at the time of tripling actually match the present day observed areas quite well.

The largest increases in decadal mean surface air temperature from the last decade of constant CO_2 to the first decade following CO_2 doubling are in high latitudes (Figure 7). The tropical and subtropical temperatures increase fairly uniformly by $\sim$1 K, while the high latitude increases reach 8 K. The increases in both hemispheres are also larger in winter than in summer and are associated with reductions in both the sea ice area and thickness. As shown above, the sea ice area decreases as the CO_2 concentration increases. The largest temperature increases, in the North Pacific and North Atlantic, are associated with retreats of the sea ice. Unfortunately, these regions have much too extensive sea ice in the control simulation, so that these effects are probably exaggerated.

The vertical profile of temperature and humidity response to CO_2 increase is given in Figure 8. The largest temperature increases ($\sim$4 K) are at the surface in the high latitude Northern Hemisphere. A secondary maximum is also found in the tropical upper troposphere, associated with a significant increase in the

438

Surface Air Temperature Difference (K)

2xCO$_2$ (yrs 70-79) - control   annual

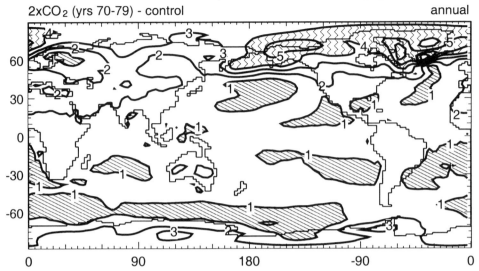

Figure 7.   The difference in annual average surface air temperature near the time of CO$_2$ doubling (years 70-79), compared to the control simulation (contour interval, 1 K).

specific humidity. In fact, the actual water vapor concentration is quite low in that region, because the temperatures are low. However, the radiative impact of the increase in water vapor is quite important. Similar increases in water vapor are found in virtually all models, and are responsible for a substantial part of the surface warming. The issue of whether water vapor will actually increase in the tropical upper troposphere is at the heart of the debate over the magnitude of the global warming expected from increasing greenhouse gases (e.g., [9]).

7. Computer Hardware

Atmospheric general circulation modeling is a computationally demanding problem and climate modeling is even more challenging, involving additional components and long time scales. Climate models must make efficient use of the most powerful computers available at any time, but the development of a successful climate model is usually an evolutionary process over many years. Therefore, evolving computer technology is of considerable interest to climate modelers, affecting the type of problems that can be addressed, the types of numerical algorithms that can be implemented efficiently, and the design of the code that implements the mathematical model.

Ideally, one would like a single very fast scalar processor with an enormous amount of memory. In reality, high speed in computing is always achieved by some form of parallelism, exploiting either multiple processors, vector processors,

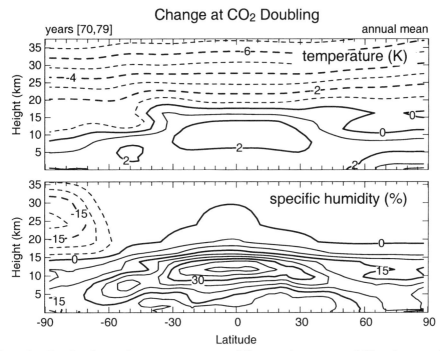

Figure 8.   Zonal and annual average temperature differences (top, contour 1 K) and percent differences in specific humidity (bottom, contour 5%) near the time of CO_2 doubling (years 70-79), compared to the control simulation.

or both. For most of the last twenty years, the most powerful computers available have used vector processors, which execute the same set of instructions sequentially over a long list (vector) of data values. The programming constructs used for these processors conform naturally to the equations solved in the fluid mechanics portions of climate models, but usually require significant effort to use for the physical parameterizations. On scalar processors it is not necessary to apply an identical sequence of instructions to a list of points, so they are generally easier to program, but offer much slower execution rates.

In high speed computers, multiple processors are used to increase the computing capacity beyond what can be obtained by a single processor. Although scalar processors are slower than vector processors, they are much cheaper, and consequently many more processors are typically used in parallel scalar machines compared to parallel vector machines. In either case, present compiler technology is not adequate for automatically achieving high performance. Both the programming difficulty and the restrictions on types of numerical algorithms that can be implemented efficiently generally increase with the number of processors that must be used.

The layout of the memory in high speed computers further complicates the modeling problem. Today's computers typically have large enough memories that many of the tricks used to conserve memory usage in the past are no longer necessary. However, many of today's (and tomorrow's) computers actually have multiple independent memories that must be managed, introducing a different kind of parallelism and requiring new programming constructs.

A portable GCM, such as CCM3, must be able to efficiently use multiple processors, on either vector or scalar machines with either shared or distributed memory. This makes the design and implementation of such a model much more complex than would be the case if the model could be targeted at a single computer system. While a model may perform optimally on a particular system architecture (shared memory parallel vector, for CCM3), its numerical algorithms cannot generally be implemented on all architectures with equal efficiency. The choice of numerical algorithms in a model must then be partly determined by the primary target machine architecture. Unfortunately, the present trend is toward large numbers of processors with groups of several processors sharing a section of a distributed memory. This architecture is one of the most difficult to program efficiently and is being implemented in various ways by different manufacturers, some using scalar processors, others using vector processors. Maintaining efficient portable models will be an ongoing (and expensive) challenge.

8. Summary

Understanding the behavior of the climate system on time scales of years to millennia requires understanding the physical, chemical, and biogeochemical interactions among climate subsystems. Climate system models are now being developed to address these problems and to understand the response of the climate system to anthropogenic influences. These models are still in an early stage of development and primarily encompass the physical climate system at present, although progress is being made at including chemical and biogeochemical components. Even a physical climate model is exceedingly complex, involving communication between models for the atmosphere, ocean, land surface, and sea ice. Such models have been in use for about 20 years and are now becoming more mature, with reasonably accurate and stable simulations being obtained without resorting to ad hoc fixes. However, much work remains to be done on improving physical climate models, while other climate system components are being incorporated.

An AGCM intended for use in a coupled model must closely satisfy constraints on energy conservation and accuracy of the surface fluxes in order to be useful. In principle OGCMs are quite similar to AGCMs although some of the physical parameterization problems are quite different. Sea ice models are quite different because, under many circumstances, ice does not behave as a fluid. The long term behavior of the climate system is sensitive to all of the components and accurate models for each are required.

Coupled climate models are more computationally demanding than AGCMs and are typically run for centuries rather than 15 years or less. The computational demands will continue to increase rapidly, as chemical and biogeochemical components are added. Making efficient use of the available computer systems is essential. Maintaining portability across systems is also important, both to allow simultaneous use of different computing platforms and to allow migration to new platforms as computing technology evolves. Efficiency and portability are often conflicting requirements that make the design and maintenance of both AGCMs and coupled models difficult and expensive.

References

1. Boville, B.A., and Gent, P.R. (1998) The NCAR climate system model, version one, *J. Climate* **11**, 1115–1130.
2. Boville, B.A., and Hurrell, J. (1998) A comparison of the atmospheric circulations in CCM3 and CSM1, *J. Climate* **11**, 1327–1341.
3. Boning, C.W., Holland, W.R., Bryan, F.O., Danabasoglu, G., and McWilliams, J.C., (1995) An overlooked problem in model simulations of the thermohaline circulation and heat transports in the Atlantic Ocean, *J. Climate* **8**, 515–523.
4. Bryan, K. (1969) A numerical method for the study of the circulation of the world ocean, *J. Comput. Phys.* **4**, 347–376.
5. Bryan, F.O. (1998) Climate drift in a multi-century integration of the NCAR Climate System Model, *J. Climate* **11**, 1455–1471.
6. Bryan, F.O., Kauffman, B.G., Large, W.G., and Gent, P.R. (1996) *The NCAR CSM Flux Coupler*, NCAR Technical Note No. 424, Boulder.
7. Covey, C. (1995) Global ocean circulation and equator–pole heat transport as a function of ocean GCM resolution, *Climate Dyn.* **11**, 425–437.
8. Danabasoglu, G., McWilliams, J.C., and Gent, P.R., (1994) The role of mesoscale tracer transports in the global ocean circulation, *Science* **264**, 1123–1126.
9. Dickinson, R.E., Meleshko, V., Randall, D., Sarachik, E., Silva-Dias, P., and Slingo, A. (1996) Climate Processes, in J.T. Houghton, L.G. Meira Filho, B.A. Callander, N. Harris, A. Kattenberg, and K. Maskell (eds.), *Climate Change 1995: The Science of Climate Change,* Cambridge University Press, Cambridge.
10. Flato, G.M., and Hibler, W.D. (1992) Modeling ice pack as a cavitating fluid, *J. Phys. Oceanogr.* **22**, 626–651.
11. Gates, W.L., Henderson-Sellers, A., Boer, G.J., Folland, C.K., Kitoh, A., McAvaney, B.J., Semazzi, F., Smith, N., Weaver, A.J., and Zeng, Q.C. (1996) Climate models — evaluation, in J.T. Houghton, L.G. Meira Filho, B.A. Callander, N. Harris, A. Kattenberg, and K. Maskell (eds.), *Climate Change 1995: The Science of Climate Change,* Cambridge University Press, Cambridge.
12. Gent, P.R., Bryan, F.O., Danabasoglu, G., Doncy, S.C., Holland, W.R., Large, W.G., and McWilliams, J.C. (1998) The NCAR Climate System Model global ocean component, *J. Climate* **11**, 1287–1306.
13. Gent, P.R., and McWilliams, J.C. (1990) Isopycnal mixing in ocean circulation models, *J. Phys. Oceanogr.* **20**, 150–155.
14. Gill, A.E. (1982) *Atmosphere-ocean dynamics,* Academic Press, New York.
15. Gleckler, P.J., Randall, D.A., Boer, G., Colman, R., Dix, M., Galin, V., Helfand, M., Kiehl, J., Kitoh, A., Lau, W., Liang, X.-Y., Lykossov, V., McAvaney, B., Miyakoda, K., Planton, S., and

442

Stern, W. (1995) Cloud–radiative effects on implied oceanic energy transports as simulated by atmospheric general circulation models, *Geophysical Research Letters* **22**, 791–794.

16.   Gordon, C., Cooper, C., Senior, C.A., Banks, H., Gregory, J.M., Johns, T.C., Mitchell, J.F.B., and Wood, R.A. (1999) The simulation of SST, sea ice extents and ocean heat transports in a version of the Hadley Centre coupled model without flux adjustments. *Climate Dyn.*, submitted.

17.   Hack, J.J. (1998) An analysis of the improvement in implied meridional ocean energy transport as simulated by the NCAR CCM3, *J. Climate* **11**, 1237–1244.

18.   Hibler, W.D. (1979) A dynamic thermodynamic sea ice model, *J. Phys. Oceanogr.* **9**, 815–846.

19.   Hibler, W.D., and Flato, G.M. (1992) Sea Ice Models, in K.E. Trenberth (ed), *Climate System Modeling*, Cambridge University Press, Cambridge.

20.   Johns, T.C, Carnell, R.E., Crossley, J.F., Gregory, J.M., Mitchell, J.F.B., Senior, C.A., Tett, S.F.B., and Wood, R.A. (1997) The second Hadley Centre coupled ocean-atmosphere GCM: model description, spinup and validation, *Climate Dyn.* **13**, 103–134.

21.   Kattenberg, A., Giorgi, F., Grassl, H., Meehl, G.A., Mitchell, J.F.B., Stouffer, R.J., Tokioka, T., Weaver, A.J., and Wigley, T.M.L. (1996) Climate models — projections of future climate, in J.T. Houghton, L.G. Meira Filho, B.A. Callander, N. Harris, A. Kattenberg, and K. Maskell (eds.), *Climate Change 1995: The Science of Climate Change,* Cambridge University Press, Cambridge.

22.   Kiehl, J.T., Hack, J.J., and Hurrell, J. (1998) The energy budget of the NCAR Community Climate Model: CCM3, *J. Climate* **11**, 1151–1178.

23.   Large, W.G., Danabasoglu, G., Doney, S.C., and McWilliams, J.C. (1997) Sensitivity to surface forcing and boundary layer mixing in the NCAR CSM ocean model: annual-mean climatology, *J. Phys. Oceanogr.* **27**, 2418–2447.

24.   Madec, G., and Imbard, M. (1996) A global ocean mesh to overcome the North Pole singularity, *Climate Dyn.* **12**, 381–388.

25.   Meehl, G.A., Boer, G.J., Covey, C., Latif, M., and Stouffer, R. (1997) Analyzing and intercomparing global coupled atmosphere-ocean-sea ice climate models, *Eos* **78**, 445–451.

26.   Oberhuber, J.M. (1995) Simulation of the Atlantic circulation with a coupled sea ice-mixed layer-isopycnal general circulation model. Part I: Model description, *J. Phys. Oceanogr.* **23**, 808–829.

27.   Pacanowski, R.C., Dixon, K., and Rosati, A. (1993) The GFDL Modular Ocean Model users guide, GFDL Ocean Group Tech. Rep. 2. [Available from GFDL, Princeton University, Princeton, NJ, USA 08542.]

28.   Shine, K.P., Fouquart, Y., Ramaswamy, V., Solomon, S., Srinivasan, J. (1994) Radiative Forcing, in J.T. Houghton, L.G. Meira Filho, J. Bruce, B.A. Callander, E. Haites, N. Harris, and K. Maskell (eds.), *Climate Change 1994: Radiative Forcing of Climate Change and An Evaluation of the IPCC IS92 Emission Scenarios*, Cambridge University Press, Cambridge.

29.   Stouffer, R.J., and Dixon, K.W. (1998) Initialization of coupled models for use in climate studies: A review, in Report of the first session of the JSC/CLIVAR working group on coupled modelling, WCRP Informal Report No. 3/1998.

30.   Trenberth, K.E. (1997) The heat budget of the atmosphere and ocean, *Proc. First Int. Conf. on Reanalysis* WMO/WCRP.

31.   Trenberth, K.E., and Solomon, A. (1994) The global heat balance: Heat transports in the atmosphere and ocean, *Climate Dyn.* **10**, 107–134.

32.   Weatherly, J.W., Briegleb, B.P., Large, W.G., and Maslanik, J.A. (1998) Sea ice and polar climate in the NCAR CSM, *J. Climate* **11**, 1472–1486.

CLIMATE MODEL INTERCOMPARISON

G. J. BOER
Canadian Centre for Climate Modelling and Analysis
Atmospheric Environment Service
University of Victoria
Victoria, B.C., V8W 2Y2   CANADA

1. Introduction

Global climate models (GCMs) are the accepted tools for investigating the behavior of the climate system and for simulating potential climate change due to increases in greenhouse gases or other changes to the forcing of the system. GCMs attempt to obtain the climate from first principles by solving the complex coupled partial differential equations that embody the physical principles that control the behavior of the atmosphere, ocean, cryosphere, and land surface. The equations are solved by numerical methods and the solutions take the form of evolving values of the quantities that determine the state of the system, namely the winds, temperatures, moisture, clouds, precipitation, evaporation, flows of radiative energy, and so on. New values of these quantities are calculated on a global grid of points and at many levels in the atmosphere and ocean every half an hour or less of simulated time. The GCM produces *much* more complete information than is available from observations (especially so in the case of the ocean).

The enormous amount of data produced by a climate model makes the "analysis and validation" of model output a complex and difficult task as discussed in Chapter 3. Simulated climates will differ from the observed climate but, while the analysis and verification of model climate can reveal model deficiencies, this does not necessarily lead directly to model improvement. Model improvement is made difficult by the non-linearity of the system which masks the connection between cause and effect, by the long simulations that are required to test if model changes lead to improvements in simulated climate, and by the lack of general techniques to isolate and eliminate the causes of model deficiencies.

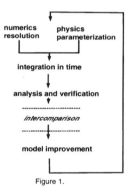

Figure 1.

Figure 1 parallels the similar figure in Chapter 3 but now incorporates an *intercomparison* step which is, in effect, a *community* analysis/verification process where results from an array of models are compared among themselves as well as with the observed climate. While it may seem somewhat incestuous to compare models with themselves, there are several reasons to do so. The purposes of model intercomparison are (1) to document the ability of models to simulate the current climate; (2) to compare the ability of models to simulate perturbed climates; (3) to identify such *common deficiencies* as exist in simulated climates; (4) on the basis of these deficiencies to formulate hypotheses con-

P. Mote and A. O'Neill (eds.), Numerical Modeling of the Global Atmosphere in the Climate System, 443–464.

444

cerning their causes; (5) to give some attention to, and draw inferences concerning, the effects of numerical methods, model resolution, and parameterizations employed in models; (6) to suggest numerical experiments to clarify the causes and potential solution to model deficiencies; and (7) to document model evolution and improvement.

One of the most compelling reasons for intercomparison is to assess the ability of models to simulate current climate as a justification for their use for simulating potential anthropogenic climate change. The Intergovernmental Panel on Climate Change (IPCC) incorporates chapters on model evaluation in its reports ([17],[18],[19]) and these include aspects of model intercomparison.

2. Model Intercomparison

Model intercomparison has a fairly lengthy history consisting of roughly four phases: (1) *ad hoc* juxtaposition of diagrams of model results (e.g., [10]); (2) the compilation and intercomparison of a restricted set of model climate statistics which are displayed on common diagrams (e.g., mean sea-level pressure and precipitation [11], 500mb height [25], atmosphere-ocean heat fluxes and stresses [20]); (3) a combination of the two (e.g., [3], [4]); and (4) purpose-designed intercomparisons where data are supplied in electronic form for analysis and intercomparison (e.g., [8], [9], the Atmospheric Model Intercomparison Program (AMIP) [12], Gates *et al.* [14], and other recent model intercomparisons discussed below).

2.1 FLAVORS OF INTERCOMPARISON

There are numerous "flavors" or "streams" of intercomparison depending on how the process is viewed:

(1) The *"pure" numerics* stream, deals with the treatment of the resolved dynamics on the "left-hand-side" of the model equations as embodied in model numerics and resolution (see Section 3.3 of Chapter 3). The intent is to separate out numerical/resolution aspects from those involving the physical parameterization as they affect model climate and to treat only the first. One approach uses a simple specified forcing in a way that minimizes the complicated interplay between numerics/resolution and physical parameterization that occurs in a full GCM. Two somewhat different forcing specifications have been suggested by Held and Suarez [16] and by Boer and Denis [6], and there is an informal "dynamical cores" intercomparison (see *http://www-pcmdi.llnl.gov/dc*).

(2) The *"pure" physics* or parameterization stream deals with the difficult problem of the parameterization of subgrid-scale physical processes in GCMs. These are the "right-hand-side" terms in the equations as discussed in Chapter 3. Many different physical processes are parameterized in a GCM, as discussed in Chapters 8-13, each of which could potentially be investigated via intercomparison. The "stand alone" approach has been fairly common where the parameterized behavior of a particular process is studied in a column model. Examples include an earlier Intercomparison of Radiation Codes in Climate Models and current intercomparisons such as the Project for Intercomparison of Land-surface Parameterization Schemes. The stand-alone approach has the considerable virtue that conditions are well controlled. It does not, however, give information on the

interactions and feedbacks that arise when a particular parameterization is used in a full model and in conjunction with other parameterizations and the numerical treatment and resolution of that model.

The effect of different parameterizations of a particular process can be tested, and typically are tested in model development mode, in one's own model, but this does not provide the flavor of a full comparison. There have not been any "in-model" intercomparisons where a specific parameterization has been tested in an array of full GCMs. There are cases where a particular kind (but not implementation) of a parameterized process is tested in several models and shows similar effects on simulated climate. An example of this is the gravity wave drag parameterization (Chapter 12) which has been adopted in most GCMs, including those used for numerical weather prediction.

(3) The *full model* approach is adopted by most Model Intercomparison Projects (MIPs) but it is also the most difficult since the effects of different numerics, resolution, and physical parameterizations are all intertwined in the results. It is an attractive option in a practical sense since all modelers perform a range of simulations as part of the development/improvement cycle of Figure 1, and an intercomparison can be part of that process.

Some standardization in full model intercomparisons is possible as *controlled* rather than *as available* simulations. Here the control is not on the model itself but on the "external" conditions such as the solar constant, CO_2 concentration, sea-surface temperatures and so on; control does not extend to the initial state, the resolution, or other features of the model and simulation. Finally, full model intercomparisons can be further subdivided into investigations of (1) *current climate* and its variations, (2) *model sensitivity* and feedbacks (e.g., cloud feedback, climate sensitivity), and (3) *perturbed climate* (paleoclimate, global warming).

2.2 ADVANTAGES AND DISADVANTAGES OF INTERCOMPARISON

Intercomparison has the advantages of (1) focusing attention on a wide array of model results and highlighting common problems or *systematic deficiencies* in order to understand and correct them; (2) providing some documentation of the evolutionary improvement of models and of their current capabilities; (3) providing information, although largely implicit, on the consequences of numerics, resolution, parameterization and other features of model formulation; (4) potentially spreading the cost of investigating many different types and formulations of climate model over the modeling community; and (5) entraining non-modelers to apply their expertise to the analysis of many aspects of model behavior.

Model intercomparison is not, however, an unalloyed good since (1) analysis may be superficial in its approach or in its focus on only a few variables; (2) identifying common difficulties does not identify causes nor necessarily suggest remedies; (3) it is usually difficult to infer general consequences from the heterogeneous model results; (4) an increasing number of MIPs and the desire to include a greater number of variables and a larger amount of data can be costly in time and effort; (5) observed climatological information may be lacking or of unknown quality for making comparison with some model results; and, in particular, (6) MIPs do not necessarily point investigators to focused

446

numerical experiments that improve models in clear and immediate ways.

Nevertheless, model intercomparison is an important *community* aspect of the evolution and improvement of climate models which acts to focus attention on the current strengths and weaknesses of models and on needed model improvement.

2.3 INTERCOMPARISON PROJECTS

There are a variety of formal and informal climate model intercomparison projects, organized by various international bodies. A number of them are listed below:

  a. AMIP1 and *AMIP2* - Atmospheric Model Intercomparison Projects - organized by the CAS/JSC Working Group on Numerical Experimentation (WGNE) of the World Climate Research Programme (WCRP)

  b. CMIP - the Coupled Model Intercomparison Project - organized by the JSC/CLIVAR Working Group on Coupled Modelling (WGCM)

  c. *ENSIP* - ENSO Intercomparison Project (in coupled models) - organized by the GOALS Numerical Experimentation Group (NEG1)

  d. *GRIPS* - GCM Reality Intercomparison Project for SPARC (Stratospheric Processes And their Role in Climate) - organized by the International Council of Scientific Unions (ICSU) and representing, perhaps, the most egregious use of acronyms.

  e. *PMIP* - Paleoclimate Model Intercomparison Project - organized by International Geosphere-Biosphere Program / Past Global Changes (IGBP/PAGES) and WCRP/WGNE.

  f. *SIMIP* - Sea-Ice Model Intercomparison Project - organized by the Sea Ice Ocean Modelling (SIOM) Panel of the Arctic Climate System Study (ACSYS).

  Other intercomparison projects include TRANSCOM (transport intercomparison in models), a Dynamical Cores Intercomparison (currently without acronym), a regional climate model intercomparison (PIRCS). Intercomparison is a widespread, acronym-laden, but clearly flourishing activity of the various modeling communities. The Program for Climate Model Diagnosis and Intercomparison (PCMDI) provides infrastructure support to many of these intercomparison projects, and their web site (*http://www-pcmdi.llnl.gov/PCMDI.html*) provides information on, and links to, many of these projects.

3. Atmospheric Models

Atmospheric models, as their name only partially suggests, simulate the behavior of the atmosphere and the land surface. Ocean processes are not simulated explicitly, and specified ocean surface temperatures and sea-ice distributions provide boundary conditions for the AGCM. The long-term climatological annual cycle of sea surface temperature (SST), or an evolving sequence of monthly values from the observational record, are used as boundary conditions. The specified SSTs constrain low level temperatures over the oceans but temperatures evolve freely over land and sea ice. All other variables evolve freely.

  The specification of an observed sequence of sea-surface temperatures allows the investigation of the atmospheric variability forced by SSTs and of the models' ability to reproduce it. The atmospheric response to El Niño/La Niña SSTs is a particular example. One early *ad hoc* intercomparison effort arose from the growing realization, following the 1982-83 El Niño, that tropical Pacific SSTs had a notable influence in both tropical and extratropical regions. An attempt to understand and model this behavior using specified El Niño-like SST anomalies was encouraged by the WGNE, and various modeling groups presented and informally intercompared their results at the International Liege Colloquium [22]. The importance of ENSO as a large-scale climate perturbation is now manifestly clear, and the ability of AGCMs (AMIP) and coupled GCMs (ENSIP) to simulate this mechanism is under continuing study.

3.1 SOME RESULTS FOR BASIC PARAMETERS

The earliest intercomparisons were visual displays of one or two basic parameters indicating the "state-of-the-art" at the time. We may trace the simulation of mean sea level pressure, precipitation, and temperature in atmospheric models over several decades.

3.1.1. *Mean Sea Level Pressure*
Figure 2 gives the zonally averaged distribution of northern winter mean sea level pressure from Gates [11] for January, and from Boer *et al.* [4] and Gates *et al.* [14] for the December-February season. The diagrams have been resized to make their aspect ratios similar to aid in comparison and, while the details of the behavior of individual models is not readily apparent from the diagrams, they do give an overall indication of the range of modeled values. The diagrams also indicates that there has been a remarkable increase in the number of AGCMs from 8 (the group listed in the upper right of the top panel) in the early 1980s, to 14 in the middle panel in the early 1990s, to over 30 models in the bottom panel in the recent AMIP1 intercomparison. Some models have been through the iterative process of Figure 1 several times, while other models are recent additions.

The mean sea-level pressure distribution represents a balance between dynamic and thermodynamic aspects of the flow and is a sensitive measure of a model's ability to simulate surface and near-surface climate. The upper panel of Figure 2 indicates some of the difficulties that beset early attempts to simulate this variable. The scatter of model results is dramatic and this is especially so at higher latitudes although, of course, averages are over smaller areas than at equatorward latitudes. Some uncertainty in the reduction to sea level is also involved but this cannot explain the remarkable range in simulated pressures. This diagram could be said to be from the early childhood of atmospheric climate modeling.

The middle panel of Figure 2 shows that some improvement had been achieved in the simulation of mean sea-level pressure by the beginning of the 1990s, although the scatter of the results is still impressive. At first glance the most recent simulations in the bottom panel of Figure 2 display as much scatter as the earlier results in the middle panel. This, however, is a consequence of the larger number of models in the intercomparison and reflects the initial attempts of newer modeling groups. Although numerical values are not available, it is clear that the sea-level pressure is closer to observed values in more recent simulations for the majority of models.

448

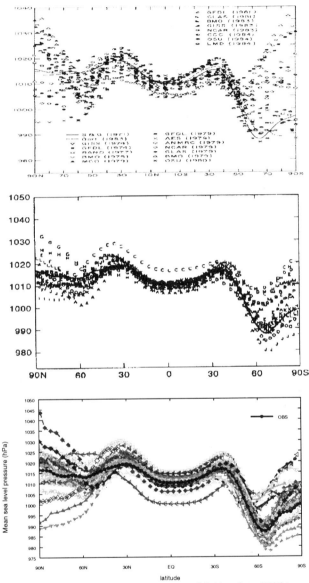

Figure 2. Modelled mean sea-level pressure (hPa) from Gates (1987) top,
Boer et al., (1991) middle, and Gates et al., (1999) bottom.

3.1.2 *Precipitation*

Precipitation is a basic climate variable and a measure of the vigor of the hydrological cycle in the model. Since the atmosphere gains energy via the release of latent heat, precipitation is also involved in the energy budget and thermodynamic balance. Figure 3 provides the same information for precipitation as does Figure 2 for the mean sea level

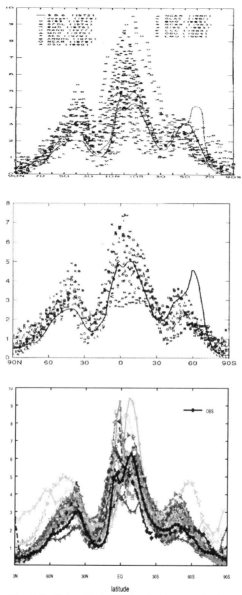

Figure 3. Modelled precipitation rates (mm/day) from Gates (1978) top,
Boer et al., (1991) middle, and Gates et al., (1999) bottom.

pressure. Uncertainty in the "observed" distribution of precipitation is illustrated by the
differences between these curves in earlier and later diagrams. The panels in the figure
indicate that (1) an earlier tendency for excessive precipitation in polar regions remains
but has generally been reduced in more recent models, (2) apparently excessive precipitation in northern mid latitudes remains a feature of most models, and (3) tropical pre-

450

cipitation rates range from the notably deficient to the notably excessive. Once again, despite the rather large scatter in the latest results, the majority of model results have moved toward the observations (and the "observations" have moved toward the modeled result in the southern hemisphere).

3.1.3 *Temperature*

Temperature is one of the most fundamental climate variables. It is of immense practical importance and has a long observational record, which provides increasing evidence of anthropogenic global warming (Chapter 21). The scatter among models in simulating temperature gives an indication of the consistency of models in treating this important quantity.

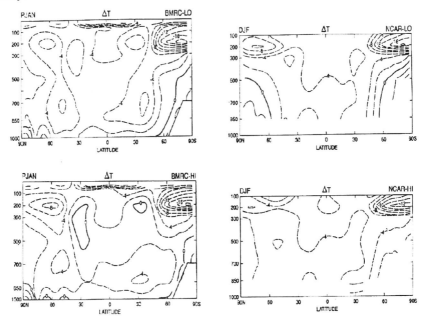

Figure 4. Differences between modelled and observed temperatures (degrees C) for northern winter for versions of the BMRC and NCAR models at different resolutions

Although modeled surface air temperature over the oceans is largely controlled by the specified SSTs, temperature is free to evolve over land, over sea ice, and away from the surface. Figure 4 gives examples of the difference between modeled and observed temperature, as a function of latitude and height, for earlier versions of the Bureau of Meteorology Research Centre (BMRC) and the National Center for Atmospheric Research (NCAR) models from Boer *et al.* [3], [4]. The diagram illustrates characteristic deficiencies in the simulation of the temperature distribution and gives an early indication of the effect of a modest increase in horizontal model resolution. The resolution increase is from spherical harmonic resolutions R15 to R31 (with corresponding horizontal grids of 48x40 and 96x80) in the case of the BMRC model and from R15 to T42 (horizontal grids of 48x40 and 128x64) in the case of the NCAR model. Figure 5 from Gates *et al.*

[14] summarizes recent results for the AMIP1 models in terms of the average model value or *model mean* $\bar{T}$ and its difference from observations $\Delta T = \bar{T} - T_o$, together with the *intermodel standard deviation*, $\sigma_T = \sqrt{\overline{(T - \bar{T})^2}}$, which is a measure of the scatter or disagreement of model results.

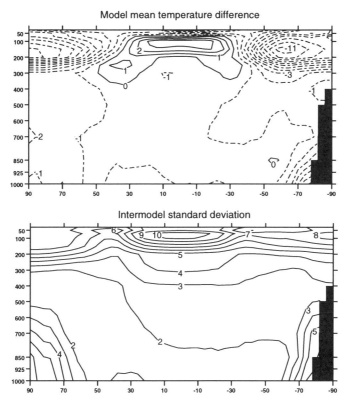

Figure 5. Difference between average of AMIP1 modelled temperatures and observations (upper) and the standard deviation of modelled temperatures (lower). Units are degrees C.

Figures 4 and 5 both indicate a well-known systematic model deficiency namely that temperatures are generally too cold, especially in the polar upper troposphere and lower stratosphere. Note that increasing horizontal resolution by itself does not necessarily improve the situation in Figure 4. The systematic nature of this feature across a wide variety of models suggests a common deficiency in some aspect of model formulation. The temperature difference of Figure 5 and that of Figure 11 of Chapter 3 show that this error is diminishing, but still exists, in recent models. It is not clear what aspect of model improvement has most influenced this. Improved resolution in the horizontal and vertical, improved treatment of clouds, radiation, and other parameterizations may all contribute in part to the improvement. The lower panel of Figure 5 measures the scatter among model results and indicates that model simulation of temperature is less consistent near the tropopause and also near the surface at polar latitudes. Apparently the models do not consistently treat the balance between convective processes, radiation, and transport that governs the temperature and location of the tropopause, nor the rather different processes governing polar temperatures.

One of the difficulties with both model validation and intercomparison is that no generally accepted measure of "model goodness" is available. If such a measure existed, it could be used to show more clearly the nature and rate of improvement of individual models and of models as a whole. As it is, the often-used root mean square difference and correlation measures of agreement between model and observed values are of only limited use since different models show different levels of agreement for different variables and no one model is "best" for all variables.

3.2 DIAGNOSTIC SUBPROJECTS

There are, of course, many other critical AGCM variables and many other statistics to intercompare and analyze beyond those illustrated above. The wealth of data available to modern intercomparisons such as AMIP demands also a wealth of analysis methods and approaches. *Diagnostic Subprojects*, initiated through the AMIP1 Panel, are sophisticated analyses of AMIP results by investigators who may be outside the general modeling community that produced the data. The intent is to widen the range of model aspects that are intercompared and to apply specialist expertise to the investigation of the modeled (and real) system. There are at least 26 AMIP1 diagnostic subprojects treating a wide range of atmospheric behavior. The PCMDI website lists information on AMIP1 and AMIP2 diagnostic subprojects and provides a bibliography of AMIP-related publications. CMIP and other intercomparisons have also adopted diagnostic subprojects as a means to expand the interest in, and the investigation of, model behavior.

4. Coupled Models

Coupled global climate models (CGCMs) add ocean and sea-ice components to the atmosphere and land-surface components of an AGCM. The term "coupled model", in this context, applies to a fully three-dimensional AGCM coupled to a fully three-dimensional OGCM. Global coupled modeling for climate is a relatively recent development for the modeling community. Despite the cost and complexity, fully coupled models are being developed and used in increasing numbers, largely fuelled by the need for improved and more sophisticated predictions of climate variation and simulations of climate change. Coupled models may be used to simulate past climate change and potential future climate change due to changes in the forcing of the climate system. One of the first intercomparisons of fully coupled models [13] investigates selected features of the control climates of four then existing models. A recent basic intercomparison (by Lambert and Boer in Chapter 5 of IPCC 1995) expands this to 11 models and deals mainly with surface climate features of coupled models which encompass a range of resolutions, initialization procedures, and coupling methods. The study is expanded in CMIP1 to include results from 19 models.

The question of *flux adjustment* looms large in the verification and intercomparison of coupled models. In a coupled climate model, the atmospheric and oceanic components typically exchange information at least once a day. The ocean component provides ocean surface temperatures to the atmosphere component which, in turn, provides fluxes of energy, fresh water, and momentum to the ocean component. These are used to calculate new ocean temperatures and so on. Symbolically $H \Rightarrow H_o$ represents the fluxes from the atmosphere passing to the ocean. Flux adjustment or "anomaly coupling" is a method whereby the anomalies from the respective climatological averages (indicated by an overbar) are exchanged between components, rather than the raw fluxes themselves. Here $\bar{H} + H' \Rightarrow \bar{H}_o + H'$ or alternatively $H_o = H + (\bar{H}_o - \bar{H}) = H + \delta\bar{H}$ and the atmospheric flux H is modified by the "flux adjustment" $\delta\bar{H}$, which arises because the climatological fluxes that the AGCM produces given observed climatological ocean surface temperatures are not the same as the climatological fluxes that the ocean model requires to produce these temperatures. Without flux adjustment the coupled model will "drift" toward a new state with a different climate where the atmospheric and oceanic fluxes are in balance. The justification for flux adjustment is to obtain a control climate that is near the observed climate so that the *climate feedback* processes in the system are operating in a reasonable range and the consequences of changes to the forcing of the system (e.g., increasing greenhouse gas concentrations) will be properly simulated. The counterargument is that flux adjustments indicate model deficiencies and may affect the way the coupled system responds to changes in forcing.

Approximately half of the coupled models considered in IPCC 1995 and in CMIP use flux adjustment, while the other half avoid flux adjustment at the cost of some climate drift (which can be large in some cases). As models improve, flux adjustments automatically become smaller so that ultimately they disappear. The generation of coupled models currently under development all aim to eliminate flux adjustment while minimizing climate drift so that flux-adjusted models are expected to become a threatened species.

Coupled models must be initialized and ideally would be integrated for many years in order to come into equilibrium with their forcing but, since oceanic timescales are so long (some thousands of years), this is computational very expensive. Alternative methods of initialization are to (1) "spin up" the two components independently, usually with acceleration techniques applied to the ocean, before coupling (however, the coupled model may subsequently drift because the separate states are not in balance); (2) initialize the components based on observations, in which case the result will not be in equilibrium and will certainly drift to some extent; or (3) integrate the coupled model until the upper ocean has come into equilibrium, although the deep ocean will not have had time to come into equilibrium. All of these approaches are represented in the current generation of coupled models.

In a coupled model, ocean temperatures evolve in response to energy fluxes at the surface (perhaps modified by flux adjustment) and ocean transport processes, as does the sea ice in polar regions. Precipitation over the oceans affects salinity and precipitation occurrence will be affected by evolving SSTs and so on. The interaction between ocean and atmosphere gives rises to a new class of "coupled modes of variation" in the models.

The climate of a coupled model is less constrained than that of an AGCM so achieving realistic simulations is more complex and the scatter among model results would be expected to be larger. Figure 6 gives the model average and the intermodel standard deviation of temperature and precipitation from the 11 models in the IPCC 1995 intercomparison. Coupled model results for these variables have the same general flavor as

454

those for AGCMs but with a number of characteristic differences. Surface air temperatures differ by a greater degree from observations and, not unexpectedly, differences are

Observed surface air temperature (°C)   DJF

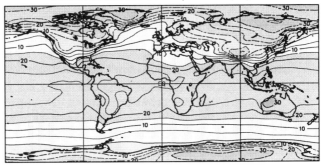

Difference of model average from observed

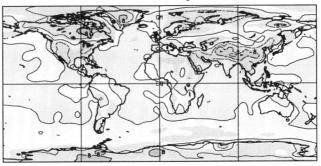

Intermodel standard deviation of surface air temperature

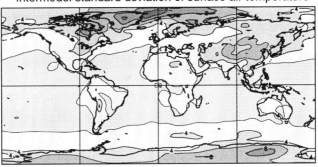

Figure 6

larger for models that do not use flux adjustment. Differences are largest over land with its smaller heat capacity and lack of flux adjustment than over the oceans. Differences are also comparatively large over polar land and sea-ice areas.

Observed precipitation rate (mm d⁻¹)   DJF

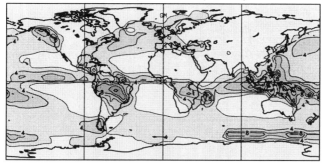

Model average precipitation rate

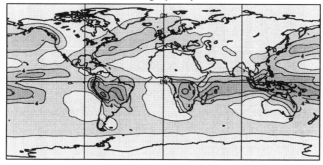

Intermodel standard deviation of precipitation rate

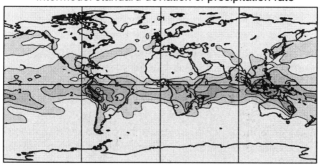

Figure 7

Precipitation is concentrated in the "storm tracks" at middle latitudes and especially in the intertropical convergence zone near the equator. Models are certainly capable of reproducing the overall distribution of precipitation. Model scatter is roughly proportional to the mean precipitation values. The middle-latitude excess of modeled precipitation in Figure 3 is seen to be associated with the oceanic winter storm tracks in Figure 7 and the scatter of tropical precipitation values in Figure 3 are associated with inconsist-

456

encies of model representation of the tropical band of large precipitation amounts.

There are many other variables and parameters of interest for climate which have been given greater or lesser attention in intercomparison studies. The ability of models to simulate cryospheric behavior in terms of the distribution and variation of snow cover and ice amount is important because the "albedo feedback" acts to enhance climate warming caused by greenhouse gases or other forcing. A warmer temperature melts ice and snow thereby reducing the albedo and the amount of incoming solar radiation

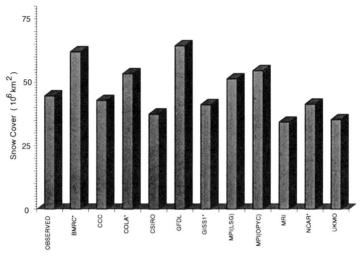

Figure 8. Northern Hemisphere winter snow covered area.

reflected back to space. The extra energy that is absorbed serves to increase temperatures and melt more ice and snow in a positive feedback loop. Figure 8 illustrates the range of winter snow covered area simulated by the models and illustrates also that, although non-flux adjusted models (indicated by an asterisk) are generally less accurate in their simulations of snow cover, this is certainly not always the case.

There can be marked differences in simulated sea-ice cover among models, as shown in Figure 9. The ice cover in the southern hemisphere is not confined by the continental boundaries, as is the case in the northern hemisphere, and the variation of ice covered area from winter to summer is quite large. Some models in this sample have a weak annual cycle of sea-ice area, some have little sea ice at all, and some have a very reasonable distribution of sea ice. Once again, flux-adjusted models tend to have distributions that agree better with observations than do non-flux-adjusted models but this is nevertheless not always the case. The virtual disappearance of ice in some models is an extreme example of climate drift and, together with lesser differences in snow cover, affects the ice-albedo feedback process mentioned above. These kinds of deficiencies in cryospheric aspects of the first generation of coupled models are being considerably overcome in the second generation of such models.

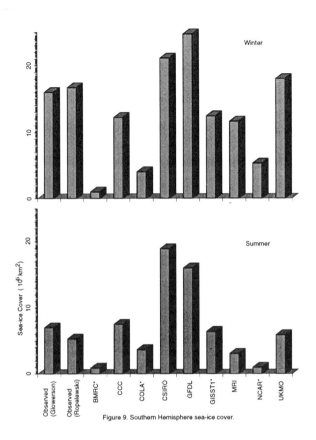

Figure 9. Southern Hemisphere sea-ice cover.

5. Model Sensitivity

Various "flavors" of intercomparison are discussed in Section 2.1 and the distinction between "pure" numerics/resolution and "pure" physical parameterization is contrasted with "full model" intercomparisons. The usual reductionist idea is that an understanding of the whole may be obtained by isolating its components. The alternative is to study model sensitivity to various features by analyzing the output of full models. This is difficult, of course, in the complicated non-linear world of climate simulation. We briefly discuss model sensitivity to resolution, gravity wave drag, and the representation of cloudiness.

5.1 RESOLUTION

Interest in the effect of model resolution on the simulation of climate has been a constant of climate modeling. Resolution studies have a long history from early models [26] to more recent studies in individual models ([2], [7], [24]), and include studies of the effects of resolution on various physical processes [23] and numerical convergence of the dynamics of a GCM as a function of resolution ([16], [6]).

Resolution effects may operate directly by reducing numerical errors or indirectly through the effect of increasing resolution on model parameterizations. Increasing resolution to reduce numerical errors may not reduce, and may even increase, parameterization errors (see the discussion in Chapter 3). Boer *et al.* [4] attempt to deduce some consequences of increasing resolution in the atmospheric models available in that intercomparison. Model climate tends to be comparatively insensitive to model resolution, as suggested by Figure 4, where the systematic errors in temperature in the same model codes do not change greatly with increased resolution. The general conclusion from these kinds of studies is that increased resolution *by itself* is seldom effective in improving the simulated climate but that increased resolution in conjunction with improved parameterizations does lead to improved climate simulation.

5.2 SURFACE STRESS AND GRAVITY WAVE DRAG

The difficulty in identifying the causes and cures of systematic model deficiencies is illustrated by the mean sea level pressure distributions of Figure 2. Although it is not immediately apparent, the early results in the upper panel of Figure 2 fall roughly into two groups. The Type A mean sea-level pressure distribution has a reasonable Antarctic trough but pressures are too low in the northern extratropics. The Type B distribution has reasonable pressures in the northern extratropics but the Antarctic trough is much too shallow. In other words, the early versions of many atmospheric models were able to simulate reasonable December-February mean sea-level pressure distributions in the extratropics of one or the other hemispheres but not both.

This puzzling situation may be understood [4] in the context of the vertically integrated and zonally averaged angular momentum budget (Chapter 3) written here as

$$\frac{1}{a\cos\varphi}\frac{\partial}{\partial\varphi}A\cos\varphi = \tau_m + \tau_s$$

where $A = a\cos\varphi\int[\overline{uv}]dp/g$ is the north-south angular momentum flux and the right-hand-side source/sink terms are the mountain torque τ_m and the (dominant) surface stress torque $\tau_s = a\cos\varphi[\tau_\lambda]$. The atmosphere gains angular momentum in the region of tropical easterlies and transports it poleward in both hemispheres to its sink, via the stress torque, in the extratropical westerlies.

The surface stress τ_λ is represented by a drag law and has an approximate geostrophic relationship to the surface pressure gradient as

$$\tau_\lambda \approx C_D|V|u \approx -C_D\frac{|V|}{fa}\frac{\partial p_s}{\partial\varphi}$$

and this equation indicates that the required sink for angular momentum depends on the value of the drag C_D and on the pressure gradient. The observed mean sea-level pressure in Figure 2 is almost flat in the northern extratropics where gradients are weak, but pressure is low and gradients are strong in the southern hemisphere extratropical Antarctic trough. If C_D is comparatively large, the equation indicates that the required stress can be attained with a relatively weak pressure gradient, but that if C_D is comparatively small then the pressure gradient must be large in order to provide the required stress torque.

A small value of C_D leads to a Type A distribution of pressure where the Antarctic trough is deep enough but pressures are low and gradients are too large in the northern extratropics. A large value of C_D leads to a Type B distribution where the Antarctic trough that is too weak but the northern extratropical pressure distribution is reasonable. While either case can be obtained by adjusting C_D in the model, the correct distribution of mean sea-level pressure in both hemispheres cannot be obtained simultaneously. This rather puzzling difficulty was finally overcome by introducing a new sink of momentum in the northern hemisphere in the Type A case by adding the "missing physics" of gravity wave drag.

Gravity-wave drag ([1], [21], Chapter 12) attempts to parameterize the effects of atmospheric internal gravity waves which are launched by sub-grid scale topographic features. The waves transport momentum vertically and are dissipated at some higher region in the atmosphere. They thereby connect the free atmosphere with the surface and transfer momentum to the surface by this mechanism. The new equation for the surface stress then becomes

$$\hat{\tau}_\lambda = \tau_{gwd} + \tau_\lambda \approx \tau_{gwd} + C_D|V|u \approx \tau_{gwd} - \frac{C_D|V|}{fa}\frac{\partial p_s}{\partial \varphi}$$

where the gravity wave drag now helps to balance the budget. The frictional stress required to balance the angular momentum budget in the northern extratropics decreases and hence the pressure distribution is improved, as is the distribution of zonal wind and temperature in the troposphere.

Most AGCMs now include a representation of gravity-wave drag and, because the transport of angular momentum tends to increase with resolution for reasons which are not completely clear, this is especially so for higher resolution GCMs. The example of gravity-wave drag illustrates the difficulty of recognizing, understanding, and parameterizing the physical processes that affect climate in models.

5.3 CLOUDS

It is an accepted truth that the treatment of clouds in climate models is a major area of uncertainty [19]. The cloud parameterization in models affects the simulation of current climate but, perhaps more importantly, can also affect the simulated response to changes in forcing. Uncertainties in cloud treatment thereby translate into uncertainties in simulated climate change. For questions of global warming, it is important to know model "sensitivity", that is, the model response to a given change in forcing.

Cess et al. ([8], [9]) investigate one aspect of model sensitivity by intercomparing the "cloud/radiative responses" of a group of 19 AGCMs to an imposed change in SST. If the heating of the global climate system is represented by the flow of radiation across the top of the atmosphere $H = S - F$ where S is the net solar into and F the net longwave out of the system and if $H_a = S_a - F_a$ is the corresponding value for the same atmosphere but in the absence of clouds (the subscript indicates the quantities for the *atmosphere only*, i.e., not including the effect of clouds) then the cloud radiative forcing is

460

defined as $CRF = H - H_a = (S - S_a) - (F - F_a)$. The cloud radiative forcing is an attempt to avoid difficulties with the description and classification of clouds by going directly to their radiative effects. Clouds will generally act to cool the system in the shortwave (by reflecting solar radiation) and to warm it in the longwave by radiating less from the tops of clouds than would be the case from the surface in the absence of clouds.

The change in cloud forcing ΔCRF is termed the *cloud feedback* and gives an indication of how clouds affect the response of the system to a change in forcing. If the cloud feedback is positive, changes in clouds act to enhance the effects of a change in forcing (i.e. GHG-warming is amplified) and the reverse if the feedback is negative (GHG warming is partially compensated). In the intercomparison, the change in the forcing of the system is not particularly realistic since it is an imposed SST change, but it is straightforward for purposes of intercomparison.

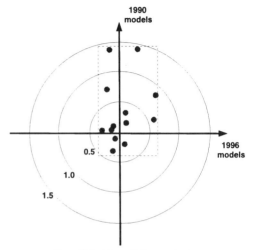

Figure 10. Evolution and intercomparison of modelled "cloud feedback"

Figure 10 replots the results from the 1990 and 1996 intercomparisons of cloud feedback in evolving versions of a set of GCMs [9]. In 1990, cloud feedbacks in models varied from modest negative to strong positive values as indicated by the height of the box enclosing the points in Figure 10. The 1990 intercomparison very dramatically focused attention on cloud effects in models and led to modifications and changes in cloud formulations. The more recent results from the 1996 intercomparison span a smaller range, as indicated by the width of the box, and are as likely to be positive as negative. This convergence in cloud feedback behavior in models means that the effects of clouds in climate change, at least measured in this way, should be more uniform among models. Nevertheless, IPCC [19] cautions that the convergence of results may not reflect a true reduction in the uncertainty of cloud processes in models since the reasons for the reduction are not always well understood.

5.4 SENSITIVITY OF SURFACE MOISTURE AND ENERGY BUDGETS

Gutowski *et al.* [15] perform a mini-intercomparison of the surface moisture and energy balance of a number of GCMs which is extended in Boer [5]. The models in this case are full AGCMs coupled to mixed layer ocean components. Such models have been used to give a first order estimate of the climate change expected with a change in external forcing. One standard climate change experiment consists of a simulation of the current or $1 \times CO_2$ climate together with a simulation of an equilibrium $2 \times CO_2$ climate change where the concentration of CO_2 in the model atmosphere has been doubled. The equilibrium global mean temperature change $\delta \langle T \rangle$ is termed the "climate sensitivity" and is a straightforward measure of the response of the modeled climate to this change in forcing. As summarized in IPCC ([17], [19], simulated values are of the order of 4°C but with a considerable range depending on a variety of aspects of the models and of their $1xCO_2$ climates. Simulated temperature changes show characteristic features with, for instance, the largest warming found at high latitudes, over the continents, and in winter.

An associated "hydro-logical sensitivity" meas-ures the change in the vigor of the hydrological cycle as the fractional change in the global mean precipitation and evaporation rate $\delta\langle P\rangle/\langle P\rangle = \delta\langle E\rangle/\langle E\rangle$. A warmer climate implies a more vigorous hydrologi-cal cycle and the propor-tionality between the temperature and hydrologi-cal sensitivity is measured

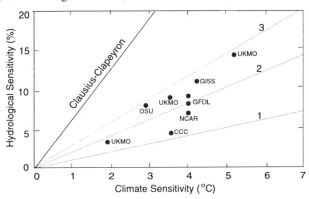

Figure 11. The relationship between surface warming and the acceleration of the hydrological cycle

by α in $\delta\langle P\rangle/\langle P\rangle = \alpha\delta\langle T\rangle$. It is plausible that α is a constant in models (and the real system) as some fraction of the limiting value given by the Clausius-Clapeyron relationship. Figure 11 shows results from a (now somewhat dated) group of models indicating that this is not the case and that different models, and different versions of the same model, show a range of differences in climate sensitivity, hydrological sensitivity, and α (the slanted lines in the diagram).

In the models, the relation between climate and hydrological sensitivity apparently depends on the nature of the control of the surface energy budget, especially as it is affected by cloud properties. This is inferred from changes in the averaged surface energy budget

$$S + F + LE + H + A = N \approx 0$$

where the terms are respectively the solar and longwave radiative fluxes, latent and sensible heat fluxes, and a surrogate for advection in the upper layer of the ocean. All flux terms are measured as positive if they act to warm the surface and the result should average to zero. Table 1 gives the changes in the energy balance $\delta R + L\delta E + \delta H \approx 0$ between the $1xCO_2$ and $2xCO_2$ climates where $\delta R = \delta S + \delta F$ is the net radiative change at the surface.

Changes in δR and δH both act to warm the surface and this is balanced by the latent heat flux term $L\delta E$ which acts to cool it. The difference in the representation of cloudiness in the model is seen most dramatically in δS, the change of solar absorbed by the surface where the CCC model shows a decrease as a consequence of the parameterization of cloud optical parameters in that model while other models show an increase. The model climate sensitivity, and especially the hydrological sensitivity, is seen from Figure 11 to be related to this cloud/radiative treatment. Because of the importance of model "sensitivity" for climate change simulations, an expanded and improved

462

intercomparison along these lines is currently being undertaken by B. MacAvaney and H. leTreut for the WGCM (section 2.3).

Table 1: CO$_2$ induced changes in surface energy balance terms

	NCAR	GFDL	GISS	CCC
δR	5.5	6.4	7.0	1.4
LδE	-7.1	-7.1	-10.4	-3.0
δH	2.7	0.6	3.3	1.6
δS	1.3	1.5	2.9	-2.4
δF	4.2	4.9	4.1	3.8

6. Concluding Remarks

There is a great wealth of model data available for intercomparison and analysis, and an ongoing gusher of new results of ever greater sophistication and ever increasing resolution encompassing ever more completely all components of the climate system from the ocean depths to the stratospheric heights. Model intercomparison projects have grown to match the breadth of modeling interests and have illuminated and documented modeling abilities. It must be realized, however, that intercomparisons provide a somewhat fuzzy snapshot of the state-of-the-art at a particular time since models are never static and continue to evolve and improve. An intercomparison can deal with only a limited subset of variables and processes at a particular time.

Model intercomparison "points to divergence and encourages convergence" of model results and this is entirely healthy provided that the resulting convergence is attained by improving the physical representativeness of the models. Intercomparison results are many and wide-ranging, and associated diagnostic studies may point attention in directions and toward processes that modeling groups might not themselves analyze and validate. The complex non-linear interconnected nature of the climate system makes itself felt in the results of intercomparison studies, however, and there is not a simple relationship between a model's ability to represent a particular climate variable or process and its ability to represent others. Intercomparison results thereby reiterate the great complexity and non-linear interconnections of the real system and of the modeled system.

In brief, model intercomparisons (1) provide information on model behavior, (2) document model abilities, (3) illustrate aspects of model sensitivity, (4) demonstrate model evolution, (4) encourage convergence of model results, (5) suggest reasons for some model deficiencies, and (6) encourage better analysis of model output. Model intercomparison must strive to "complete the cycle" of Figure 1 and devise methodologies for linking model deficiencies to model numerics, resolution, and physical parameterizations in a way that leads directly to model improvement and a better understanding of the climate system.

Acknowledgements

Thanks to Allan Frei, Jay Hnilo, and Jouni Räisänen for their comments on an earlier version of this chapter and to Steve Lambert for his helpful recommendations.

References

1. Boer, G.J., McFarlane, N.A., and Laprise, R. (1984) The climatology of the Canadian Climate Centre general circulation model as obtained from a five-year simulation, *Atmos.- Ocean* **22**, 430-475.

2. Boer, G.J., and Lazare, M. (1988) Some results concerning the effect of horizontal resolution and gravity-wave drag on simulated climate, *J. Climate* **1**, 789-806.

3. Boer, G.J., et al. (1991) An intercomparison of the climates simulated by 14 atmospheric general circulation models, CAS/JSC Working Group on Numerical Experimentation, WCRP-58, WMO/TD-No. 425, World Meteorological Organization, Geneva.

4. Boer, G.J., et al. (1992) Some results from an intercomparison of climates simulated by 14 atmospheric general circulation models, *J. Geophys. Res.* **97**, 12, 771-12,786.

5. Boer, G.J. (1993) Climate change and the regulation of the surface moisture and energy budgets, *Clim. Dyn.* **8**, 225-239.

6. Boer, G.J. and Denis, B. (1997) Numerical convergence of the dynamics of a GCM, *Clim. Dyn.* **13**, 359-374.

7. Boville, B.A. (1991) Sensitivity of simulated climate to model resolution, *J. Climate* **4**, 469-485.

8. Cess, R.D., et al. (1990) Interpretation and interpretation of climate feedback processes in 19 atmospheric general circulation models, *J. Geophys. Res.* **95**, 16,601-16,615.

9. Cess, R.D., et al. (1996) Cloud feedback processes in atmospheric general circulation models: An update, *J. Geophys. Res.* **101**, 12791-12794.

10. Gates, W.L. (ed.) (1979) Report on the HOC Study Conference on Climate Models: Performance, intercomparison and sensitivity studies, *GARP 22*, **1**, World Meteorological Organization.

11. Gates, W.L. (1987) Problems and prospects in climate modeling, in U. Radok (ed.), *Toward Understanding Climate Change*, Westview Press, Boulder, pp 5-34.

12. Gates, W.L. (1992) AMIP: the Atmospheric Model Intercomparison Project, *Bull. Amer. Meteor. Soc.* **73**, 1962-1970.

13. Gates, W.L., Cubash, U., Meehl, G.A., Mitchell, J.F.B., and Stouffer, R.J. (1993) An intercomparison of selected features of the control climates simulated by coupled ocean-atmosphere general circulation models, *World Climate Research Programme,* WCRP-82. WMO/TD No. 574, WMO, Geneva.

14. Gates, W.L., et al. (1999) An overview of the results of the Atmospheric Model Intercomparison Project (AMIP), *Bull. Amer. Meteor. Soc.* **80**, 29-56

15. Gutowski. W.J, Gutzler, D.S., and Wang, W.-C. (1991) Surface energy balances of three general circulation models: implications for simulating regional climate change, *J. Clim.* **4**, 121-134.

16. Held, I.M and Suarez, M.J. (1994) A proposal for the intercomparison of the dynamical cores of atmospheric general circulation models, *Bull. Am. Met. Soc.* **75**, 1825-1830.

17. IPCC (1990) *Climate Change, the IPCC Scientific Assessment* (eds. J.T.Houghton, G.J. Jenkins, and J. J. Ephraums), Cambridge University Press, Cambridge.

18. IPCC (1992) *Climate Change 1992, the Supplementary Report to the IPCC Scientific Assessment* (eds. J.T.Houghton, B.A. Callander, and S.K. Varney), Cambridge University Press, Cambridge.

19. IPCC (1995) *Climate Change 1995, the Science of Climate Change* (eds J.T.Houghton, L.G. Meira Filho, B.A. Callander, N. Harris, A. Kattenberg, and K. Maskell), Cambridge University Press, Cambridge.

20. Lambert, S.J., and Boer, G.J. (1989) Atmosphere-ocean heat fluxes and stresses in general circulation models, *Atmos.-Ocean* **27**, 692-715.

21. McFarlane, N.A. (1987) The effects of orographically excited gravity wave drag on the general circulation of the lower stratosphere and troposphere, *J. Atmos. Sci.* **44**, 1775-1800.

22. Nihoul, J.C.J. (ed) (1985) *Coupled ocean-atmosphere models*, Elsevier Oceanography Series, Elsevier.

23. Phillips, T.J., Corsetti, L.C., and Grotch, S.L. (1995) The impact of horizontal resolution on moist processes in the ECMWF model, *Clim. Dyn.* **11**, 85-102.

24. Stendel, M., and Roeckner, E. (1999) Impacts of horizontal resolution on simulated climate statistics in ECHAM 4. Max-Planck-Institut für Meteorologie *HTML* Report. Available at *http://www.mpimet.mpg.de.*

25. von Storch, H., Roeckner, E., and Cubasch, U. (1985) Intercomparison of extended-range January simulations with general circulation models: Statistical assessment of ensemble properties, *Beitr. Phys. Atmosph.* **58**, 477-497.

26. Wellck, R.E., Kasahara, A., Washington, W.M., and Santo, G.D. (1971) Effect of horizontal resolution in a finite-difference model of the general circulation, *Mon. Wea. Rev.* **99**, 673-683.

PALEOCLIMATE MODELING

P. VALDES
Department of Meteorology
University of Reading
Earley Gate
Reading RG6 6BB
UK

1. Introduction

Numerical modeling of the atmospheric and ocean system is motivated by the need for predictions. In the past, this has been dominated by the need for weather forecasts, but over the last decade or two the need to predict seasonal and climate change has greatly increased in importance. There is now great public concern about possible future anthropogenic climate change, and the only practical way of addressing these concerns is through numerical experimentation. However, any numerical model of the climate system requires a profound understanding of the physical processes which influence climate, and inevitably this is potentially biased towards the present day climate regime. The role of paleoclimate studies is to test this understanding of climate processes by examining past periods that are radically different from the present. In addition, when good data are available, we can use the past to test computer models in climate regimes very different from the present. If the models perform well for the present and the past, then this gives us increased confidence in their predictions for the future.

In the next section, we will briefly consider the geological evidence for past climate change and discuss the particular challenges that these present to numerical models. In the following section, a hierarchy of paleoclimate models will be presented and their strengths and weaknesses will be evaluated. Two examples of the particular challenges and insights provided by paleoclimate modeling will then be described.

P. Mote and A. O'Neill (eds.), Numerical Modeling of the Global Atmosphere in the Climate System, 465–488.
© 2000 Kluwer Academic Publishers. Printed in the Netherlands.

2. Motivation

One of the most common public misconceptions about climate is that our present climate is fairly typical and that climate would have been unchanging if mankind had not been altering the atmospheric composition. Geological evidence clearly shows that this is not the case. Climate is never constant, and the geological record shows that there have been periods in the past which have been a lot colder or a lot warmer than the present. Moreover, on a regional basis, even the rate of change of climate in the past can be comparable to that expected for the next few hundred years. Figure 1 shows a generalized curve for global mean temperature and global mean precipitation for the history of the Earth. Data for this curve come from a huge variety of indirect, proxy indicators of climate. These include the distribution of flora and fauna, sedimentary rocks such as coals, evaporites, bauxites and aeolian deposits, and glacial deposits. Flora and fauna can be quantitative (see [62], [61], [40]), whereas sedimentary deposits are more qualitative. For instance, bauxites are known to form only within certain ranges of temperature and moisture [50].

The pre-Cambrian climates (older than 570 million years) are very uncertain, although there are a few important climate features. There is evidence for running water as early as 3.8 billion years, despite a solar output that was only 70% of present conditions [21]. Energy balance models (see the next section) would suggest that such a large reduction in solar output would lead to a totally ice-covered planet. This is called the "cool Sun paradox", and it has been suggested that extremely high atmospheric CO_2 and CH_4 concentrations could have compensated for the low output [27].

Equally puzzling is the cool period in the late pre-Cambrian (570 million years ago). Data are very patchy for this period, but there are a number of sites where glacial deposits are found at extremely low latitudes. It has been suggested that there were fairly extensive ice sheets, even at low latitudes, caused by a substantial change in the Earth's obliquity [44] or decreased levels of CO_2 (for further discussion, see [15]).

Data begin to get more reliable in the Mesozoic era (225 to 65 million years ago). Throughout this period, there is no evidence for ice at either pole. Further evidence for extreme high latitude warmth comes from flora, which suggests that Northern Hemisphere continental interior temperatures were near or above zero, even in winter. Such warm periods extended for many millions of years, and the advent of glaciation at the poles seems to have been a relatively recent event. Such a fundamentally different climate regime from the present dramatically illustrates how much climate can vary. The causes for this warmth are still unclear. High atmospheric CO_2 concentrations have been suggested, but modeling studies (see later) cannot fully simulate the warmth of winter continental interiors.

Since the Mesozoic, the climate has gradually cooled. The cooling started

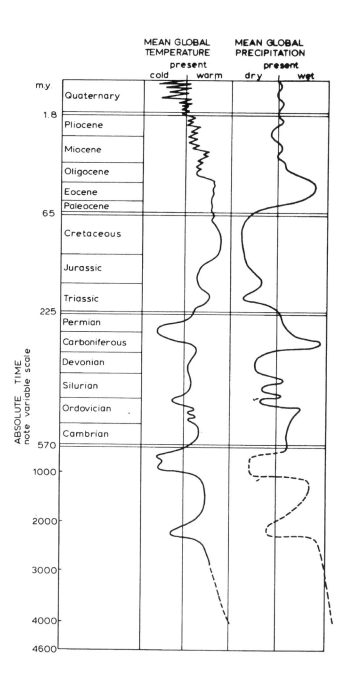

Figure 1. Generalized curve for mean annual temperature and mean annual precipitation. In practice, the geological indicators for moisture are land based and many of the indicators (e.g., evaporites) record precipitation minus evaporation. This figure is reproduced by kind permission of Professor L. Frakes [22].

in the Eocene and has continued to the present. Extensive ice sheets appear to have developed over Antartica in the Oligocene (37 to 26 million years ago), and Northern Hemisphere glaciation did not occur until the Miocene (26 to 5 million years ago). However, by the time we reach the Quarternary (the last 2 million years), glaciation existed at both poles. For this latter period, we have much more extensive and high resolution data. Evidence from sediment cores (Figure 2a) shows that the climate of the Quaternary has been oscillating. From 2.5 to 0.7 million years ago, the oscillation was characterized by periods of about 40,000 years with a relatively small amplitude. However, for the last 700,000 years, the characteristic of the oscillations are different. The amplitude is much larger and the period lengthened to about 100,000 years, although there is also evidence for a 20,000 years oscillation.

Ice cores can give more detailed climate information for the last glacial cycle (Figure 2b). This reveals that 120,000 years ago temperatures were similar to, or even slightly warmer than, the present day. The temperature then rapidly decreased for about 10,000 years and then gradually continued to cool, reaching a minimum about 21,000 years ago (called the Last Glacial Maximum, LGM). At this time, ice sheets covered large regions of the Northern Hemisphere continents to a depth of 2.5 km or more. Carbon dioxide variations are also shown (based on analysis of air bubbles trapped in the ice), and it can be seen that they vary in a similar manner to temperature. This correlation does not, however, determine cause and effect.

After this period, climate rapidly warmed but this was punctuated by rapid and substantial variations [35]. These variations occur on many time scales from thousand of years, to decadal. Finally, we reach the last few thousand years (the Holocene). The mid-Holocene was characterized by warm summers in the Northern Hemisphere continents.

The basic cause for the 100,000, 40,000 and 21,000 year oscillations appears now to be well established. The period of the oscillations corresponds to changes in the eccentricity, obliquity, and precession of the Earth's orbit, respectively ([31], [32]). However, calculations of the resulting changes in incoming solar radiation show that the main changes are seasonal, and at the precession period (i.e., 21,000 years). Furthermore, during the LGM, the orbital parameters were almost identical to the present, but the climate was radically different. Thus climate feedback mechanisms have to be invoked to explain the observed variation, as will be discussed in Section 4.

This brief survey of the paleoclimate data is an attempt to illustrate that our climate has never been constant. It naturally varies on all time scales from decades to millions of years. This provides an important and challenging test of our understanding of climate and climate processes. It also gives us an opportunity to test climate models in climate regimes very different from the present. Periods such as the Holocene, the LGM, or the mid-Cretaceous, represent an extreme range

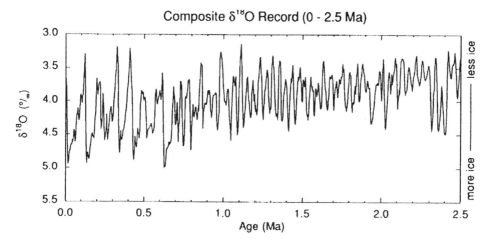

S. HEMISPHERE CLIMATE RECORDS

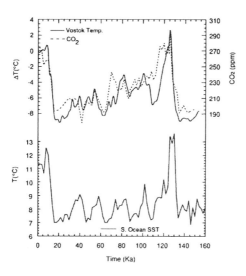

Figure 2. (a) Climate variations through the last 2.5 million years based on data from sediment cores. (From *Paleoclimatology* by Thomas J. Crowley and Gerald R. North. Copyright (c) 1991 by Oxford University Press, Inc. Used by permission of Oxford University Press, Inc.) (b) Climate variations through the last 160,000 years based on data from sediment and ice cores.

of climate variability. If models can successively simulate these periods, then we can be more confident that they will correctly respond to much smaller climate change, such as that predicted for man-made climate change caused by increasing concentrations of radiatively active gases.

It is frequently stated that pre-Quaternary climates (those older than 2 million years) may provide an interesting test of our climate understanding, but an inadequate period for testing numerical models, because the boundary conditions are uncertain. The basis for this statement is that for those more distant climates, uncertainities in data interpretation and the boundary conditions required for climate models (such as mountain heights) are so large that quantative testing of models is inappropriate. In my opinion, this is an over simplified argument. We do indeed have more information on Quarternary climates, but in general the magnitude of climate change is much smaller than some pre-Quaternary periods. This is especially true for past warm periods. The last interglacial period (125,000 years ago) was probably one of the warmest within the last million years but it was only 1-2°C warmer than present. This should be compared to the 5-6°C warming in the mid-Cretaceous. Thus the uncertainities (i.e., noise) for the distant past may be large, but so is the climate change (i.e., signal). Thus for some pre-Quaternary climates, the signal-noise ratio may be as large as the more recent periods and still represent an important test of climate models.

3. Types of Climate Model

General circulation models are the main tools for climate and paleoclimate studies, but they are not the only climate models that exist (see Chapter 5 of this volume). Some problems are better tackled with simpler calculations. Often the best course of research is to utilize a hierarchy of climate models of various degrees of sophistication.

3.1. QUALITATIVE

In their most crude form, qualitative climate models are just a conceptual description of a (hopefully) reasonable climate scenario. Parrish [46] and Parrish and Curtis [47] extended this concept by estimating climate from a set of principles based on present day conditions. They took paleogeographic reconstructions for a number of different periods throughout the Paleozoic, and sketched in the surface pressure distribution based on understanding present climate, i.e., they note that for present day conditions, there are lows over the mid-latitude oceans during winter. They attribute this to the thermal inertia of the oceans (warm in winter compared with the land) and thus sketch low pressures over oceanic areas for previous periods.

The advantage of this method is that it is quick and easy. The disadvantages are that the arguments used to place the pressure centres are oversimplified. In addition, such models cannot easily include the interactions between different components of climate. For instance, the present day Indian monsoon is strongly influenced by the Himalayan Mountains and the Ethiopian Highlands. Without using numerical models, it is almost impossible to separate the different processes. Thus predictions for monsoons in the past may have been incorrect.

Qualitative climate models are also very difficult to evaluate. Predictions of surface temperature, snow fall, soil moisture, etc. are impossible. Even the strength and exact position of the pressure centres are arguable. Different researchers can produce different answers for the same paleogeography. It is almost impossible to refine this method. To improve the paleoclimate predictions we must resort to more quantitative methods, even though at times they may not result in significant improvements.

3.2. ENERGY BALANCE MODELS

Many paleoclimate problems are associated with the temporal variability of climate. One key aspect of Quaternary climate change is to explain the reasons for the amplitude, periods, and phases of the variability seen in sediment and ice cores. Such studies go well beyond the demands for predicting the next 100 years, and the climate models used for such future climate change predictions cannot be run for 100,000 years. Instead, we must resort to a much simpler type of model which can be integrated for many thousands of years (e.g., [16], [52], [23]). Ideally such models need to include sub-models for the atmosphere, oceans, cryosphere, and carbon cycle. To include all such components in a model that can be run for several glacial/interglacial cycles requires some form of simplification.

Perhaps the simplest type of truly numerical model is that based upon energy balance considerations. The idea is that for climate to be in equilibrium the incoming solar energy must be balanced by outgoing longwave radiation. The very simplest of these models calculates the global mean surface temperature using

$$S_0 \times (1 - \alpha_p) = \varepsilon_p \sigma T_s^4$$

Where S_0 is the solar constant (currently about 1370 Wm^{-2}), α_p is the planetary albedo (proportion of solar radiation reflected back to space, currently approximately 0.30), ε_p is proportional to the infrared emissivity, and is currently approximately 0.61. σ is the Stefan-Boltzman constant (5.67 x 10-8 $Wm^{-2}K^{-1}$) and T_s is the surface temperature (in Kelvin).

This type of 0-dimensional model can be used to predict global mean temperature for changes in the albedo or atmospheric CO_2 (through changes in the emissivity). It also has some interesting characteristics. If the albedo depends on temperatures, it can be shown that there may be at least two perfectly stable, equilibrium states of climate for the same solar forcing (see [15]). One solution corresponds to an ice-covered Earth with a high albedo, the other to a climate similar to today with a relatively low albedo. This property of multiple equilibria (caused by ice albedo feedback) shows that climate may change abruptly for a small change in forcing. Multiple equilibria have also been seen in GCMs, for ice albedo feedback, vegetation feedback (see later), and North Atlantic Ocean circulation ([39], [51]). Multiple equilibria present an interesting challenge to

472

paleoclimate modeling. Their existence implies that to explain the reasons for any particular climate regime, we may have to consider the history of climate, not just the current state.

Energy balance models can also be developed for more dimensions. Gallée *et al.* [23, 24], Berger *et al.* [6] have developed a 2.5 dimensional model. The two main dimensions are height and latitude, and the model also includes different sectors of the land/ocean, so that they have a zonally averaged Atlantic, Pacific and Indian Ocean. The ocean model includes the deep thermohaline circulation, and the land surface model includes a detailed representation of the cryosphere. The model is forced by the changes in the distribution of incoming solar radiation and the observed atmospheric CO_2 variation. The resulting prediction of surface temperature and ice volume shows a remarkably close fit to paleoclimate data. The dominant process is ice albedo feedback. In addition, ice albedo ageing helps amplify the rapid deglaciation at the end of the last ice age (ice albedo ageing occurs because the albedo of snow decreases as it gets older, mainly because of changes in the crystal structure). As climate begins to warm, the albedo of ice is reduced (mainly due to the melting and refreezing of the ice) and this then acts to amplify the initial warming. The changes in radiatively active gases also amplify the response, but the basic variation is still present if the gases are held constant.

The strength of this type of model is that it can tackle problems that are impossible to consider using conventional general circulation models. The disadvantage is that some potentially important climate processes have to be treated extremely crudely. For instance, clouds are specified and do not change when climate changes. Furthermore, the major Northern Hemisphere ice sheets have considerable longitudinal variations, but these cannot be included in a 2.5D model. Some of these disadvantages are not present in GCMs but these models can only ever be used for "snapshot" type simulations. Although some parts of the climate system have to be prescribed rather than predicted, there are fewer approximations of the climate system processes themselves. In the following section, we will consider what aspects of the paleoclimate system have to be specified for GCM studies.

3.3. PALEOCLIMATE GCMS

To perform a paleoclimate simulation, the general circulation model requires some basic information about the solar output, the Earth's orbital parameters, atmospheric composition, coastlines, orographic relief, and surface albedo. In addition, an atmospheric general circulation model must include sea surface temperature data. We shall refer to all of these quantities as boundary conditions. They must be provided from the geological record; estimates for most of these quantities do exist or can be derived.

3.3.1. *Solar Output*

Current astrophysical theory predicts that the solar constant has increased by approximately 2% over the last 200 million years (see [21]). However, there is the possibility of changes by 5% or more on time scales of 10 million years [26]. For the Quaternary, solar output is assumed constant.

3.3.2. *Earth's Orbital Parameters*

The Earth's rotation rate has changed by approximately 3% over the last 200 million years. For the Mesozoic, the periods associated with the obliquity and precession of the Earth's orbit have been calculated [5], but the amplitude/phase cannot be computed. The periods of the eccentricity cycle are thought not to have changed. Thus for the Mesozoic period, models are typically either run with present orbital parameters, or using extreme limits for the orbital parameters ([43], [58]). For the Quaternary, orbital forcing is one of the main driving mechanisms and is well known from astronomical calculation.

3.3.3. *Atmospheric Composition*

This is one of the most difficult aspects of paleoclimate modeling (but see 3.3.6 below). The most important naturally occurring radiatively active gases are ozone, CO_2, water vapour, and methane. Water vapor is predicted in the model. Ozone abundance can be directly linked to oxygen. For the late Quaternary, ice cores show the concentration of CO_2 and CH_4, but for the distant past, the situation is more difficult. Berner [8] (and references therein), using a geochemical model, predicted that, 100 million years ago, CO_2 levels were up to 8 times present day levels, and O_2 levels were 20% higher that present. Different photosynthetic systems (C3, C4, or CAM) may also give a guide to O_2 and CO_2 levels [12].

The concentrations of other radiatively active gases, such as methane, are completely unknown except for data from ice cores that cover the last 100,000 years. However, there is evidence to suggest that there may have been catastrophic releases of methane from oceanic reservoirs of methane hydrates [18]. The resulting methane is oxidized to CO_2 and thus results in long term warming of the climate system. Similar releases of methane have been proposed for future climate change [28].

3.3.4. *Continental Position, coastlines and orography*

For the Quaternary, the continents have not changed in position, but sea level changes can result in changes in the land sea mask of a typical climate model. At the Last Glacial Maximum (21,000 years ago), sea level was depressed by more than 100 m. In addition, the shallowing may have an important impact on seaways and hence ocean heat and salinity transport. In particular, there were major changes in the Indonesian region which may have affected the connection between the Pacific warm pool region, and the Indian ocean. Such changes require considerable care when setting up an ocean model simulation of the LGM.

In addition, the Quaternary ice sheets have to be added to the model. These represent extra "mountains" of ice which are believed to be up to 3 km thick. The areal extents of these ice sheets are relatively well known from the geological record, but their height is less certain. The most famous reconstructions were by CLIMAP [14], but more recent reconstructions by Peltier [48] suggested changes which in some regions could be as much as 1000 m. However, the broad pattern is unchanged, with major ice sheets over the North American continent and over Scandinavia. Greenland and Antarctica were less dramatically effected.

For pre-Quaternary climates, atmospheric models can use paleogeographic reconstructions. These are relatively well known for the Mesozoic onwards, but are somewhat less certain for earlier periods. Orography/bathymetry are less well known. The position of major relief can be deduced from the movements of the plates, but the absolute elevations are highly uncertain. Indeed, there is considerable speculation about the exact elevation history of Tibet and the Himalayas [41].

3.3.5. *Land Surface Properties*

For many studies of present day climate, vegetation and permanent ice sheets are prescribed from observations. However, for past and future climate change studies, these are potentially important interactive components of the climate system. Vegetation changes the surface albedo, and the transfer of moisture from the land to the atmosphere, and has an important impact on the carbon cycle (e.g., [4]). In addition, vegetation should not be thought of as static. It is influenced by climate and should really be a predicted part of the climate system. Early attempts at predicting vegetation used simple empirical relationships between climate and vegetation type or used observed values from the past (e.g., [20], [45]). More recent studies have used more complex vegetation models which attempt to simulate fundamental processes of the vegetation/climate interactions (e.g., [42], [59]). These models are suitable for predicting the equilibrium vegetation distributions, given a suitable climate as forcing. Such models are very suited for paleoclimate studies, but for future climate change the key issue is the transient nature of response. For instance, the time scale for a forest to move is a century or more. Hence for the next century of two, vegetation will not be in equilibrium and so called "dynamic vegetation" models are being developed to simulate the potentially gradual movement of vegetation belts.

The interaction between vegetation and climate is two-way and can lead to new examples of multiple equilibria. Claussen [13] showed that if a biosphere model is coupled to a climate model and initialized with no vegetation over the Sahara, then the coupled model will keep this region unvegetated. However, if the model is initialized with vegetation over the Saharan region, the resulting circulation change is sufficient to maintain a "green Sahara". Further studies have shown that these multiple equilibria depend on the time period being considered. For the present day and the LGM, two solutions exist. However, for 6000 years ago, there is only one solution corresponding to a wet, vegetated Sahara. This is broadly consistent with the geological record that suggests that this region was much wetter during the mid-Holocene.

High latitudes are also thought to have important vegetation-climate interactions. When tundra is covered with snow, the surface albedo is high. However in a boreal forest, snow will fall between the trees and even a thick layer of snow will not result in a high albedo. Thus a boreal forest will be warmer than tundra, and this acts as a positive feedback mechanism [11].

It has been suggested that such a mechanism is important for periods such as the inception of the ice sheets, about 115,000 years ago. This was a period when the orbital parameters resulted in less incoming solar radiation during Northern hemisphere summers, and this resulted in cooler summers and the potential for winter snows to survive through the summer. This would then lead to a build-up of snow and ice. Ice albedo feedback then acts to amplify the changes and the great ice sheets start to grow (this is one of the key mechanisms for explaining the link between orbital changes and climate variability). Energy balance models successfully reproduce this mechanism, but only one GCM has managed to maintain snow during the summer season (Dong and Valdes [19]) without any other changes, although Sytkus et al. [57] showed that ocean changes could also help). Recently, Gallimore and Kutzbach [25] and deNoblet et al. [17] have suggested that the cooler summers result in an expansion of tundra, resulting in much higher surface albedo and thus amplifying the cooling caused by orbital changes only.

Land ice is another important aspect of paleoclimate modeling. The time scale for the major ice sheets to grow is measured in thousands of years. Hence from a GCM perspective, these are a boundary condition yet they clearly strongly depend on climate. In this case, we cannot truly predict the ice sheet but we can examine if the resulting climate is consistent with the imposed ice sheet. This is best done in offline models that calculate the ice sheet surface mass balance on a finer grid than the GCM, in order to ensure that the narrow zone of melting at the edge of the ice sheet is properly represented. Such calculations are also important for future climate change studies, where the surface mass balance is important for calculating the potential effects of the ice sheets on future sea level change.

3.3.6. Ocean Circulation

A perfect climate model would incorporate a fully dynamic coupled ocean-atmosphere system (see Chapter 18 of this volume). However, this is challenging even for present day conditions, and there are further challenges for paleoclimate simulations because of the problem of how to initialize the model and how long is needed to reach equilibrium. So this type of model has not been widely used for paleoclimate studies (but see [38], [9], [10], [3]). There are two common alternatives. The first is to prescribe sea surface temperature. Data for the Quaternary (CLIMAP [14]) and the pre-Quaternary (e.g., [54], [63]) exist and can be used in atmospheric-only model simulations. The benefit of this approach is that the integrations can be fairly short and a large number of sensitivity experiments can be performed to examine the importance of various changes. The disadvantage is that the model can be validated only against land datasets, and the resulting climate simulation cannot really be thought of as a prediction because a significant component of the climate system has been imposed. However, after completing the simulation, the global energy balance can be evaluated to test whether the model is consistent with the imposed SST.

The other alternative is to use a simple thermodynamic slab ocean model. The advantage of this approach is that the model predicts sea surface temperature, but there has to be an assumption about the horizontal transport of heat in the ocean. For present day simulations, this can either be imposed from observations or empirically deduced from the model to ensure that the resulting simulation of the present day climate is realistic. The problem for paleoclimate studies is that there are no observations of ocean heat transport in the past and so it is impossible to constrain this assumption for periods other than the present. The most common assumption is that it is constant (widely used for Quaternary studies), but some studies have examined the sensitivity of the model by doubling or halving the transport (e.g., [56], [53]).

Having provided the model with the appropriate boundary conditions, the model is run for a suitable period to ensure that it has reached a new climate equilibrium. This effectively produces a "snapshot" of climate for the chosen forcing (boundary conditions). In some cases, the simulations may be designed to test the sensitivity of the climate system to changes in a particular component (e.g., Tibetan uplift). In such cases, testing of the model results against observed climate should be handled with care because the model is not truly simulating a particular period. In addition, model simulations should be tested for consistency. Does the incoming energy equal the outgoing energy? Is the model truly in equilibrium or are there longer-term tendencies in some variables?

4. Case Studies

4.1. MID-HOLOCENE (6000 YEARS AGO)

The mid-Holocene is being extensively studied as part of the Paleoclimate Model Intercomparison Project (PMIP, [33]). The PMIP was designed to answer what aspects of past climate change simulations are robust, and to thoroughly validate these changes against past climate data. Its aim is to compare the results from climate models, driven by exactly the same paleoclimate forcing. This will help us to understand better the mechanisms of climate change and to test the ability of the models to reproduce past climatic conditions. The project complements the Atmospheric Modeling Intercomparison Project (AMIP, see Chapter 19 of this volume) which is evaluating the ability of climate models to simulate correctly our current climate. A good simulation of present day climate is a pre-requisite for reliable model predictions, but it is also important to test their sensitivity to changes. PMIP will be able to evaluate this question, using simulations for the LGM and for the mid-Holocene (6,000 years ago).

The seasonal variation of climate was different in the mid-Holocene. The Northern Hemisphere summer was warmer than at present, but Northern Hemisphere winters were colder. This is thought to have been caused by changes in the Earth's orbital parameters. The PMIP experiment changed the orbital parameters appropriately, and the atmospheric CO_2 concentrations were reduced to pre-industrial levels. However, an important limitation of the simulation is that the sea surface temperatures are prescribed to be the same as at present. Thus this simulation is a simple experiment to examine the effect of changes in the seasonal and latitudinal distribution of insolation. The emphasis is on model-model comparisons, although model data validation is also being performed.

All models show that the land mean, annual mean temperature and precipitation barely changes, but that the seasonal cycle is enhanced, with warm (wet) summers and cooler (drier) winters. However, there is a fair scatter in the amplitude of the changes (Figures 3 and 4), especially for precipitation.

Regional climate change predictions are more variable. For instance, all models show that enhanced summer insolation during the mid-Holocene results in increased summer warming of the Northern Hemisphere land. Owing to this amplified land-ocean temperature contrast, the monsoon circulation is enhanced and penetrates further north into the continents. As a result, the monsoon rainfall over South Asia and North Africa is increased. However, there are some large variations in the magnitude of the change (Figure 5). Additionally, in the North African region all of the models show too little variation in the hydrological conditions, compared to lake level and pollen data [34]. An additional further corollary is that the extent of model-model differences cannot be easily linked to the skill of the models in simulating present climate. This suggests that the ability to correctly simulate present climate is not necessarily a good indicator of the ability of climate models to simulate climate change. A good simulation of present climate is necessary, but NOT sufficient for good climate change simulations.

The lack of agreement between models and data may be due to the limitations of the simulations because the sea surface temperature and the land surface conditions have been held fixed. New simulations with changes in land surface (e.g., [7]) have suggested that there is a substantial feedback between the atmospheric hydrological cycle and the land surface. This is consistent with the ideas of Claussen [13] and Kubatzki and Claussen [36], who found that for 6000 years ago there was only one (vegetated) equilibrium solution for the coupled

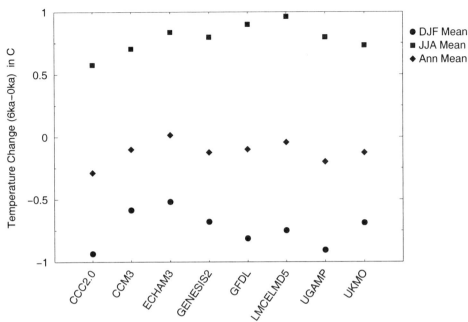

Figure 3. Changes in land mean temperatures for a selection of 8 different models participating in PMIP. Note that the annual mean is relatively little changed, but that seasonally the changes can be appreciable.

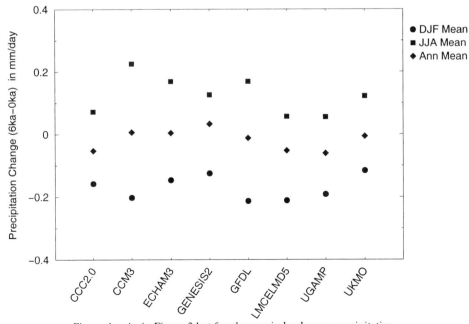

Figure 4. As in Figure 3 but for changes in land mean precipitation.

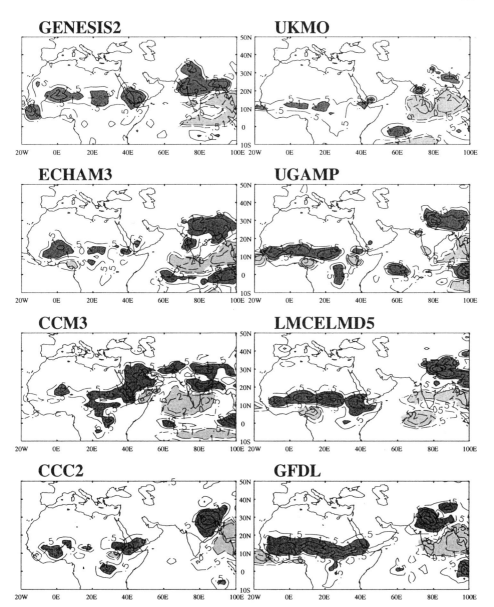

Figure 5. Changes in summer monsoon precipitation (June-July-August-September mean) for a selection of 8 different models participating in PMIP. The plot shows the change in precipitation (6000 years BP – present day). The contours are at −10, −5, −2 − 1, −0.50.5, 1, 2, 5, 10 mm/day and dark shading indicates increased precipitation (greater than 1mm/day) and the lighter shading indicates decreased precipitation (less than −1 mm/day).

atmosphere-vegetation model.

Kutzbach and Liu [37] and Hewitt and Mitchell [30], using a coupled ocean atmosphere model, have also found sensitivity of the African monsoon to ocean temperatures. Seasonal lags introduced by coupling the ocean result in an earlier and hence longer monsoon which greatly amplifies the direct response to orbital forcing. However, currently no model has reproduced the full extent of the changes seen in the data.

4.2. MID-CRETACEOUS

As mentioned in the introduction, the mid-Cretaceous is a period which was significantly warmer than the present, especially at high latitudes. There have been a number of attempts at modeling these warm periods and the results have been remarkably consistent. Figures 7 and 8 show the simulation of surface air temperature from the UGAMP model and the UKMO model for DJF and JJA means. Both models are using the same boundary conditions, which are shown in Figure 6.

Both models show that continental temperatures in winter fall well below zero. This result has been seen in all other model simulations for the period (e.g., [1], [2]) and presents a major mis-match between model and data. The latter is based on flora and faunal estimates and suggests that for both the American and Eurasian regions, continental interior temperatures could not have dropped significantly below zero during winter (e.g., [29]). This lack of cold extremes has led to the notion that the climate was "equable". The model-data disagreement is 20°C or more, and no reasonable change in the boundary conditions has significantly reduced this mis-match. This has prompted some authors ([55] and [60]) to suggest that either the data are wrong, or that we have a fundamental lack of understanding of the processes that influence climate, or that we have a fundamental problem in numerically simulating these climate processes.

Other aspects of the simulations are more reasonable. The mean annual temperatures are in better agreement with the data, and the warm summer temperatures are sufficient to melt winter snow. Thus the models do not predict permanent land ice, although seasonal snow and small mountain glaciers are possible.

Figure 9 shows that there are some very large model-model differences in the moisture distribution. The UGAMP model has much less rainfall over the continents and this results in very dry summer soil moisture (not shown). This then leads to very high summer temperatures that are further reinforced by lack of cloud cover in the UGAMP model. By comparison, the UKMO model predicts more rainfall over the landmasses in summer (and associated cloud cover), and this results in much lower summer temperatures.

One of the most likely candidates for the cause of the differences is the convective parameterisation scheme. The UGAMP model uses the Betts-Miller con-

Paleogeography

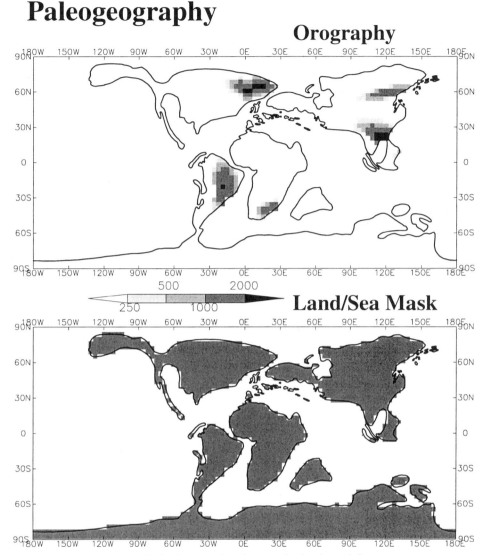

Figure 6. Model land-sea mask and orography for the mid-Cretaceous.

vection scheme whereas the UKMO model uses a penetrative convection scheme. These two schemes are potentially sensitive to different processes and hence respond very differently in radically different climates.

The magnitude of the model differences in precipitation and temperature for the mid-Cretaceous is much larger than the equivalent present day simulations (although the same characteristics can be seen). This may suggest that both models have had some "tuning" to achieve the best possible simulation for present day. This is a good example of where the "testing" of these models in very different

DJF Mean Surface Air Temperature
UGAMP Model

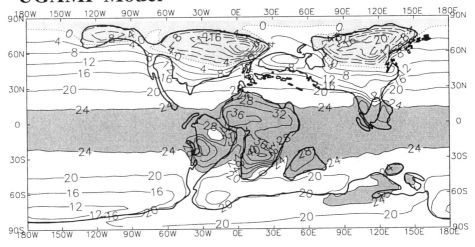

UKMO Model

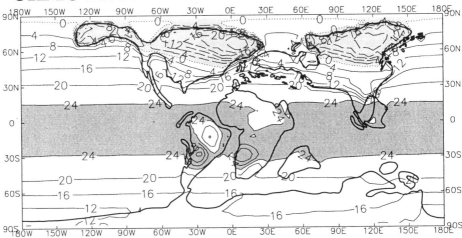

Figure 7. DJF mean surface air temperature as predicted by the UGAMP model and the UKMO model. The contour interval is 4°C and the zero line is dotted and negative contours are dashed. Areas colder than −20°C are lightly shaded and areas warmer than 24°C are darkly shaded.

climate regimes helps us to better evaluate our ability to model some climate processes.

It should also be added that the UKMO model is probably better at simulating present day precipitation patterns. However, it seems to do a poorer job (compared with the UGAMP model) at simulating the mid-Cretaceous moisture patterns [49]. There are examples of aeolian deposits in regions that the UKMO model is pre-

JJA Mean Surface Air Temperature
UGAMP Model

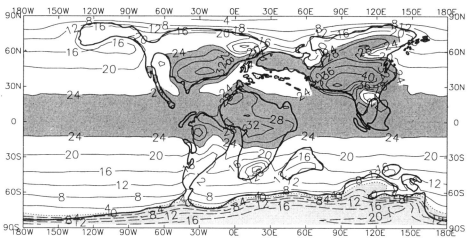

UKMO Model

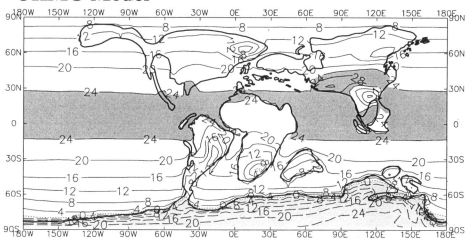

Figure 8.   As in Figure 7 but for the JJA season.

dicting to be wet. The interpretation could be that the UKMO moisture simulations have incorrect climate sensitivity, but it should be cautioned that the precipitation fields are much more influenced by some of the uncertainties in the boundary condition. Thus it may be that the UGAMP model is getting it right for the wrong reasons.

JJA Total Precipitation (in mm/day)
UGAMP Model

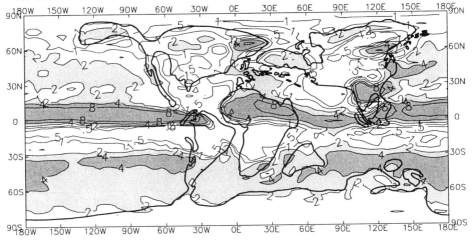

UKMO Model

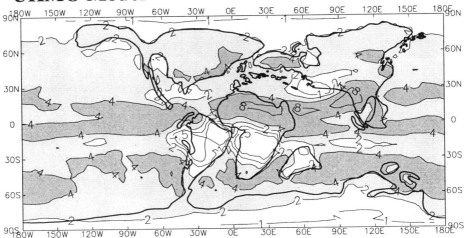

Figure 9. As in Figure 8 but for total precipitation. Units are mm/day and the contours are at 0.5, 1, 2, 5, and 10 mm/day. Areas greater than 4mm/day are shaded.

5. Summary

This brief chapter has attempted to discuss a huge subject and inevitably it has had to cover many important areas in a very shallow way; for more information, the reader should refer to Frakes [22] and Crowley and North [15]. The most important point is that paleoclimate studies are one of the few ways in which we can seriously test our understanding of climate processes and our ability to model them in climate regimes significantly different from the present. No past time period is a direct analogy for possible anthropogenic-induced future climate change, but by thoroughly understanding the mechanisms of climate change, we can hope to improve our confidence in future climate change predictions.

In many ways, paleoclimate modeling is just an example of the application of the current generation of climate models. However, for a complete understanding of past climate change, we need to develop models of the whole "climate system". This is an ambitious task that will necessitate extra components that are not generally incorporated into climate models designed for future climate change predictions (e.g., ice sheet growth and lithospheric rebound are not generally part of future climate change models). These longer time scale processes are important to include in an interactive model, although they do not always need to be coupled in a completely synchronous fashion.

Finally, I should emphasize that paleoclimate studies clearly show that climate has always varied and will continue to vary into the future. Any predictions of man-made future climate change must compare the simulation to the natural variability of the system, and this requires us to expand our knowledge and ability to model past climate change. This is not just "testing the models". It is testing our fundamental understanding of how the climate system works. We cannot claim to have a complete understanding of this climate system unless we can explain past climate variability.

Acknowledgements

I thank the many students from the ASI who added their valuable comments in a draft version of these lecture notes. The data from the PMIP simulations were kindly provided by the PMIP group members.

References

1. Barron, E.J. and Washington, W.M. (1984) The role of geographic variables in explaining paleoclimates: results from a Cretaceous climate model sensitivity studies, *J. Geophys. Res.* **89**, 1267-1279.
2. Barron, E.J., Fawcett, P.J., Pollard, D. and Thompson, S. (1993) Model simulations of Cretaceous climates: the role of geography and carbon dioxide, *Phil. Trans. Roy. Soc.* **B341**, 307-316.
3. Barron, E.J. and Peterson, W.H. (1990) Model simulation of the Cretaceous ocean circulation, *Science* **244**, 684-686.
4. Beerling, D.J., Woodward, F.I. and Valdes, P.J. (1998) Global terrestrial productivity in the mid-Cretaceous (100Ma): model simulations and data, *Geol. Soc. Am. Spec. Publ.* (In press).
5. Berger, A., Loutre, M.F. and Dehant, V. (1989) Influence of the changing lunar orbit on the astronomical frequencies of pre-Quaternary insolation patterns, *Paleoceanography* **4**, 555-564.

486

6. Berger, A., Loutre, M.F. and Gallée, H. (1998) Sensitivity of the LLN climate model to the astronomical and CO_2 forcings over the last 200 ky, *Clim. Dyn.* **14**, 615-629.

7. Brostrom, A., Coe, M., Harrison, S.P., Gallimore, R., Kutzbach, J.E., Foley, J., Prentice, I.C. and Behling, P. (1998) Land surface feedbacks and palaeomonsoons in northern Africa, *Geophys. Res. Lett.* **25**, 3615-3618.

8. Berner, R.A. (1994) GEOCARB II: A revised model of atmospheric CO_2 over Phanerozoic time, *Amer. J. Sci.* **294**, 56-91.

9. Bice, K.L. (1997) An investigation of Early Eocene Deep Water Warmth using uncoupled atmosphere and ocean general circulation models: Model sensitivity to geography, initial temperatures. Atmospheric forcing and continental runoff, *Technical Report 97-002*. Earth System Science Center, Pennsylvanian State University, University Park.

10. Bush, A.B.G., and Philander, S.G.H. (1997) The late Cretaceous simulation with a coupled atmosphere-ocean general circulation model, *Paleoceanography* **12**, 495-516.

11. Bonan, G.B., Pollard, D. and Thompson, S.L. (1992) Effects of boreal forest vegetation on global climate, *Nature* **359**, 716-718.

12. Cerling, T.E. (1991) Carbon dioxide in the atmosphere: Evidence from Cenozoic and Mesozoic paleosols, *Amer. J. Sci.* **291**, 377-400.

13. Claussen M. (1994) On coupling global biome models with climate models, *Climate Research* **4**, 203-221.

14. CLIMAP Project Members (1981) Seasonal reconstruction of the Earth's surface at the last glacial maximum, *Geol. Soc. Am. Map Chart Ser.* **MC-36**.

15. Crowley, T.J. and North, G.R. (1991) *Paleoclimatology*, Oxford University Press, New York.

16. Deblonde, G. and Peltier, W.R. (1991) Simulations of continental ice sheet growth over the last glacial-interglacial cycle: experiments with a one-level seasonal energy balance model including realistic topography, *J. Geophys. Res.* **96**, 9189-9215.

17. deNoblet, N., Prentice, I.C, Joussaume, S., Texier, D., Botta, A. and Haxeltine, A. (1996) Possible role of atmosphere-biosphere interactions in triggering the last glaciation, *Geophys. Res. Lett.* **23**, 3191-3194.

18. Dickens, G.R., Castillo, M.M. and Walker, J.C.G. (1997) A blast of gas in the latest Paleocene: simulating first-order effects of massive dissociation of oceanic methane hydrate, *Geology* **25**, 259-262.

19. Dong, B. and Valdes, P.J. (1995) Sensitivity studies of Northern Hemisphere glaciation using an atmospheric general circulation model, *J. Climate* **8**, 2471-2496.

20. Dutton, J.F. and Barron, E.J. (1996) Genesis sensitivity to changes in past vegetation, *Palaeoclimates* **1**, 325-354.

21. Endal, A.S. and Sofia, S. (1981) Rotation in solar-type stars. I. Evolutionary models for the spin-down of the Sun, *Astrophys. Jour.* **243**, 625-640.

22. Frakes, L. (1979) *Climates Throughout Geological Time*, Elsevier, Amsterdam.

23. Gallée, H., van Ypersele, J.P., Fichefet, T., Tricot, C. and Berger, A. (1991) Simulation of the last glacial cycle by a coupled sectorially averaged climate-ice-sheet model. I. The Climate Model, *J. Geophys. Res.* **96**, 13,139-13,161.

24. Gallée, H., van Ypersele, J.P., Fichefet, T., Marsiat, I., Tricot, C. Berger, A. (1992) Simulation of the last glacial cycle by a coupled sectorially averaged climate-ice-sheet model. II. Response to insolation and CO_2 variations, *J. Geophys. Res.* **97**, 15,713-15,740.

25. Gallimore, R.G. and Kutzbach, J.E. (1996) Role of orbitally induced changes in tundra area in the onset of glaciation, *Nature* **381**, 503-505.

26. Gough, D. (1977) Theoretical predictions of variations in solar output, in O. White (ed.), *The solar Output and Its Variation*, Colorado Assoc. Univ. Press.

27. Hart, M.H. (1978) The evolution of the atmosphere of the earth, *Icarus* **33**, 23-39.

28. Harvey, L.D.D. and Huang, Z. (1995) Evaluation of the potential impact of methane clathrate destabilisation on future global warming, *J. Geophys. Res.* **100**, 2905-2926.

29. Herman, A. B. and Spicer, R. A. (1997) New quantitative palaeoclimate data for the Late Cretaceous Arctic: evidence for a warm polar ocean, *Paleogeography, Paleoclimatology, Paleoecology* **128**, 227-251.

30. Hewitt, C.D. and Mitchell, J.F.B. (1998) A fully coupled GCM simulation of the climate of the mid-Holocene, *Geophys. Res. Lett.* **25**, 361-364.

31. Imbrie, J., Boyle, E.A., Clemens, S.C., Duffy, A., Howards, W.R., Kukla, G., Kutzbach, J., Martinson, D.G. McIntye, A., Mix, A.C., Molfino, B. Morley, J.J., Peterson, L.C., Pisias, N.G., Prell, W.L., Raymo, M.E. Shackleton, N.J. and Toggweiler, J.R. (1992) On the structure of major glaciation cycles, *Palaeoceanography* **7**, 701-738.

32. Imbrie, J., Berger, A., Boyle, E.A., Clemens, S.C., Duffy, A., Howards, W.R., Kukla, G., Kutzbach, J., Martinson, D.G. McIntye, A., Mix, A.C., Molfino, B. Morley, J.J., Peterson, L.C., Pisias, N.G., Prell, W.L., Raymo, M.E. Shackleton, N.J. and Toggweiler, J.R. (1993) On the structure and origin of major glaciation cycles. 2. The 100,000 year cycle, *Palaeoceanography* **8**, 699-735.

33. Joussaume, S. and Taylor, K. (1995) Status of the Paleoclimate Modeling Intercomparison Project (PMIP), in *Proceedings of the First International AMIP Scientific Conference*, Monterrey, California, USA, 15-19 May 1995, **WCRP-92**, 425-430.

34. Joussaume, S., Taylor, K.E., Bracannot, P., Mitchell, J., Kutzbach, J., Harrison, S.P., Prentice, I.C., Abe-Ouchi, A., Bartlein, P., Bonfils, C., Broccoli, A.J., Dong, B., Guiot, J., Herterich, K., Hewitt, C., Jolly, D., Kim, J.W., Kisolov, A., Kitoh, A., Masson, V., McAvaney, B., McFarlane, N., de Noblet, N., Peterschmitt, J.Y., Pollard, D., Rind, D., Royer, J.F., Schlesinger, M., Syktus, J., Thompson, S., Valdes, P.J., Vettoretti, G., Webb, R.S. and Wyputta, U. (1999) Monsoon changes for 6000 years ago: results of 18 simulations from the Paleoclimate Modeling Intercomparison Project, *Geophys. Res. Lett.* (In press).

35. Kapsner, W.R., Alley, R.B., Shuman, C.A., Anandakrishnan, S. and Grootes, P.M. (1995) Dominant influence of atmospheric circulation on snow accumulation in Greenland over the past 18,000 years, *Nature* **373**, 52-54.

36. Kubatzki, C. and Claussen, M. (1998) Simulation of the global bio-geophysical interactions during the Last Glacial Maximum, *Clim. Dyn.* **14**, 461-471.

37. Kutzbach, J.E. and Liu, Z. (1997) Response of the African monsoon to orbital forcing and ocean feedbacks in the middle Holocene, *Science* **278**, 440-443.

38. Maier-Reimer, E.K., Mikolajewicz, U. and Crowley, T.J. (1990) Ocean GCM sensitivity experiments with an open central American isthmus, *Paleoceanog.* **5**, 349-366.

39. Manabe, S. and Stouffer, R. (1988) Two stable equilibria of a coupled ocean-atmosphere model, *J. Climate* **1**, 841-866.

40. Markwick, P.J. (1994) 'Equability', continentality, and Tertiary "climate": the crocodilian perspective, *Geology* **22**, 613-616.

41. Molnar, P., England, P. and Martinod, J. (1993) Mantle Dynamics, Uplift of the Tibetan Plateau, and the Indian Monsoon, *Reviews of Geophysics* **31**, 357-396.

42. Neilson, R.P. (1995) A model for predicting continental-scale vegetation distribution and water balance, *Ecological Applications* **5**, p362-385.

43. Oglesby, R.J. and Park, J. (1989) The effect of precessional insolation changes on Cretaceous climate and cyclic sedimentation, *J. Geophys. Res.* **94**, 14,793-14,816.

44. Oglesby, R.J. and Ogg, J.G. (1998) The effect of large fluctuations in obliquity on climates of the Late Proterozoic, *Palaeoclimates: Data and Modelling* **2**, 293-316.

45. Otto-Bliesner, B.L and Upchurch, G.R. (1997) Vegetation-induced warming of high latitude regions during the Late Cretaceous period, *Nature* **385**, 804-807.

46. Parrish, J.T. (1982) Upwelling and petroleum source beds, with reference to the Paleozoic, *Am. Assoc. Petroleum Geologists Bull.* **66**, 750-774.

47. Parrish, J.T. and Curtis, R.L. (1982) Atmospheric circulation, upwelling, and organic-rich rocks in the Mesozoic and Cenozoic eras, *Paleogeography, Paleoclimatology, Paleoecology* **40**, 31-66.

48. Peltier,W.R. (1994) Ice Age paleotopography, *Science* **265**, 195-201.

49. Price, G.D., Sellwood, B.W. and Valdes, P.J. (1995) Sedimentological Evaluation of General Circulation Model Simulations for the "Greenhouse" Earth: Cretaceous and Jurassic Case Studies, *Sedimentary Geology* **100**, 159-180.

50. Price, G.P., Valdes, P.J. and Sellwood, B.W. (1997) Prediction of modern Bauxite occurrence: implications for climate reconstructions, *Paleogeography, Paleoclimatology, Paleoecology* **131**, 1-13.

51. Rahmstorf, S. (1996) On the freshwater forcing and transport of the Atlantic thermohaline circulation, *Climate Dynamics* **12**, 799-811.

52. Saltzman, B. and Maasch, K.A. (1991) A first order global model of late Cenozoic climatic change. II. Further analysis based on a simplification of CO_2 dynamics, *Climate Dynamics* **5**, 201-210.

53. Schmidt, G.A. and Mysak, L.A. (1996) Can increased poleward oceanic heat flux explain the warm Cretaceous climate? *Paleoceanography* **11**, 579-593.

54. Sellwood, B.W., Price, G.D. and Valdes, P.J. (1994) Cooler estimates of Cretaceous temperatures, *Nature* **370**, 453-455.

55. Sloan, L.C. and Barron, E.J. (1990) Equable climates during Earth History? *Geology* **18**, 489-492.

56. Sloan, L.C., Walker, J.C.G. and Moore, T.C. (1995) Possible role of oceanic heat-transport in early Eocene climate, *Paleoceanography* **10**, 347-356.

57. Sytkus, J., Gordon, H. and Chappell, J. (1994) Sensitivity of a coupled atmosphere dynamic upper ocean GCM to variations of CO_2 , solar constant, and orbital forcing, *Geophys. Res. Lett.* **21**, 1599-1602.

58. Valdes, P.J., Sellwood , B.W. and Price, G.D (1995) Modelling Late Jurassic Milankovitch Climate Variations, in M.R. House and A.S. Gale (eds.), *Orbital Forcing Timescales and Cyclostratigraphy*, Geological Society Special Publication No 85, pp 115-132.

59. Woodward, F.I., Smith, T.M. and Emanuel, W.R. (1995) A global land primary productivity and phytogeography model, *Global Biogechem. Cycles* **9**, 471-490.

60. Wing, S.L. (1991) Equable climates during Earth history? Comment, *Geology* **19**, 539-540.

61. Wing, S.L. and Greenwood, D.R. (1993) Fossils and fossil climate: the case for equable continental interiors in the Eocene, *Phil. Trans. R. Soc. Lond. B* **341**, 243-252.

62. Wolfe, J.A. (1978) A paleobotanical interpretation of Tertiary climates in the Northern Hemisphere, *American Scientist* **66**, 691-703.

63. Zachos, J.C., L.D. Stott and Lohmann, K.C. (1994) Evolution of early Cenozoic marine temperatures, *Paleoceanography* **9**, 353-387.

SIMULATING FUTURE CLIMATE

G. J. BOER
Canadian Centre for Climate Modelling and Analysis
Atmospheric Environment Service
University of Victoria
Victoria, B.C., V8W 2Y2   CANADA

1. Introduction

General circulation models (GCMs) are tools for investigating the behaviour of the climate system and for simulating climate change. GCMs embody the physical principles that govern this system and, to the extent that they correctly and completely do so, are able to obtain the current climate from basic principles. Moreover, since they are physically based, climate models provide a way to investigate other climates (such as that of Mars for instance) or, in particular, that of the Earth in the past and in the future.

The large changes in climate that have occurred on geological timescales are revealed by indirect and proxy measures of temperature and other climate quantities as discussed in Chapter 20. Climate is not fixed and has varied over a considerable range. The causes of past climate change are not completely understood but several lines of evidence indicate that changes of solar input associated with the earth's orbital variations are a primary cause, although mediated by the internal working of the climate system. Evidence of recent climate change is based on the instrumental record of the past 150 years and indicates a global mean climate warming from the preindustrial period of the 1800s on the order of 0.5°C. There is evidence that the rate of warming is accelerating and this has drawn attention to possible causes of recent climate change and also to the potential for human-induced climate change of greater magnitude in the future.

Human activity has directly and indirectly altered the concentration of trace gases and particulates in the atmosphere and these may disproportionately affect climate. The earth's climate is the result of a complex balance where the powerful but unevenly distributed input of solar energy is modulated, converted to other forms, transported by the atmosphere and ocean, and ultimately released back to space as thermal radiation. The particulates in the atmosphere are important because they interact strongly with the incoming solar beam and because they may also change the distribution and optical properties of clouds by acting as cloud condensation nuclei. The trace gases in the atmosphere are correspondingly important because they interact with the balancing output of thermal radiation to space. In particular, CO_2, CH_4, and other "greenhouse gases" (GHGs) act to warm the surface by impeding this output to space.

Our ability to simulate climate and its changes has grown rapidly during the last two decades because of increases in computer power and the improvement and elaboration of the complex coupled atmosphere/ocean/land/sea-ice models that form the basis for projections of future climate.

P. Mote and A. O'Neill (eds.), Numerical Modeling of the Global Atmosphere in the Climate System, 489–504.

490

2. International Aspects

The possibility that human activity is already changing the earth's climate, and that this may accelerate in the future, is a scientific, economic, and political challenge of considerable scope. The Intergovernmental Panel on Climate Change (the IPCC) has been established by the World Meteorological Organization (WMO) and the United Nations Environment Programme (UNEP) and charged with: (1) assessing the available scientific information on climate change; (2) assessing the environmental and socio-economic impacts of climate change; and (3) formulating strategies with which to respond to climate change. In order to respond to these tasks, the IPCC has formed three working groups, often referred to by number. They are: Working Group I - The Science of Climate Change; Working Group II - Impacts, Adaptation, and Mitigation; and Working Group III - Socio-economic and Cross-cutting Issues.

The IPCC through its working groups has undertaken to produce a number of "assessments". The WGI reports (referred to here as IPCC 1990, IPCC 1992, IPCC 1995) on the science of climate change are of particular interest and comprise the FAR and the SAR (the first and second assessment reports in 1990 and 1995) as well as an interim report. The third assessment is currently underway and will culminate in the TAR (third assessment report) early in the year 2001. Climate model results have been, and continue to be, basic to the WGI task of assessing the scientific basis of climate change. Model results are also basic to the efforts of the other working groups who use model-produced "climate change scenarios" as input to an array of applications models that attempt to understand and quantify the social and economic costs of different courses of action relating to climate change.

The chain of climate studies, that is paralleled to some extent by the IPCC structure, is indicated in Figure 1. Here the socio-economic studies of WGII form the basis for a set of "emission scenarios" for CO_2 and other trace gases. These scenarios are based on estimates of population growth, economic growth, and energy supply as well as other considerations such as international efforts to limit deforestation and the emissions of various trace gases. According to WGII, the purpose of emission scenarios is: (1) to evaluate the consequences of "no intervention" to reduce GHG emissions as an input for climate models of various kinds; (2) to evaluate the consequences of interventions to reduce GHG emissions and as input to climate models; (3) to examine the feasibility and costs of mitigating GHG emissions from different regions and economic sectors in a variety of ways; and (4) as input for negotiating possible emission reductions for countries and geographic regions. Future emissions depend on a considerable number of assumptions that are themselves uncertain. The effort is therefore to provide a range of emission scenarios that reasonably spans this range of uncertainties.

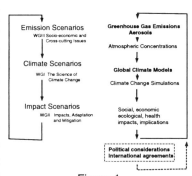

Figure 1

The global climate models that are used to simulate future climate change generally do not use emissions directly since radiative and other calculations depend on atmospheric concentrations of trace gases and on the concentrations and optical properties of particulates. The intermediate step in which the emissions are converted to the required concentrations may be comparatively simple if the sources and sinks of the trace gases are themselves straightforward but otherwise can be complicated and add an increment of uncertainty to the problem (as discussed in [11] for instance).

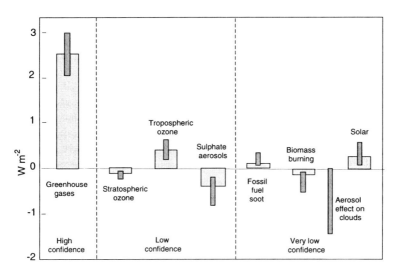

Figure 2. Global radiative forcing 1850-1992. Thin bars
indicate the range of uncertainty.

The resulting "concentration scenarios" provide the basis on which to estimate the change in the radiative forcing of the climate system, that is, the amount by which the gases and particles interfere with the radiative balance of the system and hence act to warm or cool it. Figure 2, redrawn from [11], estimates the "historical" change in radiative forcing from 1850 to 1992 due to a variety of causes. Note particularly that the confidence in these results ranges from high for the GHGs to very low for some tropospheric aerosol effects. Although CO_2 has the largest radiative effect it is by no means the only GHG in the atmosphere. It has been common to represent the effect of all GHGs in terms of an "equivalent" amount of CO_2 and thereby simplify the concentration scenarios for global models. In particular, IPCC 1995 and recent simulations of future climate change have typically used scenarios based on equivalent CO_2 concentrations and the sulphate "direct" effect (as opposed to the uncertain indirect effect of the particles as they affect clouds). The effects of ozone have generally been neglected as being uncertain into the future and also tending to cancel. The remaining forcing effects are known with very low confidence so are not included. More sophisticated simulations, underway for IPCC 2000, treat individual greenhouse gases and may include an interactive sulphur cycle for determining the distribution and effect of sulphate aerosols.

492

The concentration scenarios are input to global climate models which simulate the global distribution of climate change. Temperature and precipitation are of primary interest but values of many other quantities are also available as described in Chapter 3. These "climate change scenarios" are themselves input into a wide variety of studies of potential economic, social, health and other impacts and consequences. These may lead to political considerations, arrangements, and agreements which, in turn, influence estimates of future emissions so that the circuit is complete.

3. Simulating Historical and Future Climate

Recent climate studies have attempted to simulate climate from the preindustrial period through to the present using historical GHG concentrations and aerosol loadings, and thence to the end of the 21st century using projected GHG and aerosol amounts. The simulations for the period 1900 to the present allow a comparison with available observations. However, the change in climate forcing for this period, and hence the expected climate change, is small compared projections for the next 100 years. Simulations of this kind have been undertaken by several modelling groups; CCCma ([2], [3]), the Hadley Centre ([16], [17]), MPI [4], and GFDL [6]. Selected results from these simulations have been collected by the IPCC Data Centre for use in impact and assessment studies (*http://ipcc-ddc.cru.uea.ac.uk*) and results from other models may be added. The availability of simulations from different models using the same GHG and aerosol scenarios provides some information on the internal consistency of the results and hence an indication of their robustness.

3.1 INITIALIZATION, ADAPTION AND THE CONTROL SIMULATION

Figure 3 shows schematically how a typical climate change simulation is performed. The simulations described here are performed with the CCCma coupled climate model [5]. The atmospheric component of this model is described by McFarlane *et al.* [15], the results of an equilibrium $2\times CO_2$ simulation with a simplified ocean model by Boer *et al.* [1], and aerosol effects in this context by Reader and Boer [20]. The atmospheric model uses spectral numerics at resolution T32L10 with a 96x48 grid and 10 levels in the vertical. The three-dimensional oceanic component of the full coupled model is based on the Modular Ocean Model [18] and operates on a 192x96 grid with 29 levels in the vertical. Thus, four ocean grid squares underlie an atmospheric grid square.

Because of the very long timescales of the ocean compared to the atmosphere, the two component models are spun-up separately to approximately equilibrium states and then coupled using "flux adjustment" or "anomaly coupling". There is a period of adaptation which may include further adjustment. A *control* simulation is performed with fixed GHG and aerosol forcing. This provides a reference against which to compare a variety of *experimental* simulations in which GHG concentrations and aerosol loadings increase with time.

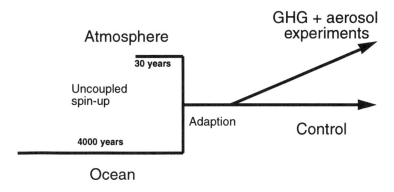

Figure 3. Coupled model initialization and simulation.

3.2 GREENHOUSE GAS AND AEROSOL FORCING

The effects of GHGs are included in these simulations by specifying an equivalent CO_2 concentration in the model atmosphere. The GHGs are well mixed in the atmosphere and the radiative effects of changing concentrations are calculated within the model as part of the longwave radiation calculation which itself depends on the evolving and interactive moisture and cloud distributions in the model atmosphere. The direct effect of sulphate aerosols is to reflect some of the incoming solar radiation away from the surface and this effect is included in the model by altering the surface albedo based on the prescribed aerosol loading. This affects the shortwave radiative calculation based on the time of day and year and on the nature and distribution of clouds. While the GHGs are well mixed throughout the atmosphere, the sulphate aerosols are produced by industrial activity and are most abundant near their sources owing to their relatively short atmospheric lifetimes. The aerosol loading pattern for the present, and the projected pattern for 2050, are shown in Figure 4. The global mean change in radiative forcing in the model due to GHGs and aerosols is plotted at 50 year intervals for the period from 1900 to 2100 in Figure 5. The GHG and aerosol specification for these simulations follows that of Mitchell *et al.* [16] and is based on the IPCC WG II "1992a emission scenario" described in IPCC 1995 [11] and elsewhere.

3.3 AVAILABLE SIMULATIONS AND DATA

The suite of simulations available with this model for the period 1900-2100 include: (1) a control run (now extended to 1000 years), (2) three independent simulations with the GHG+aerosol forcing discussed above, (3) a simulation with GHG-only forcing, (4) a "stabilization" simulation where the forcing increases to 2050 and then is kept constant at that level and, (5) a similar stabilization simulation but at the year 2100 forcing level.

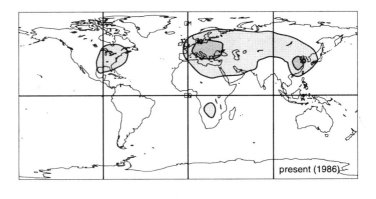

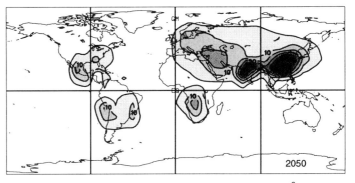

Figure 4. Anthropogenic sulphate loading   (mg m^{-2}).

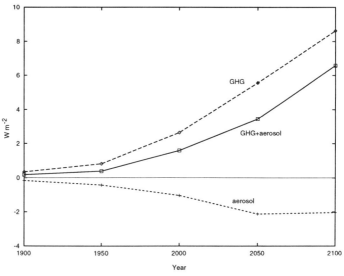

Figure 5. Global mean radiative forcing.

Data from simulations (1-3) have been made available to the IPCC data centre for use in impact and assessment studies and are also available on the CCCma website at *http://www.cccma.bc.ec.gc.ca.*

As noted in Chapter 3, a global climate model produces data for an array of climate variables evolving in time on a three-dimensional grid of points in the atmosphere and ocean. We concentrate here on the simulated changes to the mean climate for a small number of variables of basic interest, especially surface air temperature. A climate variable is "robust" for climate change purposes if its signal to noise ratio is high; that is, if its response to a change in forcing is relatively large compared to the natural variability found in the control simulation. Temperature is robust in this sense and, moreover, is the variable with the best historical observational coverage. Precipitation is a basic climate variable which is not robust in the same sense as temperature. Nevertheless, it is of immense practical importance and must be considered. Changes in the surface moisture balance affect soil moisture which, although is not well measured, has important practical implications. Changes to the cryosphere are reflected in the area covered by snow and sea-ice.

4. Climate Change in the 20th Century

The observed climate record is uneven as to coverage and sparse as to variables. For the period from 1900 to the present, which is of interest here, only two suitable data sets are available. These are the surface air temperature data set of Parker and Jones ([14], [19]) which provides gridded data over the globe (albeit with areas and periods of missing data) and the precipitation data set of Hulme ([7], [8]) which provides gridded data over land (also with areas and periods of missing data). These data sets are based on instrumental measurements of temperature and precipitation which have been processed in various ways to enhance their uniformity and representativeness.

4.1 TEMPERATURE

Figure 6 shows the global annual mean temperature anomaly with respect to the 1900-1930 average for the Parker/Jones data set, the average of the three independent model simulations with GHG+aerosol forcing, and the model simulation with GHG-only forcing. The observation-based curve displays a long-term increasing trend on which is superimposed shorter term variations. The presumption is that the temperature record has the form

$$T(t) = T_o + T_f(t) + T'(t)$$

where the terms on the right hand side are, respectively, the unperturbed global mean climate T_o (corresponding to the long-term average temperature of the control run), a deterministic "forced" climate change component T_f which varies smoothly in response to the forcing changes illustrated in Figure 5, and the remaining natural variability T' which is internally generated and non-deterministic. A perfect climate model would simulate the deterministic part $T_o + T_f$ of the temperature curve but not the details of the non-deterministic natural variability.

496

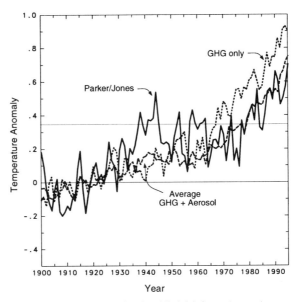

Figure 6. Observed and modelled global mean temperature.

The average of the three GHG+aerosol simulations provides a better estimate of the signal than a single realization since the random natural variability tends to average out (this is apparent in the comparative smoothness of this curve).

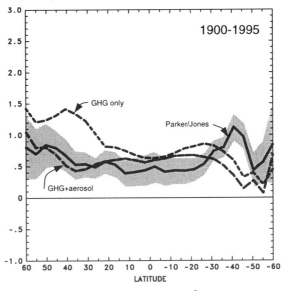

Figure 7. Linear trend in temperature ($^{\circ}$C per 100 years).

Figure 6 shows that: (1) the model is able to reproduce the long-term trend in observed temperature for the GHG+aerosol case, (2) the GHG-only simulation displays excessive warming compared to the observations, and (3) the simulations and the observations display superimposed "natural variability" (partially averaged out in the mean of the three GHG+aerosol simulations) which does not, and is not expected to, coincide.

Figure 7 compares the observed and simulated linear trend in temperature at each latitude. The observed trend, in °C per century, is plotted together with a shaded tube which should contain the trend estimate approximately 95% of the time in the presence of sampling error due to climate variability. The trend estimates from the GHG+aerosol simulations generally fall within this band indicating that the simulated and observed temperature trends cannot generally be statistically distinguished from one another. This is not the case for the GHG-only simulation which has a trend distinct from both the observed and the GHG+aerosol values.

The geographic patterns of climate change during the 1900-1995 period are not as distinct as the globally or longitudinally averaged values since the natural variability T' is not spatially averaged in this case and so is not reduced in comparison to the signal. Some evidence of the model's ability to simulate aspects of the geographic pattern of historical climate change is discussed by Boer *et al.* [2].

4.2 PRECIPITATION

Precipitation is the other primary climate variable for which an observation-based record is available, albeit only over land. Figure 8 displays the average over land for the observations (where data exists) and the same average for the GHG+aerosol and the GHG-only simulations (masking out model data where no observed data exists). The diagram indicates that: (1) there is a general increase in precipitation over land in both the observations and the simulations; (2) observed increases are slightly larger than simulated increases; and (3) both observations and simulations indicate a recent decrease in land precipitation.

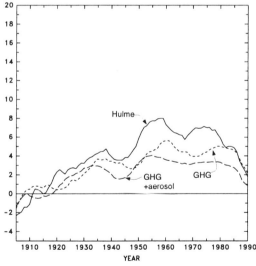

Figure 8. Percentage change in observed and modelled
precipitation over land.

Land precipitation is not the same as globally averaged precipitation and the results are further confounded by the problem of missing data. Masking the control run results in the same way as the data (not shown) suggests that at least part of the apparent recent decline in the observed and simulated values is associated with changes in sampling. The results indicate that land precipitation rates have both risen and fallen in concert for the observations and in the forced simulations but not for the unforced control simulation. The implications are twofold, namely that the observed climate change from 1900 to the present is consistent with the forcing changes due to GHGs and aerosols and that coupled models are capable of reproducing at least the main features of this change.

5. Simulating Future Climate Change

Simulations of future climate change depend on the projected changes in GHG concentrations and aerosol loading that lead to the changes in forcing shown in Figure 5. The forcing is expected to increase markedly from the present into the 21st century and the rate of climate change is expected to accelerate also.

5.1 TEMPERATURE

Figure 9 gives the simulated global annual mean temperature for the control, the three GHG+aerosol and the GHG-only simulations. The projected temperature change from the present to 2050 is approximately 1.8°C with a further 2.5°C to the end of the century. The GHG-only simulation gives even larger temperature increases. The magnitude of the temperature change depends, among other things, on the model's "climate sensitiv-

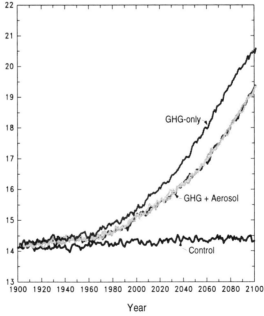

Figure 9. Global mean annual average temperature (°C).

ity" which measures the equilibrium response to a given change in radiative forcing. The typical measure of climate sensitivity is the equilibrium temperature change under a doubling of CO_2 concentration. The CCCma model's climate sensitivity is 3.5°C, which falls within the IPCC estimate of the probable range of climate sensitivity of from 1.5 to 4.5°C albeit in the upper half of the range. The evolution of temperature with

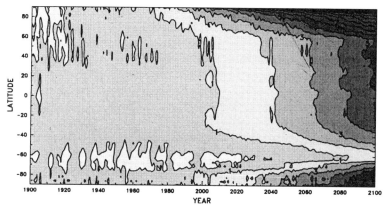

Figure 10. Temperature evolution under GHG + aerosol forcing (°C).

latitude is shown in Figure 10 which indicates: (1) a modest increase in simulated temperature to the present; (2) a rapidly accelerating warming from the present; and (3) a marked asymmetry between northern and southern hemispheres in the rate of warming (associated with ocean interactions).

Figure 11 shows the geographical pattern of the warming to 2050 and indicates: (1) the asymmetry of the warming between hemispheres and between land and ocean; (2) regions of minor cooling in certain ocean areas despite the overall warming; and (3) some indication of an "El Niño-like" warming pattern in the tropical Pacific (which is more apparent in its effects on precipitation).

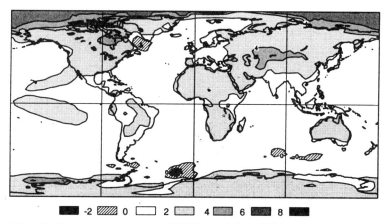

Figure 11. Temperature change from 1975-95 to 2040-60 for GHG+aerosol forcing (°C).

5.2 PRECIPITATION

The acceleration of the hydrological cycle is a general feature of a warmer climate. Global annual average changes in precipitation rate are shown in Figure 12. The increase in

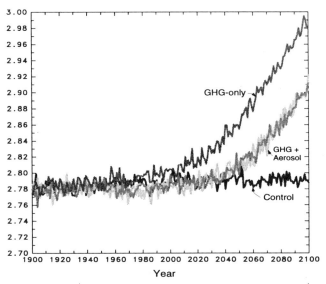

Figure 12. Global mean annual average precipitation rate (mm day⁻¹).

precipitation is 4% to the end of 2100 for the GHG+aerosol simulations and 7% for the GHG-only case. The acceleration of the increase in precipitation rate is delayed compared to that for temperature. The percentage change in precipitation with latitude is displayed in Figure 13 which shows a banded structure indicating both shifts in pattern and changes in rates. The largest percentage increases are seen in middle and high latitudes although the largest absolute increases are found at lower latitudes.

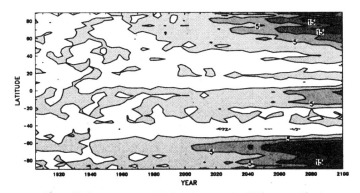

Figure 13. Percentage precipitation change under GHG+aerosol forcing.

Figure 14 gives the geographical pattern of precipitation change and shows that the general tendency for increased precipitation is not uniform, with some regions of decreased precipitation seen over both land and ocean. The decrease of tropical precipitation in the eastern, and increase of precipitation in the central and western Pacific, resembles that which occurs periodically in association with the El Niño. Other regions of decreased precipitation are seen over comparatively low latitude land areas with increases general over higher latitude land areas.

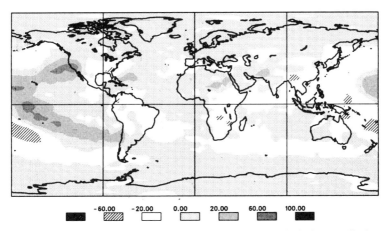

Figure 14. Percentage precipitation change from 1975-95 to 2040-2060 for GHG+aerosol forcing.

5.3 LAND HYDROLOGY AND THE CRYOSPHERE

Changes in temperature and precipitation affect other aspects of the climate system. Changes in land hydrology are largest in the northern hemisphere summer where precipitation decreases, evaporation increases and, as a consequence, soil moisture availability and runoff decrease markedly. The simulated percentage decreases are large and, if realistic, would be expected to have important agricultural and other implications.

The rate and magnitude of temperature and other changes depend on the opposing positive and negative feedbacks in the climate system whereby an initial radiative perturbation is either enhanced or counteracted. One of the most powerful positive feedbacks is the albedo feedback where an increase in temperature results in decreasing amounts of snow and ice whence the albedo of the system decreases, more solar radiation is absorbed, and the temperature increases farther. The simulations show a major retreat of snow and ice cover that is associated with, and partially responsible for, the enhanced temperature changes seen at high northern latitudes.

6. Climate Impact, Adaptation, and Mitigation

The results of climate change simulations such as these are used as "input scenarios" for subsequent impact, adaptation and mitigation studies. Two recent publications *Climate Change 1995: Impacts, Adaptations and Mitigation* [12] and *Regional Impact of Cli-*

502

mate Change: an assessment of vulnerability [13] summarize a range of studies in these areas. In particular the first of these reports concludes that: (1) human-induced climate change adds an important new stress to ecological and socioeconomic systems; (2) most systems are sensitive to climate change (including ecological, socioeconomic and human health systems); (3) the impact of climate change is difficult to quantify; (4) successful adaptation depends on a mix of technological advances and institutional arrangements which depend also on financing and information exchange; (5) vulnerability increases as adaptive capacity decreases; (6) detection of the effects of climate change on ecological and social systems will be difficult and unexpected changes are possible; and (7) enhanced research and monitoring are essential.

Changes in the "climate" of socioeconomic, ecological, and health systems may in turn affect the physical climate system if they lead to reductions (or increases) in anthropogenic GHG emissions and aerosol loadings or to changes in land use, rates of deforestation and so on. These actual and potential interactions between the human activities and the climate system close the chain of studies indicated in Figure 1.

7. Summary

Modern climate models attempt to represent the physical processes governing the behaviour of the atmosphere, ocean, land, and cryosphere. They are used to simulate both past and future climates using greenhouse gas concentrations and aerosol loadings which alter climate forcing. Information on GHG and aerosol concentrations is available for the recent past but future values depend on projected population levels and economic and industrial activity.

An ensemble of simulations with the CCCma global coupled model, with prescribed changes to GHG concentration and aerosol loading, reproduces basic features of observed climate change from 1900 to the present and provides also an indication of the climate change that is possible in the next century. The rate of climate warming is simulated to accelerate from the present but to be non-uniform as to hemisphere, season and underlying surface. Snow and sea-ice cover retreat in association with the warming.

The warming is accompanied by an overall acceleration of the hydrological cycle giving an increase in globally averaged precipitation and evaporation. There are associated regional patterns of increase and decrease including an El Niño-like shift of precipitation in the tropical Pacific. Land hydrology changes are largest in the northern hemisphere in summer where precipitation decreases, evaporation increases and, as a consequence, soil moisture and runoff decrease markedly.

Climate model results of this kind provide input for ecological and socio-economic studies of the possible effects of climate change. These may, through the choices made by political and economic decision makers, feedback on the climate system through reduced (or increased) GHG emissions and other climate forcing changes associated with human activity.

Acknowledgement

Thanks to Greg Flato for helpful comments.

References

1. Boer, G.J., McFarlane, N.A., and Lazare, M. (1992) Greenhouse gas-induced climate change simulated with the CCC second-generation general circulation model *J. Climate* **5**, 1045-1077.

2. Boer, G.J., Flato, G., Reader, M.C., and Ramsden, D. (1999) A transient climate change simulation with greenhouse gas and aerosol forcing: experimental design and comparison with the instrumental record for the 20th century. Submitted to *Clim. Dyn.*

3. Boer, G.J., Flato, G., and Ramsden, D. (1999) A transient climate change simulation with greenhouse gas and aerosol forcing: projected climate change for the 21st century. Submitted to *Clim. Dyn.*

4. Cubash, U. (ed), Caneill, J.-Y., Filiberti, M.A., Hegerl, G., Johns, T.C., Keen, A., Parey, S., Thual, O., Ulbrich, U., Voss, R., Waszkewitz, J., Wild, M., and van Ypersele, J.P. (1997) *Anthropogenic climate change,* European commission Report EUR 17466 EN.

5. Flato, G.M., Boer, G.J., Lee, W., McFarlane, N., Ramsden, D., Reader, M., and Weaver, A. (1999) The CCCma global coupled model and its climate. Submitted to *Clim. Dyn.*

6. Haywood, J.M., Stouffer, R.J., Wetherald, R.T., Manabe, S., and Ramaswamy, V. (1997) Transient response of a coupled model to estimated changes in greenhouse gas and sulphate concentrations, *Geophys. Res. Lett.* **24**, 1335-1338.

7. Hulme, M. (1992) Global land precipitation climatology for the evaluation of general circulation models, *Clim. Dyn.* **7**, 57-72.

8. Hulme, M. (1994) Validation of large-scale precipitation fields in general circulation models, in Desbois, M., and Desalmand, F. (eds), *Global Precipitation and Climate Change*, NATO ASI Series, Springer-Verlage, Berlin.

9. IPCC (1990) *Climate Change, the IPCC Scientific Assessment,* Cambridge University Press, Cambridge.

10. IPCC (1992) *Climate Change 1992, the Supplementary Report to the IPCC Scientific Assessment*, Cambridge University Press, Cambridge.

11. IPCC (1995) *Climate Change 1995, the Science of Climate Change,* Cambridge University Press, Cambridge.

12. IPCC (1995) *Climate Change 1995, Impacts, Adaptations and Mitigation of Climate Change,* Cambridge University Press, Cambridge.

504

13. IPCC (1998) *The Regional Impacts of Climate Change, An Assessment of Vulnerability,* Cambridge University Press, Cambridge.

14. Jones, P.D. (1994) Hemispheric surface air temperature variations: a reanalysis and an update to 1993, *J. Climate* **7**, 1794-1802.

15. McFarlane, N.A., Boer, G.J., Blanchet, J.-P., and Lazare, M. (1992) The Canadian Climate Centre second generation general circulation model and its equilibrium climate, *J. Climate* **5**, 1013-1044.

16. Mitchell, J.F.B., Johns, T.C., Gregory, J.M., and Tett, S.F.B. (1995) Climate response to increasing levels of greenhouse gases and sulphate aerosols, *Nature* **376**, 501-504.

17. Mitchell, J.F.B. and Johns, T.C. (1997) On the modification of global warming by sulphate aerosols, *J. Climate* **10**, 245-267.

18. Pacanowski, R.C., Dixon, K., and Rosati, A. (1993) The GFDL modular ocean model users guide, GFDL Ocean Group Tech. Rep. No. 2. Geophysical Fluid Dynamics Laboratory, Princeton.

19. Parker, D.E., Folland, C.K., and Jackson, M. (1995) Marine surface temperature: observed variations and data requirements, *Climatic Change* **31**, 559-600.

20. Reader, M.C. and Boer, G.J. (1998) The modification of greenhouse gas warming by the direct effect of sulphate aerosols, *Clim. Dynam.* **14**, 593-608.

Index

accuracy
  of implicit diffusion scheme, 153–154
  of orographic forcing in semi-Lagrangian scheme, 192
  of semi-Lagrangian method, 174–177
adjoint methods, 122
advection equation, 1-D, 130
advection over the pole, 199
advection schemes
  centered differences, 131
  explicit, 131–140
  forward approximation, 139
  implicit, 140–142
  leapfrog, 131
advection-diffusion approximations, 155–158
aerodynamic resistance, 327, 337
aerosol dynamics, 353–371
aerosol forcing, 493, 494
age of air, 355
albedo, 71, 323, 335
  and energy balance models, 471
  ice-albedo feedback, 436, 475, 501
  of snow, 340
aliasing, 132, 162, 169
analysis of model climate, 59–81, 443
anelastic approximation, 251, 298
anisotropy of gravity waves, 303
anomaly correlation, 388–390, 398–400
Arakawa Jacobian, 3–4
Arakawa-Schubert mass flux scheme, 251–255, 258
assimilated data, to verify forecasts, 388
atmosphere, structure of, 76
Atmospheric Model Intercomparison Project, 47, 128, 423, 431, 446, 477
atmospheric models, simplified, 105–117
Atmospheric Radiation Measurement program, 282

attractor, 31, 37
backward implicit scheme, 141
  for diffusion, 153, 156
baroclinic instability theory, 2
baroclinic instability, in simple GCM, 112–114
Betts-Miller convective scheme, 248–249
Blackadar mixing length, 231
blocking, 14, 101
body force, 297, 308
boundary conditions, 29, 30, 34–36, 51
  and seasonal predictions, 392, 393
  changes in, 123
  equations, 61
  for gravity waves, 304, 315
  for paleoclimate simulations, 472–476
  in NCAR CCM, 204
  in weather prediction, 391
  land surface, 346
  lateral, in nested models, 408–410
  list of, for climate model, 46–47
  lower, in stratosphere-mesosphere models, 114–115
  sensitivity experiments, 41, 49
boundary layer, 9
  and clouds, 274
  and convection, 222, 249
  structure, 222–223
  vertical diffusion as damping in, 155
  vertical diffusion in, 150
  winds in, 222
boundary layer processes, 221–236
boundary layer surface stress
  and gravity-wave drag, 314
bucket model, 329–331
budget equations, 64
  hierarchy, 66, 69
budgets of trace constituents, 354–355
bulk exchange formulas, 227, 326

bulk Richardson number, 227, 229, 327
bulk transfer functions, 227, 231
buoyancy sorting, 260–261

Canadian Centre for Climate Modelling
    and Analysis, results for GCM
    of, 71–79, 492
Canadian Regional Climate Model, 185
canonical analysis, 100–101
canonical correlation analysis, 100
capillarity of soil, 333
carbon dioxide, 34, 337, 421
    and longwave radiation, 283
    and past climate, 466, 469, 472,
        473, 477
cascade of energy, 150
centered difference operators, 135, 146
centered differences, 151, 157
changing a model, 120–121
Charney, Jule, 2
chemistry, 353–371
Clausius-Clapeyron relationship, 461
climate
    as distinct from weather, 66
    balance equations for, 70
    of the past, 466–470
climate analysis, 60
climate change, 11, 19, 391, 392, 435–
        438, 489
    and changes in land cover, 347
    and chemical processes, 353
    and land ice, 475
    and radiative balance, 71
    and runoff, 346
    and spinup period, 33
    compared to natural variability, 495
    international aspects, 490–492
    lessons from past climates, 481
    modeled, compared with observed,
        495–498
    past, 489
    simulation of, 492, 498–501
climate change scenarios
    and policy implications, 490
    as inputs to studies of impacts, 490,
        492, 501, 502

climate drift, 423–424, 434
climate feedback, 453
climate impacts studies, 501, 502
climate model, see general circulation
        model
climate models
    energy balance, 471–472
    qualitative, 470–471
Climate Prediction Center, 388, 392
climate scenarios, 49
climate sensitivity, 460–462, 498, 499
climate simulations, categories of, 46–
        50
climate statistics, 60, 66, 69
climate system, 29
    and radiative balance, 71
    boundary conditions for atmospheric
        model, 36
    studying features of, 121
climate system model, 392, 419–420,
        430–438, 481
climatology of model, 84
closure in convective schemes, 258
cloud feedback, 460
cloud fraction, 264–266, 276
cloud ice, 267
cloud optical properties, 284
cloud radiative forcing, 71, 84, 460
cloud scheme
    diagnostic, 265
    prognostic, 265–266
cloud water, 267, 270, 277
clouds
    and model intercomparison, 459,
        460
    and radiation, 292–293
    in boundary layer, 221, 235–236
    modeled, compared with observed,
        73
    role in climate system, 263–264
cluster analysis, 102
Community Climate Model, see National
        Center for Atmospheric Re-
        search
complex EOF analysis, 95–99
computational instability, 3, 133

computational mode, 137–138, 142–144
computer hardware, 438–440
computer time, for radiative calculation, 285, 291
concentration scenarios of greenhouse gases, 491
condensation, 239
continuity equation
    discrete form, 208
    for climate analysis, 64
    in NCAR CCM, 208
    integral form, 205
control parameters, 34, 40–43
control run, 85, 119–120
    and climate change simulation, 493
convection, 239–261
    and boundary layer, 222
    deep, in Betts-Miller scheme, 248
    shallow, in Betts-Miller scheme, 248–249
convective adjustment, 245–248
convective available potential energy, 241, 243, 258
convective instability, 245
    and gravity-wave breaking, 300
convective parameterizations, 245–261
    and simulation of mid-Cretaceous, 480–481
    Betts-Miller, 248–249
    buoyancy sorting, 260–261
    convective adjustment, 245–248
    mass flux, 250–259
    moisture convergence, 259–260
convective velocity scale, 228
cool Sun paradox, 466
Coriolis parameter, 146
counter-gradient heat flux, 233
Coupled Model Intercomparison Project, 420, 435, 446, 452, 453
coupled models
    freshwater input in, 344
    future climate, 392, 489
    importance of surface fluxes in, 74
    intercomparison of, 452–457
coupling constraints, 421
Courant number, 133

for grids with variable longitude, 145
orographic flow with semi-Lagrangian scheme, 196
typical value in atmospheric models, 137
Courant-Friedrichs-Lewy (CFL) condition, 133, 134
covariance, 88
Crank-Nicholson scheme, 141
    for diffusion, 151–152, 155, 156
cryosphere, simulation of future changes in, 501

Darcy's Law, 334
data assimilation, 375
    advantages, 377–378
    description, 376–377
    four-dimensional variational, 383–384
    three-dimensional variational, 381–383
decomposition of quantities into mean and eddy components, 66
deforestation and climate, 322, 345
degeneracy of EOFs, 91–92, 95
degrees of freedom, number of, 31
designing a GCM experiment, 119–125
detrainment, 251–256
diagnostic test, 120–121
diagnostic variables, 31, 36
diagnostics, 59–81, 121
diffusion
    approximations, 150–156
    higher order, 150
    horizontal, 155
    horizontal, in NCAR CCM, 206, 207, 212
    implicit form, 151
    in spectral transform models, 150
    second order, 150
    simplified in spherical harmonics, 166
    vertical, 155
    vertical, in NCAR CCM, 205
discrete operator notation, 135

508

dissipation of gravity waves, 300–301
divergence, 166
divergence equation, 167
    discrete form, 211–213
Doppler spread parameterization of gravity waves, 306
doubly spatially-averaged Eulerian treatment of orography, 196, 197
down-gradient representation of turbulent transfer, 230
downdrafts, 252, 256–257
downscaling, 414–415
downward control of circulation, 311–317
drainage of water into ground, 324, 333
dry static energy, 243–244
dynamical core model, 112–114
dynamical schemes, evaluating differences between, 120–121
dynamical system, 29

Earth Radiation Budget Experiment, 66, 72, 288
eddy diffusivity, 111, 230, 234
eddy flux in boundary layer, 224
eddy kinetic energy, 77, 404
    dependence on resolution, 404–405
eddy transport, 77–78, 426
effective CO_2, 435, 491
effects of ENSO on midlatitudes, 395, 396
El Niño Southern Oscillation (ENSO), 16–20, 427
    forecasts of, 18, 394
    key to seasonal predictions, 393–397
Eliassen-Palm flux, 312
emission inventories, 360–362
emission scenarios, 490
emissivity, 324
emissivity method, 283–284
empirical orthogonal functions, 38, 86, 90–100
energy balance
    global, 421
    local, 422
        of snow, 340
energy balance models, 471–472
energy budget
    at the surface, 322–331
    illustrated, 70
    intercomparison, 460–462
    of snow, 340
energy conservation, 421, 440
    discrete form, 213, 214
energy cycle, 80, 81
energy fluxes, surface, 70, 73, 74, 321
energy transport, 77–79
ensemble forecasts
    for seasonal predictions, 388–390, 394, 397
    of ENSO, 394
ensemble forecasts for seasonal predictions, 398
ensemble of convective plumes, 251, 254–257
ensemble runs, 14, 37, 42, 120, 122–123
ENSO, see El Niño Southern Oscillation
entraining plume, 250–256, 260
equable climate, 480
equilibrium
    of chemical reservoirs, 354–355
    of coupled models, 453
    of modeled climate, 423–424
    of paleoclimate simulations, 476
equivalent CO_2, 435, 491
equivalent potential temperature, 245
Eulerian formulation of GCM, 207–215
Eulerian treatment of orography, 192, 196, 197
European Centre for Medium-Range Weather Forecasts, 7
    and seasonal predictions, 394, 396
    cloud schemes in model of, 265–275
    radiative scheme used by, 282–292
    reanalysis data set, 385
evaporation, 239, 322, 328, 336
evaporation, potential, 326, 328
experiment "tree", 121, 125

experiment design, 51, 52, 119–125
explicit scheme, 131–140
  for diffusion, 156
  in shallow-water model, 167
extended EOF, 97
external forcing, 30, 41, 42, 49
external variables, 35

feedbacks, 436, 501
  ice albedo, 436, 475, 501
finite difference method, 3–4, 128, 130–158
finite element methods, 198
Fisher test, 85–86
fixed dynamical heating model, 122
flux adjustment, 18, 424–425, 433, 452–457, 492
flux correction, *see* flux adjustment
flux coupler, 419, 420, 431
flux of trace constituents at sea surface, 361
flux profiles, 225, 226
forced-oscillator equation, 185–191
  analytical solution, 187, 188
  numerical solution, 188–191
forecast
  seasonal, *see* seasonal prediction
  weather, *see* weather prediction
forecast error of tropical Pacific, 395
forward approximations
  for diffusion, 154
forward explicit scheme
  for diffusion, 156
forward scheme, 139–140
four-dimensional variational data assimilation, 383–384

Galerkin method, 198
Gaussian grid, 164, 165
general circulation model
  "ten commandments" of running, 125
  as dynamical system, 29–32
  early development of, 2–13
  example of, 204–215
  ocean, 13, 425–428, 440

Phillips' pioneering experiment, 2–3
  running, 50–53
  simple alternatives to, 105–117, 122
Geophysical Fluid Dynamics Laboratory, 1, 4, 8, 14, 21
Gibbs phenomenon, 171
gravity waves
  Doppler shifted in semi-Lagrangian framework, 184, 185
  in semi-implicit approximations, 146
  parameterized in zonally averaged models, 111
  with semi-implicit scheme, 207
gravity-wave breaking, 300, 306
gravity-wave drag, 297–319
  and model intercomparison, 458
  effect on temperature, 310, 312, 313
  effect on wind, 309, 312, 313
  uncertainties in, 317
gravity-wave dynamics, 298–301
gravity-wave saturation, 301
greenhouse gases, 353, 435, 489–494

Hadley Cell, 357
heat budget
  convective, 250
  surface, 325–326, 331–334
heat capacity of soil, 333
heat conduction, 332
heat flux
  and baroclinic instability, 113, 115
  in boundary layer, 223, 231, 233, 234
  in convection, 239
  in surface layer, 225, 229
heat transport, 426
  ocean, 422, 424, 431, 476
Hermite cubic polynomial, 182, 183
hierarchy of models, 106, 117, 470
Hilbert Transform, 96
hindcasts in seasonal predictions, 397
horizontal derivatives in spherical coordinates, 165
horizontal diffusion, 212

representing boundary layer processes, 224
Hotelling test, 86
Hovmoeller diagram, 90
hybrid vertical coordinate, 201–204, 307
    and reference atmosphere, 207
hydraulic conductivity, 334
hydrological sensitivity, 461, 462
hydrology, 339–346
    simulation of future changes in, 501
hydrostatic equation
    discrete form, 211
    in NCAR CCM, 211
    integral form, 205

ice cores, 468
ice rheology, 429
ice sheets at Last Glacial Maximum, 474
ice-albedo feedback, 436, 475, 501
implicit scheme
    for advection, 140–142
    for diffusion, 151–154, 156, 157
inertial-gravity waves, 146, 186
    effect on solution, 147
    filtering used to damp, 149
infiltration of water into ground, 324, 331–333, 342–343
initial conditions, 33, 38–40, 51
    and seasonal predictions, 394, 397, 399
    and sensitivity experiments, 33
    changes in, 123
    climate change simulation, 492–493
    dynamical systems perspective, 32–34
    transient phase of simulation, 33
    uncertainty in, 401
instability
    computational, 3, 133
    conditional, 240–245
    convective, 245
    Kelvin-Helmholtz, 301
interannual variations, 19–20

interception of rainfall by vegetation, 335–337
intercomparison, 59, 123, 443–464
    advantages and disadvantages, 445, 446
    and seasonal predictions, 397
    phases, 444
    purposes, 443, 444
    results for basic quantities, 447–452
    types, 444, 445
intercomparison of land surface models, 345
Intercomparison of Radiation Codes for Climate Models, 282
intercomparison projects, 445, 446
interdecadal variations, 20–21
Intergovernmental Panel on Climate Change, 18, 59, 444, 490, 491
International Research Institute, 392
International Satellite Cloud Climatology Project, 66, 72
interpolation, 33
    in semi-Lagrangian method, 173–178, 182–184
isentropic coordinates, 109–111
isotropy in boundary layer, 233

Kalman filter, 384
Kelvin-Helmholtz instability, 301
kinetic energy, 64, 231–232
Kuo convective scheme, 259–260

land ice
    for paleoclimate simulations, 475
land surface model, 322
land surface processes, 321–347
land surface properties
    for paleoclimate simulations, 474–475, 477
Laplacian
    in spherical coordinates, 165
    spherical harmonics as eigenfunctions of, 166
lapse rate, 241–242, 246, 247
large eddy simulation, 11

Last Glacial Maximum, 468, 473–474
latent heat release, 235, 246
latitude-longitude grid, 129, 130
leaf area index, 336, 338
Legendre functions, 162–164, 168
level of free convection, 221, 240–242
level of neutral buoyancy, 241–242
lifting condensation level, 221, 241–242, 246
limit of predictability, 388, 391, 395
limited area models, 408–409
line-by-line models, 282
linear drag, 150
linear forcing, 45
long-range forecasts, 13–16
longwave radiation, 283–287
    at surface, 288, 290
    outgoing, 286–287

Météo-France stretched-grid model, 410–414
Manabe, Syukuro, 21
mass conservation, 169
mass flux convective scheme, 250–259
mass transport, 84
Maximum Entropy Method, 89
mean quantities
    vs. transient or eddy quantities, 66
    with different levels of averaging, 69
Mellor, George, 9
melting of snow, 340
meridional mean, 67
mesosphere, 356
Mesozoic era, climate of, 466, 467
mid-Cretaceous, climate of, 468, 480
mid-Holocene, climate of, 467, 468, 476–480
middle atmosphere
    and gravity waves, 305–317
mixing length
    in boundary layer, 231
model
    climate system, see climate system model

general circulation, see general circulation model
    limited area, 408–409
    line-by-line, 282
    shallow water, see shallow water model
    two-dimensional, 107–112
    zonally averaged, 109–112
model climatology, 84
model improvement
    through data assimilation, 377
    through intercomparison, 443
model intercomparison, see intercomparison
model, changing, 120–121
moist convective adjustment, 245–248
moist static energy, 243–244
moisture budget, 64, 70, 250, 460–462
moisture convergence, 259–260
moisture diffusion in soil, 333–334
moisture flux
    in boundary layer, 223
    in convection, 239
    in surface layer, 225, 229
moisture profile, reference, 248
momentum equation
    discrete form, 210
    for climate analysis, 64
    in NCAR CCM, 210
    integral form, 204
momentum flux, 298, 314
    and baroclinic instability, 113, 115
Monin-Obukhov length scale, 226
monsoon circulation, 17, 19, 477, 479–480
mountains, see orography
multiple equilibria in climate, 471, 474–475

National Center for Atmospheric Research (NCAR), 11
    climate system model, 419–420, 430–438
    Community Climate Model, vii–viii, 204–215
nested models, 408–409

non-linear diffusion parameterization of gravity waves, 306
non-local representation of boundary layer, 231–235
non-orographic gravity waves, 305–306, 317
nonlinear viscosity, 4, 8
North Atlantic Oscillation, 91, 93
numerical approximations, 127–215
numerical weather prediction, *see* weather prediction

observations and data assimilation, 376
observations compared with model output, 72–79, 447–457
    clouds, 73
    eddy kinetic energy, 77
    energy fluxes at surface, 74
    energy transport, 79
    meridional streamfunction, 76
    precipitation, 448–450, 455
    sea ice, 456–457
    sea level pressure, 75, 447–448
    sea surface temperature, 433
    snow, 456
    specific humidity, 76
    temperature, 76, 450–454
    temperature at surface, 74
    top-of-atmosphere radiation, 72
    zonal wind, 76
ocean circulation and paleoclimate, 475–476
ocean general circulation models, 13, 425–428, 440
ocean heat transport, 422, 424, 431, 476
ocean model, slab, 476
optimal interpolation, 378
orbital changes and climate, 468, 473
orographic and other stationary forcing, 184–197
orographic gravity-wave drag, 302–305
    effects of, 306–317
orography
    dependence on resolution, 404, 406
    effect on modeled surface temperature, 84

    in mid-Cretaceous, 481
    in semi-Lagrangian scheme, 191–193
    in spectral model, 46
    of stretched-grid model, 413
    sensitivity experiments, 41, 49
outgoing longwave radiation, 71, 286–287
    and energy balance models, 471
output variables, 37
ozone depletion, 353

paleoclimate, 12, 49, 423, 465–482
Paleoclimate Model Intercomparison Project 446, 476–480
parameter space, 42, 43
parameterization
    boundary layer processes, 221–236
    clouds, 263–278
    convection, 245–261
    gravity wave drag, 297–317
    land surface processes and hydrology, 321–347
    radiation, 281–293
parameterization errors, 63
parcel method for determining conditional instability, 240
perturbations, 39, 41
    ideal size of, 123–124
phase error of advection schemes
    centered differences, 136
    explicit and implicit schemes, 137
    semi-Lagrangian method, 174–178
    spectral method (linear advection equation), 160
phase space, 31, 32
    trajectory in, 37
Phillips' GCM experiment, 2
photosynthesis, 337
physical laws, 30
physical mode, compared with computational mode, 138
planetary-scale waves, 107–109, 111, 396
polar filter, 169
pole problem, 144–145, 198

as economical problem, 134
description, 129
filtering as solution for, 144, 145
grid with variable longitude as solution for, 145
illustrated, 134, 199
in early GCMs, 4
in finite difference models, 128
in semi-implicit schemes, 149
in semi-Lagrangian method, 181
solved by hexagonal-icosahedral grid, 108
solved by spherical harmonic functions, 162
porosity of soil, 331
postprocessing, 53
potential evaporation, 326, 328, 336
potential temperature, 245
potential vorticity, 107
    as prognostic variable, 115
    illustration, 116
pre-Cambrian era, climate of, 466, 467
precipitation
    and convective adjustment, 246
    and runoff, 324–325
    dependence on resolution, 405, 407
    in mid-Cretaceous, 480
    in mid-Holocene, 477–479
    in stretched-grid model, 412, 414
    in Zhang-McFarlane convective scheme, 257
    model intercomparison, 448–450, 455
    simulation of future changes in, 500–501
    simulation of observed trends in, 497–498
    spatial variability of, 341–342
    spurious, associated with sigma coordinates, 201, 202
    spurious, associated with spectral method, 171
precipitation recycling, 322, 343
predictability, 14, 388, 391, 395, 398
predictability of ENSO, 395
predictive skill, 388, 393

of seasonal predictions, 397–399
pressure coordinate, 62, 201
pressure gradient force
    in hybrid coordinates, 202
    in sigma coordinates, 201
principal component analysis, 90–100
principal oscillation pattern analysis, 101
principles of modeling, 119–125
prognostic variables, 30
Program for Climate Model Diagnosis and Intercomparison, 65

qualitative climate models, 470–471
Quaternary period
    climate, 468, 476
    ice sheets during, 474
    sea surface temperature during, 476
quick fixes, 120

radiation, 71, 281–293
    and climate change, 489
    and clouds, 84, 263, 460
    and energy balance models, 471
    as diffusive process, 150
    balance, 322, 324
    longwave, 263, 281, 471
    outgoing longwave, 71
    separation between shortwave and longwave, 281
    shortwave, 263, 281
    surface, 322–324
radiative forcing
    by clouds, 71, 84, 460
    by different radiative absorbers, 491
radiative-convective model, 122
reanalyses, 65, 385, 397, 401
reduced grids, 130, 198–200
reference atmosphere, 207, 209
reference profiles, 248
regional models, 403–415
reservoirs of trace constituents, 354–355
residence time, 355, 356
resolution
    and clouds, 272–274
    and parameterization errors, 63

and quality of model climate, 63
and seasonal predictions, 398
and speed of radiative calculation, 285
climate models compared with weather prediction models, 170
effect on modeled climate, 404–408, 457–458
in ocean GCMs, 427
typical, 128
resolved and unresolved quantities in climate analysis, 62
response function
defined, 135–136
for advection-diffusion equation, 155–156
for diffusion, centered difference, 151
for explicit scheme, 135
for implicit scheme, 140
restart, 36, 50, 51
Richardson number, 227, 229
river routing, 344
Rossby waves, 145, 146
roughness height, 326, 327
roughness length, 226, 334, 335, 340
running a GCM, 119–125
runoff, 324, 329, 340–342

saturation
and cloud formation, 264, 266
of gravity waves, 302
sea ice, 438
model intercomparison, 456–457
models, 428–430
sea level pressure
model intercomparison, 447, 448
sea surface temperature, 397–401
and extended forecasts, 14
and paleoclimate, 476
during El Niño, 17
tropical, effects on midlatitude climate, 18
seasonal predictions, 16–18, 35, 387–401
seasonal variations, 47

second-order moments, 87–88
semi-implicit approximations, 146–149
in NCAR CCM, 206, 207, 214, 215
semi-implicit semi-Lagrangian scheme, 179–180, 184–197
semi-Lagrangian advection
combined with other methods, 184
semi-Lagrangian method, 7–8, 128, 171–184
sensitivity
and intercomparison, 457–462
defining, 42–46
experiments, 35, 40–42, 48, 49
land surface, 345–346
paleoclimate, 476
linear, 44
of model to longwave radiation, 283, 286
tests, 119
shallow water model, 6, 106–110, 116
equations for, 146, 167, 187
semi-Lagrangian scheme and orography, 193–196
used to illustrate spectral method, 166–169
shortwave radiation, 71, 281, 291
sigma coordinate, 169
advantages and disadvantages, 201
and reference atmosphere, 207
illustrated, 62
similarity theory for boundary layer, 225
simplified models, 105–117, 122
singular value decomposition, 100
sinks of trace constituents, 360–367
skill of prediction, see predictive skill
Smagorinsky, Joseph, 1, 8
snow, 339, 340
albedo of, 323
as reservoir of water, 324
model intercomparison, 456
observed compared with modeled, 456
snowmelt, 340
soil
heat capacity of, 333

moisture, 323, 331, 334
moisture diffusion in, 333, 334
porosity, 331
thermal properties of, 332–333
water content of, 330, 334
water vapor flux from, 328
soil hydrology, 340–345
solar radiation, 71, 281, 291
changes over geologic time, 473
during Last Glacial Maximum, 475
sources of trace constituents, 360–367
space-time spectra, 88–89
spatially-averaged Eulerian treatment of
orography, 192, 196, 197
spectral method, 4–8, 158–171
accuracy compared with finite difference, 170
and spherical geometry, 128
coefficients, 159
computational efficiency compared
with finite difference, 170
disadvantages, 170, 171
time step restriction, 160
spectral transform method, 161–162
in NCAR CCM, 207
spherical coordinates, 127–129
with semi-Lagrangian method, 181
spherical harmonic functions, 162–166
combined with semi-implicit time
scheme, 166
eigenfunctions of Laplacian, 166
history, 5
illustrated, 6, 164
solution to pole problem, 162
truncation, 163
spinup, 431, 433–434, 453, 492, 493
stability condition
Courant-Friedrichs-Lewy, 133, 134
for advection-diffusion equation,
156
for explicit scheme, 132
leapfrog solution to forced-oscillator
equation, 188, 189
semi-Lagrangian solution to forced-
oscillator equation, 189, 190
von Neumann, 133

standing eddy component, 67, 68
state space, *see* phase space
state variables, 30
stationary forcing
numerical schemes, 184
statistical downscaling, 414–415
statistical estimation, 39
statistical methods, 42
statistical significance, 42, 122–123, 125
statistical tests, 85–86
Fisher test, 85–86
Hotelling test, 86
Student t-test, 85
stomatal conductance, 338
stomatal resistance, 337, 338
storage capacity of soil, 326
stratosphere, 356, 359–360
and gravity waves, 305–317
shallow water model of, 107–109
stratosphere-mesosphere models, 114–
116
stratosphere-troposphere exchange, 359–
360
stretched-grid model, 410–414
structure of the atmosphere, zonal mean,
76
Student t-test, 85
subgrid-scale processes, 34
sublimation, 340
subsurface temperature, 332
surface
albedo, 340
climate, 74, 75
energy budget, intercomparison, 460–
462
moisture budget, intercomparison,
460–462
radiation budget, 322–324
surface flux, 326, 422–423, 431
surface layer, 225–229
heat flux in, 225
moisture flux in, 225
thermally disconnected from atmosphere, 228
wind stress in, 225
surface roughness, 327

surface stress, 458, 459

teleconnection
    and ENSO, 19
    and model analysis, 101
    and seasonal predictions, 17, 396
    definition, 14–15
temperature
    effect of gravity-wave drag on, 310, 312, 313
    in mid-Cretaceous, 479–482
    in mid-Holocene, 477–478
    model intercomparison, 450–454
    of soil, 332
    reference profile for deep convection, 248
    simulation of future changes in, 498–499
    simulation of observed trends in, 495–497
temporal difference operator, 139
thermal properties of soil, 332–333
thermodynamic equation, 64, 205, 210
thermohaline circulation, 21
three-dimensional variational data assimilation, 381–383
time differencing
    in NCAR CCM, 205–207
    in semi-Lagrangian method, 178–179
time filter, 142–144, 206
time split approximations, 157–158
    in NCAR CCM, 205, 207
time step, 131
    and pole problem, 144, 145
    in semi-implicit semi-Lagrangian scheme, 191
    stability constraint on, 133–134, 136, 141, 142, 146–149, 160, 174
timescales of transport, 358
top of atmosphere radiative fluxes, 71, 286–287
topography and runoff, 343, 344
total energy, 65
trace constituents, 357

and data assimilation, 378, 381, 384
sinks, 360–367
sources, 360–367
trade wind cumulus, 249
trajectory, parcel
    in semi-Lagrangian method, 171–173, 208
transformed-Eulerian mean, 311–312
transient eddy component, 67, 68
transit time, 358
transpiration, 337
transport
    of trace species, 356–360
trends, observed
    precipitation, 497–498
    temperature, 495–497
tropical Pacific ocean
    and seasonal predictions, 393
tropopause folding, 360
truncation
    restriction on to avoid aliasing in sigma coordinates, 169
    spherical harmonic, 163
    triangular, 163, 169
truncation error, 3, 8
    avoided with spectral method, 158
tuning parameters, 35
turbulence closure scheme, 9–11
turbulence, in boundary layer, 221
turbulent eddies, 325
turbulent fluxes, 230–235, 326–327
turbulent kinetic energy, 231–232
turnover time, 355
two-dimensional models
    horizontal, 107–109
    latitude-height, 109–112
two-stream approximation for radiation, 281, 283, 285

uncentering as solution to Doppler shifting, 185
unresolved motions, 297
updrafts, 250–256

validation experiments, 46–48

validation of model climate, 59–81, 83,
  443
    clouds, 267–275, 277, 278
    in paleoclimate simulations, 477
    observations used for, 65
    radiative schemes, 287–291
    regional models, 405, 407, 409
variability of model climate, 87–102
variable-resolution model, 410–414
variational methods, 33
VARIMAX rotation of EOFS, 95
vegetation, 323, 334–338
vegetation resistance, 337
vegetation schemes, 331
verification, *see* validation
vertical coordinates, 61, 200–203
    height, 200, 201
    hybrid (sigma-altitude), 202
    hybrid (sigma-potential tempera-
      ture), 203
    isentropic, 202, 203
    pressure, 201
    sigma, 201, 202
    step-mountain, 202
vertical diffusion, in NCAR CCM, 205
vertical momentum flux, 298, 300, 314
vertical velocity, 209, 210, 250–254
virtual temperature, 205
volcanic eruptions, 354
von Neumann stability condition, 133
vorticity, 166, 167, 211–213

water
    budget at surface, 321–331
    budget, and snow, 340
    in soil, 331–334
    storage in land surface, 325, 331,
      334
water table and runoff, 344
water vapor, 420, 438, 439
    advection errors in spectral method,
      170, 171
    and longwave radiation, 283
    continuity equation, 205, 207
    flux from soil, 328
    flux from vegetation, 337

wave breaking, 300, 306
weather prediction, 1, 2, 14, 33
    adjoint methods used in, 122
    and data assimilation, 376
    compared with seasonal prediction,
      387, 390, 391
    with stretched-grid model, 414
weather regimes, 102
wetness ratio, 326
wind
    dependence on resolution, 405–406,
      408
    effect of gravity-wave drag on, 309,
      312, 313
    in boundary layer, 222
wind stress
    in surface layer, 225
WKBJ approximation, 298–299

Zhang-McFarlane convective scheme,
  256–259
zonal mean, 67
zonally averaged models, 109–112